MW00844816

Black Hole Gravitohydromagnetics

Astrophysics and Space Science Library

Black Hole Gravitohydromagnetics

Second Edition

by

Brian Punsly

Brian Punsly
4014 Emerald St. 116
Torrance, CA
USA
and
ICRANet
Piazza della Repubblica 10
65100, Pescara
Italy

Cover Illustration: A plot of the magnetic field (the white curves) around a rapidly rotating black hole (the blue disk) in a 3-D MHD numerical simulation called "KDJ" that is discussed in detail in Chapter 11 (at time step 10,000 M in geometrized units). The false color contour plot depicts the density of turbulent accreting gas. The gravito-hydromagnetic dynamo (the subject of this book), in the inner regions of the accretion flow, twists the magnetic field azimuthally. These rotating magnetic twists propagate vertically as powerful, bipolar jets of electromagnetic energy (Poynting flux). This plot was provided courtesy of Shigenobu Hirose using the freeware, Paraview. Note that the North and South poles in the Paraview plot are reversed relative to the plots in Chapter 11, therefore the stronger jet is in the Northern hemisphere.

ISBN: 978-3-540-76955-2 e-ISBN: 978-3-540-76957-6
DOI: 10.1007/978-3-540-76957-6

Library of Congress Control Number: 2008929587

Cover design: eStudio Calamar S.L.

Printed on acid-free paper

9 8 7 6 5 4 3 2 1

springer.com

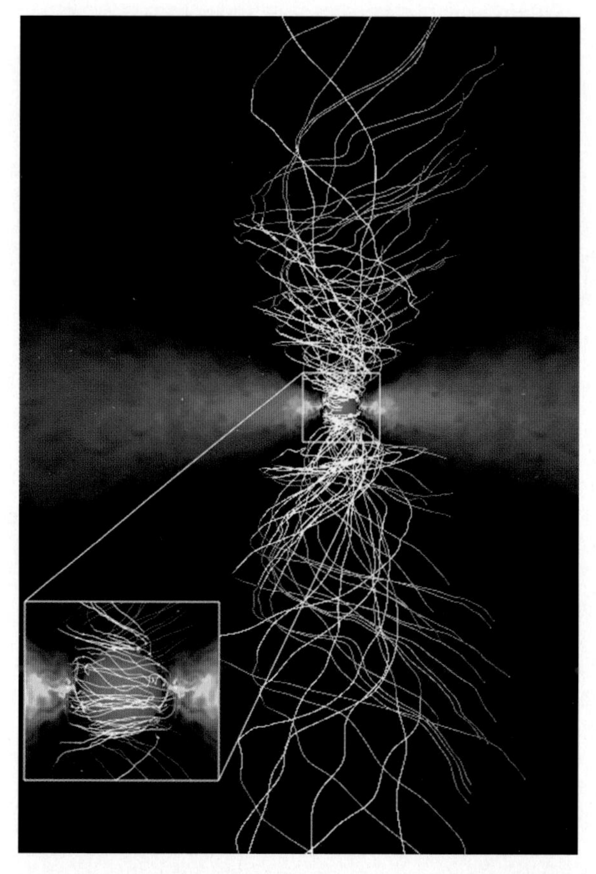

A plot of the magnetic field (the white curves) around a rapidly rotating black hole (the blue disk) in a 3-D MHD numerical simulation called "KDJ" that is discussed in detail in Chapter 11 (at time step 10,000 M in geometrized units). The false color contour plot depicts the density of turbulent accreting gas. This is an expanded view of the cover illustration (in the inset) that highlights the large scale relativistic bipolar jets that are driven by the gravitohydromagnetic dynamo. The low pitch angle magnetic helices in the outer regions of the nested set of field lines correspond to the regions of strongest Poynting (electromagnetic energy) flux in the jet. A strong flare is just beginning in the Southern hemisphere. Note that the North and South poles in the Paraview plot are reversed relative to the plots in Chapter 11. This plot was provided courtesy of Shigenobu Hirose.

Preface

"Black hole gravitohydromagnetics," or simply black hole GHM, is the study of the physical interactions of highly magnetized plasmas in the context of the differential spacetime rotation (known colloquially as the dragging of inertial frames) that is induced by the gravitational field of a rapidly rotating (Kerr) black hole. The strong large-scale magnetic field limit is essential for the external Universe to be significantly coupled to the black hole. In fact, it allows for a physical realization of the Christodoulou/Ruffini or Penrose/Floyd conceptualizations of black hole energy extraction. This is a concept that is often abused in the astrophysical community as most processes that torque a black hole only do so if the internal energy of the black hole is increased as well. A true "Penrose process" actually decreases the internal energy of the black hole. This tight constraint on the inflowing plasma state is fundamental to black hole GHM. It is the ability of GHM to describe the underlying physics behind the extraction of rotational inertia from a black hole that is of interest in astronomy. Specifically, the relevance of supermassive black hole central engines in powerful extragalactic radio sources is strongly suggested by the most modern observational evidence that is presented in Chapters 1 and 10. The theory is likely applicable to galactic black holes and "collapsars" in certain circumstances that have yet to be explored.

The main advances in the study of GHM since the first edition of this book are the advent of perfect magnetohydrodynamic simulations of black hole magnetospheres. Chapter 11 is a study of numerical simulations performed during the last seven years that are relevant to GHM. All theoretical treatments of black hole driven jets of plasma are predicated on certain assumptions. The most significant of these is the poloidal magnetic field distribution in the black magnetosphere. The bottom line on this topic is that we still know very little about what the magnetic field "usually" looks like around a supermassive black hole in an active galactic nucleus. For example, the ergospheric disk in Chapter 8 of this book assumes a large scale vertical magnetic flux in the equatorial plane near the black hole event horizon in the ergosphere. So far, no simulation has shown this to occur; however, only a very small subset of the astrophysically possible magnetospheric environments have been explored to date. The simulations can teach us about new possibilities that were not previously envisioned. For example, it was surprising to see that a strong GHM interaction in 3-D simulations drives powerful, episodic flares of relativistic plasma from the ergospheric accretion flow near the equatorial plane. In a time averaged sense, the GHM driven flares mimic many of the fundamental features of the idealized ergospheric disk. The flares of electromagnetic energy flux are coincident with flares in vertical magnetic flux that permeate the inner regions of the ergospheric equatorial accretion flow. The random local plasma physics that produces these flares in vertical flux is the ultimate trigger for the GHM driven jets. In summary, the simulations teach us about field configurations that were never before visualized. Since the poloidal magnetic flux distribution near a supermassive black hole is not known, this is fertile ground for expanding our understanding of black hole driven jets. It is likely that this will be the focal point of most numerical work in the near future.

A massive endeavor like the second edition of this book does not happen in isolation. I wish to thank Jean-Pierre DeVilliers, John Hawley and Julian Krolik for sharing their state of the art 3-D simulations of black hole accretion flows. These were extensive supercomputing efforts involving many hundreds of CPUs. I would never have been able to generate one of these complex numerical simulations and I am lucky that they were generous with their knowledge and data. I am also indebted to Vladimir Semenov and Sergey Dyadechkin who, working tirelessly with a small 2 GHz processor, generated the beautiful movies of the GHM interaction and subsequent jet production for Science magazine. These results are summarized in Section 11.2. Finally, this effort was facilitated by the support of ICRANET over the years. They have supported the page charges for the numerous peer reviewed papers that led to this second edition. The intellectual support and friendship of Remo Ruffini has been instrumental in the long-term pursuit of these topics.

Los Angeles, July 2008 *Brian Punsly*

Contents

Chapter 1
Introduction

1.1 Introductory Physical Perspective

The importance of magnetized plasma in astrophysical objects manifests itself in a wide range of phenomena from solar flares and the Aurora Borealis to pulsar winds and extragalactic radio jets. It is now widely accepted that $1–20\,M_\odot$ black holes populate the galaxies that fill the Universe. Furthermore, water maser mappings with the VLBA (Very Long Baseline Array) and kinematical gas analysis with HST (The Hubble Space Telescope) make a convincing argument that central black holes in galaxies are commonplace with masses of $10^5\,M_\odot$ to more than $10^9\,M_\odot$ (for a good review of these data see [1,2]). Thus, one expects that the interactions of magnetized plasma with the gravitational fields of black holes permeate the Universe and the astrophysical consequences should be spectacular. This monograph is the first text designed to be a formal study of this new branch of physics which is such a fundamental part of our Universe.

Why a new branch of physics? Fluid mechanics is well described by hydrodynamics. However, the introduction of magnetic fields into the flow of fluids, gases and plasmas creates forces unknown to hydrodynamics. This produces a much higher level of complexity requiring the development of a new subject, magnetohydrodynamics (MHD), or as it is sometimes called, hydromagnetics. Similarly, plasmas interacting with both magnetic fields and the near field gravitational forces of a black hole are an order of magnitude more complex than pure MHD flows. Consequently, a new formalism is required that synthesizes general relativity and plasma physics. In particular, we are interested in plasma effects induced by black hole gravity that cannot be found in more commonly studied astrophysical environments. For example, accretion disks could be found around any compact object. The combination of both magnetic and gravitational interactions with plasma flows is encompassed in the expression gravitohydromagnetics (GHM).

This is neither an easy subject to present nor to be absorbed by the student or reader. The interaction is reasonably complex by physical standards; however, the real impediment is that the subject spans two disparate areas of physics. Relativists are typically uncomfortable with the sophisticated plasma physics required to

B. Punsly, *Black Hole Gravitohydromagnetics, 2nd. ed.*,

Astrophysics and Space Science Library 355, doi: 10/1007/978-3-540-76957-6 1, describe the dynamical response of a magneto-fluid to the immense gravitational force applied by a black hole. Similarly, astrophysicists, who are well versed in standard plasma calculations, are generally unfamiliar with the advanced level of relativistic formalism necessary to probe the underlying physical interaction. This book unfortunately compromises on the review of background material for the sake of cogency. As such, a reader who has not taken a course in general relativity most likely will be overwhelmed and may never appreciate the rigorous nature of many of the basic physical concepts presented.

In spite of historical efforts to simplify the subject, the dynamics of accreting plasma in strong magnetic fields near a black hole are far from trivial. The existence of two strong forces (electromagnetic and gravitational) is incompatible with a simple description of either the plasma being in a perfect MHD state (vanishing proper electric field) everywhere or flowing along geodesic trajectories of the gravitational field. This incompatibility is manifest in the notion that an accreting flow that is dominated by strong magnetic fields (perfect MHD) in the vicinity of the hole must eventually transition to a flow state determined entirely by the gravitational field as it propagates even closer to the event horizon. This dramatic change in character of the flow does not happen gracefully and it is likely to be one of the most intense interactions attainable in the known Universe. The goal of this treatment is to supply the tools (black hole GHM) necessary for developing an intuition for the role of the black hole in this significant astrophysical context. Astrophysically, the most interesting consequence of black hole GHM is the possibility that a wind of magnetized plasma (a jet) can be driven by the interaction of the black hole gravitational field and a plasma filled magnetosphere. The main purpose of this book is to illustrate, through simplified models, the fundamental physical process that allows a rotating black hole to power a magnetized wind.

1.2 Evidence for Astrophysical Black Holes

A black hole has never been seen directly by definition. Yet, it is commonly accepted that astrophysical black holes exist. Black holes are "seen" only indirectly through their interactions with nearby matter. Because the gravitational field of a black hole is the most intense of any compact object, one expects unique signatures of their effects on the surrounding environment. For more than two decades, astronomers have been detecting what appears to be the physical manifestations of these theoretically predicted effects.

The most basic reasoning suggests that there is no known subatomic physics that can prevent a large enough mass from catastrophically collapsing to a black hole. The discovery of asymptotic freedom in Quantum Chromodynamics showed that as baryonic matter becomes more compressed, the interaction between constituent quarks becomes weaker (they essentially become Feynman's partons). Thus, our most advanced knowledge of particle physics implies the inevitability of catastrophic collapse through an event horizon if the gravitationally bound mass is large enough in a collapsing star. Of course, there could be unknown physics that could provide a positive pressure in condensed matter, halting the collapse, but it does not show up in experiments to date.

Based on this, the first evidence for black holes was a direct consequence of the maximum mass allowable for neutron stars (values $\sim$2–3 $M_\odot$ are typically estimated). Beyond this maximum mass, a stellar remnant cannot be supported by degeneracy pressure and collapse to a black hole is inevitable. Thus, the discovery of invisible companions in binary stellar systems with dynamical masses (mass functions) greater than 3 $M_\odot$ was the first evidence for black holes. To date, approximately two dozen black hole candidates are known in binary systems [3,4].

Clearly, the most famous black hole candidate is Cygnus X-1. It possesses the first predicted indirect feature of a black hole's interaction with surrounding gas. As gas is sucked into a black hole from far away, it gets crammed into smaller volumes. The viscous friction of the accreting gas should produce large amounts of heating and thermal radiation, as modified by electron scattering (see [5] for a discussion of Schwarzschild black holes and [6] for a discussion that includes Kerr black holes). For a black hole of a few solar masses, the thermal temperature of the accreting gas should be very high compared to O-stars. In fact, Cygnus X-1 was discovered by detecting this predicted X-ray emission.

In order to differentiate the spectrum of radiation emanating from accretion as to whether it originates near a black hole or neutron star requires the subjective parametric modeling of the flow. A more convincing argument that differentiates black hole accretion from neutron star accretion has been provided in the context of compact sources that are transient emitters of X-rays and γ-rays. Differences have been noted in quiescent states produced by advection dominated accretion [7]. For a black hole, the accreting gas must eventually approach the event horizon. Thus, the thermal energy is trapped in the advection dominated flow and becomes redshifted away and the flare ends in a whimper. For a neutron star, the accretion flow terminates on the stellar surface. Thus, the flare can end in a bang of thermonuclear burning or, at a minimum, the thermal energy in the advection dominated flow must be radiated from the heated up star. Consequently, the post flare quiescent states in black holes (transients from compact objects with masses $>3\,M_\odot$) have been observed to be much fainter than for neutron stars [7].

The next strongest circumstantial case for black holes is in the nuclei of galaxies. By observing the kinematics of nuclear gas and stars (by means of the Doppler shift), one can find evidence of simple Keplerian motion in some objects. This provides a straightforward dynamical estimate of the central mass [1]. The greater the spatial resolution, the more convincing the central black hole estimate. The greatest resolution is with high frequency very long baseline interferometry (VLBI). Water maser mappings at 22 GHz with the VLBA resolve subparsec scale structures orbiting the nuclei of nearby galaxies. A handful of mass estimates have been made for central black holes in nearby galaxies [2]. Most of these have insufficient data or an ambiguous interpretation. One measurement is clearly representative of a thin disk with a small warp. Molecular gas appears to be spiralling about a $4.2\times10^7\,M_\odot$ black hole in NGC 4258.

Table 1.1 Dynamically estimated central black hole masses

Source	Type	M_{BH} ($M_\odot$)	Method
Milky Way	Sbc	2.8×10^6	SD = stellar dynamics
NGC 0221 (M 32)	E2	3.4×10^6	SD
NGC 0224 (M 31)	Sb	3.0×10^7	SD
NGC 3115	S0	2.0×10^9	SD
NGC 3377	E5	1.8×10^8	SD
NGC 3379 (M 105)	E1	6.7×10^7	SD
NGC 4342	S0	3.0×10^8	SD
NGC 4486B	E1	5.7×10^8	SD
NGC 4594 (M 104)	Sa	1.0×10^9	SD
NGC 4374 (M 84)	E1	1.4×10^9	GD = gas dynamics
NGC 4486 (M 87)	E0	3.3×10^9	GD
NGC 4261	E2	4.5×10^8	GD
NGC 7052	E4	3.3×10^8	GD
NGC 6251[a]	E2	4.8×10^8	GD
NGC 1068 (M 77)	Sb	1.0×10^7	MD = maser dynamics
NGC 4258 (M 106)	Sbc	4.2×10^7	MD
NGC 4945	Scd	1.4×10^6	MD

[a] Data from [8]; all other data from [2]

A cruder but similar kinematical analysis involves studying the motion of ionized gas with HST. Clear examples of central disks of gas on the order of 100 pc across have been observed in a few galaxies and numerous black hole masses have been estimated from 10^5–$10^9\,M_\odot$. Table 1.1 lists various black hole masses determined in nearby galaxies [2] including NGC 6251 [8].

Figure 1.1 shows an HST image of the nucleus of NGC 4261, an E2 galaxy. Note the light from the active nucleus penetrating the center of the disk. Inserts (Fig. 1.2) show a radio jet emanating from the active nucleus. A similar HST image of the central disk in NGC 7052, an E4 radio galaxy, is shown in Fig. 1.3. Note that the jet axes are tied to a central engine axis that is not the same as the axis of the disk of orbiting gas.

The earliest indirect evidence for supermassive black holes in galactic nuclei came from the study of active galactic nuclei (AGN). As with galactic black holes in binary systems, the rapid accretion of gas onto the central black hole in a galaxy will have viscous dissipation and a thermal spectrum as modified by electron scattering, a "modified black body" spectrum. Due to the much larger size of the nuclear black holes, the viscous dissipation occurs on much larger scales, producing more radiation with most of the energy emitted at lower frequency (optical/UV as opposed to X-rays). Based on simplified accretion disk models [6, 9], a blue/UV excess was found in Seyfert I galaxy and quasar spectra indicating accretion onto supermassive black holes [10]. This blue/UV excess or "big blue bump" is an ubiquitous property of quasar spectra. In [11], a sample of Seyfert galaxy and quasar spectra were fit by the emission from accretion disks about central black holes with masses ranging from $10^7\,M_\odot$ to $10^9\,M_\odot$ (compare to the dynamical masses in Table 1.1).

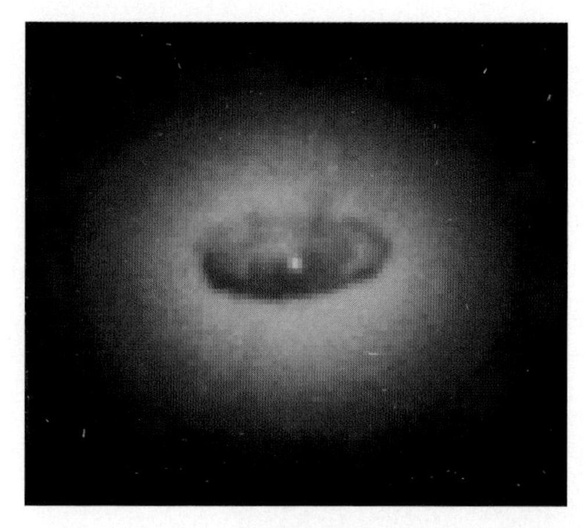

Fig. 1.1 An HST image of the central disk in the elliptical galaxy NGC 4261. Note the bright central feature, possibly accretion disk radiation from the region of the active nucleus shining through the dusty gaseous disk (alternatively, it could be the high frequency tail of the synchrotron emission form the base of the jet that is shown in Fig. 1.2). The disk is approximately 250 pc across and a gas kinematical estimate of the central black hole mass is $4.5 \times 10^8\,M_\odot$. Photograph provided courtesy of Laura Ferrarese

To this point we have talked about evidence for black holes through their effect on the nearby environment. In no instance was there cause to invoke magnetized plasma interactions near the hole to explain the data. An interesting case is the class of AGN that are giant elliptical galaxies, often residing at the center of a cluster. These have the most massive central black holes measured to date, $\sim 10^9\,M_\odot$ (see Table 1.1). They also seem to be the hosts of the radio loud class of AGN as indicated in the examples in Figs. 1.1–1.3 (this seems to hold true for more distant quasars as well [12, 13]). The best studied source of this type is M87, long suspected of harboring a supermassive central black hole. It is not a particularly strong radio source intrinsically, but it is so nearby (in a cosmological sense) that its radio emission can be well studied. It is one of the few radio sources that emits a jet that is optically detected as well. HST images reveal a jet propagating off axis from the center of a disk of nuclear gas on the order of 20 pc in diameter (see Fig. 1.4). Doppler measurements of the disk emission yield a dynamical estimate of a central black hole mass (see Table 1.1). Optically, it appears that the jet is emerging from the environs of the black hole, a finding that is supported by higher resolution radio data as well. The emission from the jet is nonthermal in origin and is well described by synchrotron radiation from hot plasma in a magnetic field that permeates the jet. This is an ubiquitous characteristic of extragalactic radio sources. Astrophysics requires an explanation as to how a jet of magnetized plasma can be generated by the environs of a black hole. The principal application of black hole GHM is the physical explanation of the central engines of extragalactic radio source that it provides.

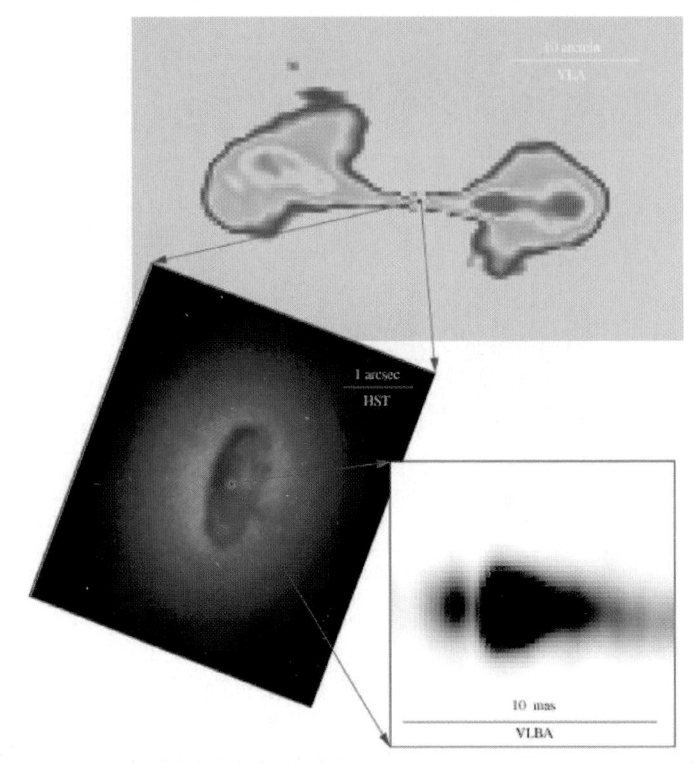

Fig. 1.2 Inserts of the large scale FR I radio structure of NGC 4261 and the small parsec scale VLBA jet that appears to emanate from the bright spot in the center of the disk (which is featured more prominently in Fig. 1.1). Photograph provided courtesy of Laura Ferrarese

1.3 Extragalactic Radio Sources

The most powerful extragalactic radio sources are associated with AGN that produce quasar emission. As stated in the last section, the bright nuclei in AGN are commonly believed to be the light produced by viscous dissipation of an accretion flow onto a black hole. The highest accretion rates produce quasar emission in galactic nuclei. It is estimated in [11] that black holes in quasars accrete mass at a rate sufficient to produce a disk luminosity $L_D > 0.1 L_{Edd}$, where L_{Edd} is the Eddington luminosity at which radiation pressure balances gravity. Seyfert galaxies have accretion rates that are lower, $L_D \sim 0.01 L_{Edd}$, with the most luminous Seyfert galaxies having $L_D \sim 0.1 - 0.2 L_{Edd}$. All of this is model dependent, but it is supported by the few known dynamical black hole mass estimates of AGN as well [14].

A magnetized accretion disk can exist, in principle, around the central black hole of either a Seyfert galaxy or quasar. Theoretically, electrodynamic luminosity (a possible energy source for a jet) is associated with large scale torques applied by a magnetic field and is independent of the gas dynamical viscous losses producing the optical/UV excess comprising the bulk of L_D. Thus, if a magnetized accretion

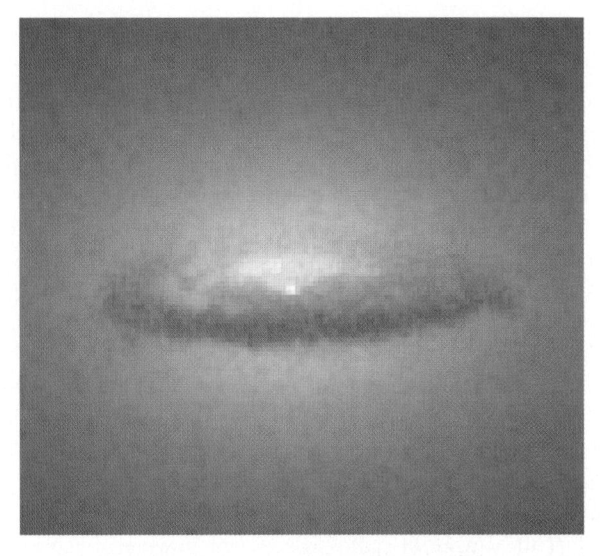

Fig. 1.3 The central disk of the elliptical galaxy NGC 7052 is revealed in this HST image. The disk is 1,000 pc in diameter and the orbital kinematics imply a central black hole mass of $3 \times 10^8\,M_\odot$. Notice the bright central region that shines through the disk as in NGC 4261. This is a weak radio source and the VLA jet is misaligned with the symmetry axis of the disk, as in NGC 4261. The photograph is provided courtesy of Roeland van der Marel

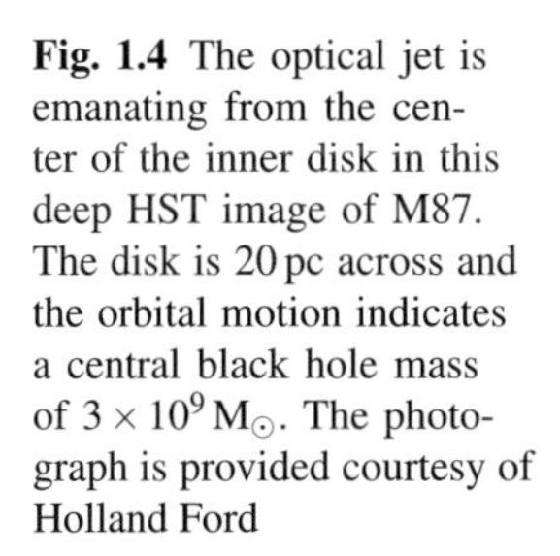

Fig. 1.4 The optical jet is emanating from the center of the inner disk in this deep HST image of M87. The disk is 20 pc across and the orbital motion indicates a central black hole mass of $3 \times 10^9\,M_\odot$. The photograph is provided courtesy of Holland Ford

disk powers the jet there is no reason why the strongest radio sources should be associated with the highest viscous losses in a flow, quasars.

However, if the quasar phenomenon is associated with the largest influx of angular momentum (largest accretion rates) then one would expect nuclear black holes in quasars to rotate more rapidly in quasars than in Seyfert galaxies or normal galaxies. It is the rotational energy that makes a black hole alive as there is extractable energy or reducible mass [15, 16] (see Sect. 1.4 for a discussion). This suggests the relevance of rapidly rotating black holes in the central engines of strong extragalactic radio sources.

Furthermore, the accretion flow alone does not seem to power the radio emission as 90% of quasars are radio quiet. The energy radiated from the central engine to the radio lobes can equal or exceed the optical/UV luminosity of a quasar. If accretion powers the radio lobes, then one would expect radio loud quasars to have highly modified accretion with a different optical/UV character than radio quiet quasars. This is contrary to observation. Radio loud quasars have indistinguishable UV broad emission lines, optical/UV luminosities and optical/UV continuum spectra from radio quiet quasars [17, 18]. Note, we did not include all of the optical broad emission lines because of an apparent difference in the optical Fe II complex between radio quiet quasars and lobe dominated radio loud quasars [19]. This strongly suggests that a central engine other than accretion seems to power FR II radio jets and lobes. This is strong circumstantial evidence supporting the hypothesis that black hole energy extraction and GHM is important in these objects.

It seems plausible that the large reducible mass of central black holes in quasars is the reason that the strongest radio sources are powered by the central engine in quasars. Thus, it is the ability of magnetized plasma interactions to extract the rotational energy of a rapidly rotating black hole that is an important application of black hole GHM.

1.3.1 Unified Scheme for Radio Loud AGN

Diagnostics of the central engine can be ascertained by understanding the connections amongst various types of radio loud AGNs. Before the mid-1980s, there appeared to be a zoo of unrelated radio loud AGN morphological types. The interpretation of radio loud AGNs within a unified scheme [20–22] revolutionized our perspective of the central engine (see [23] for a comprehensive review).

Powerful extragalactic radio sources generally occur in one of six categories:

1. FR I (Fanaroff–Riley Type I) radio galaxies
2. FR II radio galaxies
3. Lobe dominated radio loud (FR II) quasars
4. Core dominated radio loud quasars
5. Steep spectrum compact radio cores
6. BL Lac objects

In this section we will briefly describe each class and note its place in the unified scheme.

1.3.1.1 FR I Radio Galaxies

FR I radio galaxies have intrinsic extended radio luminosities (integrated from 10 MHz to 250 GHz in the galaxy's rest frame) less than $\sim 10^{43}$ ergs s^{-1} (assuming $H_0 = 55\,\mathrm{km\,s^{-1}\,Mpc^{-1}}$ and $q_0 = 0$ which are used throughout the text un-

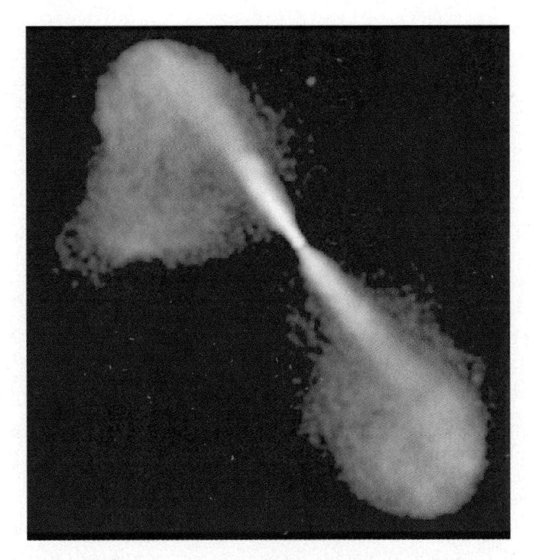

Fig. 1.5 A 5 GHz deep VLA image of a prototypical FR I radio source 3C 296. The jets are very bright compared to the diffuse lobe emission. Image provided courtesy of Alan Bridle

less otherwise stated). The luminosity is distributed in diffuse plume-like structures extended over a few hundred kiloparsecs. Although there is a wide range of morphologies for FR I radio sources [24], a classical example is given by 3C 296 (see Fig. 1.5). The radio structures in FR I sources do not have concentrated regions of emission called knots and the radio lobes are "edge darkened."

1.3.1.2 FR II Radio Sources

FR II radio galaxies have extended radio luminosities ranging from 10^{43} to 10^{47} ergs s^{-1}. The jets feeding the lobes are more collimated than FR I radio jets on kiloparsec scales. Also, by contrast, most of the radio luminosity emanates from knots in the jet or particularly strong knots in the lobes (called "hot spots") that produce an "edge brightened" appearance. Radio lobes can be separated by distances as large as a few Mpc, implying enormous amounts of stored energy.

Lobe dominated radio loud quasars have lobe and jet luminosities similar to these radio galaxies and are classified as FR II radio sources. Compare the deep VLA maps of the nearby radio galaxy Cygnus A (Fig. 1.6) and the radio loud quasar 3C 175 (Fig. 1.7); the morphology is very similar. FR II quasar counterjet/jet luminosity ratios are significantly less than for FR II radio galaxies. Only one jet is detectable in general, even for the deepest VLA maps (see [25] for a very detailed study). The strongest 3C catalog radio sources tend to be quasars rather than radio galaxies. The main distinction from FR II radio galaxies is the quasar optical/UV emission and broad emission lines from the nucleus. Host galaxy identifications of both FR II radio galaxies and radio loud lobe dominated quasars are always elliptical galaxies or irregular shaped interacting galaxies [13]. This is illustrated in Fig. 1.8 for the FR II radio galaxy 3C 219. Note the strong knot at the base of the counter jet.

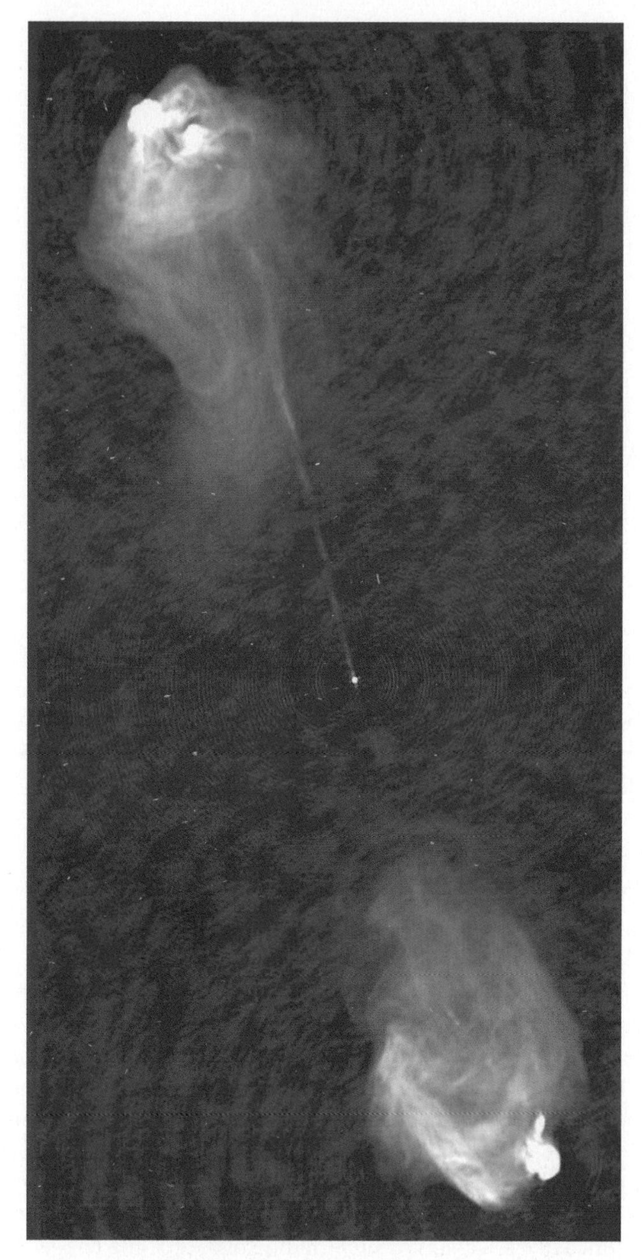

Fig. 1.6 A deep VLA image of Cygnus A at 5 GHz. The lobes are separated by 180 kpc ($H_0 = 55\,\mathrm{km\,s^{-1}\,Mpc^{-1}}$, $q_0 = 0$). Notice the strong hot spots at the end of each lobe where most of the luminosity resides. A highly collimated low surface brightness jet extends into the eastern lobe from a faint radio core. There are suggestions of a counter jet in the image. The counter jet is more pronounced in Fig. 1.10. The VLA image was provided courtesy of Rick Perley

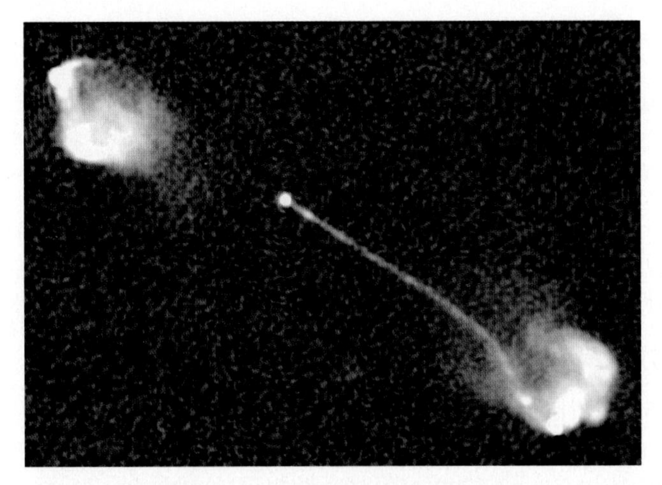

Fig. 1.7 A deep 5 GHz VLA image of the radio loud quasar 3C 175. Notice the morphological similarity to Cygnus A. The jet is more pronounced relative to the lobe emission than Cygnus A, and there is no hint of a counter jet. This is anecdotal evidence for mildly relativistic flows in kiloparsec scale jets. Image provided courtesy of Alan Bridle

Fig. 1.8 This deep 5 GHz VLA image of the FR II radio galaxy 3C 219 shows a strong jet and a knot in a counter jet. It is overlaid on the diffuse (blue) optical image of the host elliptical galaxy. Image provided courtesy of Alan Bridle

1.3.1.3 Core Dominated Quasars

Certain quasars have very strong radio cores that appear more dominant at high radio frequency (i.e., 5 GHz or above). The cores are merely unresolved radio jets. Current VLBI imaging at high frequency, 43 and 86 GHz, resolves jets to a fraction of a milliarcsecond. The jets look very similar on all scales from an arcsecond down. Higher resolution always reveals more knots of radio emission in the jet. Most of these core dominated quasars have microwave spectral indices that are much flatter than the steep spectral emission found in lobes. If one defines the spectral flux, F_ν, then a power law often approximates the spectrum in a band of frequencies, $F_\nu = F_0 \nu^{-\alpha}$. For lobes, $\alpha \approx 1.0$, and kiloparsec jets have $\alpha \approx 0.65$, while core dominated quasars generally have radio core spectral indices of $\alpha < 0.5$ (in the frequency band 1–5GHz). The spectral flux of the radio core typically turns over (steepens) between 50 and 250 GHz in the quasar rest frame [26, 27]. The radio emission from the flat spectrum radio core is often variable.

The spectral energy distribution, νF_ν, of the radio core typically peaks in the mid-infrared [28]. The high frequency optical tail is steep spectrum, variable and polarized. Radio loud quasars, whether core or lobe dominated, seem to have broad emission line regions similar to those in radio quiet quasars. Furthermore, subtracting any optical/UV emission from the high energy tail of the core spectra yields spectra and luminosities similar to radio quiet quasars.

1.3.1.4 Compact Radio Cores

There is also a panoply of compact radio sources (i.e., the total emission at 5 GHz is dominated by a "core" of radio emission that is radiated from a region less than 10 kpc across) that includes both flat and steep spectrum galaxies and quasars. The most common are compact steep spectrum quasars typically with emission on the order of 1 kpc in a twisted jet. Some of these objects have an inverted spectrum that peaks around 1 GHz and are referred to as "Gigahertz Peaked Radio Sources." Compact steep spectrum quasars rarely have significant emission on the scale of 100 kpc. The quasars 3C 286, 3C 287, and 3C 298 are well-known representatives of powerful, compact, steep spectrum objects with virtually no extended luminosity. The significant extended emission in 3C 380 is very unusual. Radio maps of compact steep spectrum cores can be found in [29] and a deep map of 3C 380 is published in [22].

Given sufficient resolution, a flat spectrum radio core usually can be found buried at the base of the jet within a steep spectrum radio core. This is consistent with a synchrotron radiation source of jet emission that becomes self-absorbed in the more compact inner regions. Every indication is that these sources have jets whose propagation is blocked by a dense intergalactic medium, or they are young radio sources with jets in the process of blasting out of the galaxy and will eventually become FR II radio sources.

1.3.1.5 BL Lac Objects

BL Lac objects have strong flat spectrum radio cores and are extremely core dominated. However, they produce very weak signals of accretion phenomena. Their broad emission lines are intrinsically weaker than core dominated quasars. The optical luminosity is dominated by the high frequency tail of the radiation from the unresolved jet comprising the core. The optical luminosity is highly variable, steep spectrum, and polarized. Subtracting the optical/UV emission from the high frequency tail of the jet in a quiescent state of core emission leaves a residual luminosity too weak to be considered a quasar (i.e., the accretion disk luminosity is weak). BL Lac objects are more common at low redshift than core dominated quasars and the opposite is true at high redshifts.

1.3.1.6 Unification

A deep connection between the various classes of objects emerged when the VLA (Very Large Array) was put into complete operation in the early 1980s. Many BL Lac objects were then shown to have diffuse halos in the high dynamic range VLA images at 1.4 GHz [21]. Later, it was shown even more conclusively that BL Lacs usually have a halo that resembles an FR I radio lobe viewed face on (i.e., same lobe luminosity and a morphology that is plume-like with no knots) [22,30]. In fact, it was shown that most BL Lacs are FR I radio galaxies viewed along the axis of the jet [21]. The radio cores of BL Lacs reveal themselves in VLBI maps to be the relativistic subkiloparsec base of jets seen nearly end on and are therefore Doppler enhanced.

Similarly, the core dominated quasars are FR II quasars seen nearly end on (i.e., looking down the jet axis) [21, 22]. The radio core is the relativistic subkiloparsec scale jet approaching the earth. Furthermore, it was argued statistically that FR II radio galaxies are actually radio loud quasars in which the quasar emission is obscured by a surrounding dusty molecular torus [20] (see Fig. 1.9). This is supported by the morphological similarity of the radio galaxy Cygnus A (Fig. 1.7) and the quasar 3C 175 (Fig. 1.8). Often, the obscuring dusty molecular gas is assumed to have a toroidal distribution, but it need not be in such a symmetric, ordered configuration for the argument to hold. An equatorial obscuring torus is consistent with the fact that FR II radio galaxies have lower jet speeds than FR II quasar jets observed with the VLBI [31]. In the unified scheme, the jets in radio galaxies are more in the sky plane than the jets in quasars, and therefore they have lower Doppler factors. Direct evidence of obscured quasars in FR II radio galaxies is sparse as only a few objects show a hidden broad line region that is characteristic of a quasar in scattered (polarized) light.

One should note that the idea that radio loud quasars are physically distinct from radio quiet quasars is statistically robust. Efforts to construct a unified scheme in which radio loud quasars are just quasars seen from a preferred angle (i.e., pole on) assume all quasars are strong radio emitters. The fundamental flaw with this

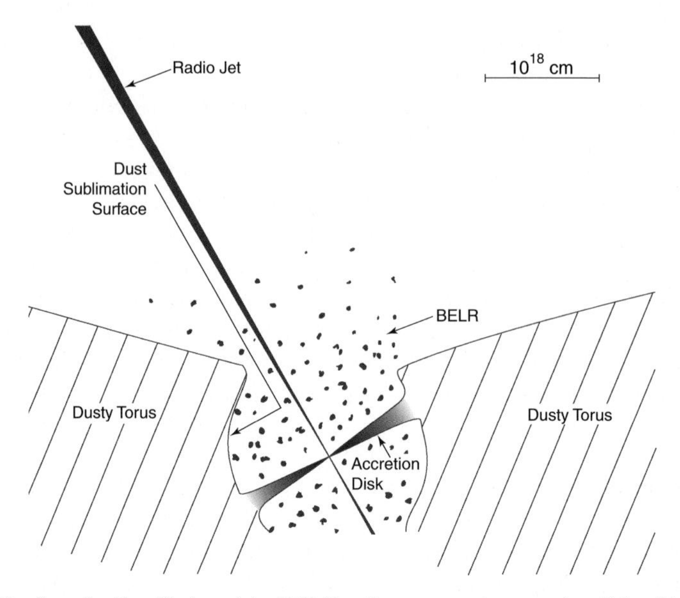

Fig. 1.9 The "standard" unified model of FR II radio sources. An accretion disk orbits a central black hole with a fiducial mass of $10^9\,M_\odot$. The optical/UV emission can be obscured by a large scale distribution of dusty molecular gas (the dusty torus) for certain lines of sight; these objects are known as FR II radio galaxies. The broad emission line region (BELR) is a distribution of clouds that are photoionized by accretion disk radiation. The dusty torus also attenuates the BELR emission in FRII radio galaxies. By contrast, in FR II quasars the line of sight reveals both the accretion disk and the BELR. Core dominated radio loud quasars are viewed along lines of sight almost colinear with the radio jet axis

idea in that the radio lobe emission is at most mildly Doppler shifted (as exemplified by 3C 175 in Fig. 1.7). The lobe emission is essentially isotropic. When core dominated quasars are observed with high dynamic range, lobe emission is usually detected [22]. In this model, radio quiet quasars are those viewed off axis. Therefore, in a single quasar population model, we should usually see the isotropic lobe emission when a quasar is viewed off axis. Thus, most quasars should be (steep spectrum) radio loud on the basis of their extended structures. However, only $\sim$10% of quasars have detectable radio lobes. Furthermore, the distribution of radio luminosities in quasars is very close to being bimodal. Consequently the statistics of radio emissivity of quasars implied by such a unified scheme was shown to conflict strongly with observation [23].

1.3.1.7 The Central Engine in the Unified Scheme

The unified scheme for extragalactic radio sources implies that the intrinsically weaker radio sources, FR I radio galaxies (or BL Lac objects when viewed end on), also have lower accretion rates (small thermal optical/UV emission). The most luminous radio sources, quasars with FR II morphology (which appear as core dominated quasars when viewed end on, or FR II radio galaxies when viewed near the equatorial plane), are associated with large accretion rates. This is interesting in the context of black hole central engines (as accretion does not seem to power strong FR II radio emission in accord with the bimodal distribution of quasar radio luminosities noted above) as it implies that rapidly rotating black holes, spun up by rapid accretion, drive the most powerful jet/lobe emission in AGNs. This makes sense from a basic principle that the amount of reducible mass (extractable energy) scales with the rotational inertia of a black hole (see Sect. 1.4).

The physics of extragalactic radio source central engines is the primary motivation for studying black hole GHM. In general it is impossible to extract the rotational energy of a black hole in any type of reasonable physical process unless there is a black hole magnetosphere. The dynamics of such a magnetosphere are governed by GHM.

The numbers are suggestive of black hole rotation as a central engine as well. Take the example of the only nearby powerful FR II radio source, Cygnus A (to be discussed in detail in the next section). The data are still somewhat debatable as to whether a hidden quasar has been found [32]. The energy supplied to the lobes in particles and fields is on the order of $\gtrsim 10^{46}$ ergs s^{-1}, (see 1.23). Cygnus A emanates from a large elliptical galaxy as is typical of radio loud AGN. Dynamical estimates of central black hole masses in nearby weaker radio loud AGN in elliptical galaxies are typically $\sim 10^8 M_\odot$–$10^9 M_\odot$ (see Table 1.1). It will be shown in this monograph that electrodynamically $\sim 10\%$ of the mass–energy of a black hole is extractable. Thus, a black hole in an large elliptical galaxy could power Cygnus A for $\sim 10^8$ years, if it were initially rapidly rotating and a significant magnetosphere were present.

1.3.2 Quantifying the Power of Extragalactic Radio Sources

In order to interpret the central engine in radio loud AGNs as supermassive black holes, one needs to quantify the power supplied to the extended radio structures. Although this is not a problem in relativistic astrophysics (as is the spirit of this book), it is of great significance for understanding the enormous power generated by the central engine and it is often not fully appreciated.

The power transported by the radio jets into the radio lobes usually far exceeds the radio luminosity. In this section, we describe the physical justification of this statement and estimate the power emitted by the central engine in Cygnus A. The observed synchrotron emission from the lobes requires both hot particles and magnetic fields. These particles and fields are advected outward as the lobes propagate into the intracluster medium and this is the dominant component of the energy supplied by the central engine. The ram pressure of the intracluster medium passing through the expanding boundary of the lobe is in balance with the internal pressure of the hot spot at the end of the lobe. This relation allows one to determine the lobe advance speed if the density of the intracluster medium is known from X-ray observations. In turn, the size of the lobes, the internal pressures and lobe advance speed yield the kinematical power of the lobe.

The synchrotron spectral luminosity is a function of the magnetic field strength and the thermal energy spectrum of the radiating electrons. There is a component of energy density from both the magnetic field and the particles. A minimum value of the total energy for a given spectral luminosity can be estimated from the radio spectra and radio maps (which give the volume of the emitting region), E_{min}. Pressure balance near the hot spots in the lobes, yields the lobe advance speed and therefore the minimum energy flux supplied by the central engine, $Q_{min} = \mathrm{d}/\mathrm{d}t\, E_{min}$. When radio spectra, deep radio maps, and X-ray spectra are available, good estimates of Q_{min} can be made, and such is the case for Cygnus A.

One can obtain an improved estimate of lobe energy compared to assuming a minimum energy plasma by studying the variation in the radio spectrum as a function of position within the lobe. This gradient in spectral index is known as "spectral aging." Spectral aging is a reference to the curvature that a power law spectrum attains over time because the high energy electrons radiate away their thermal energy faster than the low energy electrons. Thus, the high energy emission diminishes before the low energy emission causing a frequency dependent steepening (curvature) of the spectrum at high frequencies. Often the lobe advance speed disagrees with estimates derived from a spectral aging analysis that assumes a minimum energy configuration. The advantage of considering the spectral aging data in conjunction with X-ray observations of the intracluster medium is that it allows one to estimate how much the lobe energy exceeds the minimum energy value.

The power spectrum of a thermal electron gyrating in a magnetic field, B, simplifies for ultrarelativistic particles [33],

$$P(\nu) = \frac{4\sqrt{3}\pi}{3c} e^2 \gamma^{-2} \nu \int_{\nu/f_c}^{\infty} K_{5/3}(y)\, \mathrm{d}y \,\mathrm{ergs\,s^{-1}\,Hz^{-1}}; \tag{1.1}$$

where the electron cyclotron frequency is

$$\nu_B = \frac{eB}{2\pi m_e c}, \tag{1.2a}$$

the critical frequency is

$$f_c = (3/2)\,\nu_B \gamma^2, \tag{1.2b}$$

$K_{5/3}(y)$ is a modified Bessel function and γ is the thermal Lorentz factor.

Consider a power law energy distribution of electrons in which the total number of electrons in the source N_e is given by

$$N_e = \int\int N_0 \gamma^{-n}\, \mathrm{d}\gamma \mathrm{d}V\,. \tag{1.3}$$

Combining (1.1) and (1.3) and assuming an isotropic distribution of pitch angles, one finds an emissivity of

$$j(\nu) = \left(\frac{8\pi^2 N_0 e^2}{c}\right) \nu_B a(n) \left(\frac{3\nu_B}{2\nu}\right)^{\alpha} \text{ergs cm}^{-3}\,\text{s}^{-1}\,, \tag{1.4}$$

where α is the spectral index of the radiation,

$$\alpha = \frac{n-1}{2}\,, \tag{1.5}$$

and

$$a(n) = \frac{\left(2^{\frac{n-1}{2}}\sqrt{3}\right)\Gamma\left(\frac{3n-1}{12}\right)\Gamma\left(\frac{3n+19}{12}\right)\Gamma\left(\frac{n+5}{4}\right)}{8\sqrt{\pi}(n+1)\Gamma\left(\frac{n+7}{4}\right)}\,. \tag{1.6}$$

For a uniform source in a volume, V, the total number of particles contributing to the synchrotron radiation in the frequency interval $\nu_1 \leq \nu \leq \nu_2$ is found from (1.3) to be

$$N_r = N_0 V \int_{\gamma_1}^{\gamma_2} \gamma^{-n}\, d\gamma\,, \tag{1.7}$$

and the energy content stored in electrons is

$$U_e = m_e c^2 N_0 V \int_{\gamma_1}^{\gamma_2} \gamma^{(1-n)}\, d\gamma\,. \tag{1.8}$$

It has been shown [34] that by integrating the modified Bessel function spectrum of the individual electrons over the energy distribution in (1.3), that it is accurate to use (1.4) to within 10% if one chooses

$$\gamma_1 = \left[\frac{2\nu_1 y_1(n)}{3\nu_B}\right]^{\frac{1}{2}}\,, \tag{1.9a}$$

$$\gamma_2 = \left[\frac{2\nu_2 y_2(n)}{3\nu_B}\right]^{\frac{1}{2}}\,, \tag{1.9b}$$

where

$$y_1(n) = 2.2,\qquad y_2(n) = 0.10,\qquad n = 2.5, \tag{1.9c}$$

$$y_1(n) = 2.7,\qquad y_2(n) = 0.18,\qquad n = 3.0. \tag{1.9d}$$

Then, integrating $j(\nu)$ in (1.4) over frequency we find, in terms of the spectral luminosity; $L(\nu) = \int dV\, j(\nu)$, values of the total number of electrons in the source and total electron energy from (1.7) and (1.8),

$$N_r \approx \frac{6\times 10^{20} L(\nu_1) B^{-1}}{(n-1)a(n)} [y_1(n)]^{\frac{n-1}{2}} \left[1 - \left(\frac{y_2(n)\nu_1}{y_1(n)\nu_2}\right)^{\frac{n-1}{2}}\right] , \tag{1.10}$$

$$U_e \approx \frac{2\times 10^{11} B^{-\frac{3}{2}}}{a(n)(n-2)} L(\nu_1)\nu_1^{\frac{1}{2}} [y_1(n)]^{\frac{n-1}{2}} \times \left[1 - \left(\frac{y_2(n)\nu_1}{y_1(n)\nu_2}\right)^{\frac{n-1}{2}}\right] . \tag{1.11}$$

The total magnetic energy density is

$$U_B = \int \frac{B^2}{8\pi} \mathrm{d}V . \tag{1.12}$$

and there is also energy stored in protonic matter, U_{prot}. Most of the energy is probably contained in cool protons, however this is probably entrained matter and is not related to the subject of the energy flux from the central engine. We expect charge neutrality so $N_p \approx N_e$. Most of the electrons are probably cool in a steep power law, as in (1.3), with $\gamma \gtrsim 1$. Low energy electrons radiate below 10 MHz so their emission is not measured by radio telescopes. Secondly, even if it were measured, it would be highly suppressed due to absorption by thermal galactic gas as evidenced by the spectral turnover below 20 MHz in Cygnus A [35]. The energy content in protonic matter is often described by the parametric relation $U_{prot} \approx aU_e$ [36]. So the total energy density is

$$U_{TOT} \approx U_B + (a+1)U_e . \tag{1.13a}$$

Typical values of "a" used in the past are of the order of 100 based on cosmic ray energies [36]. However, all of the data described below is interpretable with $a = 0$. If one assumes that U_{prot} scales with U_e, minimizing U_{TOT} in (1.13a) with respect to B, implies by (1.11) and (1.12) that

$$U_B = 3/4\,(a+1)U_e . \tag{1.13b}$$

This is commonly known as the minimum energy state and is close to equipartition if $a = 0$.

1.3.2.1 The Radio Lobes of Cygnus A

Now consider the case of Cygnus A. The spatial dimensions of the radio lobes in Cygnus A are approximately 45 kpc by 25 kpc in each lobe. Thus, the total lobe volume is $5\times 10^{69}\,\mathrm{cm}^3$. The spectral index is $\alpha = 0.66$ from 38 to 750 MHz and the two point spectral index from 38 MHz to 5 GHz is $\alpha = 0.84$ [37]. For simplicity, we choose a homogeneous volume, even though the emissivity of the hot spots at the end of the lobes in Fig. 1.6 is greatly enhanced (we treat the hot spots separately in the next section). The integrated radio luminosity is $P_E \approx 10^{45}$ ergs s^{-1}, below 100 GHz. Again, for simplicity, we choose a single power law, $\alpha = 0.84$, or $n = 2.68$ in the electron energy distribution according to (1.5). We assume that most of the electrons are cool, so $\gamma_{min} \gtrsim 1$. Thus, from (1.9)

$$\nu_1 \approx 3.6\nu_B \,. \tag{1.14}$$

Notice that the average electron thermal Lorentz factor of the ensemble is

$$\bar{\gamma} = \frac{\alpha - 1}{\alpha - 2} = 2.5 \,. \tag{1.15}$$

Setting $E_{min} = (a + \frac{7}{4})U_e$ minimizes the energy in the lobes according to (1.13). Using (1.11) for U_e, one can solve for B in the minimum energy state. We require an expression for $L(\nu_1)$ in (1.11) which is defined through the value of ν_1 in (1.14) and this value must agree with our final computed value of B (see 1.2). Anticipating a final value of $B \approx 10^{-4}$ G, from (1.14) we get $\nu_1 \approx 10^4$ Hz. As we mentioned above, the low frequency emission is not detected at earth, so we really don't know the actual value of ν_1. The spectral luminosity is therefore approximated as

$$L(\nu) = 3 \times 10^{42} \nu^{-0.84} \mathrm{ergs\,s^{-1}\,Hz^{-1}} , \quad 10^4 \mathrm{Hz} < \nu < 10^{11} \mathrm{Hz} \,. \tag{1.16}$$

Combining, (1.11)–(1.13) with (1.17) the minimum energy magnetic field strength in the lobes is

$$B \approx (1 + a)^{2/7}\, 7.5 \times 10^{-5} \mathrm{G} \,. \tag{1.17}$$

Similarly, the total energy in the lobes is found from (1.13),

$$E_{min} = 2.5 \times 10^{60} \mathrm{ergs} \,, \tag{1.18}$$

where we take $a = 0$ in the remainder of the discussion.

This number makes sense as a minimum value based on the X-ray data on the thermal bremsstrahlung emission from the enveloping intracluster gas. It was found in [39] that $n_e \approx 10^{-2} \mathrm{cm}^{-3}$ and $T \approx 4 \times 10^7\,^\circ$K in the gas surrounding the lobes. In order to inflate the lobes into the intracluster medium requires $P\Delta V$ work. If we take the properties of the intracluster medium from the X-ray data above, $P\Delta V \approx 3 \times 10^{59}$ergs. Thus, (1.18) exceeds the minimum requirement that there needs to be at least enough internal energy to inflate the radio lobes against the pressure of the enveloping medium.

1.3.2.2 The Hot Spots in Cygnus A

The jets in Cygnus A apparently terminate in luminous hot spots at the end of the radio lobes as indicated in Fig. 1.6. The radio luminosity of the hot spots is emitted primarily from two disjoint regions that are roughly spherical with a diameter of 3.5 kpc (see Fig. 1.6, noting that the overall length of the source is about 170 kpc). Therefore, the combined volume of the hot spots is only about $10^{66}\,\mathrm{cm}^3$, yet the total hot spot radio luminosity is on the order of $5\times10^{44}\,\mathrm{ergs\,s^{-1}}$ [40]. The radio spectrum radiated from the hot spots has a spectral index of approximately $\alpha = 0.6$ [38]. Thus, we very crudely estimate the spectral luminosity of the hot spots as we did for the radio lobes (anticipating a final computed magnetic field strength of roughly 2×10^{-4}G):

$$L(\nu) = 8\times10^{39}\nu^{-0.6}\,\mathrm{ergs\,s^{-1}\,Hz^{-1}}\,, \quad 2\times10^{4}\,\mathrm{Hz} < \nu < 10^{11}\,\mathrm{Hz}\,, \tag{1.19a}$$

$$L(\nu_1) = 3.2\times10^{37}\,\mathrm{ergs\,s^{-1}\,Hz^{-1}}\,. \tag{1.19b}$$

Combining (1.11)–(1.13) with the value of $L(\nu_1)$ in (1.19b), we find the minimum energy magnetic field strength in the lobes,

$$B \approx 2.3\times10^{-4}\mathrm{G}\,. \tag{1.20}$$

This value agrees with the soft X-ray, ROSAT, observations of the lobes that is consistent with a scenario in which the hot spot material is in a minimum energy state. The radio synchrotron emission is considered to be upscattered to X-ray frequencies through inverse Compton scattering by the same hot electron population that produced the seed synchrotron radio luminosity (this is known as an SSC process and is discussed at length in Chaps. 10 and 11). It was deduced in [41] that the broadband radio to X-ray spectrum of the hot spots is consistent with the SSC spectrum of a minimum energy hot electron plasma that is characterized by a magnetic field strength of approximately 2×10^{-4} G as was found in (1.20).

1.3.2.3 The Lobe Advance Speed and Spectral Aging in Cygnus A

Inspection of Fig. 1.6 reveals that the highly collimated jet enters the lobe where it becomes destabilized and begins to decollimate. A reduced jet pressure necessarily results from this expansion. This makes the jet susceptible to the forces of the external gas pressure of the intracluster medium. The flaring jet terminates in a shock at the end of the lobe that energizes the hot spots. The terminus of the jet forms a working surface that clears out the intracluster medium replacing it with lobe material. Apparently, the flow of jet material does not disappear at the hot spot, but creates a diffuse back flow that is the radio lobe. The strongest evidence for this circumstance is spectral aging. The farther that the lobe gas is from the hot spot, the longer it has been since it exited the jet at the hot spot and was diverted into this back flow. Thus, the back flow has a correspondingly longer time to radiate away thermal energy as synchrotron photons the farther one is from the hot spot. This creates a correlation between the depletion of high energy electrons and distance within the lobe from the hot spot. Thus, the high frequency spectral turnover that this paucity of high energy electrons produces is found to occur at lower and lower frequencies the farther from the hot spots that one observes the lobe. This is known as spectral aging. The phenomenon is clearly exemplified by the fact that the 70 kpc gap between the two radio lobes seen in the 5 GHz image in Fig. 1.6 is filled in at low frequency (327 MHz) making for a continuous ovaloid of radio emission [38]. In this subsection, we describe the dynamics of the working surface at the hot spots and incorporate our knowledge of spectral aging in the lobes to estimate the age of Cygnus A.

Pressure balance at the working surface of the lobe/intracluster medium interface yields

$$\rho_{int} v_{adv}^2 = P \geq P_{min} \approx \left[\frac{U_B + U_e}{3}\right], \tag{1.21}$$

where ρ_{int} is the value of the mass density of the intracluster medium adjacent to the lobes and v_{adv} is the speed at which the lobes propagate into the intracluster medium. Using the previously quoted value from [39] of $\rho_{int} \approx 10^{-2}\, m_p\, \mathrm{cm}^{-3}$, (1.21) can be evaluated in conjunction with (1.13) and (1.20) to yield a lower bound on v_{adv},

$$v_{adv} \geq 0.0095c\,. \tag{1.22}$$

Spectral aging in the minimum energy magnetic fields in the lobes yields an estimate of lobe age, $t_{sa} \approx 1.8 \times 10^{14}$ s [38, 40]. Compare this with the lobe separation time based on v_{adv} in (1.22), $t_{sep} \leq 9 \times 10^{14}$s. Thus, the two numbers are consistent if either v_{adv} has slowed over time or the lobes are not in the minimum energy state. Consider, the first possibility that implies that t_{sa} is the relevant age of the radio source. Combining this age with the minimum energy stored in the lobes from (1.18) and the radio luminosity, P_E, yields a lower bound on the energy flux supplied by the central engine, Q,

$$Q \geq \frac{\mathrm{d}E_{min}}{\mathrm{d}t} \approx 1.5 \times 10^{46} \mathrm{ergs\,s}^{-1}\,. \tag{1.23a}$$

Alternatively, [38] find that one can equate t_{sa} and t_{sep} if B is less than its minimum energy value within the lobes. To see how this occurs, note that a relativistically hot electron radiates most of its energy near the critical frequency, $\nu_m = \frac{2}{3} f_c$ (where f_c was defined in 1.2b). Thus, for a given magnetic field strength, particles of thermal Lorentz factor, γ, radiate predominantly at a frequency $\nu = \nu_m(\gamma)$. When a break in the radio spectrum occurs at $\nu = \nu_m(\gamma)$, then the implication is that the plasma has existed for a sufficient length of time within the lobe for particles of thermal Lorentz factor, γ, to have radiated away their energy. This synchrotron radiation time scale is given by $t_{sy} = (\gamma m_e c^2)/P$, where the synchrotron power, P, can be found by integrating (1.1) over frequency. From [33] we have (using cgs units)

$$t_{sy}(\gamma) = \frac{5 \times 10^{11}}{\sqrt{\nu_m} B^{\frac{3}{2}}} \mathrm{s}\,. \tag{1.23b}$$

Thus, t_{sep} will agree with t_{sa} in the lobe if B is decreased by a factor of about three in the lobes below its equipartition value. From (1.11), this will increase the electron thermal energy by a factor of about 5 and the magnetic energy will decrease by a factor of 9 as a consequence of (1.12). Thus, the energy stored in the lobes, E_{stored} will increase by a factor of three over its minimum energy value given by (1.18),

$$E_{stored} \approx 7.5 \times 10^{60} \text{ergs} . \tag{1.23c}$$

We want to estimate the total energy delivered by the central engine, so we must also consider the prodigious radiation losses during the lifetime of the source:

$$P_E \, t_{sep} \approx 9 \times 10^{59} \text{ergs} . \tag{1.23d}$$

Secondly, we must also determine the bulk kinetic energy of the protons passing through the hot spots. From the average thermal Lorentz factor of the electrons in (1.15) and the lobe advance speed in (1.22), the bulk kinetic energy of the protons represents 7% of the total energy flux provided by the central engine. Combining this with (1.23cd) we can estimate the total energy flux delivered by the central engine, Q, in this nonequipartition scenario,

$$Q = 1.07 \left(\frac{E_{stored} + P_E t_{sep}}{t_{sep}} \right) \gtrsim 10^{46} \text{ergs s}^{-1} . \tag{1.23e}$$

1.3.2.4 The Central Engine of Cygnus A

There is only one large sample of estimates of Q_{min} for radio loud AGN [42]. We compare Q_{min} with the extended luminosity, P_E, for the stronger radio galaxies in their sample in Table 1.2. Notice how the value of $Q_{min} \gtrsim 10 P_E$ computed for Cygnus A in (1.23a) is typical of the results contained in Table 1.2.

Furthermore, these estimates ignore the contribution from hot protons (i.e., $a \gg 0$ in (1.13)). If hot protons contribute to the internal energy of the lobes then the values of Q_{min} in (1.23a) and Table 1.2 are under estimated.

Table 1.2 Kinetic powers of FR II radio galaxies

Source	Q_{min} (ergs s^{-1})	P_E (ergs s^{-1})
3C 33	9×10^{44}	1×10^{43}
3C 42	3×10^{45}	3×10^{44}
3C 79	6×10^{45}	2×10^{44}
3C 109	6×10^{45}	2×10^{44}
3C 123	5×10^{45}	1×10^{45}
3C 173.1	5×10^{45}	2×10^{44}
3C 219	1×10^{45}	8×10^{43}
3C 244.1	3×10^{45}	5×10^{44}
3C 295	7×10^{45}	3×10^{45}
3C 300	1×10^{45}	1×10^{44}
3C 341	2×10^{46}	3×10^{44}
3C 438	3×10^{45}	7×10^{44}

According to the calculations of the last subsection and (1.23e), in particular, the central engine has already delivered at least 10^{61} ergs $\approx 5 \times 10^6\,\mathrm{M}_\odot c^2$ to the radio lobes during the past lifetime of Cygnus A. The future lifetime of the source and past source evolution can only increase the total energy supplied by the central engine. Combining this with the fact that no known physically reasonable process (we dismiss annihilation energy as a reasonable choice for the central engine that produces a steady long term source of collimated energy flux in the radio band with no measurable γ-rays) is close to converting 100% of its rest mass to energy and we have only a minimum energy bound for Q, we conclude that the mass of the central engine in Cygnus A $\gg 10^7\,\mathrm{M}_\odot$.

The reason for studying Cygnus A in such detail, as opposed to another strong radio galaxy from Table 1.2, is that Cygnus A is a much stronger radio source (as measured at earth) due to its cosmological proximity. This allows high resolution (high frequency) VLBA detection of the radio jet that provides an upper bound to the physical size of the central engine. We cannot attain as tight a constraint on more distant, fainter, strong FR II radio sources. Consider the VLBI data on Cygnus A in Fig. 1.10 [43]. The 43 GHz VLBA map shows the jet emanating from a region less than one light year across. This upper limit is almost certainly not physical, but would decrease with increasing resolution. Considering that we estimated the mass of the central engine to be $\gg 10^7\,\mathrm{M}_\odot$, it is hard to understand how anything except a black hole could exist with such a large mass in such a small volume. A discussion as to how such a mass concentration would have to coalesce to form a black hole can be found in [44] and references therein. Similar arguments have been made for the central mass concentration in the Milky Way in [45], where it was found that coalescence to a black hole would occur in less than 10^7 years, which is less than the lifetime of the radio source in Cygnus A.

1.3.3 Summary of Evidence of a Black Hole Central Engine in Radio Loud AGN

It was demonstrated in Sect. 1.3.2, through a very crude analysis, that the central engine of a radio loud AGN supplies far more power to the radio lobes than is indicated directly from the radio luminosity. The large volume from which the radiation is emitted requires enormous amounts of hot particles and magnetic flux. The stored energy, $E_{min} \sim V^{3/7}$ in the minimum energy analysis of (1.13). The total energy supplied to a powerful FR II radio source (a quasar, "hidden" or otherwise in the unified scheme) would require the complete conversion of 10^6–$10^9\,\mathrm{M}_\odot$ of rest mass to energy in the lobes (see maps [25] and repeat the analysis of Sect. 1.3.2 to see that 3C 9 requires more than $10^8\,\mathrm{M}_\odot c^2$ of energy to power the radio source). VLBA maps show that radio jets emerge from regions smaller than one light year. The fact that at least $\sim 10^8\,\mathrm{M}_\odot$ exists in a central engine less than one light year across yields the unavoidable conclusion that a supermassive black hole resides in the central engine. This leaves two possible energy sources, the black hole or the accretion flow onto the black hole.

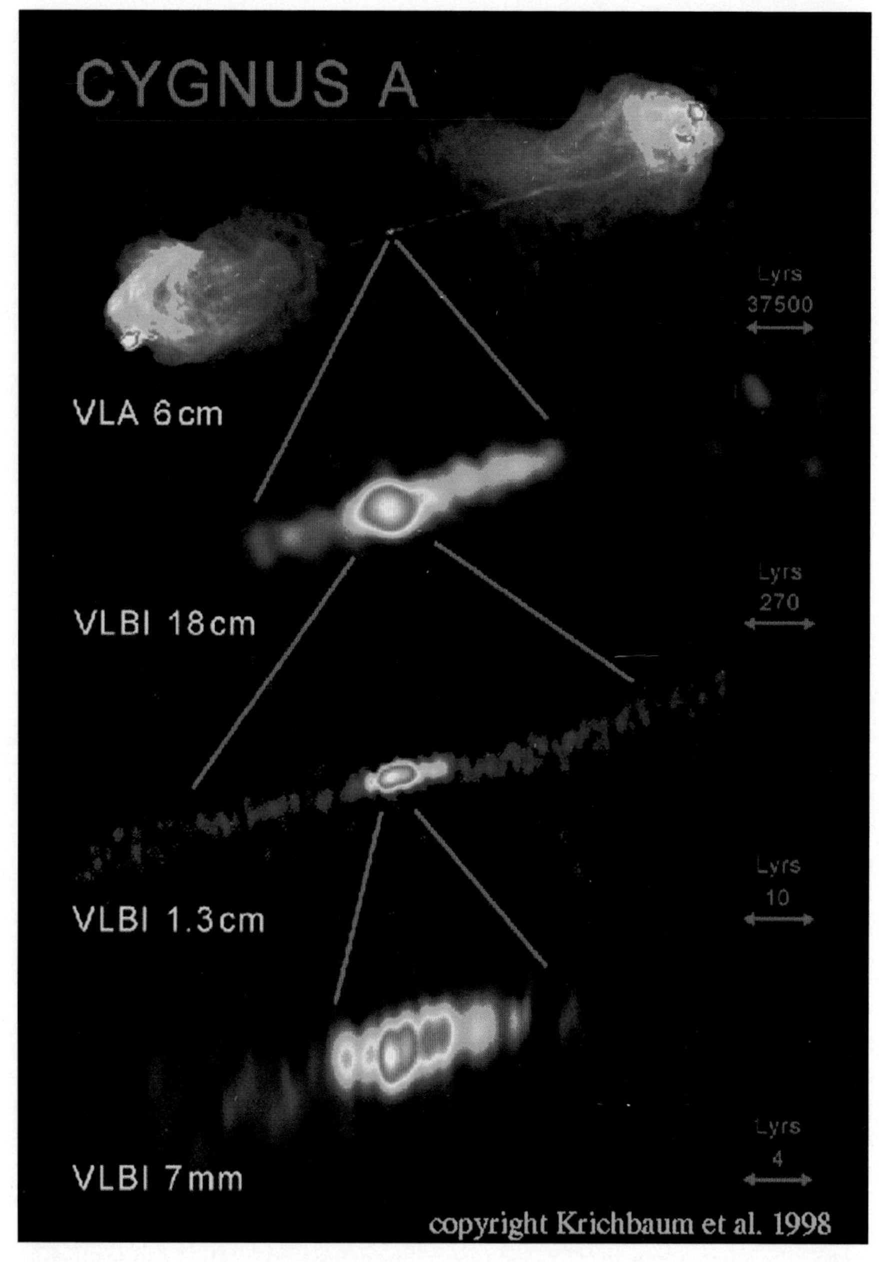

Fig. 1.10 The jet in Cygnus A is mapped from scales on the order of 50 kpc down to less than a light year in this series of inserts. The VLBI maps indicate that the central engine is less than a light year in diameter. The images are from Krichbaum et al. 1998

The bimodal distribution of quasar radio luminosity and the lack of strong Doppler beaming in the lobes (as discussed in Sect. 1.3.1) implies that radio loudness is not primarily a consequence of line of sight effects relative to the intrinsic geometry of the quasar. Radio loud quasars seem to be physically different than most quasars. Even when there is more power radiated into the large scale radio structures than is released from viscous dissipation in the accretion flow (optical/UV luminosity) the broad UV emission line profiles as well as the optical/UV continuum spectrum and luminosity are indistinguishable from those found in the radio quiet quasar population. Thus, the accretion flow does not seem to power FR II radio emission, otherwise these extremely powerful radio sources would have distinct optical/UV signatures due to their vastly different dissipation mechanisms.

The strongest radio sources in the unified scheme are associated with quasar activity. The quasar in the host galaxy is now commonly believed to be the result of large viscous losses in an accretion flow that scales with the accretion rate. Thus, within the radio loud population, large accretion rates yield strong radio structures; yet these are not powered by accretion. However, large accretion rates tend to spin up the central black hole. The central engines of strong FR II radio sources would seem to be rapidly rotating supermassive black holes.

The conclusion of this section is currently the subject of debate in the astrophysical community. However, at the time of publication, rapidly rotating supermassive black holes are the most viable known power sources for explaining all of the properties of the radio loud AGN population.

1.4 Extracting Energy from a Black Hole

In the last section, it was demonstrated that based on our current knowledge of physics that supermassive black holes are located in the central engines of extragalactic radio sources. Furthermore, comparing the radio loud and radio quiet quasar populations implies that accretion flows yield the quasar emission and the physical state of the central black hole is determinant for the existence of bipolar radio jets. Yet, the signature of a black hole is the manner in which it sucks mass–energy inescapably toward the event horizon. Thus, how can energy be extracted from the central black holes of radio loud AGN?

Mathematically it has been shown that rotating or charged black holes are “alive” and that some of their energy is extractable [15]. However, no practical physical realization of an efficient process has been available to theorists until the advent of black hole GHM.

The most general electro-vac black hole solution is that of Kerr–Newman. The axisymmetric, time stationary space–time metric is uniquely determined by three quantities, M, a, and Q, the mass, angular momentum per unit mass, and the charge of the hole respectively. In this book we treat the physics as occurring on a background Kerr–Newman space–time (i.e., the energy density of external plasma and fields are too small to affect the metric). In Boyer–Lindquist coordinates the metric,

$g_{\mu\nu}$, is given by the line element

$$\mathrm{d}s^2 \equiv g_{\mu\nu}\,\mathrm{d}x^\mu \mathrm{d}x^\nu = -\left(1-\frac{2Mr-Q^2}{\rho^2}\right)\mathrm{d}t^2+\rho^2\mathrm{d}\theta^2+\left(\frac{\rho^2}{\Delta}\right)\mathrm{d}r^2$$
$$-\frac{(4Mr-2Q^2)a}{\rho^2}\sin^2\theta\,\mathrm{d}\phi\,\mathrm{d}t+\left[(r^2+a^2)+\frac{(2Mr-Q^2)a^2}{\rho^2}\sin^2\theta\right]\sin^2\theta\,\mathrm{d}\phi^2\,, \tag{1.24a}$$

where

$$\rho^2 = r^2+a^2\cos^2\theta, \tag{1.24b}$$
$$\Delta = r^2-2Mr+a^2+Q^2 \equiv (r-r_+)(r-r_-)\,. \tag{1.24c}$$

There are two event horizons given by the roots of the equation $\Delta = 0$. The outer horizon at r_+ is of physical interest

$$r_+ = M+\sqrt{M^2-Q^2-a^2}\,. \tag{1.25}$$

The surface area of the event horizon, A, is given by

$$A = 4\pi(r_+^2+a^2) = 16\pi M_{ir}^2\,, \tag{1.26}$$

where M_{ir} is identified as the irreducible or rest mass of the hole [15]. Combining (1.25) and (1.26), the mass of the black hole decomposes into its rest mass, the electromagnetic energy and the rotational energy as follows:

$$M^2 = \left(M_{ir}+\frac{Q^2}{4M_{ir}}\right)^2+\left(\frac{Ma}{2M_{ir}}\right)^2\,. \tag{1.27}$$

Now consider the variation of $M_{ir}, \delta M_{ir}$ in (1.26) from the capture of a particle of energy, $\mathcal{E}$, angular momentum about symmetry axis of the hole, $\mathcal{L}_\phi$ and electric charge, q,

$$\frac{\delta M_{ir}}{M_{ir}} = \frac{1}{\sqrt{M^2-a^2-Q^2}}\left[\mathcal{E}-\frac{\Omega_H}{c}\mathcal{L}_\phi\right]\,. \tag{1.28}$$

In (1.28) the generalized four momentum $\pi^\mu = P^\mu + q/cA^\mu$ is used to define $\mathcal{E}$ and $\mathcal{L}_\phi$. The mechanical four momentum is P^μ and A^μ is the Kerr–Newman vector potential. In Boyer–Lindquist coordinates Ω_H is the angular velocity of the horizon as viewed from asymptotic infinity ($r \to +\infty$),

$$\Omega_H = \frac{a}{r_+^2+a^2}\,, \tag{1.29}$$

$$\mathcal{E} \equiv -\pi\cdot\partial/\partial t\,, \tag{1.30a}$$
$$\mathcal{L}_\phi \equiv \pi\cdot\partial/\partial\phi\,. \tag{1.30b}$$

The values of $\mathcal{E}$ and $\mathcal{L}_\phi$ cannot be assigned arbitrarily to a particle near the event horizon. Namely, in a physical frame near the horizon (i.e., one with a time-like world line defined through the four velocity normalization, $u \cdot u = -1$) at fixed "r" coordinate, the velocity of the particle must be inward at almost the speed of light. This will be proven rigorously in Chap. 3, but it follows from the defining characteristic of the event horizon. Light cannot escape outward through the horizon, thus massive particles must move inward with a divergently large four velocity relative to a noninertial observer at fixed coordinate $r \gtrsim r_+$. In this frame π^0 is therefore dominated by the inertial piece, P^0, from the inward ultrarelativistic motion.

Consider such a frame at fixed radial coordinate, given by the Zero Angular Momentum Observers (ZAMOs) with four velocity,

$$u^\mu = \frac{1}{\sqrt{g_{tt} - g_{\phi\phi}(g_{\phi t})^2}}\left[\frac{\partial}{\partial t} - \frac{g_{\phi t}}{g_{\phi\phi}}\frac{\partial}{\partial \phi}\right], \tag{1.31}$$

where $g_{\mu\nu}$ are the Boyer–Lindquist metric coefficients of (1.24) and $u \cdot u = -1$. In this frame, $\pi^0 = -\pi \cdot u$. Noting that in the expression for the ZAMO four velocity in (1.31),

$$\lim_{r \to r_+} \left(-g_{\phi t}/g_{\phi\phi}\right) = \Omega_H\,, \tag{1.32}$$

we can simplify the expression for $\delta M_{ir}/M_{ir}$ in (1.28):

$$\frac{\delta M_{ir}}{M_{ir}} = \lim_{r \to r_+}\left[\sqrt{g_{tt} - g_{\phi\phi}(g_{\phi t})^2}\left(\pi^0/\sqrt{M^2 - a^2 - Q^2}\right)\right], \tag{1.33a}$$

$$\approx \lim_{r \to r_+}\left[\frac{\sqrt{2}(r - r_+)^{1/2}P^0 \sin\theta}{(g_{\phi\phi})^{1/2}}\right] > 0\,. \tag{1.33b}$$

The mechanical energy of the particle is necessarily positive, $P^0 > 0$, in the physical ZAMO frame and outside the horizon, $r > r_+$, thus $\delta M_{ir} > 0$ in (1.33b). In fact it will be shown in Chap. 3 that $(r - r_+)^{1/2}P^0$ is a positive constant in the vicinity of the horizon. The inequality in (1.33) proves the result that the irreducible mass of a black hole must always increase when a classical particle is absorbed by the horizon.

Differentiating (1.27) yields the first law of black hole thermodynamics.

$$\mathrm{d}M = \frac{\kappa}{8\pi}\mathrm{d}A + \Omega_H \mathrm{d}(Ma) + \left(A_t - \frac{\Omega_H}{c}A_\phi\right)\mathrm{d}Q\,, \tag{1.34}$$

where κ is the surface gravity at the horizon,

$$\kappa = \frac{\sqrt{M^2 - a^2 - Q^2}}{r_+^2 + a^2}\,. \tag{1.35}$$

In an astrophysical context it is difficult to get a charge on the hole large enough so that $|Q| \sim M$. Note that this upper bound ($Q^2 \ll M^2$) does not preclude charges and their fields that would be enormous even by astrophysical standards (see the discussion of Chap. 11 for example). Thus, in order to understand the consequences of the first law of black hole thermodynamics in an astrophysical context, we can ignore the charge on the hole in (1.34). In analogy to (1.30a) and (1.30b) consider the absorption of a particle with mechanical energy, ω, and angular momentum, m;

$$\omega = -P \cdot \partial/\partial t\,, \tag{1.36a}$$

$$m = P \cdot \partial/\partial \phi\,. \tag{1.36b}$$

Because $\delta M_{ir} > 0$ during the absorption of a classical particle, $dA > 0$ in (1.34). This is known as the second law of black hole thermodynamics. The second law and (1.34) imply that for a particle to be absorbed by the horizon,

$$\omega - \frac{\Omega_H}{c} m > 0\,. \tag{1.37}$$

From (1.30) this also ensures that $P^0 > 0$ in the ZAMO frames as required for a physical particle.

Consider the case $\omega < 0$ (negative energy as viewed from asymptotic infinity) and $m < 0$ for particle capture in the context of the condition (1.37). The surface area of the horizon will increase according to the second law of black hole thermodynamics, requiring $|\omega| < |(\Omega_H/c)m|$. Such states do not exist as $r \to +\infty$ since $\omega < 0$; however, these particles need not have originated far from the hole at large "r" coordinate. Near the horizon such states exist because the locally measured energy is positive, $P^0 > 0$, in the ZAMO frames by (1.30) and (1.37). The locally measured energy by any other physical observer (i.e., with a four velocity satisfying, $u \cdot u = -1$) at the same "r" coordinate as the ZAMO is related by a local Lorentz transformation. Thus $P^0 > 0$ in the ZAMO frame is a necessary and sufficient condition to prove positive energy as measured by all observers near the horizon. This is an allowed absorption process of physically well-defined particles that has $dM = \omega < 0$ and $d(Ma) = m < 0$. The absorption decreases the energy and angular momentum (in the decomposition of (1.27)) of the black hole and increases the surface area of the horizon in the process, M_{ir}. Essentially it is a process that extracts the rotational energy of the hole.

The available extractable energy is the reducible mass,

$$M_{red} = M - M_{ir} = M - \frac{1}{2}\sqrt{r_+^2 + a^2}\,. \tag{1.38}$$

For a Schwarzschild black hole $M_{red} = 0$. For a maximally rotating black hole

$$M_{red} = M(1 - \sqrt{2}) = 0.29M\,, \quad a = M\,. \tag{1.39}$$

In high accretion systems such as a quasar, $a \lesssim M$ seems unavoidable. However, in practice it is unrealistic to extract the maximum theoretical value of 29% of the black hole rest mass.

The process by which the reducible mass is extracted required the absorption of a particle with $\omega < 0$. Clearly there are no such particles at infinity. In Kerr black holes, $\omega < 0$ states can exist in the ergosphere, yet these particles do not come from asymptotic infinity; they must be created within the ergosphere. Any process that extracts energy from the hole must rely on $\mathcal{E} < 0$ states, where in all generality A^μ can include contributions from electromagnetic sources outside the horizon.

It has been shown mathematically how the rotational energy of a black hole can be tapped using $\mathcal{E} < 0$ states of matter. However, we still know nothing about whether there is any realizable physical process that can extract M_{red} in an astrophysical context. Our interest is to apply this concept to the central engines of radio loud AGN. The biggest clue that we have is that the jets in FR II radio sources are highly collimated. The radio image of Cygnus A has a jet with an opening angle of $\leq 1.5°$ over $\sim$35 kpc in length [47]. First consider that the intracluster medium is not pressurized enough nor sufficiently uniform to collimate the jet over this length. This is verified by the expansion at the end of the jet into the intracluster medium as an expanding lobe. Secondly, consider a fluid that contains both gas pressure and pressure from a tangled local magnetic field. The jet would be describable as a hydrodynamic fluid. Consider a hypersonic under-expanded jet emanating from a nozzle with a half opening angle ϕ_e and exit velocity u_e. The opening angle of the jet, ϕ, is accurately calculated to be [14]:

$$\tan\phi \approx \left[\frac{u_e}{u}\right]^{\frac{\Gamma+1}{2}} \frac{\sqrt{2}}{\Gamma-1}\frac{1}{M_e}\,,$$
$$\times\left[\frac{M_e^2}{2}\left(\tan^2\phi_e\right)(\Gamma-1)^2 + 1 - \left(R/R_e\right)^{-2(\Gamma-1)}\right]^{1/2}\,, \tag{1.40a}$$

where Γ is the gas adiabatic constant, u is the axial velocity of the jet, R its cross-sectional radius, and R_e the same at the nozzle exit. For large exit Mach numbers, M_e, in a decelerating jet

$$\tan\phi \approx \left[\frac{u_e}{u}\right]^{\frac{\Gamma+1}{2}} \tan\phi_e\,, \quad M_e^2\left(\tan^2\phi_e\right) \gg \frac{2}{(\Gamma-1)^2}\,. \tag{1.40b}$$

For an accelerating jet,

$$\tan\phi \approx \tan\phi_e\,, \quad M_e^2\left(\tan^2\phi_e\right) \gg \frac{2}{(\Gamma-1)^2}\,. \tag{1.40c}$$

In the limit of $\tan\phi_e = 0$, a decelerating jet asymptotically approaches an opening half angle,

$$\tan\phi \approx \left[\frac{u_e}{u}\right]^{\frac{\Gamma+1}{2}} \frac{\sqrt{2}}{\Gamma-1} M_e^{-1}\,, \tag{1.40d}$$

and an accelerating jet has an opening half angle given by

$$\tan\phi \approx \frac{\sqrt{2}}{\Gamma-1} M_e^{-1}\,. \tag{1.40e}$$

Equations (1.40d) and (1.40e) reveal the physics of collimation of an under-expanded jet in relation to (1.40). The jet is kinematically collimated. It is over-pressurized relative to the ambient medium, so it expands as fast as it can, sonically. Thus, the opening angle is approximately the ratio of the axial speed to the sonic speed (as a result of free expansion) $\sim M_e^{-1}$.

For all values of ϕ_e and equations of state, (1.40) implies a very large Mach number in a hydrodynamic jet model of the kiloparsec scale jet in Cygnus A. The Mach number would have to be so large that the jet would be highly relativistic, $1 - u^2/c^2 \ll 1$. There is no evidence to support the existence of this type of jet velocity on scales $\sim$50 kpc in any FR II source [25, 46]. It is most likely, although not rigorously verified, that kiloparsec scale jets are only mildly relativistic. If the kiloparsec scale jets were highly relativistic, it would be difficult to explain the high surface brightness jets seen in some FR II radio galaxies such as 3C 219 (see Fig. 1.9). Even at smaller scales, the VLBI maps of Cygnus A [43] indicate jet speeds of $<0.7c$ on $\sim$100 pc from the central engine. A similar result is obtained in [48].

The large scale collimation of the jet in Cygnus A can be accomplished by hoop stresses from a large scale toroidal magnetic field [49]. No one has ever successfully modeled such an electrodynamically collimated jet [50, 51]. However, a real jet requires dissipation (magnetic induction) for closure currents to flow [52]. Jet models to date have utilized only the perfect MHD assumption, and this is inconsistent with the physics of a jet. It seems unavoidable that toroidal magnetic fields collimate FR II radio jets and we simply have not developed the computational and theoretical ability to model the global plasma state of a real jet. In spite of this, the 3-D simulation of [53] shows some striking resemblance to the jet in Cygnus A. This model has a substantial toroidal magnetic field.

The prominent role of magnetic fields (collimation and synchrotron emission) in radio jets suggests that large scale poloidal magnetic flux links the jets to the central engine. In the context of radio loud AGNs, we are compelled to explore the possibility of somehow indirectly applying a large scale magnetic torque to a rapidly rotating supermassive black hole (there is no surface to torque directly as in a star). This process requires an understanding of GHM in the ergosphere where $\mathcal{E} < 0$ states can exist. Ostensibly, we are looking for a GHM interaction in which a relativistic jet is driven outward and $\mathcal{E} < 0$ plasma is simultaneously created in the ergosphere that taps the rotational energy of the hole. In a global energy conservation context, the rotational energy of the hole (reducible mass) is powering the jet. Extracting the rotational energy of the hole is consistent with the notion that large accretion systems (AGNs with the quasar phenomena, "hidden" or otherwise) have central black holes that rotate the fastest and have the most powerful radio jets (as evidenced by the powerful FR II radio lobes they support).

1.5 Historical Perspective

The discovery of apparent superluminal motion in radio loud quasars and BL Lac objects on subkiloparsec scales with VLBI in the 1970s changed our perspective on radio loud AGN. Superluminal motion and associated rapid flux variability were observed and shown to be a consequence of relativistic apparition due to the jet pointing almost directly toward earth (see [54] for a sophisticated discussion). Typical values of subkiloparsec scale jet Lorentz factors in radio loud quasars are inferred to be, $\gamma \sim 5-25$, see [31] and [55] for example. Thus, even though the plasma flows mildly relativistically in a kiloparsec scale jet and subrelativistically $\sim$100 kpc away in the lobes [25, 46], it is ejected from the central engine as a relativistic jet. For a large sample of superluminal sources see [56].

The existence of relativistic jets on parsec scales is an important constraint on possible central engines. Early attempts to drive the jets in radio loud quasars with radiation pressure in the steep accretion vortex at the center of a very thick accretion disk, radiating at super Eddington luminosity, were popular in the late 1970s. However, the constraint of both high collimation (by the shape of the funnel, see (1.40b)), and the need for relativistic velocity are incompatible. Only mildly relativistic jets are attainable [57, 58].

The other place that astrophysicists suspect that relativistic jets reside is in the pulsars. The spindown rate of a pulsar is generally attributed to radiation losses from the neutron star in the form of a tenuous, magnetically dominated plasma wind [59]. There is a convincing argument in [60] and [61] that the Crab Nebula is likely to be powered by the relativistic wind from the central pulsar. The pulsar wind is magnetically slung due to the strong neutron star magnetic field and is therefore far more relativistic than the solar wind that is initiated with far weaker magnetic stresses [62].

This connection with pulsar winds led to the idea of making a flat pulsar from the accretion disk in AGNs by introducing a large scale poloidal magnetic flux. This line of theoretical research began in the 1970s [63, 64]. This is still currently the most popular type of model, primarily because the astrophysicist does not really encounter any complicating effects of general relativity. In spite of the $\sim$100 papers on this subject, they still do not address many important issues. The accretion flow of a radio loud quasar radiates indistinguishably in the optical/UV from a radio quiet quasar even though in one case the effective viscosity is applied primarily by a large scale magnetic torque (i.e., in a powerful radio loud quasar), and in the other case the viscous dissipation is microscopic in origin. If one invokes complete independence of the two torquing effects then the scenario does not explain why low viscous loss systems are anticorrelated with strong FR II radio sources (unified scheme). Furthermore, these models ignore the role of magnetic flux in the ergosphere which is likely to dominate the energetics in rapidly rotating systems, as the accretion disk is not a sink for magnetic flux, but a pathway toward the hole (see Chap. 10 for a complete discussion). For these reasons, the perspective of this text is that magnetized accretion disks are not the primary component of the central engine of radio loud quasars.

At the same time, the idea of using the extractable rotational energy (reducible mass) of a rotating black hole to power a wind was being examined [65]. These authors showed through a simplified model that it was possible to power an outflow with an accreting plasma permeated by a large scale magnetic field near a rotating black hole. The inflow part of the problem was solved in the weak magnetic field limit, thus any paired outgoing wind that was driven would have to be very weak. This limit allowed the authors to treat the plasma motion as geodesic.

A few years later, a preliminary investigation was made into the opposite limit in which the magnetic filed dominates the dynamics in both the inflow and outflow [66]. Such a model is more compatible with a relativistic outflow. The authors found a mathematical model in which the black hole behaved analogously to the neutron star in a pulsar. In particular, the Poynting flux and poloidal currents appear to emanate from the event horizon as seen by a global observer (like the surface of the central neutron star in a pulsar). There is no significant plasma interaction anywhere in the flow. In fact, the solution exists and was discovered in the limit of no plasma interaction or inertia (the force-free limit, $\boldsymbol{J} \times \boldsymbol{B} = 0$). The mathematical solution has no causal structure and therefore one is free to impose a magnetically dominated solution everywhere. This conflicts with the fundamental notion of a black hole that all accretion flows are inertially dominated near the event horizon. This expedience circumvents the necessity to introduce black hole GHM. As such, the solution is very amenable to usage by those not versed in the relevant details of general relativity.

Unfortunately, by studying the causal structure of the Blandford–Znajek mechanism [66], this author along with Coroniti noticed that this solution is merely a mathematical exercise and is not physical [67]. The fundamental physics of electrodynamic extraction proposed in [66] with a super-radiant magnetic field (see [68] for a discussion of super-radiance), i.e., a field that "rotates" slower than the event horizon, was very novel and was not invalidated by the arguments of [67]. The point of [67] was that since the method of solution was not causal, the solution was not unique and the detailed plasma physics in the black hole magnetosphere replete with its physical boundary plasma will ultimately determine the amount of energy extracted from the black hole, including the electrodynamic contribution. One needs to introduce a plasma interaction with the magnetic field near the black hole. It is unavoidably a complicated plasma physics problem that requires a full development of black hole GHM.

1.6 Black Hole GHM

In order to understand how the rotational energy of a black hole can be tapped by the existence of a magnetosphere, one must explore black hole GHM. There is no surface to a black hole as there is in a neutron star powered pulsar. So any putative interaction region "hovers" outside the hole. This ill-defined region involves magnetized plasma dynamics in curved space–time that can create $\mathcal{E} < 0$ plasma in the ergosphere.

It is clear that any attempt to describe the ergospheric interaction must rely heavily on the properties of relativistic plasmas. In particular, the structure of the interaction region will be causally determined by relativistic plasma waves. Chapter 2 is a review of (not so well known) relativistic plasma physics with special attention to the waves that can be propagated in the plasma. This is a crucial chapter and is imperative to an understanding of the physics of black hole GHM. As such the treatment is unique in that it caters to a physicist's need to see results derived in a straightforward manner as opposed to treating this as a mathematical physics exercise in quasi-linear, hyperbolic systems as it appears elsewhere [69, 70]. The structure of relativistic waves is discussed for the first time with a particular emphasis on the differences between Alfvén and fast modes. One of the major sources of problems with the development of the Blandford–Znajek model was that the fast and Alfvén waves were treated as modes that propagate similar information with different wave speeds. This is a very inaccurate representation; the information content of an Alfvén wave packet and a fast wave packet are distinctly defined by the plasma.

The next basic concept is to understand particle motion in the ergosphere. Chapter 3 develops this concept including a description of negative energy plasma states and the horizon boundary condition of inertial dominance.

At this point, we are ready to introduce electromagnetism on the curved space–time background. The next four chapters (Chaps. 4–7) are a formulation of the physics of the GHM interaction in the ergosphere of a black hole that can create relativistic outgoing winds of magnetized plasma. The exposition of black hole GHM begins with vacuum electrodynamics that reveals the properties of Maxwell's equations on the Kerr space–time background. This treatment provides the fundamental underpinnings of the plasma physics in the ergosphere developed in Chaps. 5–7.

Chapters 8 and 9 are applications of GHM to two model ergospheric dynamos that drive large scale magnetized winds. These two models have very different magnetic flux plasma distributions, yet the fundamental physics governing GHM are virtually identical. This lends support to the claim of commonality of physics for all ergospheric dynamos as described in Chap. 7.

Chapter 10 is an attempt to incorporate the theory of winds from magnetized black holes into the theory of radio loud AGN. Even though this interpretation is not yet commonly accepted as the central engine of radio loud AGN, it certainly seems capable of consistently describing many more properties of extragalactic radio sources than any other existing theory.

The final chapter is dedicated to the fast growing field of numerical simulations of black hole magnetospheres that initiated around the time that the first edition went to press. There are some interesting 3-D simulations that show GHM at work. When the GHM dynamo is triggered, the output power is enormous as expected from theoretical considerations. However, in order to truly understand the relevance of GHM to astrophysics will require the long term development of more sophisticated simulations.

Chapter 2
Relativistic Plasma Physics

2.1 Introduction

Black hole GHM is essentially plasma physics in the magnetosphere near a black hole. Consequently, a deep understanding of plasmas in a relativistic context is essential before delineating the particulars on a black hole background. In a plasma-filled magnetosphere, information can be transmitted from one region to another only by means of the modes of propagation allowed by the plasma state. For example, a plasma distribution in the ergosphere can induce currents to flow in a wind far from the hole if, and only if, a wave packet carrying the appropriate information requiring the current flow is transmitted from the ergospheric plasma and received by the wind plasma far from the hole. The relativistic structure of plasma waves is the prime emphasis of this chapter.

Unfortunately, there is no existing treatment of relativistic plasma waves in the literature that elaborates on the properties of the wave modes, nor derives the simplest results (i.e., waves speeds in a perfect MHD plasma) in a manner accessible to theorists. Relativistic plasmas in the context of mathematical physics are described in two monographs [69, 70]. It is difficult for a physicist even to find perfect MHD wave speeds in these books, let alone more substantial results. Although these treatments are excellent for understanding the solution space for the quasilinear hyperbolic set of equations and the formulation of strong shocks in general relativity, they are of limited value for understanding the basic structure of simple plasma waves, even in perfect MHD.

Consequently, we take this opportunity to spell out the structure of waves in an MHD plasma in a manner that parallels standard texts on nonrelativistic plasma physics. The results are derived in a straightforward manner using only special relativity. This is still of great value as we will exploit the equivalence principle later on when we study black hole magnetospheres. A special relativistic treatment in a freely falling coordinate patch at any point of space–time can be used to study the plasma modes as long as the wavelength is much less than the radius of curvature of space–time ($\sim 10^{14}$ cm for an astrophysical black hole). In the freely falling

B. Punsly, *Black Hole Gravitohydromagnetics, 2nd. ed.*

Astrophysics and Space Science Library 355, doi: 10/1007/978-3-540-76957-6_2,

frame, the connection terms are small in the covariant derivatives on scales much less than the radius of curvature [71]. This is not overly restrictive as perturbations to an equilibrium situation should be interpretable in terms of local discontinuities propagating within the plasma. Considering the discussion above, it is not surprising that the Kerr background introduces nothing that changes the basic character of the short wavelength modes or the step discontinuities. As a verification (in Chap. 6), we actually compute the structure of very long wavelength modes. This is a very difficult calculation, but it is necessary to show that no mysterious physics is hidden in this limit.

We will not restrict ourselves to perfect MHD, low frequency waves. The effects of a finite, but small, resistivity are derived and the charge separation in high frequency waves is considered as well. The long section, 2.9, on the plasma-filled cylindrical waveguide is very important. This example (which could be fabricated) is a laboratory example of the physics of a relativistic MHD wind from a magnetized star. The causal structure is clearly definable and is one of the few intellectual tools that allow us to make comparisons and contrasts between a black hole magnetosphere and something we could actually experience on earth.

2.2 The Equations of Perfect MHD Plasmas

The existence of a perfect MHD plasma is defined in terms of the vanishing of the proper electric field (i.e., the electric field in the rest of the frame of the plasma). If u^{α} is the bulk four velocity of the plasma and $F^{\mu\nu}$ is the Maxwell tensor, this condition is mathematically expressed as

$$F^{\mu\nu}u_{\nu} = 0 \qquad \forall\mu \, . \tag{2.1}$$

Locally, the vanishing of the proper electric field is ensured if the constituent particles can remain threaded onto the magnetic field in their gyro-orbits. Consequently, this condition is equivalent to the magnetic field being frozen into the plasma and is also called the frozen-in condition. If the gyro-frequency or Larmor frequency $\Omega_L \gg \nu_c$, where ν_c is the collision frequency, one expects perfect MHD to hold based on local considerations (see Sect. 2.10),

$$\Omega_L = \frac{eB}{mc} \, . \tag{2.2}$$

However, globally this condition requires a more profound analysis. Global constraints can produce a ν_c that is not a kinetic term from the Boltzmann or Fokker–Planck equations. The effective collision frequency can be induced from plasma wave-plasma wave scattering that result from the global properties of the flow (this actually occurs in black hole magnetospheres).

Equation (2.1) is equivalent to the vanishing of the electric field component parallel to the magnetic field. Relativistically, this is stated

$$*F^{\mu\nu}F_{\mu\nu} = 0\,, \tag{2.3}$$

where $*F^{\mu\nu}$ is the Maxwell dual tensor defined by

$$*F^{\mu\nu} = \frac{1}{2}\varepsilon^{\mu\nu\alpha\beta}F_{\alpha\beta}\,. \tag{2.4}$$

Similarly, there is another relativistic invariant besides (2.3) that is constrained by a perfect MHD plasma state

$$F^{\alpha\beta}F_{\alpha\beta} \geq 0\,. \tag{2.5}$$

Equation (2.5) implies that the field is magnetic (i.e., $B^2 - E^2 \geq 0$).

The first law of thermodynamics for a relativistic fluid is the law of mass–energy conservation for a fluid element.

$$\mathrm{d}\rho = \frac{(\rho + P)\mathrm{d}n}{n} + nT\mathrm{d}\mathbb{S}\,, \tag{2.6}$$

where ρ is the mass–energy density, P is the pressure, T is the temperature, n is the particle number density, and $\mathbb{S}$ is the entropy per baryon. The local rate of entropy of production in the fluid $S^{\mu}{}_{;\mu}$ can be expressed in terms of the heat flow vector q^{μ} as [72]

$$S^{\mu}{}_{;\mu} = \frac{-q^{\mu}a_{\mu}}{T} - \frac{T_{,\mu}}{T^2}q^{\mu}\,, \tag{2.7}$$

where a^{μ} is the four acceleration,

$$a^{\mu} = u^{\mu}{}_{;\nu}u^{\nu}\,, \tag{2.8}$$

and S^{μ} is the entropy four vector,

$$S^{\mu} = n\mathbb{S}u^{\mu} + \frac{q^{\mu}}{T}\,. \tag{2.9}$$

The perfect MHD condition (2.1) implies that there is no Ohmic heating in the plasma,

$$u_{\mu}F^{\mu\nu}J_{\nu} = 0\,, \tag{2.10}$$

where J^{ν} is the four current density. Consequently, there is no heat flow, q^{μ}, or by (2.7) any entropy generation, so the first law of thermodynamics becomes

$$\frac{\mathrm{d}\rho}{\mathrm{d}\tau} = \mu\frac{\mathrm{d}n}{\mathrm{d}\tau}\,, \tag{2.11}$$

where $\mathrm{d}/\mathrm{d}\tau$ is the convective derivative (i.e., derivative with respect to proper time)

$$\frac{\mathrm{d}}{\mathrm{d}\tau} \equiv u^{\alpha}\frac{\partial}{\partial x^{\alpha}} , \tag{2.12}$$

and the specific enthalpy is

$$\mu = \frac{\rho + P}{n} . \tag{2.13}$$

Equations (2.1) and (2.10) allow us to write the momentum equations as that of a perfect fluid

$$n\mu u^{\alpha}u^{\beta}{}_{;\alpha} = \frac{F^{\beta\nu}J_{\nu}}{c} + \left(g^{\beta\mu} + u^{\beta}u^{\mu}\right)\frac{\partial}{\partial x^{\mu}}P , \tag{2.14}$$

where $g^{\beta\mu}$ is the metric tensor.

The first law of thermodynamics (2.11), the momentum equation (2.14), mass conservation (2.15),

$$(nu^{\mu})_{;\mu} = 0 , \tag{2.15}$$

and Maxwell's equations (2.16),

$$F^{\mu\nu}{}_{;\nu} = \frac{4\pi J^{\mu}}{c} , \tag{2.16a}$$

$$F_{\alpha\beta;\gamma} + F_{\gamma\alpha;\beta} + F_{\beta\gamma;\alpha} = 0 . \tag{2.16b}$$

form the coupled set of perfect MHD equations. For a discussion of the set of circumstances that are required to establish this hydromagnetic description of the fluid, see [73] Chap. 3.

2.3 Perfect MHD Wave Speeds in a Warm Plasma

We examine low frequency waves in a warm plasma with no dissipative effects. Our basic equations are (2.11)–(2.16) in the limit that the oscillatory frequency, ω, is much less than the cyclotron frequency of a proton, Ω_i as defined in (2.2). For a positronic plasma, this relation becomes $\omega \ll \Omega_e$ where Ω_e is the electron cyclotron frequency. In this first treatment of waves we also assume $\omega^2 \ll \omega_{pi}^2$ in a protonic plasma and $\omega^2 \ll \omega_{pe}^2$ in a positronic plasma. The plasma frequency ω_p is defined for each species as

$$\omega_p^2 = \frac{4\pi ne^2}{\mu} . \tag{2.17}$$

Three independent low frequency modes exist in a plasma. The longitudinal hydrodynamic sonic mode is coupled to the two independent transverse electromagnetic vacuum modes.

In order to simplify the analysis, consider a freely falling frame initially at rest with respect to the fluid. In this frame, the connection is zero at the origin of the coordinate patch and Maxwell's equations become as in flat space–time,

$$\nabla\times\boldsymbol{B}=\frac{1}{c}\frac{\partial\boldsymbol{E}}{\partial t}+\frac{4\pi}{c}\boldsymbol{J}\,, \tag{2.18a}$$

$$\nabla\times\boldsymbol{E}=-\frac{1}{c}\frac{\partial\boldsymbol{B}}{\partial t}\,. \tag{2.18b}$$

In (2.18), t is the time (the covector dual to the four velocity) of the local freely falling observer and the electromagnetic field components are,

$$\boldsymbol{E}\equiv E^{\alpha}=F^{\alpha\mu}u_{\mu}\,, \tag{2.19a}$$

$$\boldsymbol{B}=B^{\alpha}=*F^{\alpha\mu}u_{\mu}\,. \tag{2.19b}$$

Equations (2.18ab) can be combined to get the second order Maxwell's' equation

$$\nabla^{2}\boldsymbol{E}-\nabla(\nabla\cdot\boldsymbol{E})=\frac{1}{c^{2}}\frac{\partial^{2}\boldsymbol{E}}{\partial t^{2}}+\frac{4\pi}{c^{2}}\frac{\partial\boldsymbol{J}}{\partial t}\,. \tag{2.20}$$

We simplify the momentum equation for a linear perturbation analysis. The waves are chosen in a plane wave representation to have an oscillatory behavior $\exp\left[\mathrm{i}(\boldsymbol{k}\cdot\boldsymbol{r}-\omega t)\right]$. We define the x direction to be parallel to $\boldsymbol{k}$, so all perturbed quantities are Fourier analyzed in (x,t) space. For example, the four velocity is

$$u^{\mu}=u_{0}^{\mu}+\mathrm{e}^{\mathrm{i}(kx-\omega t)}\,\delta u^{\mu} \tag{2.21a}$$

$$=\mathrm{e}^{\mathrm{i}(kx-\omega t)}\,\delta u^{\mu}\,. \tag{2.21b}$$

The unperturbed value of the four velocity, u_{0}^{μ}, vanishes identically by our choice of frame. This simplifies the momentum equation (2.14) tremendously. The next approximation is that the waves vary on much smaller scales than the background fields. Thus, we consider the magnetic field of the unperturbed state to be a constant.

$$\frac{\partial}{\partial x^{\alpha}}\boldsymbol{B}_{0}=0\,. \tag{2.22}$$

Ignoring gradients in $\boldsymbol{B}_0$ is the same as ignoring the currents in (2.18). Thus, we ignore the $\boldsymbol{J}_0\times\delta\boldsymbol{B}$ forces in the perturbed momentum to the first order. Thus, the perturbed momentum equation becomes

$$-\mathrm{i}\omega\mu\,\delta\boldsymbol{u}=-\mathrm{i}\boldsymbol{k}\delta P+\frac{\delta\boldsymbol{J}\times\boldsymbol{B}_{0}}{c}\,. \tag{2.23}$$

Maxwell's equation (2.20) becomes with the aid of (2.1) and (2.19a) an equation of perturbed quantities only

$$k^2\,\delta\boldsymbol{E} - \boldsymbol{k}\,(\boldsymbol{k}\cdot\delta\boldsymbol{E}) = \frac{\omega^2}{c^2}\,\delta\boldsymbol{E} + \frac{\mathrm{i}4\pi\omega}{c^2}\,\delta\boldsymbol{J}\,. \tag{2.24}$$

The perturbed frozen-in condition is

$$\delta\boldsymbol{E} + \frac{\delta\boldsymbol{u}\times\boldsymbol{B}_0}{c} = 0\,. \tag{2.25}$$

In order to perturb the thermodynamic quantities, note that the sound speed, c_S is given by

$$c_S^2 = \frac{\partial P}{\partial\rho}\Big|_{\mathbb{S}=\mathrm{constant}}\,. \tag{2.26}$$

For the adiabatic version of the first law of thermodynamics (2.11), we have an adiabatic gas law with adiabatic constant, Γ,

$$Pn^{-\Gamma} = \mathrm{constant}\,, \quad c_S^2 = \frac{\Gamma P}{n\mu}\,. \tag{2.27}$$

Thus, we find after linear perturbation

$$\frac{\nabla\delta P}{P_0} = \Gamma\frac{\nabla\delta n}{n_0}\,. \tag{2.28}$$

The linearized continuity equation (2.15) yields

$$\delta n = \frac{(\boldsymbol{k}\cdot\delta\boldsymbol{u})}{\omega}\,n_0\,, \tag{2.29a}$$

or

$$\nabla\delta n = \mathrm{i}\boldsymbol{k}\frac{(\boldsymbol{k}\cdot\delta\boldsymbol{u})}{\omega}n_0\,. \tag{2.29b}$$

Equations (2.26)–(2.29) allow us to rewrite the pressure gradient force in the perturbed momentum equation (2.23) as

$$\mathrm{i}\boldsymbol{k}\delta P = \frac{\mathrm{i}n_0\mu_0 c_S^2\,(\boldsymbol{k}\cdot\delta\boldsymbol{u})}{\omega}\boldsymbol{k}\,. \tag{2.30}$$

The momentum equation (2.23) of the perturbed plasma can be rewritten in terms of (2.30) as

$$\delta\boldsymbol{u} = \frac{(\boldsymbol{k}\cdot\delta\boldsymbol{u})}{\omega^2}c_S^2\boldsymbol{k} + \mathrm{i}\frac{\delta\boldsymbol{J}\times B_0}{c\omega n_0\mu_0}\,. \tag{2.31}$$

At this point we explicitly write out the components of the perturbed equations in a local Cartesian basis. The x-axis was previously defined to be parallel to $\boldsymbol{k}$ and we define the z-axis such that $\boldsymbol{B}_0$ always lies in the x–z plane. The components of Maxwell's equation, (2.24), become

$$\delta E_y = \frac{4\pi \mathrm{i}\omega}{(c^2k^2 - \omega^2)} \delta J_y \, , \tag{2.32a}$$

$$\delta E_z = \frac{4\pi \mathrm{i}\omega}{(c^2k^2 - \omega^2)} \delta J_z \, , \tag{2.32b}$$

$$\delta E_x = -\frac{4\pi \mathrm{i}}{\omega} \delta J_x \, . \tag{2.32c}$$

Similarly, the perturbed frozen-in condition (2.25) becomes

$$\delta E_x = -\frac{\delta u_y}{c} B_0 \cos\theta \, , \tag{2.33a}$$

$$\delta E_y = -\frac{\delta u_x}{c} B_0 \cos\theta + \frac{\delta u_z}{c} B_0 \sin\theta \, , \tag{2.33b}$$

$$\delta E_z = \frac{\delta u_y}{c} B_0 \sin\theta \, , \tag{2.33c}$$

where θ is the angle between the propagation vector and the magnetic field. This can be defined covariantly as

$$\cos^2\theta = \frac{\left(u^\mu * F_{\mu\nu} k^\nu\right)^2}{\left[k^\alpha - u^\alpha \left(u^\mu k_\mu\right)\right] \left[k_\alpha - u_\alpha \left(u^\nu k_\nu\right)\right] \left(F_{\mu\nu} F^{\mu\nu}\right)} \, . \tag{2.34}$$

In (2.34), k^μ is the wave propagation four vector and the spatial three vector in the free falling frame instantaneously at rest with respect to the fluid, $\boldsymbol{k}$, is

$$\boldsymbol{k} = k^\alpha - u^\alpha \left(u^\mu k_\mu\right) \, . \tag{2.35}$$

Expanding the momentum equation (2.31) into components

$$\delta u_x = \frac{\mathrm{i} B_0 \delta J_y \sin\theta}{c\omega n_0 \mu_0 \left(1 - k^2 c_S^2/\omega^2\right)} \, , \tag{2.36a}$$

$$\delta u_y = \frac{\mathrm{i} B_0}{c\omega n_0 \mu_0} \left(\delta J_z \cos\theta - \delta J_x \sin\theta\right) \, , \tag{2.36b}$$

$$\delta u_z = -\frac{\mathrm{i} B_0}{c\omega n_0 \mu_0} \left(\delta J_y \cos\theta\right) \, . \tag{2.36c}$$

Substituting (2.36) and (2.32) into the frozen-in condition (2.33) eliminates $\boldsymbol{\delta E}$ and $\boldsymbol{\delta u}$ to give an expression entirely in terms of perturbed currents. The z-equation is

$$\left[\frac{4\pi\omega^2}{c^2k^2-\omega^2}-\frac{B_0\cos^2\theta}{c^2n_0\mu_0}\right]\delta J_z+\frac{B_0^2\sin\theta\cos\theta}{cn_0\mu_0}\delta J_x=0\,, \tag{2.37a}$$

The y component of the frozen-in equation becomes

$$\left[\frac{4\pi\mathrm{i}\omega^2}{c^2k^2-\omega^2}-\frac{B_0^2}{c^2n_0\mu_0}\left(\cos^2\theta+\frac{\sin^2\theta}{1-k^2c_S^2/\omega^2}\right)\right]\delta J_y=0\,, \tag{2.37b}$$

The x -component is

$$\left[1+\frac{B_0^2\sin^2\theta}{4\pi n_0\mu_0}\right]\delta J_x-\left[\frac{B_0^2\sin\theta\cos\theta}{4\pi n_0\mu_0}\right]\delta J_z=0\,. \tag{2.37c}$$

Equation (2.37c) shows that the perturbed currents δJ_x and δJ_z in the $\boldsymbol{k}, \boldsymbol{B}$ plane are proportional to each other. Using this fact in (2.37a) yields an equation in δJ_z alone

$$\left[\frac{4\pi\omega^2}{c^2k^2-\omega^2}-\frac{B_0^2\cos^2\theta}{c^2n_0\mu_0\left(1+B_0^2\sin^2\theta/4\pi n_0\mu_0\right)}\right]\delta J_z=0\,. \tag{2.38}$$

There are three solutions to the two independent equations (2.37b) and (2.38). These are the dispersion relations for the phase velocities of the waves. In a relativistic formalism we write the phase velocity, v_φ, as

$$v_\varphi^2=\frac{\omega^2}{|\boldsymbol{k}|^2}=\frac{\left(u^\mu k_\mu\right)^2}{\left[k^\mu-u^\alpha\left(u^\mu k_\mu\right)\right]\left[k_\alpha-u_\alpha\left(u_\mu k^\mu\right)\right]}\,, \tag{2.39}$$

and the phase four velocity u_φ is

$$\frac{u_\varphi^2}{c^2}=\frac{\left(u^\mu k_\mu\right)^2}{k^\mu k_\mu}=\frac{\omega^2}{c^2k^2-\omega^2}\,. \tag{2.40}$$

The first solution to the set of equations (2.37b) and (2.38) is found by setting $\delta J_y=0$ to solve (2.37b). Then (2.38) yields the phase four-velocity defined in (2.40)

$$u_\varphi^2=\frac{B_0\cos^2\theta}{4\pi n_0\mu_0\left(1+B_0^2\sin^2\theta/4\pi n_0\mu_0c^2\right)}=U_I^2\,. \tag{2.41}$$

The wave propagation three speed is the group velocity $v_g=d\omega/dk$. For nondispersive wave $u_g=u_\varphi$ and $v_g=v_\varphi$. Equation (2.41) represents nondispersive waves associated with the shear Alfvén mode or simply the Alfvén wave. The four speed U_I of this mode is often called the intermediate speed. From (2.41) the intermediate three speed v_I is

$$v_I^2=\frac{B_0^2\cos^2\theta}{\left(4\pi n_0\mu_0c^2+B_0^2\right)}c^2\,. \tag{2.42}$$

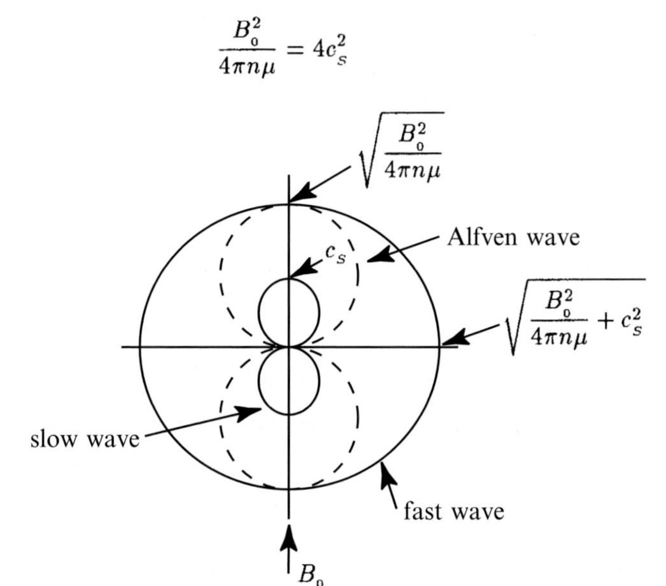

Fig. 2.1 The Friedrichs or phase polar diagram of the three plasma wave phase velocities. The polar diagram plots the velocities in the case the $B_0^2 = 16\pi n\mu_0 c_S^2$. Note that $U_F \geq U_I \geq U_{SL}$

Figure 2.1 is a Friedrichs diagram that is a polar coordinate plot of the three velocity as a function of angle of propagation relative to the magnetic field. The intermediate speed goes to zero when the wave propagates perpendicular to the field and is faster for parallel propagation.

The other solutions to the system of equations (2.37b) and (2.38) are found by setting $\delta J_z = 0$ in (2.38). Equation (2.37b) can be expanded to yield a dispersion relation for the three speeds of the two remaining waves,

$$\left(1 + \frac{B_0^2}{4\pi n_0 \mu_0 c^2}\right) v_{F,SL}^4 - v_{F,SL}^2 \left(c_S^2 + \frac{B_0^2}{4\pi n_0 \mu_0} + \frac{c_S^2 B_0^2 \cos^2\theta}{4\pi n_0 \mu_0 c^2}\right) + \frac{B_0^2 \cos^2\theta \, c_S^2}{4\pi n_0 \mu_0} = 0. \tag{2.43}$$

The two wave speed v_F and v_{SL} in (2.43) represent the compressional Alfvén wave or fast wave and slow wave propagation speeds, respectively. Due to the coupling to the sound speed these are often referred to as the two magneto-acoustic modes. Their three velocities are plotted in the polar plot of Fig. 2.1 as a function of propagation angle. The slow wave does not propagate perpendicular to the magnetic field and propagates fastest parallel to the magnetic field direction as for the Alfvén mode. The fast wave propagates at virtually the same speed in all directions with a small increase due to the sound speed for perpendicular propagation.

Equation (2.43) can be used to compute the four speed dispersion relation

$$U_{F,SL}^4 - U_{F,SL}^2 \left[\frac{c^2 c_S^2}{c^2 - c_S^2} + \frac{B_0^2}{4\pi n_0 \mu_0} \cos^2\theta + \frac{c^2}{c^2 - c_S^2} \frac{B_0^2 \sin^2\theta}{4\pi n_0 \mu_0} \right]$$

$$+ \frac{B_0^2}{4\pi n_0 \mu_0} \frac{c^2 c_S^2}{c^2 - c_S^2} \cos^2\theta = 0\,. \tag{2.44}$$

Equations (2.44) and (2.41) and Fig. 2.1 indicate the well-known result, $U_{SL}^2 \leq U_I^2 \leq U_F^2$.

2.4 Covariant Formulation of the Plasma Wave Speeds

It is of mathematical interest to rewrite the four speed dispersion relations for the plasma modes in a covariant form (with the customary choice of setting $c = 1$). For the Alfvén mode, we expand (2.42) with the covariant definitions (2.39) and (2.34) to find (note $B_0^2 = \frac{1}{2} F^{\mu\nu} F_{\mu\nu}$),

$$n_0 \mu_0 \left(1 + \frac{F_{\mu\nu} F^{\mu\nu}}{8\pi n_0 \mu_0}\right) \left(u^\mu k_\mu\right)^2 - \frac{1}{4\pi} \left(u_\mu *F^{\mu\nu} k_\nu\right)^2 = 0\,. \tag{2.45}$$

Alternatively, one can describe the wave propagation vector as the derivative of the phase, ϕ, of the wave

$$k_\lambda = \phi_{,\lambda}\,. \tag{2.46}$$

Then (2.45) can be written as

$$\left[n_0 \mu_0 \left(1 + \frac{F_{\mu\nu} F^{\mu\nu}}{8\pi n_0 \mu_0}\right) u^\alpha u^\beta - \frac{1}{4\pi} u_\mu *F^{\mu\alpha} u_\nu *F^{\nu\beta} \right] \phi_{,\alpha} \phi_{,\beta} = 0\,. \tag{2.47}$$

The surfaces ϕ = constant defined by (2.47) are the characteristic surfaces of the quasi-linear hyperbolic partial differential equations [74]. Equation (2.45) parallels the mathematical physics treatment of [70], Chap. 6.

Similarly for the magneto-acoustic modes we expand (2.43) with the covariant definitions (2.34) and (2.39):

$$\left[1 - c_S^2\right] \left[u^\mu k_\mu\right]^4 + \left[c_S^2 + \frac{F_{\mu\nu} F^{\mu\nu}}{8\pi n_0 \mu_0}\right] \left[u^\mu k_\mu\right] \left[k^\mu k_\mu\right] + \frac{c_S^2 \left[u_\mu *F^{\mu\nu} k_\nu\right]}{4\pi n_0 \mu_0} k^\mu k_\mu = 0\,. \tag{2.48}$$

In deriving (2.48) we used

$$|\boldsymbol{k}|^2 = k^\mu k_\mu + \left(u_\mu k^\mu\right)^2 = \left(g^{\lambda\mu} + u^\lambda u^\mu\right) k_\lambda k_\mu\,. \tag{2.49}$$

Expanding (2.48) with (2.49) and the definition (2.46), we can get an expression for the characteristic surfaces $\phi = \text{constant}$:

$$\left\{ \left(1 - c_S^2\right) u^\mu u^\lambda u^\nu u^\delta - \left[c_S^2 + \frac{F_{\mu\nu}F^{\mu\nu}}{8\pi n_0 \mu_0} \right] u^\mu u^\lambda g^{\nu\delta} + \frac{c_S^2}{4\pi n_0 \mu_0} \left(u_\alpha {*F^{\alpha\mu}} u_\beta {*F^{\beta\lambda}} g^{\nu\delta} \right) \right\} \phi_{,\mu} \phi_{,\lambda} \phi_{,\nu} \phi_{,g} = 0 \, . \tag{2.50}$$

Equation (2.48) reproduces the results of [70], Chap. 6. The expressions in this section are not very useful from a physical point of view, but are included for mathematical rigor.

A very useful concept is a four velocity vector for a propagating wave constructed from the four speeds (2.41) and (2.44). Note that the group four velocity (2.41) and (2.44) is a function of $\theta, u_g(\theta)$. We define

$$u_g^\alpha = \sqrt{u_g^2(\theta) + 1} \left[u^\alpha + \frac{u_g(\theta)}{u_g^2(\theta) + 1} \left(\frac{k^\alpha}{|k|} - u_g(\theta) u^\alpha \right) \right] . \tag{2.51}$$

2.5 The Perfect MHD Alfvén Mode

Equations (2.41) and (2.42) are the four and three velocities of the shear Alfvén wave. It is of substantial interest to know what information is carried by this mode. Recall that the wave speeds were found by setting $\delta J_y = 0$ in the set of equations formed by (2.37b) and (2.38),

$$\delta J_y = 0 \, . \tag{2.52a}$$

The component of current density along the propagation vector, δJ_x, is proportional to δJ_z by (2.37ac),

$$\delta J_x \neq 0 \, , \; \theta \neq 0, \pi \quad \text{and} \quad \delta J_z \neq 0, \; \theta \neq \pi \, . \tag{2.52b}$$

The current density vanishes at $\theta = \pi$ because the Alfvén wave does not propagate perpendicular to the magnetic field. In general, from (2.52b) and (2.37c)

$$\boldsymbol{k} \cdot \delta \boldsymbol{J} \neq 0 \, , \quad \theta \neq 0, \pi \, , \tag{2.52c}$$

$$\boldsymbol{B} \cdot \delta \boldsymbol{J} = \delta J_z B_0 \sin\theta \left[\frac{B_0^2 + 4\pi n_0 \mu_0 c^2}{B_0^2 \sin^2\theta + 4\pi n_0 \mu_0 c^2} \right] . \tag{2.52d}$$

In a practical sense, the only Alfvén waves of physical relevance are oblique Alfvén waves; parallel propagation is of textbook value only as a simple illustrative example of waves. Because $\boldsymbol{k} \cdot \boldsymbol{B}_0 \neq 0$, in general, (2.52cd) imply that *the oblique Alfvén wave can propagate field aligned currents.*

From (2.52a), (2.36a) and (2.36b)

$$\delta u_x = \delta u_z = 0, \quad \delta u_y \neq 0\,, \tag{2.53a}$$

$$\boldsymbol{k} \cdot \delta \boldsymbol{u} = 0\,. \tag{2.53b}$$

From (2.53b) we see that the Alfvén mode is mechanically transverse.

Inserting (2.53b) into (2.30) and the mass conservation law (2.29a), one finds that the Alfvén mode is noncompressive and is decoupled from the sonic mode

$$\delta P = 0\,, \tag{2.54a}$$

$$\delta n = 0\,. \tag{2.54b}$$

From (2.52) and (2.32) one finds that

$$\delta E_y = 0\,, \tag{2.55a}$$

$$\delta E_x \neq 0, \quad \theta \neq 0, \pi \qquad \text{and} \quad \delta E_z \neq 0\,, \quad \theta \neq \pi\,. \tag{2.55b}$$

Equation (2.55b) implies that for oblique Alfvén waves

$$\boldsymbol{k} \cdot \delta \boldsymbol{E} \neq 0\,. \tag{2.56a}$$

Then, using Gauss's Law for the charge density in the wave, ρ_e, and performing a linear perturbation, yields

$$\nabla \cdot \boldsymbol{E} = 4\pi \rho_e\,, \tag{2.56b}$$

$$\mathrm{i}\boldsymbol{k} \cdot \delta \boldsymbol{E} = 4\pi \delta \rho_e\,. \tag{2.56c}$$

Equation (2.56) states that *the oblique Alfvén wave carries a charge density and has a significant electrostatic polarization.*

Consider the induction equation

$$\frac{1}{c}\frac{\partial \boldsymbol{B}}{\partial t} = -\nabla \times \boldsymbol{E}\,, \tag{2.57a}$$

and its linearized perturbation

$$\frac{\omega}{c}\delta \boldsymbol{B} = \boldsymbol{k} \times \boldsymbol{E}\,. \tag{2.57b}$$

Applying (2.55) to (2.57b) yields for the Alfvén mode

$$\delta B_x = 0\,, \quad \delta B_z = 0\,, \quad \delta B_y \neq 0\,. \tag{2.57c}$$

The fact that $\boldsymbol{k} \cdot \delta \boldsymbol{B} = 0$ is simply the perturbed $\nabla \cdot \boldsymbol{B} = 0$ equation. In general, one can not conclude that these results must be true for wavelengths on the order of the field line curvature as a consequence of the assumption in (2.22).

2.6 The Magneto-Acoustic Waves in a Perfect MHD Plasma

The magneto-acoustic wave speeds were derived from the coupled set of (2.37b) and (2.38) by setting $\delta J_z = 0$. From (2.37c) for magneto-acoustic waves

$$\delta J_x = \delta J_z = 0\,, \quad \delta J_y \neq 0\,, \tag{2.58a}$$

$$\boldsymbol{k} \cdot \boldsymbol{J} = 0\,. \tag{2.58b}$$

Because $\boldsymbol{B}_0$ is in the $x-z$ plane, *magneto-acoustic waves carry no current along the magnetic field direction.*

Equation (2.58) combined with (2.36) implies

$$\delta u_y = 0\,, \quad \delta u_x \neq 0\,, \quad \delta u_z \neq 0\,, \tag{2.59a}$$

Thus the waves have a longitudinal mechanical component

$$\boldsymbol{k} \cdot \delta \boldsymbol{u} \neq 0\,. \tag{2.59b}$$

Inserting (2.59b) into (2.30) and the mass conservation law (2.29a) shows that the magneto-acoustic waves are compressive and coupled to the sonic mode

$$\delta \rho \neq 0\,, \tag{2.60a}$$

$$\delta n \neq 0\,. \tag{2.60b}$$

Consequently, there exist both compression and rarefaction fast and slow waves.

Inserting (2.58) into (2.32) yields

$$\delta E_x = \delta E_z = 0\,, \quad \delta E_y \neq 0\,. \tag{2.61a}$$

Thus, for magneto-acoustic waves from (2.56c)

$$\boldsymbol{k} \cdot \boldsymbol{E} = 0\,, \quad \delta \rho_e = 0\,. \tag{2.61b}$$

Equation (2.61b) states that *the magneto-acoustic waves carry no charge density and have no electrostatic polarization.* Again, one cannot conclude that these results are true for wavelengths on the order of the field line curvature as a consequence of the assumption in (2.22).

The induction equation, (2.57), combined with (2.61a) gives the perturbed magnetic fields in a magneto-acoustic wave,

$$\delta B_x = 0\,, \quad \delta B_y = 0\,, \quad \delta B_z \neq 0\,. \tag{2.62}$$

The fast and slow modes have different compressional properties. Writing the induction equation in terms of the phase velocity, one has

$$\delta E_y = \frac{v_\varphi}{c} \delta B_z\,. \tag{2.63}$$

Combining (2.32a), (2.36a) and (2.63)

$$\frac{B_0 \delta u_x}{\delta B_z} = \frac{v_\varphi}{c^2} \left[\frac{B_0^2 \sin\theta}{4\pi n_0 \mu_0} \right] \left[\frac{c^2 - v_\varphi^2}{v_\varphi^2 - c_S^2} \right] . \tag{2.64}$$

From the linearized continuity equation (2.29a) applied to (2.64)

$$\frac{\delta n}{\delta B_z} = \left[\frac{B_0 \sin\theta}{4\pi \mu_0 c^2} \right] \left[\frac{c^2 - v_\varphi^2}{v_\varphi^2 - c_S^2} \right] . \tag{2.65}$$

For the nondispersive magneto-acoustic speeds in (2.43), $c^2 > v_\varphi^2$. Thus, the sign of $\delta n / \delta B_z$ depends on the sign of $v_\varphi^2 - c_S^2$. For a slow wave, (2.43) implies $v_\varphi < c_S$, and for a fast wave, $v_\varphi > c_S$ (see Fig. 2.1). Thus, $\delta n / \delta B_z < 0$ for slow waves, and $\delta n / \delta B_z > 0$ for fast waves. The magnetic field decreases (increases) across slow compression (rarefaction) waves. The magnetic field increases (decreases) across fast compression (rarefaction) waves. These defining properties persist even in the nonlinear large amplitude versions of these waves (including shock waves).

2.7 MHD Waves in a Resistive Medium

In order to get a qualitative feel for the effects of resistivity on MHD wave propagation, we consider a scalar electrical conductivity, σ. In reality, a tensorial conductivity is important especially in the interesting case in which a strong magnetic field is present and the cross-field conduction is highly suppressed (see Sect. 2.10). We illustrate (using the simplified scalar conductivity case) that the main effect of finite conductivity is to attenuate the waves and decrease the propagation speeds.

The linearly perturbed plasma equations (2.32), (2.25) and (2.36) are the same as before except the frozen-in equation (2.25) is replaced by Ohm's law

$$\delta \boldsymbol{J} = \sigma \left(\delta \boldsymbol{E} + \frac{\delta \boldsymbol{u} \times \boldsymbol{B}_0}{c} \right) . \tag{2.66}$$

Inserting (2.32) and (2.36) into (2.66) yields three equations in analogy to (2.37),

$$\delta J_x \left[1 + \sigma \mathrm{i} \left(\frac{4\pi}{\omega} + \frac{B_0^2 \sin^2\theta}{c^2 \omega n_0 \mu_0} \right) \right] = -\frac{\sigma \mathrm{i}}{c^2 \omega n_0 \mu_0} B_0^2 \sin\theta \cos\theta \, \delta J_z , \tag{2.67a}$$

$$\delta J_y \left[1 - \frac{4\pi\sigma \mathrm{i}\omega}{c^2 k^2 - \omega^2} - \frac{\mathrm{i}\sigma B_0^2}{c^2 \omega n_0 \mu_0} \times \left(\cos^2\theta + \frac{\sin^2\theta}{1 - k^2 c_S^2 / \omega^2} \right) \right] = 0 , \tag{2.67b}$$

$$\delta J_z \left[1 - \sigma \left(\frac{4\pi \mathrm{i}\omega}{c^2 k^2 - \omega^2} + \frac{\mathrm{i} B_0^2 \cos^2\theta}{c^2 \omega n_0 \mu_0} \right) \right] + \frac{\mathrm{i}\sigma B_0^2}{c^2 \omega n_0 \mu_0} \sin\theta \cos\theta \, \delta J_x = 0 . \tag{2.67c}$$

Equation (2.67b) can be solved with $\delta J_x = \delta J_z = 0$ to give the phase velocities of the magneto-acoustic modes

$$\left\{1 + \frac{B_0^2}{4\pi n_0 \mu_0 c^2 \left[1 + \mathrm{i}\left(k^2c^2 - \omega^2\right)/4\pi\omega\sigma\right]}\right\} v_{F,SL}^4 = v_{F,SL}^2 \left\{c_S^2 + \frac{B_0^2\left(c^2 + c_S^2 \cos^2\theta\right)}{4\pi n_0 \mu_0 c^2 \left[1 + \mathrm{i}\left(k^2c^2 - \omega^2\right)/4\pi\omega\sigma\right]}\right\} - \frac{B_0^2 c_S^2 \cos^2\theta}{4\pi n_0 \mu_0 \left[1 + \mathrm{i}\left(k^2c^2 - \omega^2\right)/4\pi\omega\sigma\right]} . \tag{2.68}$$

Defining the dimensionless pure Alfvén four speed, U_A, by

$$U_A^2 = \frac{B_0^2}{4\pi n \mu_0 c^2} . \tag{2.69}$$

Then the dispersion relation (2.68) is the same as the perfect MHD magneto-acoustic relation (2.43) with the substitution $U_A^2 \to \overline{U}_A^2$:

$$\overline{U}_A^2 = \frac{U_A^2}{1 + \mathrm{i}\left(k^2c^2 - \omega^2\right)/4\pi\omega\sigma} . \tag{2.70}$$

Setting $\delta J_y = 0$ and combining (2.67a) and (2.67c) yields the dispersion relation for resistive Alfvén waves

$$\frac{U_I^2}{c^2} = \frac{\overline{U}_A^2 \cos^2\theta}{1 + U_A^2 \sin^2\theta + \omega\mathrm{i}/4\pi\sigma} . \tag{2.71}$$

Compare this with the perfect MHD analog in (2.41)

$$\frac{U_I^2}{c^2} = \frac{U_A^2 \cos^2\theta}{1 + U_A^2 \sin^2\theta} . \tag{2.72}$$

Equation (2.71) for Alfvén waves can be rewritten for large conductivities with the aid of (2.42) as

$$k \approx \frac{\omega}{v_I} + \frac{\mathrm{i}\omega^2 c^2}{8\pi\sigma v_I^3}\left[U_A^2 + 1\right]^{-1}\left(1 + \frac{v_I^2}{c^2} U_A^2 \sin^2\theta\right) . \tag{2.73}$$

Thus, the imaginary part of the propagation vector, $\boldsymbol{k}$, implies that the waves are attenuated. For smaller conductivities, the attenuation increases.

Similarly for the subrelativistic velocities, (2.71) approximates to

$$\omega = \sqrt{k^2 v_I^2 - \frac{\eta^2 k^4}{4}} - \frac{\mathrm{i}k^2\eta}{2} , \tag{2.74}$$

where η is the magnetic diffusivity

$$\eta \equiv \frac{c^2}{4\pi\sigma} . \tag{2.75}$$

Relation (2.74) shows that the resistivity slows down the waves and damps them as well. In the limit of large resistivity the slow and Alfvén waves do not propagate and the fast wave speed approaches the sound speed. Equation (2.71) represents the dispersive character of resistive Alfvén waves. For large conductivities, (2.73) shows these effects to be small and $v_g \approx v_\varphi$. A deeper discussion of resistive effects in MHD waves as well as the effects of viscosity and thermal conduction appear in [75].

As in a perfect MHD plasma, only the Alfvén mode can carry field aligned currents and only the Alfvén mode propagates a net electric charge.

2.8 High Frequency Waves in a Perfect MHD Plasma

In order to study the effects induced by high frequency oscillations, we need to decompose the plasma into the dynamics of the individual species. For the sake of simplification, we restrict this discussion to a positronic plasma as the difference in mass between the electron and proton causes added complexity (see Sect. 2.11).

Equation (2.14) is replaced by the pair of perturbed momentum equation of the individual species,

$$(n_0)_+ \mu \frac{\mathrm{d}}{\mathrm{d}t} \delta \boldsymbol{u}_+ + \nabla \delta P = -\frac{e\,(n_0)_+}{c} \left(\delta \boldsymbol{E} + \frac{\delta \boldsymbol{u}_+ \times \boldsymbol{B}_0}{c} \right) , \tag{2.76a}$$

$$(n_0)_- \mu \frac{\mathrm{d}}{\mathrm{d}t} \delta \boldsymbol{u}_- + \nabla \delta P = +\frac{e\,(n_0)_-}{c} \left(\delta \boldsymbol{E} + \frac{\delta \boldsymbol{u}_- \times \boldsymbol{B}_0}{c} \right) . \tag{2.76b}$$

Adding (2.76a) with (2.76b) gives the bulk flow momentum equation (2.23). Subtracting (2.76b) from (2.76a) yields a generalized Ohm's law that replaces the frozen-in condition (2.25)

$$\frac{\mathrm{d}\boldsymbol{J}}{\mathrm{d}t} = \frac{e^2 n_0}{\mu_0 c} \left[\delta \boldsymbol{E} + \frac{\delta \boldsymbol{u}}{c} \times \boldsymbol{B}_0 \right] , \tag{2.77a}$$

where

$$\delta \boldsymbol{u} = \frac{\delta \boldsymbol{u}_+ + \delta \boldsymbol{u}_-}{2} . \tag{2.77b}$$

The magneto-acoustic dispersion relation is found in Sect. 2.3,

$$\begin{aligned} &\left[\omega^2 - k^2 c_S^2\right] \left[\omega^4 - \omega^2 \left(c^2 k^2 + \omega_p^2\right) - \omega_p^2 U_A^2 \left(\omega^2 - c^2 k^2\right) \cos^2\theta\right] \\ &= \omega^2 \omega_p^2 U_A^2 \left(\omega^2 - c^2 k^2\right) \sin^2\theta , \end{aligned} \tag{2.78a}$$

and the Alfvén mode dispersion relation becomes

$$\left[1-\frac{\omega_p^2}{\omega^2}\right]\left[1+\frac{c^2k^2+\omega_p^2}{\omega^2}\right]-\frac{\omega_p^2U_A^2\sin^2\theta}{\omega^2}\left[1-\frac{c^2k^2+\omega_p^2}{\omega^2}\right]$$
$$=\left[1-\frac{\omega_p^2}{\omega^2}\right]\left[\frac{\omega_p^2U_A^2\cos^2\theta}{\omega^2}\right]\left[1-\frac{c^2k^2}{\omega^2}\right]. \quad (2.78b)$$

The plasma waves become dispersive and they are slowed down at high frequency.

Now consider the interesting properties of the waves. Combining the generalized Ohm's law (2.77a) with Maxwell's equations (2.24) and (2.56c), the charge density in the waves is

$$\delta\rho_e=-\frac{\omega_p^2 k}{4\pi\left[\omega_p^2-\omega^2\right]}\left[\frac{\delta u_y}{c}B_0\sin\theta\right]. \quad (2.79)$$

For high frequency waves, the Alfvén mode is the only wave that can carry field aligned currents and have an electrostatic polarization. (Note $\delta u_y \neq 0$ for Alfvén waves and $\delta u_y = 0$ for magneto-acoustic waves as in the low frequency case).

2.9 The Cylindrical Plasma-Filled Waveguide

In this lengthy section we discuss an instructive example of a cylindrical waveguide filled with plasma that is threaded by a uniform axial magnetic field supported by an external solenoid. The wave properties of this configuration have been discussed in the nonrelativistic limit in Stix [76, 77]. As with [77], for simplicity we consider a cold pressureless fluid. By contrast, we are concerned with relativistic waves. The importance of this example is that it creates a laboratory environment in which the physics of a relativistic magnetosphere about a compact object can be studied. For instance, we will discuss how attaching a rotating disk at the end of a semi-infinite magnetized cylindrical plasma-filled waveguide creates an outgoing plasma wavefront. This outgoing wave carries an energy flux associated with the Poynting vector. This is an excellent laboratory analog of the relativistic wind driven by a neutron star in the MHD theory of pulsar magnetospheres. Different variants of this configuration can be compared and contrasted with a black hole magnetosphere. The analysis greatly elucidates the role of the event horizon in black hole GHM.

2.9.1 Plasma Waves in a Cylindrical Waveguide

The discussion requires a different coordinate system than the last seven sections. Before, we singled out $\boldsymbol{k}$ as the unique direction and denoted its unit normal as $\hat{\boldsymbol{e}}_x$.

In this section, the axial direction is unique and we define

$$\boldsymbol{B}_0 = B_0 \hat{\boldsymbol{e}}_z \tag{2.80}$$

We decompose the wave vector, $\boldsymbol{k}$, into an axial component, k_z, and a radial wave number

$$\nu \equiv k_\perp = \left(k_x^2 + k_y^2\right)^{1/2} . \tag{2.81}$$

For simplicity, we consider a cold plasma, thus there is no slow mode, since $c_S \to 0$. From (2.42) and (2.34) the dispersion relation for shear Alfvén waves is

$$\omega^2 = \frac{c^2 k_z^2 B_0^2}{4\pi n_0 \mu_0 c^2 + B_0^2} . \tag{2.82}$$

Similarly from (2.43) and (2.34), the dispersion relation for the fast compressive mode is

$$\omega^2 = \frac{c^2\left(k_z^2 + \nu^2\right) B_0^2}{4\pi n_0 \mu_0 c^2 + B_0^2} \tag{2.83}$$

In order to compute the wave solutions in the waveguide, we introduce cylindrical coordinates (ρ, ϕ, z). The momentum equation (2.31) in the cold plasma limit becomes

$$\delta \boldsymbol{u} = \frac{\mathrm{i}\delta \boldsymbol{J} \times \boldsymbol{B}_0}{c\omega n_0 \mu_0} . \tag{2.84}$$

Consider the Fourier decomposition in cylindrical coordinates, for example

$$\delta \boldsymbol{E}\left(\omega, k_z, m, \rho, \phi, z, t\right) \equiv \mathrm{e}^{\mathrm{i}\left[k_z z + m\phi - \omega t\right]} \delta \boldsymbol{E} . \tag{2.85}$$

Combining the frozen-in condition (2.25) with (2.84) we find that

$$\delta E^\rho = \frac{4\pi \mathrm{i} U_A^2}{\omega} \delta J^\rho , \tag{2.86a}$$

$$\delta E^\phi = \frac{4\pi \mathrm{i} U_A^2}{\omega} \delta J^\phi , \tag{2.86b}$$

$$\delta E^z = 0 . \tag{2.86c}$$

Maxwell's equation (2.20) is complicated in cylindrical coordinates,

$$\begin{aligned}(\nabla \times \nabla \times \delta \boldsymbol{E})^\rho &= \frac{im}{\rho^2} \frac{\partial}{\partial \rho}\left(\rho\, \delta E^\phi\right) + \left(\frac{m^2}{\rho^2} + k_z^2\right) \delta E^\rho \\ &= \frac{-\omega^2}{c^2} \delta E^\rho - \frac{4\pi i \omega}{c} \delta J^\rho ,\end{aligned} \tag{2.87a}$$

$$(\nabla \times \nabla \times \delta \boldsymbol{E})^{\phi} = k_z^2 \delta E^{\phi} - \frac{\partial}{\partial \rho}\left[\frac{1}{\rho}\frac{\partial}{\partial \rho}\left(\rho\, \delta E^{\phi}\right)\right] - \mathrm{i}m\frac{\partial}{\partial \rho}\left(\frac{\delta E^{\rho}}{\rho}\right)$$
$$= \frac{-\omega^2}{c^2}\delta E^{\phi} - \frac{4\pi \mathrm{i}\omega}{c}\delta J^{\phi} \,. \tag{2.87b}$$

Since (2.87a) and (2.87b) are coupled, we must consider the various cases individually in order to solve them.

2.9.2 Fast Waves

Fast waves carry no charge density (see 2.61b). Thus,

$$\nabla \cdot \delta \boldsymbol{E} = 0 \,, \tag{2.88a}$$

or in cylindrical coordinates

$$\frac{\partial}{\partial \rho}\left(\rho\, \delta E_{\rho}\right) + \mathrm{i}m\, \delta E_{\phi} = 0 \,. \tag{2.88b}$$

First consider $m = 0$ waves. Then the divergenceless condition (2.88b) implies

$$\rho\, \delta E_{\rho} = \text{constant}, \quad m = 0 \,. \tag{2.88c}$$

By the regularity of E_{ρ} at $\rho = 0$, (2.88c) implies

$$\delta E_{\rho} = 0 \,, \quad m = 0 \,. \tag{2.88d}$$

In the axisymmetric case $m = 0$, the "ϕ" component of Maxwell's equation (2.87b) is independent of E_{ρ}. Combining (2.87b) with (2.86b) to eliminate δJ^{ϕ}, one obtains Bessel's equation and

$$\delta E^{\phi} = AJ_1(\nu\rho) + BY_1(\nu\rho) \,, \quad m = 0 \,, \tag{2.89a}$$

where J_1 is the cylindrical Bessel function and Y_1 is the Neumann function [74]. The wave number "ν" is defined through the dispersion relation (2.83)

$$\nu^2 = \left[1 + U_A^{-2}\right]\frac{\omega^2}{c^2} - k_z^2 \,. \tag{2.89b}$$

At the cylinder boundary, $\rho = R$, the tangential electric field, E^{ϕ}, must vanish in the cylindrical conductor. Thus A and B satisfy

$$\delta E^{\phi}(\rho = R) = AJ_1(\nu R) + BY_1(\nu R) = 0 \,, \quad m = 0 \,. \tag{2.89c}$$

Consider $m \neq 0$ fast waves. Substitute the divergenceless condition (2.88b) into (2.87a) to eliminate δE^{ϕ} in Maxwell's equation. Then substitute (2.86a) in the resulting equation to eliminate δJ^ρ in order to obtain Bessel's equation for $\rho\,\delta E_\rho$,

$$\delta E_\rho = \frac{A}{\rho} J_m(\nu\rho) + \frac{B}{\rho} Y_m(\nu\rho) \neq 0\,. \tag{2.90a}$$

From the divergenceless condition (2.88b) and the conductor boundary condition,

$$\delta E_\phi = \frac{\mathrm{i}}{m}\frac{\partial}{\partial\rho}\left(\rho\,\delta E_\rho\right)\,,\quad \delta E_\phi\left(\rho = R\right) = 0\,. \tag{2.90b}$$

Note that the fast mode does not propagate as $\omega \to 0$ since there is a cutoff in the dispersion relation (2.83).

2.9.3 Alfvén Waves

Combining the dispersion relation (2.82) with Maxwell's equation (2.87) yields

$$\frac{im}{\rho^2}\frac{\partial}{\partial\rho}\left(\rho\,\delta E^\phi\right) - \frac{m^2}{\rho^2}\delta E^\phi = 0\,, \tag{2.91a}$$

$$-\frac{\partial}{\partial\rho}\left[\frac{1}{\rho}\frac{\partial}{\partial\rho}\left(\rho\,\delta E^\phi\right)\right] - im\frac{\partial}{\partial\rho}\frac{\delta E_\rho}{\rho} = 0\,. \tag{2.91b}$$

From (2.57c), $\delta\boldsymbol{B}$ is orthogonal to the plane containing $\boldsymbol{k}$ and $\boldsymbol{B}_0$, so

$$\delta B_z = 0\,. \tag{2.92a}$$

From the induction equation (2.57b), it follows that

$$\left(\nabla\times\delta\boldsymbol{E}\right)_z = 0\,, \tag{2.92b}$$

which is (2.91a). Also, note that the partial derivative of (2.91a) with respect to ρ is (2.91b). Thus, there is only one independent equation.

When $m = 0$, (2.91b), along with the conductor boundary condition, $E^\phi\left(\rho = R\right) = 0$, implies that

$$\delta E^\phi = 0\,,\qquad m = 0\,. \tag{2.93a}$$

There is no equation restricting δE^ρ as it depends entirely on the charge distribution on the boundaries,

$$\delta E^\rho = \text{arbitrary}\,,\quad m = 0\,. \tag{2.93b}$$

When $m \neq 0$, we have our one independent relation

$$\delta B_z = 0\,,\quad \left(\nabla\times\delta\boldsymbol{E}\right)_z = 0\,,\quad \delta E^\phi\left(\rho = R\right) = 0\,,\quad m \neq 0\,. \tag{2.94}$$

For the $m = 0$ mode, from the frozen-in relation for δE_ρ

$$\delta \boldsymbol{u} = \delta u^\phi \hat{\boldsymbol{e}}_\phi \ . \tag{2.95}$$

From the "ϕ" component of the induction equation (2.57a)

$$\delta \mathbf{B} = \delta B^\phi \hat{\boldsymbol{e}}_\phi \ . \tag{2.96}$$

We also note that for a general value of "m" we can compute the field aligned current from the "z" component of Maxwell's equation (2.20) using (2.86c)

$$(\nabla \times \nabla \times \delta \boldsymbol{E})^z = \frac{-4\pi \mathrm{i}\omega}{c} \delta J^z \ , \tag{2.97a}$$

$$\delta J^z = \frac{c^2}{v_I} \delta \rho_e \quad , \quad v_I \equiv \omega / k_z \ . \tag{2.97b}$$

2.9.4 The Faraday Wheel

It is very instructive to attach a Faraday wheel or unipolar inductor to the end of a semi-infinite plasma-filled waveguide. We will show that the Faraday wheel is a piston for Alfvén waves in analogy to the piston for magneto-acoustic waves [78]. It is also demonstrated how the waveguide is analogous to a MHD pulsar wind and that the Faraday wheel is the analog of the neutron star. The parallels allow us to explore the fundamental role of the unipolar inductor in the relativistic MHD wind theory. The main insight that this discussion provides is that the event horizon of a Kerr black hole behaves like a Faraday wheel that is disconnected (i.e., separated from the end of the waveguide by an electrical insulator) from the plasma-filled waveguide. This lack of unipolar induction associated with the space–time near the event horizon is discussed in detail in this context in Chap. 4.

2.9.4.1 The Faraday Wheel Terminated Transmission Line

To begin with, we describe the notion of a Faraday wheel or unipolar inductor. Consider a conductive disk rotating with an angular velocity, Ω_D, about the symmetry axis (z-axis) in Fig. 2.2. The disk is immersed in the uniform magnetic field of an aligned infinite solenoid. In the laboratory frame, there is an electric field in the disk as a consequence of the frozen-in condition.

$$E^\rho = \frac{-\Omega_D \rho}{c} B_0 \ . \tag{2.98}$$

This electric field is established because the rotationally induced Lorentz force separates charge. Gauss's law applied to (2.98) yields a charge density

$$\rho_e = \frac{-\Omega_D B_0}{2\pi c} \ . \tag{2.99}$$

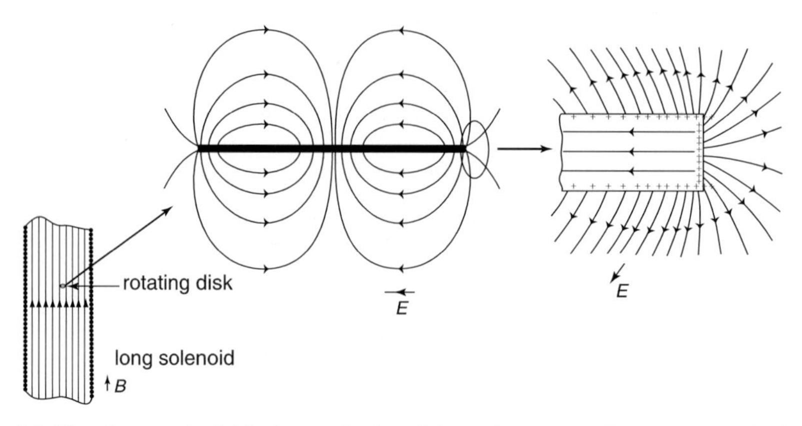

Fig. 2.2 The electrostatic field of a conducting disk rotating in a uniform magnetic field. The expanded view of the end shows the surface charge distribution

An electrostatic equilibrium is established as electrons are pushed toward the center of the disk by the Lorentz force, leaving a residual positive surface charge density at the outer edge of the disk. The resulting electrostatic field is shown in Fig. 2.2. The electrostatic force within the disk balances the rotationally induced EMF creating an equilibrium.

In order to make the rotating disk an active element, we insert it between two semi-infinite coaxial transmission lines as indicated in Fig. 2.3. The center conductor can be supported by a thin low-loss dielectric webbing such as teflon. This is known commercially as spline coaxial cable. The whole configuration is immersed inside an infinite solenoid so that a uniform axial magnetic field threads the transmission line. The vacuum electrostatic field lines are drawn as dashed curves in Fig. 2.3. When the center conductor and the outer conductor first touch the disk, the negative (positive) surface charge on the disk induces an incoming (outgoing) current in the conductors at the center (outer edge) as shown in Fig. 2.3. The presence of conduction paths at the center and edge of the disk allows charges to flow from above and below the disk that will instantaneously partially neutralize the surface charge density. This, in turn, disrupts the balance of electrostatic force with the rotationally induced EMF in the disk. This small excess EMF drives charges in an effort to replenish the lost surface charge density. However, the surface charge is again partially neutralized by electron motion in the inner and outer conductors of the transmission line, and so on. The EMF remains unbalanced and no electrostatic equilibrium can be attained. The rotational EMF drives a surface current outward radially that requires electrons to be drawn from the outer conductor and pumped outward along the center conductor. The Faraday wheel is a unipolar generator of electrical current in the transmission line circuit.

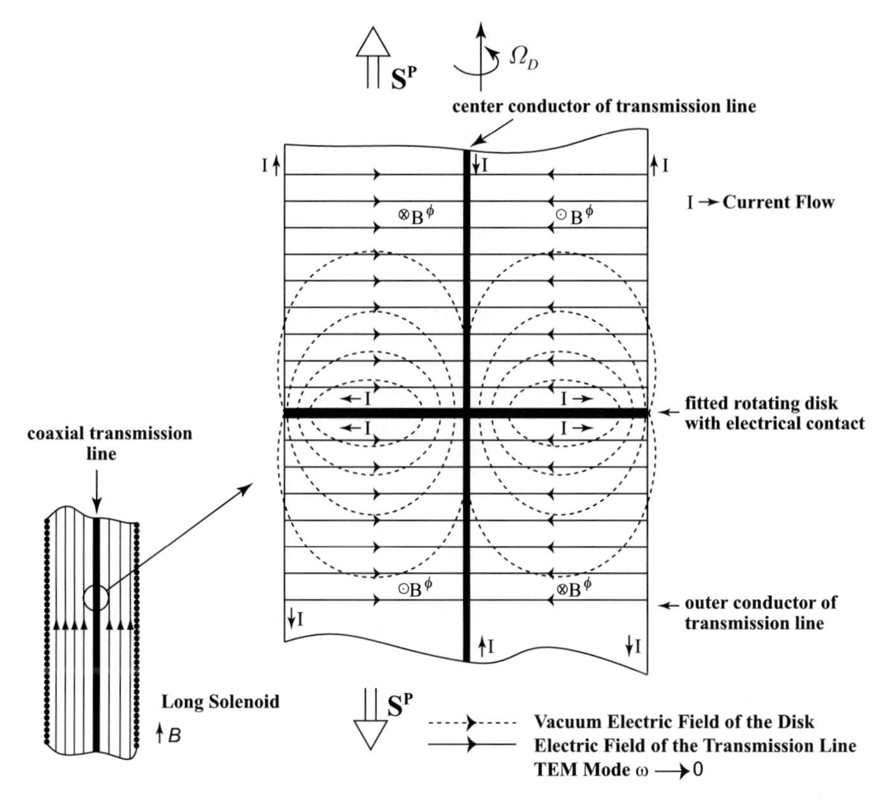

Fig. 2.3 The electrostatic field of a conducting disk rotating in a uniform magnetic field

Associated with the current flow generated by the Faraday wheel is a toroidal magnetic field, B_ϕ, as a consequence of Ampere's Law. The time stationary state of the transmission line has an electric field given by

$$\oint \boldsymbol{E} \cdot \mathrm{d}\ell = 0\,. \tag{2.100a}$$

In the disk, from (2.98), the voltage drop from ΔV, is

$$\Delta V = -\int_0^R \boldsymbol{E} \cdot \mathrm{d}\ell = \frac{+\Omega_D R^2}{2c} B_0\,. \tag{2.100b}$$

By (2.100a), the voltage drop between the center conductor to the outer conductor of the transmission line is ΔV as well. Inside the transmission line $\rho_e = 0$, so Gauss's law implies $E_\rho = E_0/\rho$ where E_0 is a constant. Combining this with (2.100b) yields

$$E_\rho = \frac{-\Omega_D R^2}{2c} \frac{B_0}{\ln\left(R/a\right)} \frac{1}{\rho}\,, \tag{2.100c}$$

where "a" is the radius of the inner conductor. Note that E_ρ in the transmission line (2.100c) does not equal E_ρ in the disk (2.98) even though $E_\parallel$ is continuous at the surface of a conductor. The fringing fields on the disk that transition (2.98) to (2.100c) are in the form of nonpropagating TM and TE modes in the transmission line [79]. The electromagnetic field in the transmission line is the $\omega \to 0$ limit of the TEM mode.

The Faraday wheel radiates a poloidal Poynting flux, S^P into the transmission line.

$$S^P = \frac{c}{4\pi}(\boldsymbol{E} \times \boldsymbol{B}) = \frac{c}{4\pi} E_\rho B_\phi \,. \tag{2.101}$$

Similarly, an angular momentum flux, S_L^P, is associated with the TEM wave

$$S_L^P = -\frac{c}{4\pi} \rho B^\phi B^z > 0 \,. \tag{2.102}$$

In order to understand the magnitude and nature of S^P and S_L^P in a time stationary problem, note that in reality the disk was attached to the transmission line at the same time $t = t_0$. Thus, there are a pair of TEM wave fronts propagating along the two semi-infinite transmission lines. The electric field in (2.100c) in the $+z$ transmission line in the actual time dependent problem is

$$E^\rho = \frac{-\Omega_D R^2 B_0}{2c \ln\left(R/a\right)} \frac{1}{\rho} \Theta\left[c\left(t-t_0\right)-z\right] \,, \tag{2.103}$$

where Θ is the Heaviside step function. The wave front has propagated a distance $c(t-t_0)$ along the z axis at time "t" (waves propagate at the speed of light in a transmission line with perfectly conducting walls and center conductors). The step function is an approximation to this complicated interface that involves fringing fields (see the appendix to this chapter). A displacement current flows at the wave front,

$$J_D^\rho = \frac{-\Omega_D R^2 B_0}{8\pi \ln\left(R/a\right)} \frac{1}{\rho} \delta\left[c\left(t-t_0\right)-z\right] \tag{2.104}$$

Applying Ampere's law to (2.104) at the flow front,

$$B^\phi = \frac{-\Omega_D R^2 B_0}{2c \ln\left(R/a\right)} \frac{1}{\rho} \Theta\left[c\left(t-t_0\right)-z\right] \,. \tag{2.105}$$

Note that one can write a fictitious Ohm's law at the flow front. Define a fictitious surface current at the flow front (a surface displacement current), $\mathcal{J}_D^\rho$, by integrating the displacement current across the flow front,

$$\mathcal{J}_D^\rho = \lim_{\varepsilon \to 0} \int_{c(t-t_0)-\varepsilon}^{c(t-t_0)+\varepsilon} J_D^\rho \mathrm{d}z \,. \tag{2.106a}$$

Then define a fictitious impedance of the flow front, Z_D, empirically as

$$Z_D \mathcal{J}_D^{\rho} \equiv E^{\rho} \,. \tag{2.106b}$$

From (2.103), (2.104) and (2.106b)

$$Z_D = 4\pi/c \,. \tag{2.106c}$$

This impedance is fictitious and is defined to elucidate the surface impedance of $4\pi/c$ of the event horizon in the membrane paradigm [80] in future discussions.

The Faraday wheel at the end of a transmission line radiates physical TEM modes of the electromagnetic field in the limit of $\omega \rightarrow 0$ (step waves). From (2.101) and (2.102), the radiation carries off energy and angular momentum. This will continue until $\boldsymbol{J} \times \boldsymbol{B}$ forces in the disk torque it to zero angular velocity. The Faraday wheel converts mechanical inertia into electromagnetic radiation. Note that (2.106c) implies that these radiation losses are equivalent mathematically to pumping currents through a surface impedance of $4\pi/c$ at infinity in a time stationary transmission line circuit.

The voltage drop across the magnetic field lines is given by (2.100b) as

$$\Delta V = \frac{-\Phi \Omega_D}{2\pi c} \,, \tag{2.107}$$

where Φ is the magnetic flux through the disk and Ω_D the disk angular velocity. ΔV is a function of disk parameters only and is determined by the Faraday wheel. Similarly, the total current flowing along the center conductor can be found from (2.105) and Ampere's law

$$I = \frac{-\Phi \Omega_D}{2\pi \, \ln\left(R/a\right)} \,. \tag{2.108}$$

The current depends only on disk parameters and is determined causally by microscopic forces within the Faraday wheel. Microscopic physics within the Faraday wheel is the causative agent producing I and ΔV in the transmission line.

2.9.4.2 The Faraday Wheel Terminated Plasma-Filled Waveguide

Consider a pair of plasma-filled, semi-infinite, cylindrical waveguides connected to a Faraday wheel as shown in Fig. 2.4. The waveguides are inside an infinite solenoid, so a uniform axial magnetic field exists within the plasma. The plasma provides a conduction path for return current, so no electrostatic equilibrium is achievable in the Faraday wheel, as was the case for the transmission line circuit. The unbalanced EMF in the Faraday wheel makes it a unipolar generator of electrical current in the waveguide circuit. The return current now flows throughout the plasma as opposed to being concentrated along the symmetry axis in the center conductor of a transmission line.

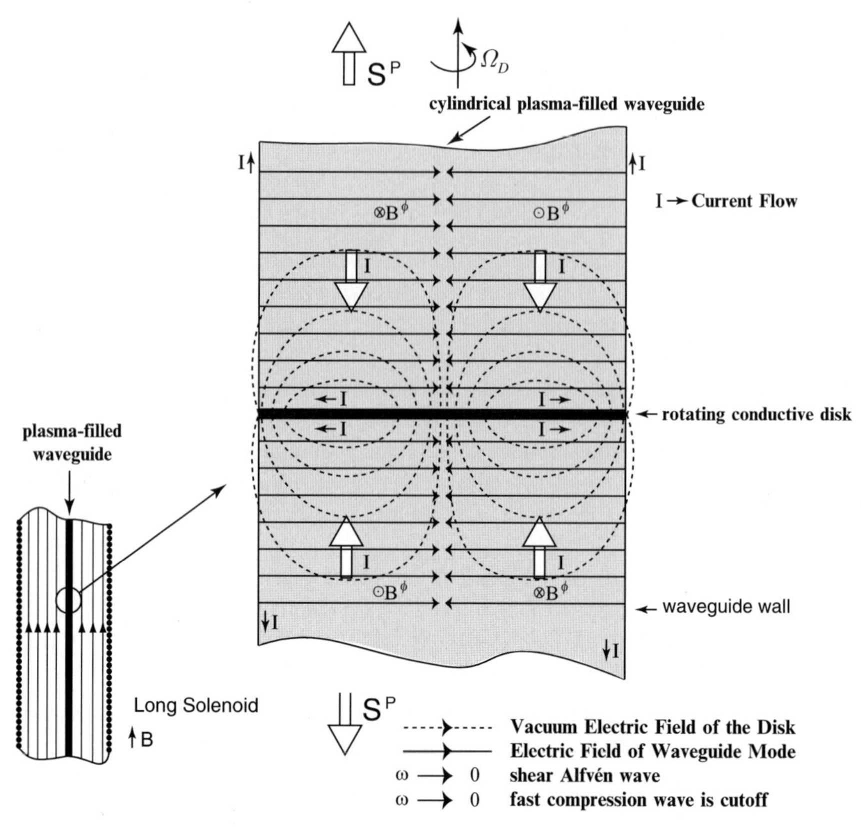

Fig. 2.4 When a rotating disk is sandwiched between two semi-infinite plasma-filled cylindrical waveguides in the presence of a uniform magnetic field, it behaves as a unipolar inductor. The Faraday wheel drives current in the waveguide wall and a distributed return current flows in the plasma (indicated by the broad arrows labeled I). The current supports a toroidal magnetic field, B^{ϕ}, in a $\omega \to 0$ shear Alfvén wave. The Faraday wheel radiates Alfvén waves in this configuration

In the plasma, E^{ρ} and ρ_e are given by (2.98) and (2.99) respectively, just as in the Faraday wheel. As with the transmission line circuit, current closure is accomplished by displacement current at the flow front as depicted in Fig. 2.5. In general, v_z, the velocity of the flow front, is not a constant, but depends on plasma parameters. Thus we have

$$J_D^{\rho} = \frac{-\Omega_D \rho}{4\pi c} B_0 v^z \delta\left[\int_{t_0}^{t} v_z \mathrm{d}t - z\right] . \tag{2.109}$$

Note that the $m = 0$ fast waves do not propagate in the cylinder as $\omega \to 0$ as discussed in Sect. 2.9.2. The Alfvén wave by contrast, exists as $\omega \to 0$ from the dispersion relation (2.82) and by (2.93b) and (2.96) it carries an E^{ρ} and B^{ϕ}. The flow can be interpreted as a step Alfvén wave in the $\omega \to 0$ limit. A formal construction of the wind as a Fourier composition of oscillatory Alfvén waves, in the extremely

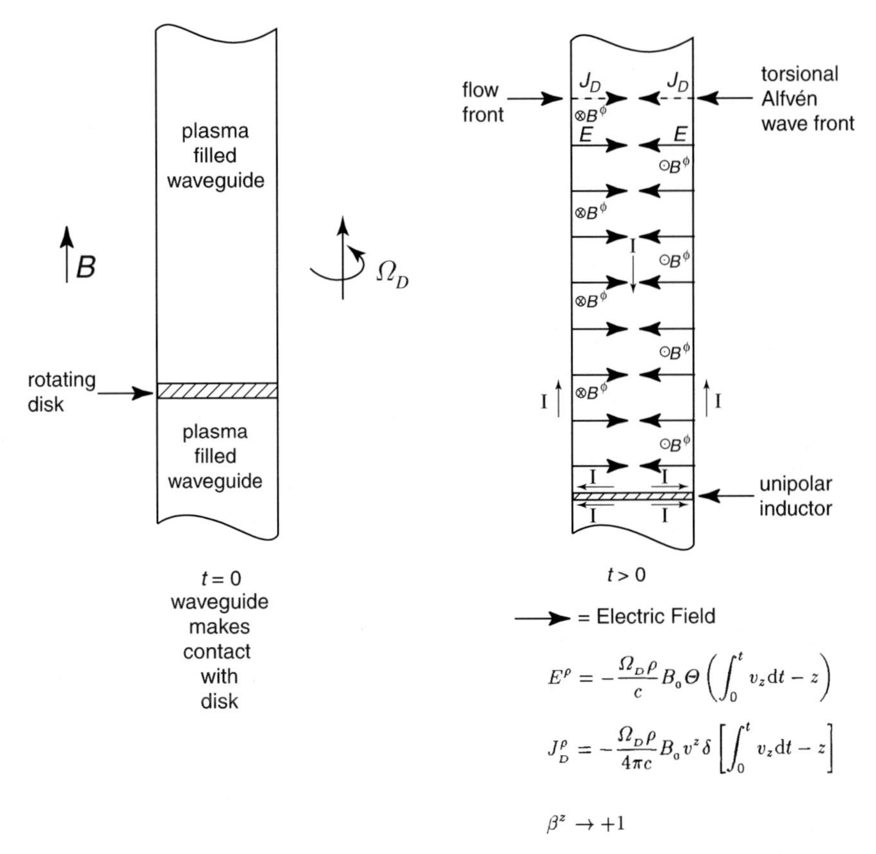

Fig. 2.5 The circuit in Fig. 2.4 closes primarily with displacement current at the flow front. In order to see this, note that the disk was attached to the pair of plasma-filled waveguides at some time, $t = t_0$ (say $t = 0$ for simplicity). At this time, the Faraday wheel begins to radiate shear Alfvén waves into the waveguides. The wave fronts propagate with a velocity v^z. At a time, t, a wave front is $\int_0^t v_z dt$ displaced from the Faraday wheel.The change in E^ρ across the flow front creates a displacement current, J_D^ρ

magnetically dominated limit, is given in the appendix to this chapter. However, it is important to note that the wind-front is complicated by fringing fields and inertial terms. Thus, the wind-front requires a fast wave precursor in order to achieve all the boundary and jump conditions. This illustrates a general principle that both the compressive modes and the Alfvén mode polarizations are required to propagate electromagnetic information in a plasma [78].

In order to solve for B^ϕ exactly in the cylinder is complicated by the effects of plasma inertia. The flow front is not purely electrodynamic, but has inertia as well. An inertial current exists in the flow front that torques the static plasma ahead of the flow to achieve the value of u^ϕ in the step wave downstream. We can solve the azimuthal momentum equation at the flow front to get the inertial current density,

$$J^{\rho} = -\frac{cn\mu u^{\phi} u^{z}}{B_0} \delta\left[\int_{t_0}^{t} v_z \mathrm{d}t - z\right] . \tag{2.110}$$

Using (2.98) and the conservation of current equation,

$$\nabla \cdot \boldsymbol{J} = -\frac{1}{4\pi}\frac{\partial}{\partial t}(\nabla \cdot \boldsymbol{E}) , \tag{2.111a}$$

we have

$$\frac{\partial J_z}{\partial z} + \frac{1}{\rho}\frac{\partial}{\partial \rho}(\rho J^{\rho}) = -\frac{\Omega_D B_0}{2\pi c} v_z \delta\left[\int_{t_0}^{t} v_z \mathrm{d}t - z\right] . \tag{2.111b}$$

Note that the term on the right hand side of (2.111a) is the divergence of the displacement current in (2.109). Inserting (2.110) into (2.111b) and integrating across the flow front implies that

$$\lim_{z \to v_z(t-t_0)} J_z = -\frac{1}{\rho}\frac{\partial}{\partial \rho}\left(\frac{\rho n\mu u^{\phi} u^{z} c}{B_0} + \frac{\Omega_D \rho^2}{4\pi c} B_0 v_z\right) , \tag{2.112}$$

where $v_z = \left(u^z/u^0\right)$. Ampere's law applied to (2.112) yields

$$B^{\phi} = -\left(\frac{4\pi n\mu u^{\phi} u^{z}}{B_0} + \frac{\Omega_D \rho}{c} B_0 \frac{v_z}{c}\right) . \tag{2.113}$$

The frozen-in condition can be written with the definition, $v^{\phi} = \left(u^{\phi}/u^0\right)$, as

$$B^{\phi} = \frac{v^{\phi} - \Omega_D \rho}{v_z} B_0 . \tag{2.114}$$

Solving (2.113) and (2.114) simultaneously yields at the flow front,

$$\lim_{z \to v_z(t-t_0)} v^{\phi} = \frac{\left(1 - v_z^2\right) \Omega_D \rho}{M^2 + 1} , \tag{2.115}$$

where M^2 is the pure Alfvén number

$$M^2 \equiv \frac{u_z^2}{U_A^2 c^2} \equiv \frac{4\pi n\mu u_z^2}{B_0^2} , \text{ and} \tag{2.116a}$$

$$U_A^2 \equiv \frac{B_0^2}{4\pi n\mu c^2} . \tag{2.116b}$$

Inserting this value of v^{ϕ} back into (2.113) gives the toroidal magnetic field just downstream of the wave front:

$$\lim_{z \to v_z(t-t_0)} B^\phi = -\frac{\Omega_D \rho B_0 v_z}{c^2}\left[\frac{U_A^2 c^2 + u_0^2}{U_A^2 c^2 + u_z^2}\right]$$
$$= E^\rho \frac{v_z}{c}\left[\frac{U_A^2 c^2 + u_0^2}{U_A^2 c^2 + u_z^2}\right] . \tag{2.117}$$

Equation (2.117) can be understood in the magnetically dominated case, $U_A^2 \gg 1$. The wind front moves at the intermediate speed of the plasma ahead of the wave front (note, $B^\phi = 0$ ahead of the wave front, so $U_I = cU_A$). Thus, by (2.82) and (2.41) $u_z^2 = c^2 U_A^2$ at the flow front. Expanding (2.117) with $u_z^2/c^2 \gg 1$, at large distances from the Faraday wheel we have

$$B^\phi \approx E^\rho \frac{v_z}{c}\left[1 + \frac{c^2 + u_\phi^2}{2u_z^2}\right] \approx \frac{E^\rho c}{v_z} . \tag{2.118}$$

This is the same result as in a split monopole MHD wind from a neutron star near the equator [67]. The field aligned poloidal current far from the Faraday wheel in (2.112) becomes

$$J_z = -\frac{1}{\rho}\frac{\Omega_D B_0}{4\pi}\frac{\partial}{\partial \rho}\left[\frac{U_A^2 c^2 + u_0^2}{U_A^2 c^2 + u_z^2}\rho^2 \frac{v_z}{c}\right] \approx \frac{c^2}{v_z}\rho_e . \tag{2.119}$$

Equation (2.119) holds in the split monopolar pulsar wind near the equator as well [67]. The charge density, ρ_e, in the waveguide is the analog of the Goldreich–Julian [81] charge density, ρ_{G-J}, of pulsar physics. Similarly (2.119) is the same axial current to charge ratio found for Alfvén waves in the cylinder (see 2.97b). This strengthens the perception that the flow is a step Alfvén wave.

Integrating (2.119) over radius yields the total axial current in the waveguide

$$I = -\frac{\Omega_D B_0 R^2}{2}\frac{v_z}{c}\left[\frac{U_A^2 c^2 + u_0^2}{U_A^2 c^2 + u_z^2}\right] \approx -\frac{\Omega_D \Phi}{2\pi}\frac{c}{v_z} . \tag{2.120}$$

The voltage drop across the magnetic field lines is the same as for the transmission line

$$\Delta V = -\frac{\Phi \Omega_D}{2\pi c} . \tag{2.121}$$

Again, the Faraday wheel establishes the global potential, ΔV, and the current, (as slightly modified by the $\left[1 - (v_z/c)^2\right](c/v_z)$ component from plasma inertia). This close analogy to an MHD pulsar establishes the essential role of a unipolar inductor in determining the current and the electrostatic potential in the wind by means of Alfvén waves emitted from the stellar surface. It also illustrates how current closure is accomplished at the flow front primarily with displacement current with a small amount of inertial current. The analogy to the pulsar magnetosphere requires a source of plasma injection in order to be complete, since plasma flows axially as a result of $J^\rho \hat{\boldsymbol{e}}_\rho \times B^\phi \hat{\boldsymbol{e}}_\phi$ forces. As plasma flows away from the disk, it must be replenished. The details of this mechanism determine the flow velocity at the disk surface and is a boundary condition necessary for solving the axial momentum equation.

2.9.4.3 The Open Circuited Transmission Line

Consider a Faraday wheel that has been disconnected from the end of a coaxial transmission line (spline) as indicated in the top view in Fig. 2.6. The circuit is open and no current flows. The electrostatic field from the surface charge on the disk terminates on the surfaces of the transmission line conductors. The voltage drop from $\oint \boldsymbol{E} \cdot \mathrm{d}\ell = 0$ exists entirely in the gap between the conductors and the disk,

$$\Delta V_{\text{gap}} = -\frac{\Omega_D \Phi}{2\pi} \,. \tag{2.122}$$

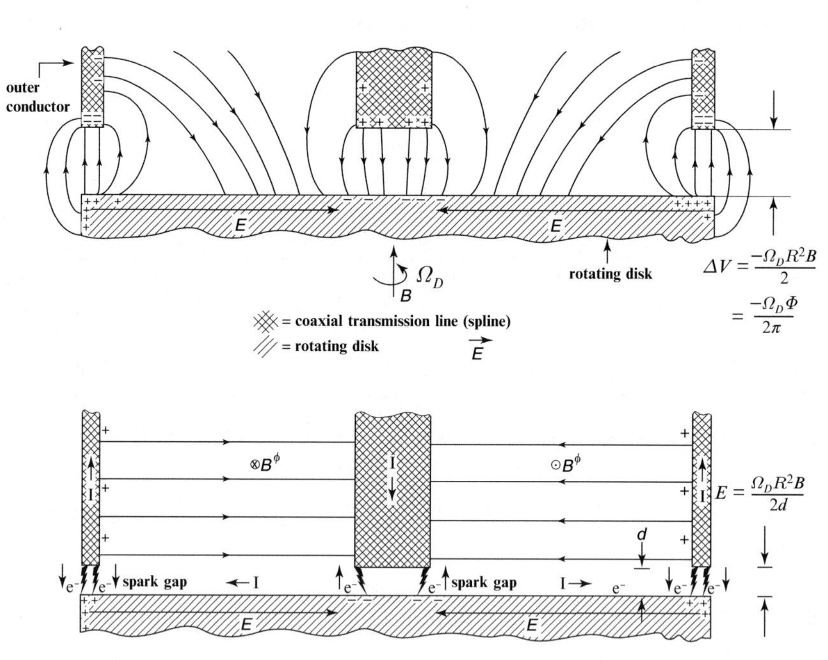

Fig. 2.6 The top of the figure shows a Faraday wheel that is physically disconnected from the pair of semi-infinite transmission lines in Fig. 2.3. An electrostatic equilibrium forms. The static charge distribution is indicated by the *plus* and *minus* signs on the conductors. Even though there is a voltage drop across the magnetic field lines at the surface of the disk, no current flows. This is the electrodynamic analog of an event horizon of a rotating black hole. At the bottom of the figure, the gap is decreased and $E_{gap} \sim 1/d$. Eventually, for small enough d, E_{gap} is large enough to create sparking. The sparks represent electron motion in the gap in the direction of the arrows adjacent to the e^-s. The current in the spark gaps flows in the opposite direction to the *arrows*. This closes the circuit and the Faraday wheel radiates $\omega \to 0$ TEM modes

There is a voltage drop across the magnetic field lines at the disk surface of the given by (2.100b), yet no current flows (i.e., a voltage drop across magnetic field lines does not imply current flow). This situation will be shown in Chap. 4 to be analogous to the electrodynamics of the event horizon.

Now let the transmission line approach the Faraday wheel as shown in the bottom of Fig. 2.6. The electric field in the gap grows inversely with the gap height, d,

$$E_{\mathrm{gap}} = -\frac{\Omega_D \Phi}{2\pi d} . \tag{2.123}$$

As the gap closes one expects arcing to occur as the electrostatic force will eventually overcome the forces that hold the electrons to the conductor. The spark gaps close the circuit allowing current and TEM waves to be emitted from the unipolar inductor as indicated in the bottom of Fig. 2.6.

2.9.4.4 The Open Circuited Plasma-Filled Waveguide

In analogy to the last section, we disconnect the Faraday from the plasma-filled waveguide as illustrated at the top of Fig. 2.7. We hold the plasma within the waveguide by a thin, low-loss dielectric sheet, such as a teflon sheet. Again, the voltage drop occurs entirely within the gap and is given by (2.122). There is a voltage across the magnetic field lines at the surface of the disk given by (2.100b). Yet, no current flows, the circuit is open. We will show in Chap. 6 that the event horizon behaves analogously to the Faraday wheel in this context. For incoming winds the horizon can have a voltage drop across the magnetic field lines, but it is not a "battery-like EMF" as it drives no current.

As the gap closes, as indicated at the bottom of Fig. 2.7, the electric field grows, as in (2.123). Eventually, for small "d" spark gaps form and the circuit is completed. This allows the Faraday wheel to behave as unipolar inductor and radiate Alfvén waves into the plasma-filled cylinder. In Chap. 4, this will be contrasted to the event horizon in which E_{gap} stays well behaved.

2.10 Anisotropic Electrical Conductivity in Strong Magnetic Fields

The plasma in black hole GHM are relativistic because they exist in magnetically dominated magnetospheres. In the strong magnetic field domain of plasmas ($U_A^2 \gg 1$), the electromagnetic field can inject very strong forces per unit mass in the tenuous plasma. This creates a potential for relativistic motion and waves in the magnetosphere.

An important aspect of strong magnetic fields in a highly conductive tenuous plasma is the anisotropic electrical conductivity tensor. The individual particles tend

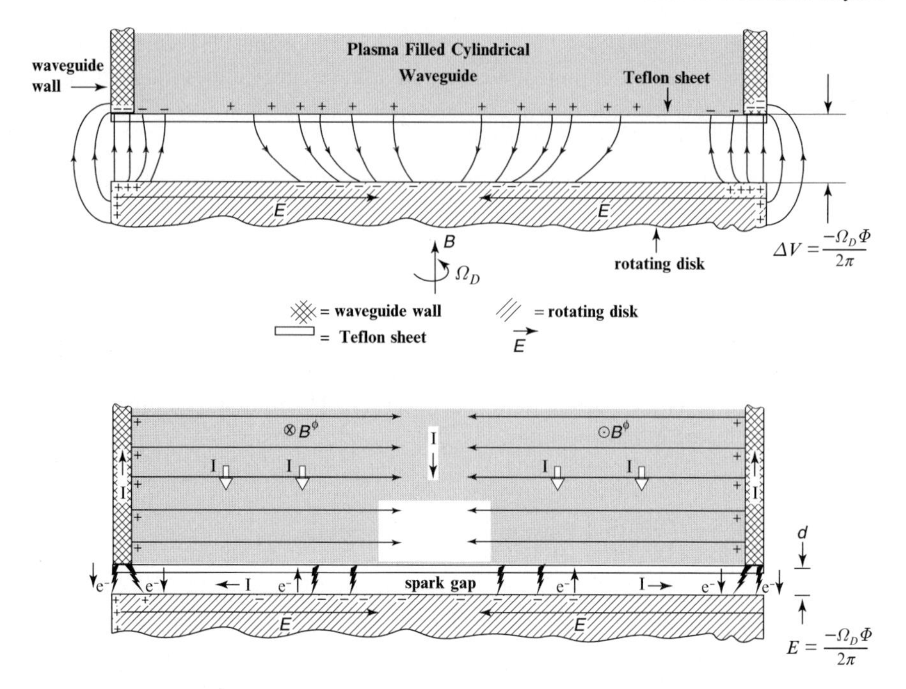

Fig. 2.7 The top of the figure shows a Faraday wheel that is physically isolated from the pair of semi-infinite plasma-filled waveguides in Fig. 2.4. An electrostatic equilibrium is established. Even though there is a voltage drop across $\boldsymbol{B}$ at the surface of the disk, no current flows. This is the analog of the event horizon of a rotating black hole in the context of MHD winds. At the bottom of the figure, the gap is decreased and $E_{gap} \sim 1/d$. Eventually, for small enough d, E_{gap} is large enough to create spark gaps. This closes the circuit and the Faraday wheel radiates $\omega \to 0$ Alfvén waves. This property contrasts a laboratory conductor with the event horizon

to be threaded on the magnetic field lines in their gyro-orbits. They can slide along the magnetic field lines freely and thus the electrical conductivity in this direction, $\sigma_{\parallel}$, is the same as for a nonmagnetized plasma. In the cross-field direction, it takes particle collisions to knock the charges out of their gyro-orbits in order for a current stream to cross the magnetic field. By the tenuous plasma/strong magnetic field assumption, these collisions occur at a rate, v_c, that is always much less than the gyro-frequency:

$$\Omega_L = \frac{qB}{\mu c} \quad , \quad v_c \ll |\Omega_L| \; , \tag{2.124}$$

where q is the charge of the particle. Consequently, electrical conduction in the cross-field direction, $\sigma_{\perp}$, is greatly impeded. We demonstrate this for both positronic and protonic plasmas in a two-fluid analysis.

We analyze the response of a resistive plasma to an imposed electric field in the rest frame (i.e., a frame moving with the bulk velocity of the flow) of the plasma. The analysis is performed with the aid of the equations of motion of the individual species of charge. We introduce electrical conductivity through a phenomenological collision term $n_0 \nu_c \mu \boldsymbol{v}_D$ in the momentum equation, where $\boldsymbol{v}_D$ is the drift velocity between the two species,

$$\boldsymbol{v}_D \equiv \frac{\boldsymbol{v}_+ - \boldsymbol{v}_-}{2} \,. \tag{2.125}$$

Because of collisions, the proper electric field, $\boldsymbol{E}$, is no longer exactly zero. Thus, momentum equations become (compare to (2.76)),

$$n_+ \mu_+ \frac{d}{d\tau} \boldsymbol{u}_+ + \nabla P_+ = \frac{e}{c} n_+ \left[\boldsymbol{E} + \frac{(\boldsymbol{v}_+ \times \boldsymbol{B})}{c} \right] - n_+ \nu_c \mu \boldsymbol{v}_D \,, \tag{2.126a}$$

$$n_- \mu_- \frac{d}{d\tau} \boldsymbol{u}_- + \nabla P_- = -\frac{e}{c} n_- \left[\boldsymbol{E} + \frac{(\boldsymbol{v}_- \times \boldsymbol{B})}{c} \right] + n_- \nu_c \mu \boldsymbol{v}_D \,. \tag{2.126b}$$

In order to evaluate electrical conductivity, we ignore the ∇P term. Note that "e" is the charge of a positron, or proton.

We can consider the current due to the motion of positive charges obtained from the equation of motion (2.126a). Let z be along the magnetic field axis; then (2.126a) yields the following matrix differential equation for the positive charge component of current density, $\boldsymbol{J}_+$

$$\begin{pmatrix} \dot{J}_x \\ \dot{J}_y \\ \dot{J}_z \end{pmatrix}_+ = \frac{(\omega_p)_+^2}{4\pi} \begin{pmatrix} E_x \\ E_y \\ E_z \end{pmatrix} + \begin{pmatrix} -\nu_c & -\Omega_L^+ & 0 \\ \Omega_L^+ & -\nu_c & 0 \\ 0 & 0 & -\nu_c \end{pmatrix} \begin{pmatrix} J_x \\ J_y \\ J_z \end{pmatrix}_+ \,, \tag{2.127a}$$

where $(\omega_p)_+$ is the plasma frequency of the positive charges. The gyro frequency Ω_L^+ is given by (2.124) where q is the positronic charge "e". Similarly, the negative charges produce a current density $\boldsymbol{J}_-$ obtained from (2.126b),

$$\begin{pmatrix} \dot{J}_x \\ \dot{J}_y \\ \dot{J}_z \end{pmatrix}_- = \frac{(\omega_p)_-^2}{4\pi} \begin{pmatrix} E_x \\ E_y \\ E_z \end{pmatrix} + \begin{pmatrix} -\nu_c & -\Omega_L^- & 0 \\ \Omega_L^- & -\nu_c & 0 \\ 0 & 0 & -\nu_c \end{pmatrix} \begin{pmatrix} J_x \\ J_y \\ J_z \end{pmatrix}_- \,. \tag{2.127b}$$

The gyro frequency, Ω_L^-, is given by (2.124) where q is now the electronic charge "$-e$".

Solving (2.127a) by matrix techniques yields the following expressions:

$$J_x^+(\tau) = \frac{(\omega_p^2)_+}{4\pi \left[\nu_c^2 + (\Omega_L^+)^2 \right]} \left\{ \left[\nu_c + e^{-\nu_c \tau} \left(\Omega_L^+ \sin \Omega_L^+ \tau - \nu \cos \Omega_L^+ \tau \right) \right] E_x \right.$$

$$\left. + \left[-\Omega_L^+ + e^{-\nu_c \tau} \left(\Omega_L^+ \cos \Omega_L^+ \tau + \nu_c \sin \Omega_L^+ \tau \right) \right] E_y \right\} \,, \tag{2.128a}$$

$$J_y^+(\tau) = \frac{(\omega_p^2)_+}{4\pi\left[\nu_c^2 + (\Omega_L^+)^2\right]}\left\{\left[\Omega_L^+ - e^{-\nu_c\tau}\left(\Omega_L^+\cos\Omega_L^+\tau + \nu\sin\Omega_L^+\tau\right)\right]E_x\right.$$

$$\left. + \left[\nu_c + e^{-\nu_c\tau}\left(\Omega_L^+\sin\Omega_L^+\tau - \nu_c\cos\Omega_L^+\tau\right)\right]E_y\right\} , \quad (2.128b)$$

$$J_z^+(\tau) = \frac{(\omega_p^2)_+}{4\pi\nu_c}\left[1 - e^{-\nu_c\tau}\right] . \quad (2.128c)$$

At late times ($\tau \to \infty$) a steady state current flow is attained in the plasma as a consequence of (2.128), from the motion of positive charge. This can be written as an Ohm's law for positive charge,

$$J_+^i = \sigma_+^{ik}E_k , \quad (2.129a)$$

$$\sigma_+^{ik} = \frac{(\omega_p^2)_+}{4\pi\left[(\Omega_L^+)^2 + \nu_c^2\right]}\begin{bmatrix} \nu_c & -\Omega_L^+ & 0 \\ \Omega_L^+ & \nu_c & 0 \\ 0 & 0 & \left[(\Omega_L^+)^2 + \nu_c^2\right]/\nu_c \end{bmatrix} . \quad (2.129b)$$

Similarly, one finds from integrating (2.127b) as $\tau \to \infty$, a current contribution from negative charges

$$J_-^i = \sigma_-^{ik}E_k , \quad (2.129c)$$

$$\sigma_-^{ik} = \frac{(\omega_p^2)_-}{4\pi\left[(\Omega_L^-) + \nu_c^2\right]}\begin{bmatrix} \nu_c & -\Omega_L^- & 0 \\ \Omega_L^- & \nu_c & 0 \\ 0 & 0 & \left[(\Omega_L^-)^2 + \nu_c^2\right]/\nu_c \end{bmatrix} . \quad (2.129d)$$

The off diagonal terms represent currents due to the charges being in gyro-orbits. Note that the positive and negative charges orbit in the opposite sense, so these currents are of opposite sign.

The total current density is given by

$$\boldsymbol{J} = \boldsymbol{J}_+ + \boldsymbol{J}_- , \quad (2.130a)$$

and the total conductivity tensor is therefore

$$\sigma^{ik} = \sigma_+^{ik} + \sigma_-^{ik} . \quad (2.130b)$$

From (2.129), for a positronic plasma, the conductivity tensor is

$$\sigma^{ik} = \frac{(\omega_{pe})^2}{4\pi\nu_c} \begin{bmatrix} \left[1+\Omega_e^2/\nu_c^2\right]^{-1} & 0 & 0 \\ 0 & \left[1+\Omega_e^2/\nu_c^2\right]^{-1} & 0 \\ 0 & 0 & 1 \end{bmatrix}, \tag{2.130c}$$

$$\Omega_e \equiv -\Omega_L^- \tag{2.130d}$$

Note that due to the equal electron and positron masses, the contribution to the currents from the gyro-motion cancel. For a protonic MHD plasma, the current is carried primarily by the lighter electrons, so (2.129) and (2.130b) imply

$$\sigma^{ik} \approx \frac{(\omega_{pe})^2}{4\pi\nu_c} \begin{bmatrix} \left[1+\Omega_e^2/\nu_c^2\right]^{-1} & +(\Omega_e/\nu_c)\left[1+\Omega_e^2/\nu_c^2\right]^{-1} & 0 \\ -(\Omega_e/\nu_c)\left[1+\Omega_e^2/\nu_c^2\right]^{-1} & \left[1+\Omega_e^2/\nu_c^2\right]^{-1} & 0 \\ 0 & 0 & 1 \end{bmatrix}. \tag{2.130e}$$

Note that the parallel conductivity along a magnetic field line is the same as in a nonmagnetic plasma,

$$\sigma_{\|} = \frac{(\omega_{pe})^2}{4\pi\nu_c}, \tag{2.131}$$

where ω_{pe} is the electron plasma frequency,

$$(\omega_{pe})^2 = \frac{4\pi n_e e^2}{\mu_e}. \tag{2.132}$$

The electrical conductivity across the magnetic field is given by (2.130) as

$$\sigma_{\perp} = \frac{(\omega_{pe})^2 \nu_c}{4\pi(\nu_c^2+\Omega_e^2)} = \frac{\sigma_{\|}}{\left[1+\Omega_e^2/\nu_c^2\right]}. \tag{2.133}$$

For a tenuous plasma in a strong magnetic field $\nu_c^2 \ll (\Omega_e)^2$, so $\sigma_{\perp} \ll \sigma_{\|}$.

In order to illustrate the poor electrical conductivity in the cross-field direction in a strong magnetosphere filled with tenuous plasma, we can maximize $\sigma_{\perp}$ with respect to ν_c (in cgs units),

$$\nu_c = \Omega_e, \tag{2.134a}$$

$$\sigma_{\perp}(\max) = \frac{(\omega_{pe})^2}{8\pi\Omega_e} = \frac{1}{2}\frac{n_e e c}{B} \approx 7\frac{n_e}{B}. \tag{2.134b}$$

The magnetically dominated assumption requires

$$U_A^2 = \frac{(\Omega_e)^2}{(\omega_{pe})^2} \gg 1, \qquad \text{positronic plasma}; \tag{2.135a}$$

$$U_A^2 = \frac{(\Omega_e)^2}{(\omega_{pi})^2} \gg 1, \qquad \text{protonic plasma}, \tag{2.135b}$$

where ω_{pi} is the proton plasma frequency defined as in (2.132). Given that the maximum magnetic field strengths in a plasma-filled magnetosphere around a $10^9\,M_\odot$ black hole are $\gtrsim 10^4$ G in an astrophysical context (see Chap. 10 for a discussion), (2.134) and (2.135) imply that

$$\sigma_\perp(\text{max}) \ll 10^{10}\text{s}^{-1} \quad (\text{positronic plasma}), \tag{2.136a}$$

$$\sigma_\perp(\text{max}) \ll 10^{7}\text{s}^{-1} \quad (\text{protonic plasma}). \tag{2.136b}$$

The value in (2.136b) is smaller because the magnetically dominated condition lowers n in a protonic plasma by three orders of magnitude. The maximal achievable $\sigma_\perp$ in a black hole magnetosphere is similar to that of pure germanium or silicon at room temperature. Thus, the magnetic field acts like an insulator for electrical currents in the cross-field direction.

This is an important aspect of magnetically dominated magnetospheres that is a large distinction to the unipolar induction of the Faraday wheel described in the last section. The magnetic field does not allow strong cross-field currents to flow in a near perfect MHD plasma. Strong cross-field currents can flow in a unipolar inductor because the large number densities in the conductive medium reverse the inequality in (2.135b). This also allows $\nu_c \gg (\Omega_e)^2$ and $\sigma_\perp \approx \sigma_\parallel$ as given by (2.131). A large n_e implies large values of $(\omega_{pe})^2$ in (2.132), which allows $\sigma_\parallel$ to be large.

In order for even a modest cross-field current to flow in a magnetically dominated magnetosphere requires strong dissipation. In order to see this, note that the effective collision frequency, ν_c, can be increased in a tenuous plasma by plasma wave scattering produced in a dissipative plasma. A larger value of ν_c increases $\sigma_\perp$ as noted in (2.134). Secondly, dissipation increases μ_e which is essentially the relativistic inertia of the plasma. This makes the flow less magnetically dominated through (2.135) since $(\omega_{pe})^2/(\Omega_e)^2 \sim \mu_e/m_e$. This also increases $\sigma_\perp$ in (2.133) as the "heavier" plasma can cross the magnetic field more easily as the large collision frequency, ν_c knocks the plasma out of its gyro-orbits. The current can flow, but $\sigma_\perp$ is still much less than $\sigma_\parallel$ and this is not like the small Ohmic dissipation in a Faraday wheel.

2.11 High Frequency Waves in Protonic Plasmas

In Sect. 2.8 we restricted the discussion of high frequency waves to positronic plasmas. In this section, we justify the generality of the results of Sect. 2.8 to protonic plasmas in the range of frequencies that are relevant to the causal structure of the GHM black hole magnetospheres. In particular we show that unless $\omega \sim \Omega_i$, (the protonic gyro frequency).

$$\Omega_i = \frac{eB}{m_p c}\,, \tag{2.137}$$

then the fast magnetoacoustic wave has negligible electrostatic polarization.

In order to modify the analysis of the positronic two fluid analysis to protonic fluids we must substitute the Ohm's law of (2.77a) with the more complicated sum of (2.127a) and (2.127b) in the limit $v_c \to 0$ (i.e., $\sigma_{\parallel} \to \infty$). The single fluid approximation breaks down at high frequencies because protons respond differently to an electric field than electrons because of their larger inertia. At "high frequencies" collisions are not frequent enough to couple the electrons and protons and charge neutrality cannot be maintained. A charge separation can occur in the magnetoacoustic waves and an analysis is required to describe quantitatively what is meant by "high frequency."

We proceed as in Sects. 2.3 and 2.8 using the Ohm's law in (2.127). This is expanded as in Sect. 2.3 with the aid of (2.32) and (2.36) which are valid in a warm plasma. We get a very complicated dispersion relation in (2.138k),

$$A^{ij}E_j = 0\,, \tag{2.138a}$$

$$A^{xx} = \frac{\left(\omega^2/\omega_{pe}\right)^2 - 1 - U_A^2 \sin^2\theta}{\left[1 - c^2k^2/\omega^2\right]}\,, \tag{2.138b}$$

$$A^{xy} = -\frac{\mathrm{i}\omega U_A^2 \sin\theta}{\Omega_i}\,, \tag{2.138c}$$

$$A^{xz} = U_A^2 \sin\theta \cos\theta\,, \tag{2.138d}$$

$$A^{yx} = \frac{\mathrm{i}\omega U_A^2 \sin\theta}{\Omega_i}\,, \tag{2.138e}$$

$$A^{yy} = \frac{\omega^2}{\left(\omega_{pe}\right)^2} + \frac{\omega^2}{c^2k^2 - \omega^2} - U_A^2\left[\cos^2\theta + \frac{\sin^2\theta}{1 - c_S^2k^2/\omega^2}\right], \tag{2.138f}$$

$$A^{yz} = -\frac{\mathrm{i}\omega U_A^2 \cos\theta}{\Omega_i}\,, \tag{2.138g}$$

$$A^{zx} = \frac{U_A^2 \sin\theta \cos\theta}{\left[1 - c^2k^2/\omega^2\right]}\,, \tag{2.138h}$$

$$A^{zy} = \frac{\mathrm{i}\omega U_A^2 \cos\theta}{\Omega_i}\,, \tag{2.138i}$$

$$A^{zz} = \frac{\omega^2}{\left(\omega_{pe}\right)^2} + \frac{\omega^2}{c^2k^2 - \omega^2} - U_A^2 \cos^2\theta\,, \tag{2.138j}$$

$$\det A = 0\,. \tag{2.138k}$$

Note that in the limit $\omega/\Omega_i \to 0$, (2.138) yields the positronic high frequency dispersion relation (2.78). Also in the subrelativistic limit, $U_A^2 \ll 1, \omega^2/c^2k^2 \ll 1$, with $(\omega_{pe})^2 \gg \omega^2$, one finds the well known dispersion relation from (2.138), [73],

$$\left[\frac{\omega^4}{k^4} - \frac{\omega^2}{k^2}\left(c_S^2 + c^2U_A^2\right) + c_S^2c^2U_A^2\cos^2\theta\right]\left[\frac{\omega^2}{k^2} - c^2U_A^2\cos^2\theta\right]$$
$$-\left[\frac{\omega U_A^2}{\Omega_i}\right]^2\left[\frac{\omega^2}{k^2} - c_S^2\right]c^4\cos^2\theta = 0\,. \tag{2.139}$$

The charge density in the wave $\sim \boldsymbol{k}\cdot\boldsymbol{E} \sim E_x$. The imaginary, off diagonal terms in (2.138) couple the magneto-acoustic modes to E_x and hence create an electrostatic polarization. However, these terms $\sim (\omega/\Omega_i)$. Consider the typical black hole magnetospheric parameters for an AGN central engine model (as discussed in relation to (2.136) and detailed in Chap. 10), $M \approx 10^9\,\mathrm{M}_\odot, B \sim 10^3\,\mathrm{G} - 10^4\,\mathrm{G}$. This implies that

$$\Omega_i > 3\mathrm{x}10^7\,\mathrm{s}^{-1}\,, \tag{2.140a}$$
$$\Omega_H \sim 10^{-4}\,\mathrm{s}^{-1}\,. \tag{2.140b}$$

We discussed in Sect. 2.9 that the charge separation in a magnetosphere is related to the rotation rate of the magnetic field (see (2.121) for example). The relevant rotation rates for a black hole magnetosphere are less than Ω_H and therefore we expect that low frequency modes $\omega \sim \Omega_H$ are needed to establish the global electrostatic potential. For these modes $(\omega/\Omega_i) \sim 10^{-12}$. Consequently, even though a charge separation can exist in a high frequency protonic fast wave, these modes play no part in determining the electrostatic potential in a black hole magnetosphere.

The analysis in this section is largely pedantic as it is generally assumed that positronic plasmas are common in the high energy environment of the black hole central engine of an AGN. This brief discussion of high frequency waves in protonic plasmas was included mainly for completeness and to strengthen our understanding MHD causality in later chapters.

2.12 Longitudinal Polarized MHD Discontinuities

To this point, the discussion of wave properties has been restricted to short wavelength 1-D waves in a homogeneous media. As such, the conclusions drawn do not necessarily follow for long wavelength modes in an inhomogeneous magnetosphere, nor for higher dimensional wavefronts. A more general technique that is valid irrespective of spatial inhomogeneities involves the propagation of abrupt discontinuities of MHD parameters, or step waves. These are the basic building blocks of perfect MHD simulations that are based on Riemann solvers. In this section, we show that one result from the oscillatory solutions remains valid: the Alfvén mode, alone, is responsible for charge propagation in the magnetically dominated limit. In this limit, it is equivalent to a pure charge discontinuity that propagates parallel to the magnetic field at the pure Alfvén speed, U_A.

Perfect MHD discontinuities are solved for by considering the continuity of the stress–energy tensor, $T^{\mu\nu}$, across the wavefront. Since ultimately, we are interested in the magnetically dominated regime and relativistic waves in the context of black hole magnetospheres, the calculation is performed in the cold limit of perfect MHD (thermal energy density is small compared to the magnetic energy density). The calculations are performed in the frame of reference of the propagating wavefront. Let x be the local normal coordinate to the wavefront and the upstream magnetic field, **B** is in the x-y plane and z lies in the wavefront surface. All that needs to be considered in order to determine the desired results is mass conservation, the frozen-in condition, which are

$$nu^x u^0 = \text{constant}\ , \ \mathbf{E} + \frac{1}{c}\mathbf{v}\times\mathbf{B} = 0\ , \tag{2.141}$$

the antisymmetric components of Maxwell's equations

$$F_{\alpha\beta;\gamma} + F_{\gamma\alpha;\beta} + F_{\beta\gamma;\alpha} = 0\ , \tag{2.142}$$

and the continuity of one component of the stress–energy tensor, T^{xz},

$$n\mu u^x u^z - \frac{1}{4\pi}(E^x E^z + B^x_u B^z) = 0\ , \tag{2.143}$$

in which all of the quantities are evaluated downstream unless there is a subscript "u" and μ is the specific enthalpy (note that by (2.141), $E^x_u = 0$). The continuity of B_x used in (2.143) follows from the $\nabla\cdot\mathbf{B} = 0$ condition in (2.142). A tremendous simplification occurs in (2.142) at the step wavefront, since to lowest order, all the singular terms must cancel (surface terms like delta functions), thus the normal covariant derivative does not depend on the connection coefficients and one has continuity of E^y and E^z at the wavefront, or

$$\frac{\partial E^y}{\partial x} = \frac{\partial E^z}{\partial x} = 0\ , \tag{2.144}$$

across the wavefront. Inserting (2.141) and (2.144) into (2.142), one gets

$$\left((u^x)^2_u\left[\frac{n_u}{n} + \frac{(b^y_u)^2}{4\pi n_u \mu c^2}\right] - \frac{(b^x_u)^2}{4\pi n_u \mu}\right)E^x = 0\ , \tag{2.145}$$

written in terms of the field in the plasma rest frame, **b**. For $n_u = n$, u^x_u is the Alfvén or intermediate wave speed of (2.41). Taking the limit of zero mass density, (2.145) has two solutions,

$$v \equiv \frac{u^x_u}{u^0_u} = \mathrm{c}\cos\theta\ , \text{or} \quad E^x = 0\ , \tag{2.146}$$

where θ is the angle between **b** and the wave normal in the plasma rest frame and v is the force-free value of the Alfvén speed in a proper frame, i.e., the zero mass limit of (2.42). Thus by (2.146), in the magnetically dominated limit, a perfect MHD discontinuity either travels at the Alfvén speed and transports $E_{\|}$, or it carries no longitudinal polarization, $E_{\|}$, or surface charge. In the magnetically dominated limit, only the Alfvén discontinuity can transport electric charge in a magnetosphere.

2.13 What is Important About This Chapter?

The primary result of this chapter is that the Alfvén wave has an electrostatic polarization and can propagate field aligned currents. The simple fast wave has no electrostatic polarization and cannot propagate field aligned currents. This result was derived for both high and low frequency waves in a perfect MHD plasma as well as in a resistive plasma. The conclusions drawn from these simple wave calculations were overstated in the first edition of this book. The properties of the fast mode do not necessarily follow in higher dimensions and in inhomogeneous media. In general, nonlinear fast waves in magnetically dominated magnetospheres, such as an abrupt strong discontinuity, might in principle transport current changes, but not physical charge discontinuities as shown in Sect. 2.12. They do not carry any information on the longitudinal polarization. If one is looking to more complicated geometries as an alternative to the simple wave analysis in this chapter, then the wavefronts are complicated and all the polarizations of the electromagnetic field are required to propagate changes in the current and the charge density. The important point is that in general both the compressive and the Alfvén MHD modes are required. The conclusion that the fast mode was not involved in establishing the Goldreich–Julian charge density and field aligned currents in an MHD magnetosphere was overstated and was off the point in the first edition. The point of this chapter is not that fast waves are incapable of affecting the wind parameters. Conversely, the causality arguments in this chapter are based on the notion that Alfvén mode must figure prominently in the determination of the global electrostatic potential and the field aligned poloidal currents that are created in a black hole driven plasma wind. The key role of the Alfven mode in the determination of the electromagnetic parameters in a relativistic jet, dominated by Poynting flux is demonstrated numerically in the simulations that are described in Sect. 11.2. In those simulations, a relativistic jet is driven from the environs of the black hole entirely with Alfvén and slow modes, no fast modes.

To illustrate the relevance of these calculations and discussions to magnetospheric physics, an example of a plasma-filled waveguide terminated by a unipolar inductor was introduced in Sect. 2.9. This example is used to illustrate the role of a unipolar inductor in a relativistic pulsar MHD wind theory. The wind equations cannot be used to determine the wind constants such as the electrostatic potential and field aligned current. These are established by Alfvén wave radiation from a neutron star that behaves as a unipolar inductor. By contrast, we use the example of a Faraday wheel that is disconnected from the end of a plasma-filled waveguide to elucidate the electrodynamic properties of the event horizon in the context of MHD winds.

2.14 Appendix. The Role of the Alfvén Wave in the Plasma-Filled Waveguide

In this appendix, it is shown by explicit construction that the propagating plasma discontinuity in the Faraday wheel terminated plasma-filled waveguide of Sect. 2.9.4 can be described almost entirely as a pure Alfvén wave packet in the magnetically dominated limit. This proof is accomplished by means of the well known Fourier decomposition of a step wave as a superposition of oscillatory waves It is found that the Alfvén wave packet does indeed transport field aligned currents and charge, not just oscillatory perturbations. In the first subsection, we compare and contrast a wave packet (a step wave) created from a linear superposition of the previously calculated oscillatory Alfvén modes in Sect. 2.9.3 to the radiation emanating from the Faraday wheel (the nature of this radiation was derived previously in terms of a formal MHD wind solution in the waveguide in Sect. 2.9.4). Finally, it is noted that the depiction of the wind as a pure Alfvén step wave becomes increasingly more accurate as one passes to the force-free limit.

2.14.1 Constructing Wave Packets

We can construct wave packets of the oscillatory solutions in Sect. 2.9.3 by taking Fourier integrals. First we need to explicitly construct the Alfvén wave oscillatory solutions.

2.14.1.1 The Electric Field

From (2.93b), the radial electric field in an axisymmetric Alfvén wave that is propagating within a cylindrical waveguide is arbitrary and is determined by the boundaries (the field lines can shear relative to each other, hence these modes are often called "shear Alfvén waves"). Considering the frozen-in condition within the Faraday wheel and the continuity of the tangential electric field at the Faraday wheel/plasma interface let us choose the radial electric field in the Alfvén wave to be given by (2.98):

$$E^{\rho} = -\frac{\Omega_D \rho}{c} B_0 e^{i(k_z z - \omega t)} , \tag{2.147}$$

i.e., the Faraday wheel boundary determines E^{ρ}. The other electric field components are given by (2.86c) and (2.93a) as

$$E^{\phi} = 0\,, \tag{2.148}$$

$$E^{z} = 0\,. \tag{2.149}$$

2.14.1.2 The Currents

From (2.141)–(2.143) and Gauss' law, the charge density is;

$$\rho_e = -\frac{\Omega_D B_0}{2\pi c} e^{i(k_z z - \omega t)}\,. \tag{2.150}$$

From the Alfvén wave dispersion relation, (2.82), and the frozen-in form of the momentum equation, (2.86a), the cross-field current density is

$$J^{\rho} = i\frac{v_I k_z \Omega_D \rho}{4\pi U_A^2 c} B_0 e^{i(k_z z - \omega t)}\,, \tag{2.151}$$

where v_I is the intermediate three speed and U_A is the pure Alfven speed. From the law of current conservation,

$$\frac{\partial \rho_e}{\partial t} + \nabla \cdot \mathbf{J} = 0\,, \tag{2.152}$$

combined with (2.150) and (2.151) yields the axial current density:

$$J^{z} = \frac{\rho_e}{v_I} c^2 = -\frac{\Omega_D B_0 c}{2\pi v_I} e^{i(k_z z - \omega t)}\,. \tag{2.153}$$

2.14.1.3 The Magnetic Field

Using the value of the axial current in (2.152) in Ampere's law yields the toroidal magnetic field strength:

$$B^{\phi} = -\frac{B_0 \Omega_D \rho}{v_I} e^{i(k_z z - \omega t)}\,. \tag{2.154}$$

The other components are given by (2.92a) and (2.96):

$$B^{z} = 0\,, \tag{2.155}$$

$$B^{\rho} = 0\,. \tag{2.156}$$

2.14.1.4 The Step Wave

We are interested in the step wave. For the Alfvén wave, one can use the dispersion relation (2.82) in the cylinder to write,

$$e^{i(k_z z - \omega t)} = e^{-ik_z(v_I t - z)}\,. \tag{2.157}$$

The step function is given by,

$$\Theta(v_I t - z) = -\frac{1}{2\pi i}\int_{-\infty}^{+\infty}\frac{e^{-ik_z(v_I t - z)}dk_z}{k_z + i\varepsilon} . \tag{2.158}$$

We construct a wave packet of Alfvén waves, Ψ, with a spectral amplitude, $A(k_z)$,

$$\Psi = \int_{-\infty}^{+\infty} A(k_z)\psi(k_z)dk_z , \tag{2.159}$$

$$A(k_z) = -\frac{1}{(2\pi i)(k_z + i\varepsilon)} , \tag{2.160}$$

where $\psi(k_z)$ is one of the oscillatory m = 0 wave function components in (2.147)–(2.156) and ε is an arbitrarily small positive number in the usual sense. The wave packet of oscillatory solutions with field and current components given by (2.147)–(2.156) and spectral amplitude given by (2.160) has the following field and current distributions (for the sake of demonstrating equivalence, the corresponding equation number from Sect. 2.9.4 that were derived by the MHD wind calculation in the cylinder is placed on the right hand side of the equals sign in these relations):

$$E^\rho = -\frac{\Omega_D \rho}{c} B_0 \Theta(v_I t - z) = (2.98) , \tag{2.161}$$

$$E^\phi = 0 , \tag{2.162}$$

$$E^z = 0 , \tag{2.163}$$

$$\rho_e = -\frac{\Omega_D B_0}{2\pi c}\Theta(v_I t - z) = (2.99) , \tag{2.164}$$

$$J^\rho = \frac{v_I \Omega_D \rho}{4\pi U_A^2 c} B_0 \delta(v_I t - z) , \tag{2.165}$$

$$J^z = \frac{\rho_e}{v_I}c^2 = -\frac{\Omega_D B_0 c}{2\pi v_I}\Theta(v_I t - z) = (2.119) , \tag{2.166}$$

$$B^\phi = -\frac{B_0 \Omega_D \rho}{v_I}\Theta(v_I t - z) = (2.118) , \tag{2.167}$$

$$B^z = 0 , \tag{2.168}$$

$$B^\rho = 0 . \tag{2.169}$$

These quantities agree with the downstream state of the plasma that was found near the flow front in Sect. 2.9.4 in the limit that $v_z = v_I$. Notice that J^ρ vanishes downstream of the wavefront and only has a surface component on the wavefront as it does in (2.110).

2.14.2 Physical Discussion

From (2.164) and (2.166), the Alfvén wave packet actually transports changes in the charge and field aligned current. The solution is not exact. There are errors associated with the small inertial terms in the magnetically dominated limit. At the wavefront, the relativistic MHD shock equations do not solve exactly. These equations are the conservation of the axial components of the stress energy tensor across the discontinuity in the frame of the wavefront. Pressure balance (magnetic pressure from the toroidal magnetic field in the frame of the wavefront) can not be achieved because the Alfvén wave, unlike the fast wave, has no compressive properties. The errors in the stress–energy balance can be found in the frame of the propagating discontinuity by using (2.161) and (2.167) to compute the proper toroidal magnetic field. The errors in the shock equations are on the order of $(\Omega_D \rho U_A^{-1})^2$. This is biquadratic in two small quantities: $\Omega_D \rho / c \ll 1$ by construction and by the magnetically dominated condition, $cU_A^{-1} \ll 1$. The violation of the shock relations are therefore a very small second order effect in the magnetically dominated limit. Notice that the errors vanish completely in the force-free limit as the Alfvén three speed approaches the speed of light. In magnetically dominated perfect MHD, these small errors are accounted for by a fast switch-on shock that creates a B^ρ and an E^ϕ downstream of the shock front. This is an MHD precursor, an infinitesimal distance upstream, to the Alfvén rotational discontinuity at the terminus of the Alfvén step wave. As demonstrated by explicit construction above, it is the Alfvén rotational discontinuity that imprints the charge and field aligned current on the waveguide plasma. The interpretation of the waveguide wind solution as an Alfvén wave is exact in the force-free limit and is extremely accurate to first order in the magnetically dominated MHD limit. However, in relativistic MHD it is important to note that both fast and Alfvén modes are typically required to match all the jump and boundary conditions at an electromagnetic discontinuity [78].

Chapter 3
Particle Trajectories in the Ergosphere

3.1 Motivation

We discussed in Sect. 1.4 that any energy extraction process involving a rotating black hole requires the absorption of matter with a negative energy, ω, as measured from asymptotic infinity. The preparation of $\omega < 0$ matter must occur within the ergosphere. This is related to the fact that the outer boundary of the ergosphere, the stationary limit at $r = r_s$ occurs when $\partial/\partial t$ changes from a timelike vector field to a spacelike vector field (see definition of ω in (1.36a)). This change happens when g_{tt} switches sign. By (1.24a) this condition is

$$r_s = M + \sqrt{M^2 - a^2 \cos^2\theta} \,. \tag{3.1}$$

Since any energy extracting process creates negative energy matter in the ergosphere, it is important to understand particle trajectories in this region with or without applied forces.

3.2 Coordinate Systems and Frames

It is useful to have different methods to describe the dynamics in general relativity in order to elucidate the physical meaning. For example, a global coordinate system such as the Boyer–Lindquist coordinates is invaluable for defining conserved quantities and the global energetics of an interaction. The Boyer–Lindquist coordinates are particularly natural because as $r \to +\infty$ they reduce to standard Minkowski space–time spherical coordinates. Thus, they represent how a distant observer would view the black hole. They are not very useful for understanding the nature of the physical interaction since they are not orthonormal or even orthogonal because of space–time curvature. Furthermore, the four velocity of the distant observers, $\partial/\partial t$, is spacelike within the ergosphere. This makes interpretations within standard physics of the earth very confusing.

B. Punsly, *Black Hole Gravitohydromagnetics, 2nd. ed.*,
Astrophysics and Space Science Library 355, doi: 10/1007/978-3-540-76957-6_3,

In practice, it is more useful to calculate in an orthonormal frame and then piece together the global geometry. The power of this is that one can exploit the equivalence principle and use Lorentz boosts to transform to and from freely falling frames where the physics is simplified because the connection coefficients vanish at the origin of these local frames. In order to fully capitalize on this property of psuedo-Riemannian manifolds, we want to find a global orthonormal frame field that has a timelike direction that is hypersurface orthogonal. Mathematically, this condition for a frame field, $\hat{e}_0$, $\hat{e}_i$ $(i=1,2,3)$ can be expressed in terms of the basis covectors ω^0, ω^i by $d\omega \wedge \omega = 0$. This is equivalent to being able to find a time coordinate ω^0 orthogonal to a spacelike 3-surface spanned by $\hat{e}_1$, $\hat{e}_2$, and $\hat{e}_3$. This procedure is known as foliating space–time with a family of 3 dimensional spacelike hypersurfaces. Thus, the local physics in an orthonormal frame can be pieced together to give the global physics in Boyer–Lindquist coordinates. The calculational process goes as follows:

1. Compute special relativistic physics in a frame (that is sometimes chosen to be instantaneously at rest with respect to a local orthonormal frame).
2. Use special relativistic transformations to find physics in local orthonormal frames.
3. Using the foliation of space–time in the orthonormal frame field yields physics in the global Boyer–Lindquist coordinates.

For most physical processes the calculational scenario in steps 1-3 is justified and will be exploited throughout the text.

The most useful hypersurface orthogonal frames are clearly ones defined at a constant r coordinate. An example of such a frame field are the ZAMO frames introduced in Sect. 1.4. They have zero angular momentum about the symmetry axis of the hole, m, as defined in (1.36b). They are the analog of static frames in the Schwarzschild geometry. The zero angular momentum condition requires that they rotate with an angular velocity as viewed from asymptotic infinity, Ω, given by

$$\mathrm{d}\phi/\mathrm{d}t = \Omega = -g_{\phi t}/g_{\phi\phi} \,. \tag{3.2}$$

The ZAMO basis vectors are determined by the transformation

$$\begin{bmatrix} \hat{e}_0 \\ \hat{e}_\phi \end{bmatrix} = \begin{bmatrix} \left|g_{tt} - \Omega^2 g_{\phi\phi}\right|^{-1/2} & \Omega \left|g_{tt} - \Omega^2 g_{\phi\phi}\right|^{-1/2} \\ 0 & 1/\sqrt{g_{\phi\phi}} \end{bmatrix} \begin{bmatrix} \tilde{e}_t \\ \tilde{e}_\phi \end{bmatrix}, \tag{3.3a}$$

$$\hat{e}_r = \left(\frac{\Delta}{\rho^2}\right)^{1/2} \tilde{e}_r = \left(\frac{\Delta^{1/2}}{\rho^2}\right) \frac{\partial}{\partial r}, \tag{3.3b}$$

$$\hat{e}_\theta = \left(\frac{1}{\rho}\right) \tilde{e}_\theta = \left(\frac{1}{\rho}\right) \frac{\partial}{\partial \theta}, \tag{3.3c}$$

where we will designate Boyer–Lindquist evaluated quantities by tildes throughout the remainder of the text.

The basis covectors are

$$\begin{bmatrix} \omega^0 \\ \omega^\phi \end{bmatrix} = \begin{bmatrix} \left| g_{tt} - \Omega^2 g_{\phi\phi} \right|^{1/2} & 0 \\ -\Omega \sqrt{g_{\phi\phi}} & \sqrt{g_{\phi\phi}} \end{bmatrix} \begin{bmatrix} \mathrm{d}t \\ \mathrm{d}\phi \end{bmatrix} , \tag{3.4a}$$

$$\omega^r = \left(\frac{\rho^2}{\Delta} \right)^{1/2} \mathrm{d}r ,$$

$$\omega^\theta = \rho \, \mathrm{d}\theta . \tag{3.4b}$$

The gravitational redshift of the ZAMO frames as viewed from asymptotic infinity is a very useful quantity, α, called the lapse function in [80]:

$$\alpha = \left| g_{tt} - \Omega^2 g_{\phi\phi} \right|^{1/2} = \frac{\Delta^{1/2} \sin\theta}{\sqrt{g_{\phi\phi}}} . \tag{3.5a}$$

Note that

$$\lim_{r \to \infty} \alpha = +1 , \tag{3.5b}$$

$$\lim_{r \to r_+} \alpha = 0 . \tag{3.5c}$$

Consequently, α is a valuable dimensionless parameter for expanding quantities in the vicinity of the event horizon, $\alpha \gtrsim 0$.

The ZAMOs are not inertial observers (otherwise the space–time would be flat). The acceleration, a^μ, is found in [72] to be

$$a^\mu = \frac{1}{\alpha} \frac{\partial}{\partial X^\mu} (\alpha) . \tag{3.6}$$

One of the computational advantages of the ZAMO orthonormal frame is that it is defined only up to a rotation in the (r, θ) plane. In the study of winds it is useful to define a rotated ZAMO basis in which the unit vector $\hat{e}_1$ is parallel to the poloidal component of the magnetic field, B^P. In terms of the Maxwell tensor in the ZAMO frames,

$$B^1 \equiv B^P = F^{2\phi} , \quad \text{and} \quad B^2 = F^{\phi 1} = 0 . \tag{3.7}$$

The basis vectors in the (r, θ) plane become

$$\begin{bmatrix} \hat{e}_1 \\ \hat{e}_2 \end{bmatrix} = \frac{1}{|B^P|} \begin{bmatrix} F^{\theta\phi} & F^{\phi r} \\ -F^{\phi r} & F^{\theta\phi} \end{bmatrix} \begin{bmatrix} \hat{e}_r \\ \hat{e}_\theta \end{bmatrix} . \tag{3.8a}$$

Using $B^r = F^{\theta\phi}$, and $B^\theta = F^{\phi r}$ the basis covectors in the rotated ZAMO frame are (note: $B^P = \sqrt{(B^\theta)^2 + (B^r)^2}$)

$$\begin{bmatrix} \omega^1 \\ \omega^2 \end{bmatrix} = \frac{1}{|B^P|} \begin{bmatrix} B^r & B^\theta \\ -B^\theta & B^r \end{bmatrix} \begin{bmatrix} \omega^r \\ \omega^\theta \end{bmatrix} . \tag{3.8b}$$

This basis essentially reduces the dimensionality by one in many calculations.

3.3 Geodesic Motion

It is of interest to understand the concept of the natural timelike trajectories of the Kerr space–time background (the geodesics) in order to understand the effects of external forces. Carter in [82] solved the geodesic equation in the Kerr space–time,

$$u^{\alpha} u^{\beta}{}_{;\alpha} = 0 \,. \tag{3.9}$$

He found four constants of motion, ω and m, the rest mass μ defined in terms of the four momentum,

$$P_{\mu} P^{\mu} = -\mu^2 \,, \tag{3.10}$$

and a fourth constant $\mathcal{K}$ called "Carter's fourth constant of motion," a general relativistic analog of the total angular momentum of the trajectory. Defining the affine parameter, λ, in terms of proper time τ as $\tau = \mu\lambda$, Carter's equations are

$$\rho^2 \left(\frac{\mathrm{d}t}{\mathrm{d}\lambda} \right) = -a \left(\omega a \sin^2\theta - m \right) + \left[\frac{r^2 + a^2}{\Delta} \right] P \,, \tag{3.11a}$$

$$\rho^2 \left(\frac{\mathrm{d}r}{\mathrm{d}\lambda} \right) = \pm\sqrt{R} \,, \tag{3.11b}$$

$$\rho^2 \left(\frac{\mathrm{d}\phi}{\mathrm{d}\lambda} \right) = -a \left(\omega a - m/\sin^2\theta \right) + aP/\Delta \,, \tag{3.11c}$$

$$\rho^2 \left(\frac{\mathrm{d}\theta}{\mathrm{d}\lambda} \right) = \pm\Theta \,, \tag{3.11d}$$

Note that

$$P = \omega \left(r^2 + a^2 \right) - ma \,, \tag{3.11e}$$

$$R = P^2 - \Delta \left(\mu^2 r^2 + \mathcal{K} \right) \,, \tag{3.11f}$$

$$\Theta = \mathcal{K} - (\omega a - m)^2 - \cos^2\theta \left[a^2 \left(\mu^2 - \omega^2 \right) + m^2/\sin^2\theta \right] . \tag{3.11g}$$

The signs chosen for (3.11b) and (3.11d) are independent.

In general, (3.11) is not of much interest in terms of an interaction of significance in the ergosphere. However, geodesics can be used to approximate any trajectory locally (i.e., like approximating a curve in Euclidean space by many short segments). Alternatively, one can describe a trajectory by (3.11) with ω, m, and $\mathcal{K}$ functions instead of constants. The timelike geodesic equations (3.11) can be used to understand non-geodesic accelerating trajectories.

Note that if $\omega \neq \Omega_H m/c$, one can approximate the geodesics near the horizon to $\mathrm{O}(\alpha^2)$ in terms of the ZAMO four momentum using (3.11). Using geometrized units and $c = 1$ in the following as $r \to r_+$, an ingoing trajectory has

$$P^0 \approx \left[\frac{\omega\left(r^2+a^2\right)}{\rho\Delta^{1/2}}\right]\left[1-\frac{\Delta a^2\sin^2\theta}{2\left(r^2+a^2\right)^2}\right]$$
$$-\left[\frac{ma}{\rho\Delta^{1/2}}\right]\left[1-\frac{\Delta\left(\rho^2+(1/2)a^2\sin^2\theta\right)}{\left(r^2+a^2\right)^2}\right], \tag{3.12a}$$

$$P^r \approx -\frac{1}{\rho\Delta^{1/2}}\left[\omega\left(r^2+a^2\right)-ma-\frac{\Delta\left(\mathcal{K}^2+\mu^2 r_+^2\right)}{2\left[\omega\left(r_+^2+a^2\right)-ma\right]}\right], \tag{3.12b}$$

$$P^\phi \approx \frac{m\rho}{(r^2+a^2)\sin\theta}, \tag{3.12c}$$

$$P^\theta \approx \frac{1}{\rho}\tilde{P}_\theta \,. \tag{3.12d}$$

The last equation is merely a consequence of the value of Carter's fourth constant of motion,

$$\mathcal{K} = \left(\tilde{P}^\theta\right)^2 + (\omega a\sin\theta - m/\sin\theta)^2 + \mu^2 a^2\cos^2\theta \,. \tag{3.13}$$

The restriction $\omega \neq m\Omega_H/c$ is not really a restriction for geodesic motion since by (3.8) and (1.26)

$$u^0 = \frac{(\omega - \Omega m/c)}{\alpha}, \tag{3.14a}$$

$$u^\phi = m/\sqrt{g_{\phi\phi}}, \tag{3.14b}$$

$$\beta^\phi = \frac{u^\phi}{u^0} = \frac{c\alpha}{\left(\Omega_H - \Omega\right)\sqrt{g_{\phi\phi}}}, \qquad \text{if } \omega = m\Omega_H/c \,. \tag{3.14c}$$

Expanding Ω, using the definition (3.2) and the metric (1.24a), gives

$$\Omega \approx \left[\frac{a}{r^2+a^2}\right]\left[1-\frac{\Delta\rho^2}{\left(r^2+a^2\right)^2}\right] = \Omega_H\left(1+\mathrm{O}(\alpha^2)\right) . \tag{3.15}$$

Using this expansion in (3.14c) yields

$$\beta^\phi \sim \alpha^{-1}, \qquad \text{if } \omega = m\Omega_H/c \,. \tag{3.16}$$

Yet $\left|\beta^\phi\right| < 1$ since $u \cdot u = -1$ for timelike trajectories. Thus, the restriction $\omega \neq m\Omega_H/c$ does not affect the asymptotic trajectories near the horizon in (3.12).

Similarly, Carter's equations (3.11) imply the completely general ingoing geodesic three-velocity near the horizon evaluated in the stationary frames is given by

$$\frac{\mathrm{d}r}{\mathrm{d}t} \approx \frac{-\Delta}{r^2+a^2}\left[1+\frac{\Delta\left(\omega a^2\sin^2\theta - ma\right)}{P\left(r^2+a^2\right)} - \frac{\Delta\left(\mathcal{K}+\mu^2 r_+^2\right)}{2P^2}\right], \tag{3.17a}$$

$$\frac{\mathrm{d}\phi}{\mathrm{d}t} \approx \frac{a}{r^2+a^2}\left[1-\frac{\Delta\left(\omega\rho^2 - ma\sin^2\theta\right)}{P\left(r^2+a^2\right)}\right] = \Omega_H\left(1+\mathrm{O}(\alpha^2)\right)\,, \tag{3.17b}$$

$$\frac{\mathrm{d}\theta}{\mathrm{d}t} \approx \frac{\pm\sqrt{\Theta}P\Delta}{r^2+a^2}\,, \tag{3.17c}$$

$$\frac{\mathrm{d}t}{\mathrm{d}\tau} \sim \alpha^{-2}\,. \tag{3.17d}$$

Note that we can consider the physical absurdity of an outgoing particle trajectory near the horizon by evaluating the energy in a freely falling frame. Define the freely falling frame to have quantum numbers:

$$\bar{\omega} = \mu\,, \tag{3.18a}$$

$$\bar{m} = 0\,, \tag{3.18b}$$

$$\mathcal{K} = \bar{\omega}a^2\sin^2\theta + \mu^2 a^2\cos^2\theta = \mu^2 a^2\,, \tag{3.18c}$$

$$\bar{u}^{\mu} = \alpha^{-1}\left(\hat{e}_0 + \sqrt{1-\alpha^2}\hat{e}_r\right)\,, \tag{3.18d}$$

then (3.11) defines a particle falling inward from rest at infinity along radial trajectories. The four velocity tangent to this trajectory is $\bar{u}^{\mu}$. The energy of a particle as viewed in this frame is $P^u = \bar{u}^{\mu}P_{\mu}$. For an outgoing trajectory near the horizon, the proper frame energy is

$$P^u \underset{r\to r_+}{\sim} \frac{2(\omega - m\Omega_H)}{\alpha^2}\,. \tag{3.19}$$

Thus, it requires a divergently large energy in a proper frame near the horizon for a particle to appear outgoing globally.

An interesting consequence of (3.11) is (3.17b), all ingoing geodesics approach the angular velocity of the horizon as $r \to r_+$. This is shown in the results of Johnston and Ruffini [83] that are reproduced in Figs. 3.1 and 3.2. The geodesics rotate faster and faster in the sense of the hole as the horizon is approached, namely as $\mathrm{d}\phi/\mathrm{d}t$ increases from 0 to Ω_H. Most of the increase occurs close to the hole in the ergosphere. This is a manifestation of the dragging of inertial frames and is an important concept for understanding the ergospheric interaction in a plasma-filled magnetosphere. The natural state of an inflowing plasma (geodesic motion) is to rotate faster and faster until $\mathrm{d}\phi/\mathrm{d}t \longrightarrow \Omega_H$ near the horizon.

3.4 The Momentum Equations of a Magneto-Fluid

In this section a tractable version of the momentum equations of a fluid is motivated. It would be desirable to find the four velocity in complete generality in the ergosphere for all possible plasma states. However, many plasma interactions are highly nonlinear kinetic effects that produce phenomena such as turbulence that are not

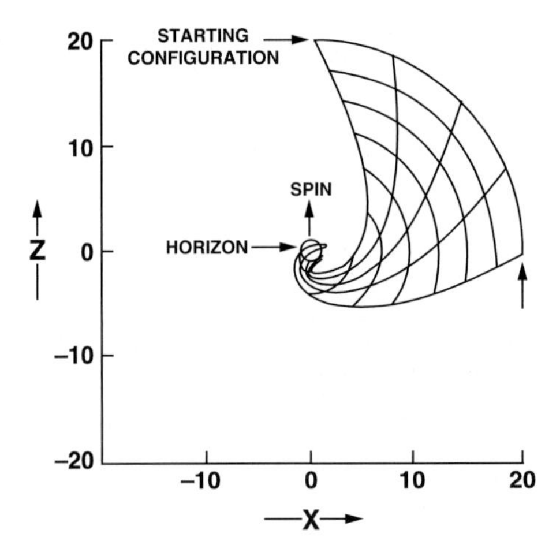

Fig. 3.1 Motion of an uncharged cloud of particles around a black hole with $a = M$. The constants of motion are $\omega = \mu, m = 0, Q \equiv \mathcal{K} - (m - \omega a)^2 = 10\mu^2$. Initially $\dot{\Theta} > 0$. The figure is from [82]

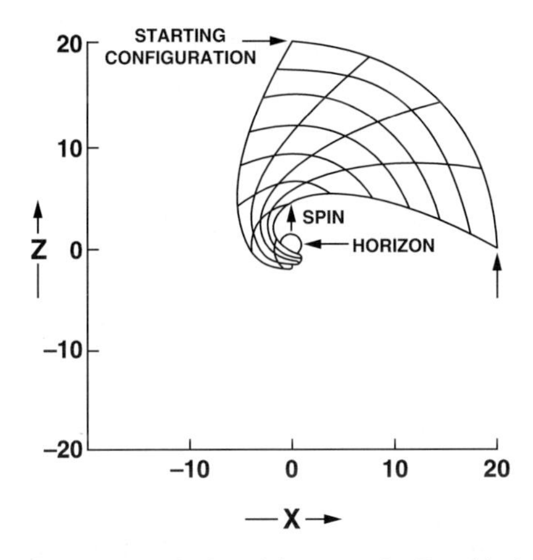

Fig. 3.2 Motion of an uncharged cloud of particles around a Kerr black hole with $a = M$. The constants of motion are $\omega = \mu, m = 0, Q = 10\mu^2$. Initially $\dot{\Theta} < 0$. The figure is from [82]

describable by a bulk four velocity. For simplicity we assume that the nonlinear discrete elements are captured within some type of bulk property of the flow, a component of internal energy for example, that is advected with an average bulk velocity, u^μ.

The next best approach would be a two-fluid analysis as is Sect. 2.9. Such an analysis yields a magneto-fluid with $\boldsymbol{J} \times \boldsymbol{B}$ forces when the equations of motion for the species are added (see (2.76) for example), yet for the individual species there is a Lorentz force. The following analysis assumes that a magneto-fluid can be described by a bulk flow four momentum

$$P^{\alpha} \equiv \mu u^{\alpha} = \frac{\mu_{+}n_{+}u_{+}^{\alpha} + \mu_{-}n_{-}u_{-}^{\alpha}}{n} , \tag{3.20a}$$

$$n = n_{+} + n_{-} . \tag{3.20b}$$

The "+" and "−" subscripts refer to the positive and negative species where n and μ are the proper density and enthalpy per particle.

To model the stress-energy tensor of the fluid we note that viscous terms are very complicated in curved space–times due to their higher derivative nature. Furthermore, we are interested in plasmas in strong magnetospheric fields so we expect electromagnetic interactions to dominate the dynamics, torque and dissipation. Thus, we ignore fluid viscosity in the fluid stress-energy tensor:

$$T_{f}^{\alpha\beta} = \rho u^{\alpha} \otimes u^{\beta} + h^{\alpha\beta} P , \tag{3.21}$$

where P is an isotropic pressure (both gas and radiation) in the plasma rest frame and $h^{\alpha\beta}$ is the projection tensor

$$h^{\alpha\beta} = g^{\alpha\beta} + u^{\alpha} \otimes u^{\beta} . \tag{3.22}$$

The electromagnetic stress-energy tensor is

$$T_{\mathrm{EM}}^{\alpha\beta} = \frac{1}{4\pi} F^{\mu\alpha} F_{\alpha}{}^{\nu} - \frac{1}{8\pi} g^{\mu\nu} F^{\gamma\delta} F_{\gamma\delta} . \tag{3.23}$$

The equations of motion are

$$T_{f}^{\alpha\beta}{}_{;\beta} + T_{\mathrm{EM}}^{\alpha\beta}{}_{;\beta} + T_{r}^{\alpha\beta}{}_{;\beta} = 0 , \tag{3.24}$$

where $T_{r}^{\alpha\beta}$ is the stress-energy tensor of the radiation field that includes Compton drag, radiation resistance and heat flow. We could have separated out the heat flow tensor as is often done, but this is not very illustrative as it leaves everything reexpressed in terms of an unknown heat flow vector. $T_{r}^{\alpha\beta}$ generally has a complicated form that is very specific to the situation under analysis, so we leave it in abstract form and expand it out when needed in the text.

In order to expand out the differential equation in (3.24) we note the correspondence between the ZAMO basis vectors and a basis of tangent vectors $\partial/\partial X^{i}$ at a point of the Kerr space–time is given by

$$\hat{e}_{0} = \frac{\partial}{\partial X^{0}} = \alpha^{-1} \left(\frac{\partial}{\partial t} + \frac{\Omega}{c} \frac{\partial}{\partial \phi} \right) , \tag{3.25a}$$

$$\hat{e}_{\phi} = \frac{\partial}{\partial X^{\phi}} = \frac{1}{\sqrt{g_{\phi\phi}}} \left(\frac{\partial}{\partial \phi} \right) , \tag{3.25b}$$

$$\hat{e}_r = \frac{\partial}{\partial X^r} = \frac{\Delta^{1/2}}{\rho}\frac{\partial}{\partial r}\,, \tag{3.25c}$$

$$\hat{e}_\theta = \frac{\partial}{\partial X^\theta} = \frac{1}{\rho}\frac{\partial}{\partial \theta}\,. \tag{3.25d}$$

The connection coefficients, $\Gamma^{\alpha}{}_{\beta\gamma}$, are found by solving the structure equations:

$$\mathrm{d}\omega^\alpha = \omega^\alpha{}_\beta \wedge \omega^\beta\,, \tag{3.26}$$

$$\omega^\alpha{}_\beta = \Gamma^\alpha{}_{\beta\gamma}\omega^\gamma\,. \tag{3.27}$$

The connection can be expressed with the help of (3.25) as [84]

$$\omega_{0r} = -\frac{1}{\alpha}\frac{\partial}{\partial X^r}(\alpha)\,\omega^0 - \frac{\sqrt{g_{\phi\phi}}}{2\alpha}\frac{\partial}{\partial X^r}(\Omega)\,\omega^\phi\,, \tag{3.28a}$$

$$\omega_{0\theta} = -\frac{1}{\alpha}\frac{\partial}{\partial X^\theta}(\alpha)\,\omega^0 - \frac{\sqrt{g_{\phi\phi}}}{2\alpha}\frac{\partial}{\partial X^\theta}(\Omega)\,\omega^\phi\,, \tag{3.28b}$$

$$\omega_{0\phi} = -\frac{\sqrt{g_{\phi\phi}}}{2\alpha}\frac{\partial}{\partial X^r}(\Omega)\,\omega^r - \frac{\sqrt{g_{\phi\phi}}}{2\alpha}\frac{\partial}{\partial X^\theta}(\Omega)\,\omega^\theta\,, \tag{3.28c}$$

$$\omega_{r\theta} = \frac{\Delta^{1/2}}{\rho}\frac{\partial}{\partial X^\theta}\left(\frac{\rho}{\Delta^{1/2}}\right)\omega^r - \frac{1}{\rho}\frac{\partial}{\partial X^r}(\rho)\,\omega^\theta\,, \tag{3.28d}$$

$$\omega_{r\phi} = -\sqrt{g_{\phi\phi}}\frac{\partial}{\partial X^r}\left(\sqrt{g_{\phi\phi}}\right)\omega^\phi + \frac{\sqrt{g_{\phi\phi}}}{2\alpha}\frac{\partial}{\partial X^r}(\Omega)\,\omega^0\,, \tag{3.28e}$$

$$\omega_{\theta\phi} = -\frac{1}{\sqrt{g_{\phi\phi}}}\frac{\partial}{\partial X^\theta}\left(\sqrt{g_{\phi\phi}}\right)\omega^0 + \frac{\sqrt{g_{\phi\phi}}}{2\alpha}\frac{\partial}{\partial X^\theta}(\Omega)\,\omega^0\,. \tag{3.28f}$$

The remaining connection forms are found by the antisymmetry condition, $\omega_{\alpha\beta} = -\omega_{\beta\alpha}$.

From (3.27) and (3.28) the connection coefficients in the ZAMO frame are

$$\Gamma^r{}_{00} = \left[M\left\{\left[\left(r^2+a^2\right)\left(r^4 - a^4\cos^2\theta + a^2r^2\sin^2\theta\right)\right] - 4Ma^2r^3\sin^2\theta\right\}\right]$$
$$\times\rho^3\left[\left(r^2+a^2\right)\rho^2 + 2Mra^2\sin^2\theta\right]\left(r^2+a^2-2Mr\right)^{1/2}\,, \tag{3.29a}$$

$$\Gamma^\theta{}_{00} = -\frac{2Mra^2\left(r^2+a^2\right)\sin\theta\cos\theta}{\rho^3\left[\left(r^2+a^2\right)\rho^2 + 2Mra^2\sin^2\theta\right]}\,, \tag{3.29b}$$

$$\Gamma^r{}_{0\phi} = -\frac{Ma\sin\theta}{\rho^3\left[\left(r^2+a^2\right)\rho^2 + 2Mra^2\sin^2\theta\right]} \times \left[\left(r^2-a^2\right)a^2\cos^2\theta + r^2\left(3r^2+a^2\right)\right]\,, \tag{3.29c}$$

$$\Gamma^\phi{}_{\theta 0} = -\frac{2Mra^3\sin^4\theta\cos\theta}{\rho^5 g_{\phi\phi}}\left(r^2+a^2-2Mr\right)^{1/2}\,, \tag{3.29d}$$

$$\Gamma^{r}{}_{\phi\phi} = -\frac{\left(r^2+a^2-2Mr\right)^{1/2}}{\rho^3\left[\rho^2\left(r^2+a^2\right)+2Mra^2\sin^2\theta\right]} \times\left\{r\left[\left(r^2+a^2\cos^2\theta\right)^2-rMa^2\sin^2\theta\right]+\frac{Ma^4}{4}\sin^2 2\theta\right\}, \quad (3.29e)$$

$$\Gamma^{\theta}{}_{\phi\phi} = -\frac{\cos\theta}{\rho^3\left[\rho^2\left(r^2+a^2\right)+2Mra^2\sin^2\theta\right]\sin\theta} \times\left\{\rho^4\left(r^2+a^2\right)+2Mra^2\sin^2\theta\left[2r^2+a^2\left(1+\cos^2\theta\right)\right]\right\}, \quad (3.29f)$$

$$\Gamma^{r}{}_{\theta\theta} = -\frac{r\left(r^2+a^2-2Mr\right)^{1/2}}{\rho^3}, \quad (3.29g)$$

$$\Gamma^{r}{}_{\theta r} = -\frac{a^2\sin\theta\cos\theta}{\rho^3}. \quad (3.29h)$$

The remaining connection coefficients are found by the antisymmetry in the connection forms and (3.27)

$$\Gamma_{(\alpha\beta)\gamma} = 0. \quad (3.30)$$

One of the main advantages of the ZAMO basis is the ability to write equations in the rotated ZAMO basis (3.8) in which $\hat{e}_i$ is aligned with the poloidal magnetic field. Connection coefficients do not transform like tensors. The transformation law from a frame with legs labeled by x^i to one with legs $\bar{x}^i$ contains an inhomogeneous term:

$$\bar{\Gamma}^{k}{}_{rl} = \frac{\partial x^t}{\partial \bar{x}^r}\frac{\partial x^n}{\partial \bar{x}^l}\frac{\partial \bar{x}^k}{\partial x^m}\Gamma^{m}{}_{tn} - \frac{\partial x^t}{\partial \bar{x}^r}\frac{\partial x^s}{\partial \bar{x}^l}\frac{\partial^2 \bar{x}^k}{\partial x^s \partial x^t}. \quad (3.31)$$

Because of time stationarity and axisymmetry of the metric, the more interesting connection coefficients transform without an inhomogenous term. From (3.8), (3.29) and (3.31) the connection in the rotated ZAMO basis is given by

$$\Gamma^{1}{}_{00} = \alpha^{-1}\frac{\partial}{\partial X^1}(\alpha), \quad (3.32a)$$

$$\Gamma^{2}{}_{00} = \alpha^{-1}\frac{\partial}{\partial X^2}(\alpha), \quad (3.32b)$$

$$\Gamma^{\phi}{}_{1\phi} = \frac{1}{\sqrt{g_{\phi\phi}}}\frac{\partial}{\partial X^1}\sqrt{g_{\phi\phi}}, \quad (3.32c)$$

$$\Gamma^{\phi}{}_{2\phi} = \frac{1}{\sqrt{g_{\phi\phi}}}\frac{\partial}{\partial X^2}\sqrt{g_{\phi\phi}}, \quad (3.32d)$$

$$\Gamma^{1}{}_{\phi 0} = \frac{\sqrt{g_{\phi\phi}}}{2\alpha}\frac{\partial}{\partial X^1}(\Omega), \quad (3.32e)$$

$$\Gamma^{2}{}_{\phi 0} = \frac{\sqrt{g_{\phi\phi}}}{2\alpha}\frac{\partial}{\partial X^2}(\Omega). \quad (3.32f)$$

Partial derivatives in the rotated ZAMO basis are found from (3.8a) to be

$$\frac{\partial}{\partial X^1} = \frac{B^r}{|B^P|}\frac{\partial}{\partial X^r} - \frac{B^\theta}{|B^P|}\frac{\partial}{\partial X^\theta}\,, \tag{3.33a}$$

$$\frac{\partial}{\partial X^2} = \frac{B^\theta}{|B^P|}\frac{\partial}{\partial X^r} + \frac{B^r}{|B^P|}\frac{\partial}{\partial X^\theta}\,. \tag{3.33b}$$

The simplified from of the poloidal connection in the rotated ZAMO basis (3.32), is the essence of the clarification and simplicity in formulating the dynamics in the rotated ZAMO basis.

The law of mass conservation is

$$(nu^\alpha)_{;\alpha} = 0\,. \tag{3.34a}$$

Expanding this out yields

$$\frac{\partial}{\partial X^\alpha}(nu^\alpha) + \Gamma^\beta{}_{\alpha\beta}u^\alpha = 0\,. \tag{3.34b}$$

From (3.29) we have

$$\Gamma^r{}_{\theta r} = \frac{1}{\sqrt{g_{rr}}}\frac{\partial}{\partial X^\theta}\sqrt{g_{rr}}\,, \tag{3.35a}$$

$$\Gamma^\theta{}_{r\theta} = \frac{1}{\sqrt{g_{\theta\theta}}}\frac{\partial}{\partial X^r}\sqrt{g_{\theta\theta}}\,. \tag{3.35b}$$

Combining these connection coefficients with the results of (3.32) and the substitutions $1 \to r$ and $2 \to \theta$, the mass conservation equation (3.34) becomes

$$\begin{aligned}&\frac{\partial}{\partial X^0}\left(nu^0\right) + \frac{\partial}{\partial X^\phi}\left(nu^\phi\right) + \\ &\frac{1}{\alpha\sqrt{g_{rr}g_{\phi\phi}}}\frac{\partial}{\partial X^\theta}\left(nu^\theta\alpha\sqrt{g_{rr}g_{\phi\phi}}\right) + \\ &\frac{1}{\alpha\sqrt{g_{\theta\theta}g_{\phi\phi}}}\frac{\partial}{\partial X^r}\left(nu^r\alpha\sqrt{g_{\theta\theta}g_{\phi\phi}}\right) = 0\,.\end{aligned} \tag{3.36a}$$

In terms of a source function, S, we have

$$S \equiv \alpha\frac{\partial}{\partial X^0}\left(nu^0\right) = \nabla^{(3)}\cdot(\alpha n\boldsymbol{u})\,, \tag{3.36b}$$

where $\nabla^{(3)}$ is the covariant derivative with respect to the connection on the spacelike hypersurface orthogonal to the ZAMO four velocity, $\hat{e}_0$. (i.e., the metric is $ds^2 = \omega^r \otimes \omega^r + \omega^\theta \otimes \omega^\theta + \omega^\phi \otimes \omega^\phi$).

Contracting the four velocity of the plasma flow into (3.24) and retaining the source term in the mass conservation law (3.36b) to account for the pair creation, we get

$$\frac{\mathrm{d}\rho}{\mathrm{d}\tau} + \mu S - \mu \frac{\mathrm{d}n}{\mathrm{d}\tau} - u \cdot \nabla T_r + u \cdot F \cdot J = 0\,. \tag{3.37}$$

This looks more like a thermodynamic relation in the entropy $\mathbb{S}$ and temperature T if we write the equation for entropy generation as

$$nT\frac{d\mathbb{S}}{d\tau} = -\mu S - u \cdot F \cdot J + u \cdot \nabla T_r\,. \tag{3.38}$$

Taking the projection (3.22) of (3.24) and implementing the thermodynamic relation (3.37), we get an equation for the bulk motion of the fluid:

$$nu^\beta \frac{\partial}{\partial X^\beta}(\mu u^\alpha) + n\mu \Gamma^\alpha{}_{\mu\nu} u^\mu u^\nu + \frac{\partial}{\partial X^\alpha}(P) = \mu u^\alpha S + \frac{F^{\mu\nu} J_\nu}{c} + (T^{\alpha\nu}{}_{;\nu})_r\,. \tag{3.39}$$

Consider the poloidal momentum equation along the magnetic field lines in (3.39):

$$n\left[\frac{\partial}{\partial\tau}(\mu u^1) + \mu\,(u^0)^2\left(\Gamma^1{}_{00} + 2\beta^\phi \Gamma^1{}_{0\phi} + (\beta^\phi)^2 \Gamma^1{}_{\phi\phi}\right)\right] + \frac{\partial}{\partial X^1}(P) = \mu u^1 S + \frac{F^{12} J^2}{c} + \left(T^{1\nu}{}_{;\nu}\right)_r\,. \tag{3.40}$$

We can define an effective gravity "g" in the ZAMO frames as

$$g \equiv (u^0)^2\left(\Gamma^1{}_{00} + 2\beta^\phi \Gamma^1{}_{0\phi} + (\beta^\phi)^2 \Gamma^1{}_{\phi\phi}\right)\,. \tag{3.41}$$

This expression has a straightforward physical interpretation. The poloidal pull of gravity is $\Gamma^1{}_{00}$ in the ZAMO frames. From the definition of $\Gamma^1{}_{0\phi}$ in (3.32e), the second term is a Coriolis force in the rotating ZAMO frame and the last term is a centrifugal force.

3.5 Frame Dragging and Negative Energy States

The ZAMO frames are a natural format for elucidating the effects of frame dragging in the ergosphere. Consider a particle rotating with an angular velocity $\Omega_p = \mathrm{d}\phi/\mathrm{d}t$ as viewed from asymptotic infinity. The azimuthal three velocity in the ZAMO frames is found from (3.8):

$$\beta^\phi = \frac{u^\phi}{u^0} = \frac{(\Omega_p - \Omega)\sqrt{g_{\phi\phi}}}{\alpha c}\,. \tag{3.42}$$

Since $-1 \leq \beta^\phi \leq 1$, we have from (3.42),

$$\Omega_{\text{min}} \leq \Omega_p \leq \Omega_{\text{max}} \,, \tag{3.43a}$$

$$\Omega_{\text{min}} = \Omega - c\alpha/\sqrt{g_{\phi\phi}} \,, \tag{3.43b}$$

$$\Omega_{\text{max}} = \Omega + c\alpha/\sqrt{g_{\phi\phi}} \,. \tag{3.43c}$$

From (3.5b) and (3.15),

$$\lim_{r \to r_+} \Omega_{\text{min}} = \Omega_H \,, \qquad \lim_{r \to r_+} \Omega_{\text{max}} = \Omega_H \,. \tag{3.44}$$

Thus, all particles must corotate with the horizon as they approach it as seen globally. This is illustrated in Figs. 3.1 and 3.2 and (3.17b).

Similarly, from (3.5b) and (1.24a) we have a purely geometrical effect of the curvilinear coordinates:

$$\lim_{r\sin\theta \to \infty} \Omega_{\text{min}} = 0 \,, \qquad \lim_{r\sin\theta \to \infty} \Omega_{\text{max}} = 0 \,. \tag{3.45}$$

More importantly, near the black hole $\Omega_{\text{min}} = 0$ when $\Omega = c\alpha/\sqrt{g_{\phi\phi}}$. Inserting (3.5a) and (3.2), the definitions of α and Ω into (3.43), this constraint yields

$$\Omega_{\text{min}} = 0 \quad \Rightarrow \quad g_{tt} = 0 \,. \tag{3.46}$$

This is the well known result that all timelike trajectories must rotate in the same sense as the horizon inside of the stationary limit at r_s (given by (3.1)) as viewed by distant observers.

Similarly, let us consider where $\omega < 0$ states exist in terms of the $-1 \leq \beta^\phi \leq 1$ condition. From (3.14) we can express β^ϕ in terms of m and ω as

$$\beta^\phi = \frac{u^\phi}{u^0} = \frac{\alpha m/\sqrt{g_{\phi\phi}}}{\omega - (\Omega/c)m} \,. \tag{3.47}$$

If $\omega < 0$, since $u^0 > 0$ for a timelike trajectory, by (3.14) $m < 0$. Thus, we are interested in the condition $0 \geq \beta^\phi \geq -1$ in (3.47), or

$$\omega - (\Omega/c)m \geq -\alpha m/\sqrt{g_{\phi\phi}} \,, \quad \omega < 0,\ m < 0. \tag{3.48a}$$

Rearranging (3.48a) we have

$$-(\Omega/c)m \geq -\alpha m/\sqrt{g_{\phi\phi}} - \omega \geq -\alpha m/\sqrt{g_{\phi\phi}} \,, \tag{3.48b}$$

or

$$\Omega/c \geq \alpha/\sqrt{g_{\phi\phi}} \,, \tag{3.48c}$$

which by the definition of Ω_{min} in (3.43b), is equivalent to the condition $\Omega_{\text{min}} \geq 0$. We showed in (3.46) that $\Omega_{\text{min}} \geq 0$ is the definition of the ergosphere. The globally defined $\omega < 0$ negative energy states can only occur in the region $r < r_s$.

It is instructive to invert (3.14) to get an expression for ω, the redshifted energy as viewed from asymptotic infinity:

$$\omega = \mu u^0 \left[\frac{\Omega_{\min}}{c} \sqrt{g_{\phi\phi}} - \left(1+\beta^{\phi}\right) \frac{\Omega}{c} \sqrt{g_{\phi\phi}} \right] . \tag{3.49}$$

The derivation of (3.48c) shows the connection between local counter rotation, $m < 0$, $\beta^{\phi} \approx -1$, and the negative energy states, $\omega < 0$. Anticipating the significance of these relativistic counter rotating states, note that

$$\omega < 0 \text{ if } \beta^{\phi} < -\frac{\alpha c}{\Omega \sqrt{g_{\phi\phi}}} \text{ and } \lim_{\beta^{\phi} \to -1} \omega = -\mu u^0 \frac{\Omega_{\min}}{c} \sqrt{g_{\phi\phi}} . \tag{3.50}$$

As $\beta^{\phi} \to -1$, u^0 becomes very large (since $u \cdot u = -1$). Thus, (3.50) shows that counter rotation in the ergosphere can create huge values of negative global specific energy, ω, if $\beta^{\phi} \approx -1$. If $\beta^{\phi} \approx -1$ when a negative energy particle is created at r_c such that $r_+ < r_c < r_s$, then

$$\omega \approx \frac{\Omega_{\min}(r_c)\, m}{c} < 0 . \tag{3.51}$$

If such a particle is created in the ergosphere and free falls into the hole, it will change the mass, M, and angular momentum, Ma, of the hole in the first law of black hole thermodynamics (1.34)

$$\delta M = \omega = \frac{\Omega_{\min}(r_c)\, m}{c} < 0 , \quad \delta(Ma) = m < 0 ; \tag{3.52a}$$

$$\delta A = \frac{8\pi}{\kappa} \left(\frac{\Omega_{\min}}{c} - \frac{\Omega_H}{c} \right) m > 0 . \tag{3.52b}$$

The inequality in (3.52b) is established by the fact that $\Omega_{\min}/c < \Omega_H/c$ if $r > r_+$. Thus, (3.52b) shows that the absorption of the negative energy matter obeys the second law of black hole thermodynamics and rotational energy is extracted by the hole.

The creation of $\beta^{\phi} \approx -1$ states of plasma in the ergosphere is precisely the relevant physics that occurs in black hole GMH processes that extract rotational energy from a black hole.

3.6 Maxwell's Equations

We expand out Maxwell's equations (2.16) in the ZAMO basis for future use with the aid of (3.28) and (3.29):

$$F^{\theta\alpha}{}_{;\alpha} = \frac{4\pi J^\theta}{c}\,, \tag{3.53a}$$

$$\frac{\partial}{\partial X^0}F^{\theta 0} + \frac{\partial}{\partial X^\phi}F^{\theta\phi} + \frac{1}{\alpha\sqrt{g_{\phi\phi}}}\frac{\partial}{\partial X^r}\left[\alpha\sqrt{g_{\phi\phi}}F^{\theta r}\right] = \frac{4\pi J^\theta}{c}\,. \tag{3.53b}$$

$$F_{\theta r;0} + F_{r0;\theta} + F_{0\theta;r} = 0\,, \tag{3.54a}$$

$$\frac{\partial}{\partial X^0}F^{\theta r} + \frac{1}{\alpha\sqrt{g_{\theta\theta}}}\frac{\partial}{\partial X^r}\left[\alpha\sqrt{g_{\theta\theta}}F^{\theta 0}\right] - \frac{1}{\alpha\sqrt{g_{rr}}}\frac{\partial}{\partial X^\theta}\left[\alpha\sqrt{g_{rr}}F^{r0}\right]$$

$$+2\left(\Gamma^\phi{}_{r0}F^{\theta\phi} + \Gamma^\phi{}_{\theta 0}F^{\phi r}\right) = 0\,. \tag{3.54b}$$

$$F_{\theta r;\phi} + F_{r\phi;\theta} + F_{\phi\theta;r} = 0\,, \tag{3.55a}$$

$$\frac{\partial}{\partial X^\phi}F^{\theta r} + \frac{1}{\sqrt{g_{\phi\phi}g_{rr}}}\frac{\partial}{\partial X^\theta}\left[\sqrt{g_{\phi\phi}g_{rr}}F^{r\phi}\right] + \frac{1}{\sqrt{g_{\phi\phi}g_{\theta\theta}}}\frac{\partial}{\partial X^r}\left[\sqrt{g_{\phi\phi}g_{\theta\theta}}F^{\phi\theta}\right] = 0\,. \tag{3.55b}$$

$$F^{r\alpha}{}_{;\alpha} = \frac{4\pi J^r}{c}\,, \tag{3.56a}$$

$$\frac{\partial}{\partial X^0}F^{r0} + \frac{\partial}{\partial X^\phi}F^{r\phi} + \frac{1}{\alpha\sqrt{g_{\phi\phi}}}\frac{\partial}{\partial X^\theta}\left[\alpha\sqrt{g_{\phi\phi}}F^{r\theta}\right] = \frac{4\pi J^r}{c}\,. \tag{3.56b}$$

$$F_{\theta 0;\phi} + F_{0\phi;\theta} + F_{\phi\theta;0} = 0\,, \tag{3.57a}$$

$$\frac{\partial}{\partial X^\phi}F^{\theta 0} - \frac{\partial}{\partial X^0}F^{\phi\theta} - \frac{1}{\alpha\sqrt{g_{\phi\phi}}}\frac{\partial}{\partial X^\theta}\left[\alpha\sqrt{g_{\phi\phi}}F^{\phi 0}\right] = 0\,. \tag{3.57b}$$

$$F_{\phi r;0} + F_{r0;\phi} + F_{0\phi;r} = 0\,, \tag{3.58a}$$

$$\frac{\partial}{\partial X^0}F^{\phi r} - \frac{\partial}{\partial X^\phi}F^{r0} + \frac{1}{\alpha\sqrt{g_{\phi\phi}}}\frac{\partial}{\partial X^r}\left[\alpha\sqrt{g_{\phi\phi}}F^{\phi 0}\right] = 0\,. \tag{3.58b}$$

$$F^{0\alpha}{}_{;\alpha} = \frac{4\pi J^0}{c}\,, \tag{3.59a}$$

$$\frac{\partial}{\partial X^\phi}F^{0\phi} + \frac{1}{\sqrt{g_{\phi\phi}g_{\theta\theta}}}\frac{\partial}{\partial X^r}\left[\sqrt{g_{\phi\phi}g_{\theta\theta}}F^{0r}\right]$$

$$+\frac{1}{\sqrt{g_{\phi\phi}g_{rr}}}\frac{\partial}{\partial X^\theta}\left[\sqrt{g_{\phi\phi}g_{rr}}F^{0\theta}\right] = \frac{4\pi J^0}{c}\,. \tag{3.59b}$$

$$F^{\phi\alpha}{}_{;\alpha} = \frac{4\pi J^{\phi}}{c}, \tag{3.60a}$$

$$\frac{\partial}{\partial X^0}F^{\phi 0} + \frac{1}{\alpha\sqrt{g_{\theta\theta}}}\frac{\partial}{\partial X^r}\left[\alpha\sqrt{g_{\theta\theta}}F^{\phi r}\right] + \frac{1}{\alpha\sqrt{g_{rr}}}\frac{\partial}{\partial X^\theta}\left[\alpha\sqrt{g_{rr}}F^{\phi\theta}\right]$$
$$+2\left(\Gamma^{\phi}{}_{0r}F^{0r} + \Gamma^{\phi}{}_{0\theta}F^{0\theta}\right) = \frac{4\pi J^{\phi}}{c}. \tag{3.60b}$$

Because of the simple form of the connection in the rotated ZAMO basis of (3.32), Ampere's law is surprisingly simple in this useful frame field:

$$F^{2\alpha}{}_{;\alpha} = \frac{4\pi J^2}{c}, \tag{3.61a}$$

$$\frac{\partial}{\partial X^0}F^{20} + \frac{\partial}{\partial X^\phi}F^{2\phi} + \frac{1}{\alpha\sqrt{g_{\phi\phi}}}\frac{\partial}{\partial X^1}\left(\alpha\sqrt{g_{\phi\phi}}F^{21}\right) = \frac{4\pi J^2}{c}, \tag{3.61b}$$

$$F^{1\alpha}{}_{;\alpha} = \frac{4\pi J^1}{c}, \tag{3.62a}$$

$$\frac{\partial}{\partial X^0}F^{10} + \frac{1}{\alpha\sqrt{g_{\phi\phi}}}\frac{\partial}{\partial X^2}\left(\alpha\sqrt{g_{\phi\phi}}F^{12}\right) = \frac{4\pi J^1}{c}. \tag{3.62b}$$

Note that (3.8) implies

$$F^{12} = F^{r\theta}. \tag{3.63}$$

We will also need Maxwell's equations in the stationary frames from time to time:

$$\frac{1}{\sqrt{-\tilde{g}}}\frac{\partial}{\partial \tilde{X}^\alpha}\left(\sqrt{-\tilde{g}}\tilde{F}^{\beta\alpha}\right) = \frac{4\pi\tilde{J}^\beta}{c}, \tag{3.64a}$$

$$\tilde{F}_{\alpha\beta,\gamma} + \tilde{F}_{\beta\gamma,\alpha} + \tilde{F}_{\gamma\alpha,\beta} = 0. \tag{3.64b}$$

Note that ordinary derivatives occur in (3.64b), a simplification of using a coordinate basis. Also $\tilde{g}$ is the determinant of the metric in (3.64a) as given by (1.24a) in Boyer–Lindquist coordinates:

$$-\tilde{g} = \rho^4 \sin^2\theta. \tag{3.64c}$$

3.7 Inviscid Hydromagnetic Horizon Boundary Conditions

In this section we examine the state of matter near the event horizon in terms of the momentum equations. We find that the plasma is always inertially dominated near the horizon, i.e., gravitational forces are the largest forces in the poloidal momentum equation (3.40). There are no quasi-stationary equilibria near the horizon.

Even when external forces are present the plasma flow approaches the asymptotic geodesic trajectories found in (3.12) and (3.17).

Note the asymptotic form of the connection in the ZAMO frames, (3.29), near the horizon. When $a \neq M$,

$$\begin{aligned}
&\Gamma^r{}_{00} \sim \alpha^{-1}, \quad \Gamma^\theta{}_{00} \sim \alpha^0, \quad \Gamma^r{}_{\phi 0} = \Gamma^r{}_{0\phi} \sim \alpha^0, \\
&\Gamma^r{}_{\theta\theta} \sim \alpha^1, \quad \Gamma^r{}_{\phi\phi} \sim \alpha^1, \quad \Gamma^\theta{}_{\phi 0} = \Gamma^\theta{}_{0\phi} \sim \alpha^1, \\
&\Gamma^\theta{}_{\phi\phi} \sim \alpha^0, \quad \Gamma^\theta{}_{rr} \sim \alpha^0,
\end{aligned} \tag{3.65}$$

and $\Gamma^r{}_{00} \sim \alpha^{-1}$ when $a = M$ except in the equatorial plane, where $\Gamma^r{}_{00} \sim \alpha^0$. Consequently, as $\alpha \to 0$, the radial gravity dominates in the effective ZAMO gravity, g, in (3.41):

$$g \approx \left(u^0\right)^2 \Gamma^r{}_{00}, \quad \alpha \to 0 . \tag{3.66}$$

We will show that g always exceeds other forces near the horizon.

For geodesic motion $\omega \neq (\Omega_H/c)m$ as shown in (3.14), the existence of an approximate force balance in the poloidal momentum equation, (3.40), (i.e., a quasi-stationary equilibrium), requires $\omega \to (\Omega_H/c)m$ in (3.12). We explore the possibility that external forces can alter ω and m so that

$$\lim_{r \to r_+} \omega = \frac{\Omega_H}{c} m . \tag{3.67}$$

We can express the general functional form of $\omega - (\Omega_H/c)m$, by noting that a physical trajectory must have finite changes in global energy, ω, and angular momentum, m, over finite proper distances:

$$\frac{1}{\sqrt{g_{rr}}} \frac{\partial}{\partial r} \left[\omega - \frac{\Omega_H}{c} m \right] < \infty . \tag{3.68}$$

Consequently, (3.68) and the condition for quasi-stationary equilibrium (3.67) imply that

$$\lim_{r \to r_+} \omega = \frac{\Omega}{c} m + \alpha \left[f_0(\theta) + \bar{f}(r, \theta) \right] , \tag{3.69a}$$

$$\lim_{r \to r_+} \bar{f}(r, \theta) = 0 , \tag{3.69b}$$

where $f_0(\theta)$ is a well-behaved function of θ. We will show that such a conjectured equilibrium can not be maintained near the horizon. Note that the functional form (3.69) for ω when inserted into (3.14) yields $\beta^\phi \sim \alpha^0$, so a timelike trajectory exists near the horizon with $\omega \to (\Omega_H/c)m$ as long as it is not geodesic (see 3.16). However, the trajectory is highly unstable to radial perturbations. We explore this instability in the next subsections by comparing ZAMO gravity g with external forces in the poloidal momentum as the equilibrium is perturbed inward. In the process the horizon boundary condition is derived.

3.7.1 Electromagnetic Forces

In order to assess the feasibility for the nongeodesic quasi-stationary equilibrium condition (3.67), we first look at the implied force balance that would be required of electromagnetic forces in (3.41). By (3.65) and (3.66) the effective gravity in the poloidal momentum equation scales near the horizon as

$$g \sim \left(u^0\right)^2 \alpha^{-1} . \tag{3.70}$$

We now compare electromagnetic forces to g as $\alpha \to 0$.

Consider the electromagnetic force term in (3.41) as $\alpha \to 0$,

$$F^{r0}J_0 + F^{r\phi}J_\phi + F^{r\theta}J_\theta = F^{r\alpha}J_\alpha . \tag{3.71}$$

The poloidal force (3.71) is analyzed term by term near the horizon.

3.7.1.1 The $F^{r0}J_0$ Force in Equilibrium

Note that the electric flux through a surface element of the horizon is expressed in terms of ZAMO covectors as

$$\Phi_E = \int F^{0r}\omega^\theta \wedge \omega^\phi \equiv \int F^{0r}\mathrm{d}X^\theta \wedge \mathrm{d}X^\phi = \int F^{0r}\sqrt{g_{\theta\theta}}\sqrt{g_{rr}}\mathrm{d}\theta\mathrm{d}\phi . \tag{3.72}$$

Since Φ_E is coordinate independent and must be regular on the horizon, F^{0r} must be finite on the horizon:

$$F^{0r} \sim \alpha^0 . \tag{3.73}$$

Now consider the current decomposition into particle drifts:

$$J^0 = e\left(n_- u^0{}_- - n_+ u^0{}_+\right) . \tag{3.74}$$

The bulk flow velocity is defined in terms of the constituent species four velocities as well as the specific enthalpies which are proper frame quantities (i.e., they scale as α^0 near the horizon) in (3.20). The equilibrium condition (3.67) combined with (3.14) imply that

$$u^0 \sim \alpha^0 . \tag{3.75a}$$

Thus, (3.75a) and the decomposition of (3.20a) in quasi-stationary equilibrium imply

$$u^0{}_+ \sim \alpha^0 \quad \text{and} \quad u^0{}_- \sim \alpha^0 . \tag{3.75b}$$

Combining (3.75b) with (3.74) yields

$$J_0 \sim \alpha^0 . \tag{3.76}$$

Then (3.76) and (3.73) provides the scaling on electromagnetic force:

$$F^{0r}J_0 \sim \alpha^0 \,. \tag{3.77a}$$

Therefore, by (3.70),

$$\frac{F^{0r}J_0}{cn\mu g} \sim \alpha \,. \tag{3.77b}$$

3.7.1.2 The $F^{r\phi}J_\phi$ and $F^{r\theta}J_\theta$ Forces in Equilibrium

In order to study this term in the poloidal momentum equation we analyze Maxwell's equation (Ampere's law), (3.56b). Decomposing the radial current into particle drifts yields

$$J^r = -e\,(n_+ u^r{}_+ - n_- u^r{}_-) \,. \tag{3.78}$$

From either (3.14a) or (3.12b) with variable ω and m, one has in quasi-stationary equilibrium by virtue of (3.67) that

$$u^r \sim \alpha^0 \,, \tag{3.79}$$

and by (3.20), this implies

$$u^r{}_+ \sim \alpha^0 \quad \text{and} \quad u^r{}_- \sim \alpha^0 \,. \tag{3.80}$$

Inserting (3.80) into (3.78) yields the asymptotic scaling,

$$J^r \sim \alpha^0 \,. \tag{3.81}$$

To analyze the time derivative in (3.56b), we note that the convective derivative of F^{r0} is well behaved in a freely falling frame near the horizon. By (3.12a), for a freely falling observer,

$$u^0 \equiv \frac{\mathrm{d}X^0}{\mathrm{d}\tau} \sim \alpha^{-1} \,, \tag{3.82a}$$

so

$$\mathrm{d}X^0 \sim \alpha^{-1}\mathrm{d}\tau \,. \tag{3.82b}$$

Both the ZAMO frame and the coordinate frame of the freely falling observer restricted to its world line are orthonormal. Thus, the natural isomorphism between vectors and covectors (when restricted to the world line) trivially states

$$\langle \mathrm{d}X^0, \frac{\partial}{\partial X^0} \rangle = 1 \,, \tag{3.83a}$$

so

$$\langle \mathrm{d}\tau, \frac{\partial}{\partial \tau} \rangle = 1 \,. \tag{3.83b}$$

which implies with the aid of (3.82b) that

$$\frac{\partial}{\partial X^0} \sim \alpha \frac{\partial}{\partial \tau} \,. \tag{3.84}$$

Combining this with (3.73) gives the desired scaling as $r \to r_+$,

$$\frac{\partial}{\partial X^0} F^{r0} \sim \alpha \,. \tag{3.85}$$

Equation (3.85) implies that there are three possible scalings of B^θ and B^ϕ with lapse function that can satisfy Ampere's law (3.56b) with the restricted form of the quasi-stationary equilibrium current J^r in (3.81):

$$F^{r\phi} \sim \alpha^0 \,, \quad F^{r\theta} \sim \alpha^n \,, \quad n > 0, \quad \text{nonaxisymmetric}, \tag{3.86a}$$

$$F^{r\phi} \sim F^{r\theta} \sim \alpha^n \,, \quad n < 0, \quad \text{nonaxisymmetric}, \tag{3.86b}$$

$$F^{r\phi} \sim \alpha \,, \quad F^{r\theta} \sim \alpha^0 \,, \quad \text{axisymmetric}, \tag{3.86c}$$

The scaling on $F^{r\phi}$ does not come from Ampere's law (3.56b) in the axisymmetric case, where it does not appear. The scaling on B^θ in (3.86c) comes from the axisymmetric form of the divergence equation (3.55b), where the regularity of $F^{\theta\phi}$ on the horizon is needed. The scaling of $F^{\theta\phi}$ is found by computing the magnetic flux through any finite area element arbitrarily near the horizon, Φ_B. This must be a well behaved quantity:

$$\Phi_B = \int F^{\theta\phi} \mathrm{d}X^\theta \wedge \mathrm{d}X^\phi \sim \alpha^0 \,. \tag{3.87a}$$

Therefore,

$$F^{\theta\phi} \sim \alpha^0 \,. \tag{3.87b}$$

Expanding the current densities J^θ and J^ϕ as in (3.74) and (3.78) gives

$$J^\theta \sim \alpha^0 \quad \text{and} \quad J^\phi \sim \alpha^0 \,. \tag{3.88a}$$

Thus, the scaling (3.86a) and (3.86c) yield forces in the poloidal momentum equation that are dominated by radial gravity:

$$\frac{F^{r\theta} J_\theta + F^{r\phi} J_\phi}{n\mu c g} \sim \alpha \,. \tag{3.88b}$$

3.7.1.3 The Ingoing Wave Fields of Condition

In order to explore the scalings (3.86b) we require $n = -1$ for an electromagnetic balancing of gravity in the poloidal momentum equation by (3.88a) and (3.70):

$$F^{r\phi} \sim F^{r\theta} \sim \alpha^{-1}\,, \quad \text{nonaxisymmetric.} \tag{3.89}$$

We will show that a quasi-stationary equilibrium that depends on relation (3.89) cannot self consistently describe the state of the plasma through the momentum equations.

Note that (3.89) would imply proper magnetic and electric fields that diverge as α^{-2} (through (3.12)) unless one also has electric fields of similar magnitude:

$$F^{\theta 0} = -F^{\theta r}\left[1+\mathrm{O}(\alpha^2)\right] \sim \alpha^{-1}\,, \tag{3.90a}$$

$$F^{\phi 0} = -F^{\phi r}\left[1+\mathrm{O}(\alpha^2)\right] \sim \alpha^{-1}\,, \quad \text{nonaxisymmetric.} \tag{3.90b}$$

These are essentially the electromagnetic fields of an ingoing wave polarized in the θ-ϕ plane. Note that near the horizon the fields in the ZAMO frames manifest themselves as waves if there is no axisymmetry. The ZAMOs see a time variation, $\partial/\partial X^0$, since they orbit at approximately the angular velocity of the horizon and see a different field strength at different ϕ coordinates. Note that in general any non-zero magnetic fields as measured in the θ–ϕ plane by a freely falling observer will result in the fields (3.90) in the ZAMO frames near the horizon. In the axisymmetric case the fields in (3.90b) vanish.

Intuitively, the idea of balancing gravitational forces with proper fields in a freely falling frame near the horizon seems flawed. To see this requires introducing the energy equation and the zero component of momentum equation near the horizon:

$$-n\mu\Gamma^{r}{}_{00}u^{r}u^{0} \approx \left[F^{0\theta}J_{\theta} + F^{0\phi}J_{\phi}\right]\left[1+\mathrm{O}(\alpha)\right]\,. \tag{3.91a}$$

Expanding out the radial momentum one more time yields

$$n\mu\Gamma^{r}{}_{00}u^{0}u^{0} = \left[F^{r\theta}J_{\theta} + F^{r\phi}J_{\phi}\right]\left[1+\mathrm{O}(\alpha)\right]\,. \tag{3.91b}$$

Combining (3.91) with the wave condition (3.90) yields a value of the ZAMO radial three velocity v^r,

$$\frac{v^r}{c} = -1\left[1+\mathrm{O}(\alpha)\right]\,. \tag{3.92}$$

Yet, this violates (3.67) and the quasi-stationary equilibrium assumption. The fields of an ingoing plasma or fluid can not balance gravity nor can ingoing electromagnetic waves balance gravity near the horizon. Equation (3.92) is the condition necessary to keep the electric fields well behaved in the fluid frame.

3.7.1.4 Electromagnetically Induced Equilibrium

The results (3.77b) and (3.88b) show that any equilibrium near the horizon induced by electromagnetic forces is highly unstable. Any finite inward perturbation produces a radial gravitational force that increases without bound. The equations of radial motion and energy quickly transition for a fluid, inward of the equilibrium position, to

$$\frac{\mathrm{d}u^r}{\mathrm{d}\tau} + \Gamma^r{}_{00}\left(u^0\right)^2 \approx 0\,, \tag{3.93a}$$

$$\frac{\mathrm{d}u^0}{\mathrm{d}\tau} - \Gamma^r{}_{00}u^0u^r \approx 0\,. \tag{3.93b}$$

These equations integrate to give the asymptotic velocities in (3.12) and (3.17), the same as a geodesic. Namely, in the ZAMO frames,

$$u^0 \sim \alpha^{-1}\,, \tag{3.94a}$$

$$v^r = \frac{u^r}{u^0} \sim -1 + \mathrm{O}(\alpha^2)\,, \tag{3.94b}$$

$$v^\phi = \frac{u^\phi}{u^0} \sim \alpha\,, \tag{3.94c}$$

$$v^\theta = \frac{u^\theta}{u^0} \sim \alpha\,. \tag{3.94d}$$

In the stationary frames,

$$\tilde{u}^t \sim \alpha^{-2}\,, \tag{3.95a}$$

$$\frac{\mathrm{d}\phi}{\mathrm{d}t} = \Omega_H\left[1 + \mathrm{O}(\alpha^2)\right]\,, \tag{3.95b}$$

$$\frac{\mathrm{d}r}{\mathrm{d}t} = -\frac{\Delta}{{r_+}^2 + a^2}\left[1 + \mathrm{O}(\alpha^2)\right]\,. \tag{3.95c}$$

Furthermore, it was shown above that there are no equilibria near the horizon associated with forces induced by the ingoing wave condition $F^{r\theta} \sim F^{r\phi} \sim \alpha^{-1}$. The equations of motion of the plasma were over constrained.

Note that due to the current decomposition, in (3.74) and (3.78) for example, into drifts between species the above analysis holds for charged separated plasmas as well in which the $\boldsymbol{J} \times \boldsymbol{B}$ force is replaced by the Lorentz force in the equation of motion (3.39). At small enough lapse function the horizon boundary conditions (3.94) and (3.95) hold for a charge separated plasma. The details of the calculation parallel the treatment above for a fluid and are left as an exercise for the reader.

3.7.2 Radiative Forces

Next we look at forces which derive from the radiation term in the equation of motion (3.40). One might suspect that Compton drag is capable of achieving a balance with gravity. Photons that appear outgoing in a global sense can transfer a component of radial momentum that also appears outgoing in a global sense to a fluid. However, the only possible sources for outgoing radiation are currents (moving charges) in the θ–ϕ plane between the point of observation and the horizon (this follows from Ampere's law (3.56b) and (3.60b)). Near the hole, one can compute a bound on the total current, $I_{\|}$, in the θ–ϕ plane between the horizon and a ZAMO located at Boyer–Lindquist coordinate $r \gtrsim r_+$ (this is the source of radiation measured by a ZAMO or, therefore, a fluid in quasi-stationary equilibrium without any additional factors of α):

$$\begin{aligned}I_{\|} &= \int_{r_{\min}}^{r} \sqrt{g_{rr}} J_{\|} \mathrm{d}r \\ &\approx 2J_{\|}\left(r_+ - r_-\right)^{-1/2}\left[\left(r - r_+\right)^{1/2} - \left(r_{\min} - r_+\right)^{1/2}\right]\rho_+ \\ &< \frac{2J_{\|}\left(r - r_-\right)^{1/2}}{\left(r_+ - r_-\right)}\rho_+ \sim \alpha\left(r\right) .\end{aligned} \tag{3.96}$$

In (3.96), $r_{\min}$ is the minimum radial coordinate of the current distribution, r_- is defined in (1.24c) and we used the current density scalings found in (3.88).

In summary, as the horizon is approached, by Ampere's law and gravitational redshifting, the flux of globally outgoing classical electromagnetic radiation becomes more feeble (it scales with lapse function), but gravity in (3.70) keeps increasing in strength. Thus, even though it is possible for there to be a Compton drag in a local inertial frame, the whole frame moves toward the horizon under the force of gravity. Near the horizon, radiation pressure will not halt the global infall of the fluid as viewed by external observers.

Another effect in the $\nabla \cdot T_r$ term is radiation resistance. But this is either a result of inertia dragging a plasma across magnetic field lines (a gravity dominated flow by definition) or accelerations in an electric field. Since we already showed in Sect. 3.7.1 that electromagnetic forces can not compete with gravity near the horizon, then certainly the induced backreaction through radiation will not either.

3.7.3 Other Possible Forces in the Equation of Motion

Note that n and μ are rest-frame-evaluated quantities so that they are well behaved and do not scale with lapse function. The pressure P is rest-frame-evaluated as well, so $\partial P/\partial X^r$ can not balance the gravity term in (3.40) with any stability:

$$\lim_{r \to r_+} \frac{\left(\partial P/\partial X^r\right)}{n\mu c g} \sim \alpha^2 \left(u_0\right)^{-2} . \tag{3.97}$$

If an anisotropic pressure tensor were chosen in (3.40), the result would be the same since the force terms that would be introduced are of a similar nature to the isotropic case.

Quantum electrodyanmic processes can induce pair creation which induces a term in the poloidal momentum equation. But, these can be discounted from preventing inertial dominance near the horizon since this is just a redistribution or creation of inertia.

It should be noted that the results of this section are more general than the title indicates since viscous forces are macroscopic models of microphysical processing involving electromagnetic interactions. The electromagnetic analysis of Sect. 3.7.1 is equally valid for microscopic as well as macroscopic fields. Thus, it follows that viscous forces will not impede the asymptotic flow.

Equations (3.94) and (3.95) are the horizon boundary conditions for classical matter in any state or form. Recall that outgoing matter states near the horizon are unphysical by (3.19). Equilibria near the horizon are unphysical when one explores the momentum equations of the charges that source the magnetic field as implied by the analysis of Sect. 1.3.7. In summary, all flows are inertially dominated near the horizon as evidenced by the boundary conditions (3.94) and (3.95) that depend only on the metric. Obviously, in in-falling coordinates such as ingoing Kerr–Schild coordinates this effect is obscured by construction: they the natural coordinates of observers in a state of relativistic in-fall relative to asymptotic infinity, near the horizon.

More applications of the horizon boundary condition that prevents stable equilibria outside of the horizon can be found in [85]. This section is based primarily on that article.

Chapter 4
Vacuum Electrodynamics

4.1 Motivation

The study of vacuum electromagnetic fields of sources for Maxwell's equations near the horizon is a fundamental starting point for understanding GHM in the ergosphere. It addresses two core issues that can determine the global flow. Firstly, Maxwell's equations can be used to determine the electrodynamic nature of space–time near the event horizon. For example, if the event horizon can impose meaningful boundary conditions electrodynamically, the GHM coupling in the ergosphere would be largely a consequence of these constraints. It will be demonstrated that contrary to many popular early treatments of black hole magnetospheres that the event horizon is an asymptotic infinity for accreting charge neutral electromagnetic sources. This is essentially a manifestation of a "no hair" theorem that is proven for rotating black holes. In particular, we show in Sect. 4.7 that the space–time near the horizon has no unipolar inductive properties like a Faraday wheel has at the end of a plasma-filled waveguide or transmission line (as discussed in Sect. 2.10). Thus, the event horizon plays no role in any GHM interaction except for being an effective sink for mass influx.

The second issue addressed by Maxwell's equations in vacuum are the structure of large scale fields near the hole and any long range interactions that are associated with electromagnetic fields. We find that the rotating geometry mixes electric and magnetic fields. All magnetic field solutions near the hole have a frame dragging induced electric field that can not be eliminated by a global coordinate transformation ($*F^{\mu\nu}F_{\mu\nu} \neq 0$). As plasma is introduced in an effort to short out this electric field, the black hole attains a net charge. Therefore, the net electromagnetic field has a component due to the Kerr–Newman black hole that results from charge accretion. The fields from the charged black hole can be substantial when $a \approx M$.

We restrict our discussion mainly to the case of axisymmetric externally imposed sources. A natural place to look for electromagnetic sources would be an accretion disk around a black hole and the accretion flow falling in from the inner edge. The black hole GHM interactions can occur on the background of the seed fields created

B. Punsly, *Black Hole Gravitohydromagnetics, 2nd. ed.*,
Astrophysics and Space Science Library 355, doi: 10/1007/978-3-540-76957-6_4,

by the equatorial accreting sources and they modify the fields in the process. We expect the inner regions of the accretion disk to experience Lens–Thirring torques and be aligned perpendicular to the rotation axis of the hole: the Bardeen–Petterson effect. A detailed examination of this issue can be found in [80] wherein it is shown that the accretion disk within $r = 100M$ is likely to have a symmetry axis aligned with that of the hole. For low accretion systems a poloidal magnetic field sourced in a disk will torque the disk into alignment with the hole's rotation axis [86]. Thus, we expect in astrophysical situations that axisymmetric fields should yield the first order effects of ergospheric vacuum electrodynamics. Furthermore, for high accretion rates as in AGN, we expect rapid neutralization of electric fields. Thus, charge neutral axisymmetric sources are of prime interest.

Even with the simplification of charge neutrality and axisymmetry, Maxwell's equations in Sect. 3.6 are very complicated. This is true in any basis of evaluation. Even though the ZAMO basis has the advantage of an integral form of Maxwell's equations due to hypersurface orthogonality (this is known as the $3+1$ split [80]), it is well known that the integral form of Ampere's law can only be used to find the magnetic field distribution in the very few cases of extreme symmetry. Furthermore, there is no Biot–Savart law in the $3+1$ split to facilitate calculation numerically. Maxwell's equations in Sect. 3.6 are intractable in a global sense because different field components are coupled in the differential equations. Thus, we can not expand the field components in an infinite series of special functions as is done in electrostatics and magnetostatics [87]. Amazingly, if the spinorial decomposition of the Maxwell tensor into Newman–Penrose coefficients [88] is performed, one finds that two of the three spin coefficients have separable differential equations [89]. The third spin coefficient is found by integrating a first order differential equation. Thus, one has the ability to solve for the Maxwell tensor as a linear combination of special functions combined with the corresponding solutions of a radial equation. The big breakthrough in the study of vacuum electrodynamics was obtained in [90] wherein the elaborate formalism of [89], designed for electromagnetic waves, was used to study the simpler case of stationary electromagnetic test fields around Kerr black holes. This chapter is primarily an exposition of this technique and the implications of the results.

4.2 Maxwell's Equations in the Newman–Penrose Formalism

There is an isomorphism between Pauli spinors, $\psi = \begin{pmatrix} \delta \\ \eta \end{pmatrix}$, and null vectors, X^{μ}, in Minkowski space–time:

$$X^{\mu} = \frac{1}{\sqrt{2}} \bar{\psi} \sigma^{\mu} \psi \,, \tag{4.1a}$$

$$X^{\mu} X_{\mu} = \frac{1}{2} \left[(\delta\bar{\delta}+\eta\bar{\eta})^2 - (\delta\bar{\eta}+\eta\bar{\delta})^2 + (\delta\bar{\eta}-\eta\bar{\delta})^2 - (\delta\bar{\delta}-\eta\bar{\eta})^2 \right] = 0. \tag{4.1b}$$

where " ‾ " means complex conjugate and σ^{μ} is a Pauli spin matrix with $\sigma^0 \equiv I$. In [91], it is shown how this relation creates a future directed null vector X^{μ} that is defined up to a phase. Inverting (4.1a) by multiplying through by σ_{μ} and summing yields

$$\frac{1}{\sqrt{2}}\begin{pmatrix} X_0+X_3 & X_1+\mathrm{i}X_2 \\ X_1-\mathrm{i}X_2 & X_0-X_3 \end{pmatrix} = \begin{pmatrix} \delta\bar{\delta} & \delta\bar{\eta} \\ \eta\bar{\delta} & \eta\bar{\eta} \end{pmatrix} = \begin{pmatrix} \delta \\ \eta \end{pmatrix} \begin{pmatrix} \bar{\delta} & \bar{\eta} \end{pmatrix} \equiv \psi^A \psi^{A'} . \quad (4.1c)$$

where $\psi^A = \begin{pmatrix} \delta \\ \eta \end{pmatrix}$ and $\psi^{A'} = \begin{pmatrix} \bar{\delta} & \bar{\eta} \end{pmatrix}$ transforms as a conjugate spinor. Note that $(X^0)^2 - (X_1)^2 - (X_2)^2 - (X_3)^2 = \delta\bar{\delta}\eta\bar{\eta} - \eta\bar{\delta}\delta\bar{\eta} = 0$. Thus, a null vector X_{μ} is represented by a bispinor $\psi^A\psi^{A'}$. This is a spinor of rank (1,1).

Based on the isomorphism between null vectors and 2-spinors, it is useful to decompose tensors in a null tetrad, $e^i_{(a)}$. There are four legs to the tetrad labeled by the subscript a. These are four independent vector fields at each point of space–time. The superscript i indicates the component representation of the vector field $e_{(a)}$ in some coordinate system. Motivated by the isomorphism (4.1) we find a tetrad of two real and two complex null vector fields. In order to generate this frame field using the isomorphism (4.1) requires more spinor degrees of freedom. We thus define a two dimensional spinorial basis with a symplectic metric:

$$\varepsilon_{AB} = -\varepsilon_{BA} \, , \quad \varepsilon_{AB}\varepsilon^{CB} = \delta^C{}_A \, . \quad (4.2a)$$

The basis spinors satisfy the normalization condition

$$\psi_A \xi^A = \varepsilon_{AB}\psi^A\xi^B = 1 \, , \quad (4.2b)$$

$$\psi_A \psi^A = \xi_A \xi^A = 0 \, . \quad (4.2c)$$

We use the spinor basis to define the null tetrad in analogy to (4.1c):

$$l^{\alpha} = \psi^A \psi^{A'} \, , \quad (4.3a)$$

$$n^{\alpha} = \xi^A \xi^{A'} \, , \quad (4.3b)$$

$$m^{\alpha} = \psi^A \xi^{A'} \, , \quad (4.3c)$$

$$\bar{m}^{\alpha} = \xi^A \psi^{A'} \, . \quad (4.3d)$$

This also implies the relations

$$X^0 = \frac{1}{\sqrt{2}}\left(\psi^A\psi^{A'} + \xi^A\xi^{A'}\right) \, , \quad (4.3e)$$

$$X_1 = \frac{1}{\sqrt{2}}\left(\psi^A\xi^{A'} + \xi^A\psi^{A'}\right) \, , \quad (4.3f)$$

$$X_2 = \frac{1}{\sqrt{2}}\left(\psi^A\xi^{A'} - \xi^A\psi^{A'}\right) , \tag{4.3g}$$

$$X_3 = \frac{1}{\sqrt{2}}\left(\psi^A\psi^{A'} - \xi^A\xi^{A'}\right) . \tag{4.3h}$$

We can generalize (4.3) to represent 4-vectors (not necessarily null) by spinors of rank (1,1), $X^{AB'}$:

$$X^\nu = -\frac{1}{2}\sigma^\nu{}_{AB'}X^{AB'} , \quad X^{AB'} = X^\mu\sigma_\mu{}^{AB'} , \tag{4.4a}$$

$$X_\nu = -\frac{1}{2}\sigma_\nu{}^{AB'}X_{AB'} , \quad X_{AB'} = X_\mu\sigma^\mu{}_{AB'} , \tag{4.4b}$$

$$\sigma^\mu{}_{AB'} = \eta^{\mu\nu}\sigma_\nu{}^{CD'}\varepsilon_{CA}\varepsilon_{D'B'} . \tag{4.4c}$$

where $\eta_{\mu\nu}$ is the Minkowski metric. Similarly, a tensor such as the Maxwell tensor decomposes as

$$F_{AB'CD'} = \sigma^\nu{}_{AB'}\sigma^\mu{}_{CD'}F_{\mu\nu} , \tag{4.4d}$$

$$F_{\mu\nu} = \frac{1}{4}\left(\sigma_\mu{}^{AB'}\sigma_\nu{}^{CD'}\right)F_{AB'CD'} . \tag{4.4e}$$

Since $F_{\mu\nu}$ is real and antisymmetric, we can expand the Maxwell tensor in terms of a symmetric rank 2 spinor ϕ_{AB},

$$F_{AB'CD'} = \phi_{AC}\varepsilon_{B'D'} + \varepsilon_{AC}\bar{\phi}_{B'D'} . \tag{4.4f}$$

The null tetrad in (4.3) obeys the orthogonality condition

$$\begin{aligned} &l \cdot m = l \cdot \bar{m} = n \cdot m = n \cdot \bar{m} = \\ &l \cdot l = n \cdot n = m \cdot m = \bar{m} \cdot \bar{m} = 0 , \end{aligned} \tag{4.5a}$$

and the normalization condition

$$l \cdot n = 1 , \quad \text{and} \quad m \cdot \bar{m} = -1 . \tag{4.5b}$$

Using the metric in the tetrad basis (4.5ab) we can write

$$e_{(1)}{}^\mu = l^\mu , \quad e_{(2)}{}^\mu = n^\mu , \quad e_{(3)}{}^\mu = m^\mu , \quad e_{(4)}{}^\mu = \bar{m}^\mu , \tag{4.5c}$$

$$\begin{aligned} &e^{(1)\,\mu} = e_{(2)}{}^\mu = n^\mu , \quad e^{(2)\,\mu} = e_{(1)}{}^\mu = l^\mu , \\ &e^{(3)\,\mu} = -e_{(4)}{}^\mu = -\bar{m}^\mu , \quad e^{(4)\,\mu} = -e_{(3)}{}^\mu = -m^\mu . \end{aligned} \tag{4.5d}$$

The advantage of a null tetrad is the possibility of simplifying the connection (see 4.12).

The covariant derivatives of the tetrad are expressed in terms of Ricci rotation coefficients, $\gamma_{(c)(a)(b)}$, as

$$e_{(a)k;i} = e^{(c)}{}_k \, \gamma_{(c)(a)(b)} \, e^{(b)}{}_i \,, \tag{4.6a}$$

$$\gamma_{(c)(a)(b)} = e_{(c)}{}^k \, e_{(a)k;i} \, e_{(b)}{}^i \,, \tag{4.6b}$$

$$\gamma_{(c)(a)(b)} = -\gamma_{(a)(c)(b)} \,. \tag{4.6c}$$

The spin coefficients are defined in terms of the Ricci rotation coefficients [88]:

$$\begin{aligned}
\kappa &= \gamma_{131} = l_{\mu;\nu} m^\mu l^\nu \,, \\
\pi &= -\gamma_{241} = -n_{\mu;\nu} \bar{m}^\mu l^\nu \,, \\
\varepsilon &= \frac{1}{2}(\gamma_{121} - \gamma_{341}) = \frac{1}{2}\left(l_{\mu;\nu} n^\mu l^\nu - m_{\mu;\nu} \bar{m}^\mu l^\nu\right) \,, \\
\tilde{\rho} &= \gamma_{134} = l_{\mu;\nu} m^\mu \bar{m}^\nu \,, \\
\lambda &= -\gamma_{244} = -n_{\mu;\nu} \bar{m}^\mu \bar{m}^\nu \,, \\
\alpha &= \frac{1}{2}(\gamma_{124} - \gamma_{344}) = \frac{1}{2}\left(l_{\mu;\nu} n^\mu \bar{m}^\nu - m_{\mu;\nu} \bar{m}^\mu \bar{m}^\nu\right) \,, \\
\sigma &= \gamma_{133} = l_{\mu;\nu} m^\mu m^\nu \,, \\
\mu &= -\gamma_{243} = -n_{\mu;\nu} \bar{m}^\mu m^\nu \,, \\
\beta &= \frac{1}{2}(\gamma_{123} - \gamma_{343}) = \frac{1}{2}\left(l_{\mu;\nu} n^\mu m^\nu - m_{\mu;\nu} \bar{m}^\mu m^\nu\right) \,, \\
\tau &= \gamma_{132} = l_{\mu;\nu} m^\mu n^\nu \,, \\
\nu &= -\gamma_{242} = -n_{\mu;\nu} \bar{m}^\mu n^\nu \,, \\
\gamma &= \frac{1}{2}(\gamma_{122} - \gamma_{342}) = \frac{1}{2}\left(l_{\mu;\nu} n^\mu n^\nu - m_{\mu;\nu} \bar{m}^\mu n^\nu\right) \,.
\end{aligned} \tag{4.7}$$

Maxwell's equations in spinorial form are [91]

$$\nabla^{A'B} \phi^A{}_B = 2\pi J^{AA'} \,, \quad J^{AA'} = \bar{J}^{AA'} \,, \tag{4.8a}$$

$$\nabla^{A'B} = (\sigma_\mu)^{A'B} \nabla^\mu \,, \tag{4.8b}$$

where ∇_μ is the covariant derivative in the tetrad basis. It is more instructive to rewrite (4.8a) using the symplectic spinor metric as

$$\frac{1}{2}\varepsilon^{BC}\left[\sigma^\mu{}_{CD'} \nabla_\mu \phi_{AB} - \sigma^\mu{}_{BD'} \nabla_\mu \phi_{AC}\right] = 2\pi J_{AD'} \,. \tag{4.8c}$$

Using (4.4f) and (4.5d), the three independent components of the Maxwell 2-spinor are defined as [89]

$$\phi_0 \equiv \phi_{00} = F_{13} = F_{ij} l^i m^j \,, \tag{4.9a}$$

$$\phi_1 \equiv \phi_{10} = \phi_{01} = \frac{1}{2}(F_{12} + F_{43}) = \frac{1}{2} F_{ij}\left(l^i n^j + \bar{m}^i m^j\right) \,, \tag{4.9b}$$

$$\phi_2 \equiv \phi_{11} = F_{42} = F_{ij} \bar{m}^i n^j \,. \tag{4.9c}$$

These equations invert to give

$$F_{\mu\nu} = 2\left[\phi_1 \left(n_{[\mu} l_{\nu]} + m_{[\mu} \bar{m}_{\nu]}\right) + \phi_2 l_{[\mu} m_{\nu]} + \phi_0 \bar{m}_{[\mu} n_{\nu]}\right] + \text{c.c.} \,. \tag{4.9d}$$

Equation (4.8c) contains covariant derivatives with respect to the tetrad defined by

$$\nabla_{(b)} A_{(a)} = e^i_{(a)} A_{i;j} e^j_{(b)} = A_{(a),(b)} - G^{(n)(m)} \gamma_{(n)(a)(b)} A_{(m)} \,, \tag{4.9e}$$

where $G^{(n)(m)}$ is the metric in the tetrad basis given by (4.5)

$$G_{(a)(b)} = G^{(a)(b)} = \begin{bmatrix} 0 & 1 & 0 & 0 \\ 1 & 0 & 0 & 0 \\ 0 & 0 & 0 & -1 \\ 0 & 0 & -1 & 0 \end{bmatrix} . \tag{4.9f}$$

Writing out the components of (4.8c), we have the spinorial Maxwell's equations in the tetrad basis:

$$\begin{aligned} \nabla_{(1)}\phi_1 - \nabla_{(4)}\phi_0 &= 2\pi J_{(1)} \,, \\ \nabla_{(1)}\phi_2 - \nabla_{(4)}\phi_1 &= 2\pi J_{(4)} \,, \\ \nabla_{(3)}\phi_1 - \nabla_{(2)}\phi_0 &= 2\pi J_{(3)} \,, \\ \nabla_{(3)}\phi_2 - \nabla_{(2)}\phi_1 &= 2\pi J_{(2)} \,. \end{aligned} \tag{4.9g}$$

Expanding the intrinsic derivatives as in [92] using (4.7) and (4.9abc), one has

$$\begin{aligned} \nabla_{(1)}\phi_1 &= \frac{1}{2}[F_{12,1} - G^{nm}(\gamma_{n11} F_{m2} + \gamma_{n21} F_{1m}) \\ &\qquad + F_{43,1} - G^{nm}(\gamma_{n41} F_{m3} + \gamma_{n31} F_{4m})] \\ &= \phi_{1,1} - (\gamma_{131} F_{42} - \gamma_{241} F_{13}) \\ &= l^\mu \frac{\partial}{\partial X^\mu}\phi_1 + \kappa\phi_2 - \pi\phi_0 \,. \end{aligned} \tag{4.9h}$$

The other intrinsic derivatives in Maxwell's equations (2.9g) have a similar form.

One can explicitly expand (4.9d) in Boyer–Lindquist coordinates as in [93]:

$$\tilde{F}^{tr} = \mathrm{Re}\left\{\frac{r^2+a^2}{\rho^2}\phi_1 + \frac{\mathrm{i}a\tilde{\rho}^*\sin\theta}{\sqrt{2}}\left(\phi_2 - \frac{\tilde{\rho}^2\Delta\phi_0}{2}\right)\right\}, \tag{4.10a}$$

$$\tilde{F}^{t\theta} = \mathrm{Re}\left\{\frac{\mathrm{i}a\sin\theta}{\rho^2}\phi_1 - \frac{(r^2+a^2)}{\sqrt{2}}\frac{\tilde{\rho}}{\Delta}\left(\phi_2 - \frac{\tilde{\rho}^2\Delta\phi_0}{2}\right)\right\}, \tag{4.10b}$$

$$\tilde{F}^{t\phi} = \mathrm{Re}\left\{-\frac{\mathrm{i}\tilde{\rho}\rho^2}{2\sqrt{2}\Delta\sin\theta}\left(\phi_2 + \frac{\tilde{\rho}^2\Delta\phi_0}{2}\right)\right\}, \tag{4.10c}$$

$$\tilde{F}^{r\theta} = \mathrm{Re}\left\{-\frac{\tilde{\rho}^*}{2\sqrt{2}}\left(\phi_2 + \frac{\tilde{\rho}^2\Delta\phi_0}{2}\right)\right\}, \tag{4.10d}$$

$$\tilde{F}^{r\phi} = \mathrm{Re}\left\{-\frac{a}{\rho^2}\phi_1 - \frac{\mathrm{i}\tilde{\rho}^*}{\sqrt{2}\sin\theta}\left(\phi_2 - \frac{\tilde{\rho}^2\Delta\phi_0}{2}\right)\right\}, \tag{4.10e}$$

$$\tilde{F}^{\theta\phi} = \mathrm{Re}\left\{-\frac{\mathrm{i}}{\rho^2\sin\theta}\phi_1 + \frac{a\tilde{\rho}^*}{\sqrt{2}\Delta}\left(\phi_2 - \frac{\tilde{\rho}^2\Delta\phi_0}{2}\right)\right\}, \tag{4.10f}$$

where the spin coefficient $\tilde{\rho}$ is given by

$$\tilde{\rho} = \frac{-1}{r - \mathrm{i}a\cos\theta}. \tag{4.11}$$

A tetrad was found by Kinnersley [94] in which the spin coefficients of (4.7) are simplified since the legs l^μ and n^ν are along the principal null directions of the Kerr space–time,

$$\kappa = \sigma = \nu = \gamma = \varepsilon = 0. \tag{4.12}$$

The legs of the tetrad are given in (t, r, θ, ϕ) Boyer–Lindquist coordinates as

$$l^\mu = \left[\frac{(r^2+a^2)}{\Delta}, 1, 0, \frac{a}{\Delta}\right], \tag{4.13a}$$

$$n^\mu = \left[(r^2+a^2), -\Delta, 0, a\right]/2\rho^2, \tag{4.13b}$$

$$m^\mu = \left[\mathrm{i}a\sin\theta, 0, 1, \frac{\mathrm{i}}{\sin\theta}\right] / \left[\sqrt{2}(r + \mathrm{i}a\cos\theta)\right]. \tag{4.13c}$$

It is shown in [89] that (4.9e) combined with (4.12) yields Maxwell's equations, (4.9g), in terms of the spin coefficients in the Kinnersley tetrad:

$$(D - 2\rho)\phi_1 - (\delta^* + \pi - 2\alpha)\phi_0 = 2\pi J_l, \tag{4.14a}$$

$$(\delta - 2\tau)\phi_1 - (\Delta + \mu - 2\lambda)\phi_0 = 2\pi J_m, \tag{4.14b}$$

$$(D - \rho)\phi_2 - (\delta^* + 2\pi)\phi_1 = 2\pi J_{\bar{m}}, \tag{4.14c}$$

$$(\delta - \tau + 2\beta)\phi_2 - (\Delta + 2\mu)\phi_1 = 2\pi J_n. \tag{4.14d}$$

The differential operators in (4.14) are defined with respect to the null tetrad as

$$D = l^{\mu} \frac{\partial}{\partial X^{\mu}} , \tag{4.15a}$$

$$\Delta = n^{\mu} \frac{\partial}{\partial X^{\mu}} , \tag{4.15b}$$

$$\delta = m^{\mu} \frac{\partial}{\partial X^{\mu}} . \tag{4.15c}$$

The spin coefficients are tabulated in the Kinnersley tetrad in [89] and [93]:

$$\begin{aligned}
\tilde{\rho} &= -\frac{1}{r - \mathrm{i}a\cos\theta} , \\
\beta &= -\frac{\tilde{\rho}^* \cot\theta}{2\sqrt{2}} , \\
\pi &= \frac{\mathrm{i}a\tilde{\rho}^2 \sin\theta}{\sqrt{2}} , \\
\tau &= -\frac{\mathrm{i}a\tilde{\rho}\tilde{\rho}^* \sin\theta}{\sqrt{2}} , \\
\mu &= \tilde{\rho}^2\tilde{\rho}^* \frac{\Delta}{2} , \\
\gamma &= \mu + \tilde{\rho}\tilde{\rho}^* \frac{(r-M)}{2} , \\
\alpha &= \pi - \beta^* .
\end{aligned} \tag{4.16}$$

The current sources in (4.14) are

$$J_l = J^{\mu} l_{\mu} , \tag{4.17a}$$

$$J_n = J^{\mu} n_{\mu} , \tag{4.17b}$$

$$J_m = J^{\mu} m_{\mu} , \tag{4.17c}$$

$$J_{\bar{m}} = J^{\mu} \bar{m}_{\mu} . \tag{4.17d}$$

A second order equation for ϕ_0 can be created by operating on (4.14b) with $D - 2\tilde{\rho} - \tilde{\rho}^*$ and (4.14a) with $\delta - \beta - \alpha^* - 2\tau + \pi^*$ and subtracting one equation from another [89]. This procedure yields

$$\begin{aligned}
&[(D - 2\tilde{\rho} - \tilde{\rho}^*)(\Delta + \mu - 2\gamma) \\
&\quad - (\delta - \beta - \alpha^* - 2\tau + \pi^*)(\delta^* + \pi - 2\alpha)]\,\phi_0 = 2\pi J_0 ,
\end{aligned} \tag{4.18}$$

$$J_0 = (\delta - \beta - \alpha^* - 2\tau + \pi^*) J_l - (D - 2\tilde{\rho} - \tilde{\rho}^*) J_m . \tag{4.19}$$

Similarly, operating on (4.14c) and (4.14d), Teukolsky [89] found a decoupled equation for ϕ_2:

$$\begin{aligned}&[(\Delta+\gamma-\gamma^*+2\mu+\mu^*)(D-\tilde{\rho})\\&\quad-(\delta^*+\alpha+\beta^*+2\pi-\tau^*)(\delta-\tau+2\beta)]\,\phi_2=2\pi J_2\,,\end{aligned} \tag{4.20}$$

$$J_2=(\Delta+\gamma-\gamma^*+2\mu+\mu^*)J_{\bar{m}}-(\delta^*+\alpha+\beta^*+2\pi-\tau^*)J_n\,. \tag{4.21}$$

Teukolsky identified spin weighted components of the Maxwell tensor:

$$\phi_{+1}=\phi_0\,, \tag{4.22a}$$
$$\phi_{-1}=\tilde{\rho}^{-2}\phi_2\,. \tag{4.22b}$$

These components have separable solutions to (4.18) and (4.20) respectively:

$$\phi_{+1}=\mathrm{e}^{-\mathrm{i}\omega t}\mathrm{e}^{im\phi}\;{}_{+1}S^m{}_l(\theta)R_{+1}(r)\,, \tag{4.23a}$$
$$\phi_{-1}=\mathrm{e}^{-\mathrm{i}\omega t}\mathrm{e}^{im\phi}\;{}_{-1}S^m{}_l(\theta)R_{-1}(r)\,, \tag{4.23b}$$

where R and S satisfy (in vacuum $G_{\pm 1}(r)=0$)

$$\Delta^{-1}\frac{\mathrm{d}}{\mathrm{d}r}\left(\Delta^2\frac{\mathrm{d}R_{+1}}{\mathrm{d}r}\right)+\left(\frac{\mathrm{P}^2-2\mathrm{i}(r-M)\mathrm{P}}{\Delta}+4\mathrm{i}\omega r-\mathrm{k}_+\right)R_{+1}=G_{+1}(r)\,, \tag{4.23c}$$

$$\Delta^{+1}\frac{\mathrm{d}^2}{\mathrm{d}^2r}R_{-1}+\left(\frac{\mathrm{P}^2+2\mathrm{i}(r-M)\mathrm{P}}{\Delta}-4\mathrm{i}\omega r-\mathrm{k}_-\right)R_{-1}=G_{-1}(r)\,, \tag{4.23d}$$

$$\begin{aligned}&\frac{1}{\sin\theta}\frac{\mathrm{d}}{\mathrm{d}\theta}\left(\sin\theta\frac{\mathrm{d}}{\mathrm{d}\theta}({}_{+1}S^m{}_l)\right)+\left(\omega^2a^2\cos^2\omega-\frac{m^2}{\sin^2\theta}\right.\\&\qquad\left.-2\omega a\cos\theta-\frac{2m\cos\theta}{\sin^2\theta}-\cot^2\theta+1+\mathrm{A}\right){}_{+1}S^m{}_l=0\,,\end{aligned} \tag{4.23e}$$

$$\begin{aligned}&\frac{1}{\sin\theta}\frac{\mathrm{d}}{\mathrm{d}\theta}\left(\sin\theta\frac{\mathrm{d}}{\mathrm{d}\theta}({}_{-1}S^m{}_l)\right)+\left(\omega^2a^2\cos^2\omega-\frac{m^2}{\sin^2\theta}\right.\\&\qquad\left.+2\omega a\cos\theta+\frac{2m\cos\theta}{\sin^2\theta}-\cot^2\theta-1+\mathrm{A}\right){}_{-1}S^m{}_l=0\,,\end{aligned} \tag{4.23f}$$

where $\mathrm{A}\equiv{}_{\pm 1}\mathrm{A}^m{}_l(\omega a)$ are separation constants and

$$\mathrm{P}\equiv\omega\left(r^2+a^2\right)-ma\,, \tag{4.23g}$$
$$\mathrm{k}_{\pm}=\mathrm{A}+\omega^2a^2-2ma\omega-(1\pm 1)\,. \tag{4.23h}$$

The functions ${}_{\pm 1}S^m{}_l(\theta)$ are spin weighted spheroidal harmonics. When $\omega a=0$ these reduce to spin weighted spherical harmonics.

When sources are present in Maxwell's equations we have

$$\phi_{\pm 1} = \int \mathrm{d}\omega \sum_{l,m} R_{\pm 1}(r)\, {}_{\pm 1}S^m{}_l(\theta)\, \mathrm{e}^{\mathrm{i}m\phi}\mathrm{e}^{-\mathrm{i}\omega t} . \tag{4.24}$$

where the source terms in the radial equations $G_{\pm 1}(r)$ are given by (4.19) and (4.21) as

$$4\pi\rho^2 J_0 = \int \mathrm{d}\omega \sum_{l,m} G_{+1}(r)\, {}_{+1}S^m{}_l(\theta)\, \mathrm{e}^{\mathrm{i}m\phi}\mathrm{e}^{-\mathrm{i}\omega t} , \tag{4.25a}$$

$$\frac{4\pi\rho^2 J_2}{(r - \mathrm{i}a\cos\theta)^2} = \int \mathrm{d}\omega \sum_{l,m} G_{-1}(r)\, {}_{-1}S^m{}_l(\theta)\, \mathrm{e}^{\mathrm{i}m\phi}\mathrm{e}^{-\mathrm{i}\omega t} . \tag{4.25b}$$

A decoupling of the equations for ϕ_1 is found in [95], but the resulting differential equation does not lend itself to separation of variables in the Kerr space–time. In general one solves the second order differential equations (4.18) and (4.20) to compute ϕ_0 and ϕ_2. This allows one to integrate the first order coupled differential equations (4.14) to find ϕ_1.

4.3 Poisson's Equations in the Kerr Space–Time

The Newman–Penrose formulation of Maxwell's equations was developed to investigate time dependent perturbations of a black hole and the proof of the no hair theorem that the most general electro-vac black hole solution is described by Q, M, and a (the Kerr–Newman black hole). A clever idea was to use those same equations to study time stationary electromagnetic test fields on the background of the Kerr metric [90, 93, 96, 97]. There is a tremendous simplification over the results of the last section: one derivative is eliminated and since $\omega a = 0$, the angular functions simplify from spin weighted spheroidal harmonics to spin weighted spherical harmonics. The treatment in [90] is particularly well developed and forms the basis for calculation throughout most of this chapter.

The spin coefficients of (4.9) are redefined in [90] as

$$\begin{aligned} \Phi_0 &= \phi_0 , \\ \Phi_1 &= \frac{(r - \mathrm{i}a\cos\theta)^2}{(r_+ - r_-)^2}\phi_1 , \\ \Phi_2 &= \frac{r - \mathrm{i}a\cos\theta}{(r_+ - r_-)^2}\phi_2 . \end{aligned} \tag{4.26}$$

Setting the time derivatives equal to zero in the first order Maxwell's equations (4.14) one has

$$\sqrt{2}\,(r_+ - r_-)^2 \left(\frac{\partial}{\partial r} + \frac{a}{\Delta}\frac{\partial}{\partial \phi} \right) \Phi_1$$
$$- (r - \mathrm{i}a\cos\theta) \left(\frac{\partial}{\partial \theta} + \cot\theta - \frac{\mathrm{i}}{\sin\theta}\frac{\partial}{\partial \phi} \right) \Phi_0 + \mathrm{i}a\sin\theta\,\Phi_0$$
$$= \sqrt{2}\,(r - \mathrm{i}a\cos\theta)^2\, 2\pi J_l\,, \tag{4.27a}$$

$$\sqrt{2}\,(r_+ - r_-)^2 \left(\frac{\partial}{\partial \theta} + \frac{\mathrm{i}}{\sin\theta}\frac{\partial}{\partial \phi} \right) \Phi_1$$
$$+ (r - \mathrm{i}a\cos\theta) \left(\frac{\partial}{\partial r} - \frac{a}{\Delta}\frac{\partial}{\partial \phi} \right) \Delta\Phi_0 - \Delta\Phi_0$$
$$= \sqrt{2}\,(r - \mathrm{i}a\cos\theta)\,\rho^2 2\pi J_m\,, \tag{4.27b}$$

$$\frac{1}{\sqrt{2}} \left(\frac{\partial}{\partial \theta} - \frac{\mathrm{i}}{\sin\theta}\frac{\partial}{\partial \phi} \right) \Phi_1$$
$$- (r - \mathrm{i}a\cos\theta) \left(\frac{\partial}{\partial r} + \frac{a}{\Delta}\frac{\partial}{\partial \phi} \right) \Phi_2 + \Phi_2$$
$$= -\sqrt{2}\frac{(r - \mathrm{i}a\cos\theta)^2}{(r_+ - r_-)^2} 2\pi J_{\bar{m}}\,, \tag{4.27c}$$

$$\frac{1}{\sqrt{2}} \left(\frac{\partial}{\partial r} - \frac{a}{\Delta}\frac{\partial}{\partial \phi} \right) \Phi_1$$
$$+ (r - \mathrm{i}a\cos\theta) \left(\frac{\partial}{\partial \theta} + \cot\theta + \frac{\mathrm{i}}{\sin\theta}\frac{\partial}{\partial \phi} \right) \frac{\Phi_2}{\Delta} - \frac{(\mathrm{i}a\sin\theta)}{\Delta}\Phi_2$$
$$= -\frac{\sqrt{2}}{\Delta}\frac{\rho^2\,(r - \mathrm{i}a\cos\theta)}{(r_+ - r_-)^2} 2\pi J_n\,. \tag{4.27d}$$

The clever aspect of the analysis in [90] is to acknowledge the conclusion of the study of the second order Maxwell's equations presented in (4.23), namely that the solutions for Φ_0 and Φ_2 are separable; then, insert the general expansion into the simpler first order equations (4.27):

$$\Phi_0 = \sum_{l=1}^{\infty} \sum_{m=-l}^{l} {}^{0}R_{lm}(r)\, {}_{+1}Y_{lm}(\theta,\phi) \equiv \sum_{l,m} {}^{0}R_{lm}(r)\, {}_{+1}Y_{lm}(\theta,\phi)\,, \tag{4.28a}$$

$$\Phi_2 = \sum_{l,m} {}^{2}R_{lm}(r)\, {}_{-1}Y_{lm}(\theta,\phi)\,. \tag{4.28b}$$

The functions ${}_{\pm 1}Y_{lm}$ are spin weighted spherical harmonics [87, 91] and the radial equation for ${}^{2}R_{lm}(r)$ is found from (4.23d) and (4.25b) to be

$$\left(r^2 - 2Mr + a^2\right)\frac{\mathrm{d}^2\left({}^{2}R_{lm}\right)}{\mathrm{d}r^2} + \left[\frac{a^2m^2 - 2\mathrm{i}am(r-M)}{r^2 - 2Mr + a^2} - l(l+1)\right]\left({}^{2}R_{lm}\right) = -4\pi\left({}^{2}J_{lm}\right) . \tag{4.29}$$

The current source ${}^{2}J_{lm}(r)$ is found in terms of J_2 in (4.21):

$${}^{2}J_{lm}(r) = \int_0^{2\pi}\int_0^{\pi}\frac{(r - \mathrm{i}a\cos\theta)^2}{\left(r_+ - r_-\right)^2}\rho^2 J_2\left({}_{-1}\bar{Y}_{lm}(\theta,\phi)\right)\sin\theta\,\mathrm{d}\theta\,\mathrm{d}\phi . \tag{4.30}$$

Using (4.21) and the Kinnersley tetrad (4.13), we expand J_2 as

$$J_2 = \frac{-\Delta}{2\sqrt{2}\rho^2\,(r - \mathrm{i}a\cos\theta)^2}\left[\sqrt{2}\left(\frac{\partial}{\partial r} - \frac{a}{\Delta}\frac{\partial}{\partial\phi} + \frac{1}{r - \mathrm{i}a\cos\theta}\right)(r - \mathrm{i}a\cos\theta)\,J_{\bar{m}} + 2\left(\frac{\partial}{\partial\theta} - \frac{\mathrm{i}}{\sin\theta}\frac{\partial}{\partial\phi} + \frac{\mathrm{i}a\sin\theta}{r - \mathrm{i}a\cos\theta}\right)\frac{\rho^2\,(r - \mathrm{i}a\cos\theta)}{\Delta}J_n\right] . \tag{4.31}$$

4.4 Laplace's Equations in the Kerr Space–Time

The radial equation (4.29) can be solved when ${}^{2}J_{lm}(r) = 0$ in terms of hypergeometric functions as can be seen by the substitution

$${}^{2}R_{lm}(X) = \left(1 - \frac{1}{X}\right)^{-\mathrm{i}Z_m}\left({}^{2}y_{lm}(X)\right) , \tag{4.32a}$$

$$X \equiv \frac{r - r_-}{r_+ - r_-} , \quad Z_m \equiv \frac{ma}{r_+ - r_-} . \tag{4.32b}$$

Equations (4.29) reduces to a hypergeometric equation:

$$X(X-1)\left({}^{2}y''_{lm}\right) - 2\mathrm{i}\left({}^{2}y'_{lm}\right) - l(l+1)\left({}^{2}y_{lm}\right) = 0 . \tag{4.33}$$

There are two linearly independent solutions in terms of hypergeometric functions, F,

$${}^{2}y_{lm}^{(I)} = \left(1 - \frac{1}{X}\right)^{2\mathrm{i}Z_m}X(X-1)F(l+2, 1-l, 2-2\mathrm{i}Z_m; X) , \tag{4.34a}$$

$${}^{2}y_{lm}^{(II)} = (-X)^{-l}F(l, l+1-2\mathrm{i}Z_m, 2l+2; X^{-1}) . \tag{4.34b}$$

It is shown in [90] how to determine ${}^{0}R_{lm}$ from ${}^{2}R_{lm}$ in (4.32) and (4.34) as well as solve for Φ_1 in vacuum, i.e., Laplace's equation. Apply the operator $\partial/\partial\theta -$

$(\mathrm{i}/\sin\theta)\partial/\partial\phi$ to (4.27a) and $\partial/\partial r + (a/\Delta)\partial/\partial\phi$ to (4.27c) with the expansions for Φ_0 and Φ_2 in (4.28) and $J_l = J_{\bar{m}} = 0$. This yields

$$ {}^0R_{lm} = \frac{2\,(r_+ - r_-)^2}{l(l+1)} \left(\frac{\mathrm{d}}{\mathrm{d}r} + \frac{\mathrm{i}am}{\Delta}\right)\left(\frac{\mathrm{d}}{\mathrm{d}r} + \frac{\mathrm{i}am}{\Delta}\right) {}^2R_{lm} \,. \tag{4.35} $$

From (4.32a) this implies that

$$ {}^0R_{lm}(X) = \left(1 - \frac{1}{X}\right)^{-\mathrm{i}Z_m} \left[\frac{2}{l(l+1)}\right] \frac{\mathrm{d}^2}{\mathrm{d}X^2}\left[{}^2y_{lm}\right] . \tag{4.36a} $$

From [98], one can show that

$$ \frac{\mathrm{d}^2}{\mathrm{d}X^2}\left[{}^2y_{lm}^{(I)}\right] = 2\mathrm{i}Z_m\,(2\mathrm{i}Z_m - 1)\left(1 - \frac{1}{X}\right)^{2\mathrm{i}Z_m} [X(X-1)]^{-1} $$
$$ \times F(l, -l-1, -2\mathrm{i}Z_m; X)\,, \qquad \text{for } Z_m \neq 0\,, \tag{4.36b} $$

$$ \frac{\mathrm{d}^2}{\mathrm{d}X^2}\left[{}^2y_{lm}^{(I)}\right] = l(l+1)F(l+2, 1-l, 2; X)\,, \qquad \text{for } Z_m = 0\,, \tag{4.36c} $$

$$ \frac{\mathrm{d}^2}{\mathrm{d}X^2}\left[{}^2y_{lm}^{(II)}\right] = l(l+1)\,(-X)^{-l-2}\,F(l+2, l+1-2\mathrm{i}Z_m, 2l+2; X^{-1})\,. \tag{4.36d} $$

Solving for Φ_1 is more difficult. By the axial symmetry of the metric we have the following expansion:

$$ \Phi_1(X,\theta,\phi) = \sum_{m=-\infty}^{\infty} \left(1 - \frac{1}{X}\right)^{-\mathrm{i}Z_m} \mathrm{e}^{\mathrm{i}m\phi}\,\Phi_{1m}(X,\theta)\,. \tag{4.37} $$

Integrating (4.27a) with $J_l = 0$ and Φ_0 given by (4.28) and (4.36) yields an expression for Φ_1 up to an undetermined function $f_m(\theta)$:

$$ \Phi_1 = \frac{\sqrt{2}}{(r_+ - r_-)} \sum_{l,m} [l(l+1)]^{-1} \left(1 - \frac{1}{X}\right)^{-\mathrm{i}Z_m} \Big\{ [l(l+1)]^{1/2} $$
$$ \times \left[(r - \mathrm{i}a\cos\theta)\frac{\mathrm{d}}{\mathrm{d}X}\left({}^2y_{lm}\right) - (r_+ - r_-)\left({}^2y_{lm}\right)\right] {}_0Y_{lm}(\theta,\phi) $$
$$ -\mathrm{i}a\sin\theta \frac{\mathrm{d}}{\mathrm{d}X}\left({}^2y_{lm}\right) {}_{+1}Y_{lm}(\theta,\phi) \Big\} $$
$$ + \sum_{m=-\infty}^{\infty} f_m(\theta)\mathrm{e}^{\mathrm{i}m\phi}\left(1 - \frac{1}{X}\right)^{-\mathrm{i}Z_m} . \tag{4.38} $$

Combining the X derivative of (4.33) with (4.27b) and, $J_m = 0$ and the expansion for Φ_1 in (4.38), yields

$$\left[\frac{\partial}{\partial\theta}-\frac{m}{\sin\theta}\right]f_m(\theta)=0\,. \tag{4.39a}$$

Similarly, (4.27c) with $J_{\bar{m}}=0$ implies

$$\left[\frac{\partial}{\partial\theta}+\frac{m}{\sin\theta}\right]f_m(\theta)=0\,. \tag{4.39b}$$

Combining (4.39b) with (4.39a) yields

$$f_m(\theta)=\mathrm{C}\delta_{m0}\,, \tag{4.39c}$$

where C is a constant. It is noted in [90] that (4.27d) with $J_n=0$ is automatically satisfied by this solution.

To complete the solution (4.38) requires the following relations from [98] and [90]:

$$\frac{\mathrm{d}}{\mathrm{d}X}\left[{}^{2}y_{lm}^{(I)}\right]=(2\mathrm{i}Z_m-1)\left(1-\frac{1}{X}\right)^{2\mathrm{i}Z_m}F(l+1,-l,1-2\mathrm{i}Z_m;X)\,, \tag{4.40a}$$

$$\frac{\mathrm{d}}{\mathrm{d}X}\left[{}^{2}y_{lm}^{(II)}\right]=l(-X)^{-l-1}F(l+1,l+1-2\mathrm{i}Z_m,2l+2;X^{-1})\,. \tag{4.40b}$$

The general solution to Laplace's equation in the Kerr space–time is found from (4.28), (4.32), (4.34), (4.36), (4.38) and (4.39). Consider a source located between r_1 and r_2, with $r_+<r_1<r_2<\infty$. In the region between the source and the horizon, we use the asymptotic form of the hypergeometric function and regularity at the horizon to find the spin components in the region $r_+<r<r_1$:

$$\Phi_0=\sum_{l,m}a_{lm}\,2[l(l+1)]^{-1}\left(1-\frac{1}{X}\right)^{-\mathrm{i}Z_m}\frac{\mathrm{d}^2}{\mathrm{d}X^2}\left[{}^{2}y_{lm}^{(I)}\right]{}_{+1}Y_{lm}(\theta,\phi)\,, \tag{4.41a}$$

$$\begin{aligned}\Phi_1=&\frac{\sqrt{2}\,(r_+-r_-)}{(r-\mathrm{i}a\cos\theta)^2}\sum_{l,m}a_{lm}\,[l(l+1)]^{-1}\left(1-\frac{1}{X}\right)^{-\mathrm{i}Z_m}\Big\{[l(l+1)]^{1/2}\\&\times\left[(r-\mathrm{i}a\cos\theta)\frac{\mathrm{d}}{\mathrm{d}X}\left({}^{2}y_{lm}^{(I)}\right)-(r_+-r_-)\left({}^{2}y_{lm}^{(I)}\right)\right]{}_{0}Y_{lm}(\theta,\phi)\\&-\mathrm{i}a\sin\theta\frac{\mathrm{d}}{\mathrm{d}X}\left({}^{2}y_{lm}^{(I)}\right){}_{+1}Y_{lm}(\theta,\phi)\Big\}+\frac{E_a}{(r-\mathrm{i}a\cos\theta)^2}\,,\end{aligned} \tag{4.41b}$$

$$\Phi_2=\frac{(r_+-r_-)^2}{(r-\mathrm{i}a\cos\theta)^2}\sum_{l,m}a_{lm}\left(1-\frac{1}{X}\right)^{-\mathrm{i}Z_m}\left({}^{2}y_{lm}^{(I)}\right){}_{-1}Y_{lm}(\theta,\phi)\,. \tag{4.41c}$$

The solutions of most interest to black hole GHM are those at $r>r_2$. Again, we require regularity at $r\longrightarrow+\infty$.

$$\Phi_0 = \sum_{l,m} b_{lm}\, 2[l(l+1)]^{-1} \left(1-\frac{1}{X}\right)^{-\mathrm{i}Z_m} \frac{\mathrm{d}^2}{\mathrm{d}X^2}\left[{}^2y_{lm}^{(II)}\right] {}_{+1}Y_{lm}(\theta,\phi)\,, \tag{4.42a}$$

$$\Phi_1 = \frac{\sqrt{2}\,(r_+ - r_-)}{(r-\mathrm{i}a\cos\theta)^2} \sum_{l,m} b_{lm}\,[l(l+1)]^{-1}\left(1-\frac{1}{X}\right)^{-\mathrm{i}Z_m}\Big\{[l(l+1)]^{1/2}$$
$$\times\left[(r-\mathrm{i}a\cos\theta)\frac{\mathrm{d}}{\mathrm{d}X}\left({}^2y_{lm}^{(II)}\right) - (r_+ - r_-)\left({}^2y_{lm}^{(II)}\right)\right] {}_0Y_{lm}(\theta,\phi)$$
$$-\mathrm{i}a\sin\theta\frac{\mathrm{d}}{\mathrm{d}X}\left({}^2y_{lm}^{(II)}\right) {}_{+1}Y_{lm}(\theta,\phi)\Big\} + \frac{E_b}{(r-\mathrm{i}a\cos\theta)^2}\,, \tag{4.42b}$$

$$\Phi_2 = \frac{(r_+ - r_-)^2}{(r-\mathrm{i}a\cos\theta)^2}\sum_{l,m} b_{lm}\left(1-\frac{1}{X}\right)^{-\mathrm{i}Z_m}\left({}^2y_{lm}^{(II)}\right) {}_{-1}Y_{lm}(\theta,\phi)\,. \tag{4.42c}$$

The constants a_{lm}, b_{lm}, E_a, and E_b are determined by the nature of the source.

The solutions for ${}^2R_{lm}^{(I)}$ and ${}^2R_{lm}^{(II)}$ found in (4.32) and (4.34) are the solution to the homogeneous radial equation (4.29). The solution to the inhomogeneous equation with ${}^2J_{lm}(r) \neq 0$ is therefore

$$ {}^2R_{lm}(r) = {}^2R_{lm}^{(I)}(X)\int \frac{4\pi\left({}^2J_{lm}(\xi)\right)\left({}^2R_{lm}^{(II)}(\xi)\right)}{\xi(\xi-1)\,W\left[\left({}^2R_{lm}^{(I)}\right),\left({}^2R_{lm}^{(II)}\right),\xi\right]}\mathrm{d}\xi$$
$$ -\ {}^2R_{lm}^{(II)}(X)\int \frac{4\pi\left({}^2J_{lm}(\xi)\right)\left({}^2R_{lm}^{(I)}(\xi)\right)}{\xi(\xi-1)\,W\left[\left({}^2R_{lm}^{(I)}\right),\left({}^2R_{lm}^{(II)}\right),\xi\right]}\mathrm{d}\xi\,. \tag{4.43}$$

where $W\left[\left({}^2R_{lm}^{(I)}\right),\left({}^2R_{lm}^{(II)}\right),\xi\right]$ is the Wronskian of ${}^2R_{lm}^{(I)}$ and ${}^2R_{lm}^{(II)}$ at the point ξ. Considering the finite support of the sources between X_1 and X_2 and regularity of the expression (4.41) and (4.42) at the horizon and infinity, respectively, and the constant value of the Wronskian computed asymptotically as $X \to \infty$, we can solve for a_{lm} and b_{lm}:

$$W\left[\left({}^2R_{lm}^{(I)}\right),\left({}^2R_{lm}^{(II)}\right),\xi\right] = \frac{(2l+1)!\ \Gamma(2-2\mathrm{i}Z_m)}{(l+1)!\ \Gamma(l+1-2\mathrm{i}Z_m)}\,. \tag{4.44}$$

$$a_{lm} = -\frac{4\pi(l+1)!\ \Gamma(l+1-2\mathrm{i}Z_m)}{(2l+1)!\ \Gamma(2-2\mathrm{i}Z_m)}\int_{X_1-\varepsilon}^{X_2+\varepsilon}\frac{\left({}^2J_{lm}(\xi)\right)\left({}^2R_{lm}^{(II)}(\xi)\right)}{\xi(\xi-1)}\mathrm{d}\xi\,, \tag{4.45a}$$

$$b_{lm} = -\frac{4\pi(l+1)!\ \Gamma(l+1-2\mathrm{i}Z_m)}{(2l+1)!\ \Gamma(2-2\mathrm{i}Z_m)}\int_{X_1-\varepsilon}^{X_2+\varepsilon}\frac{\left({}^2J_{lm}(\xi)\right)\left({}^2R_{lm}^{(I)}(\xi)\right)}{\xi(\xi-1)}\mathrm{d}\xi\,, \tag{4.45b}$$

where ε is arbitrarily small and positive.

To solve for E_a, and E_b in (4.41b) and (4.42b) we compute Gauss' law in Boyer–Lindquist coordinates for a sphere of radius r_0 with a charge of $Q(r_0)$ enclosed within,

$$\int_0^{2\pi}\int_0^{\pi}\sqrt{-\tilde{g}}\tilde{F}^{tr}\mathrm{d}\theta\mathrm{d}\phi = 4\pi Q(r_0)\,. \tag{4.46}$$

Combining (4.10a) to express $\tilde{F}^{tr}$ in terms of spin coefficients, with (4.41) and (4.42), one finds

$$E_a + \bar{E}_a = Q(r_0)\,, \quad r_0 < r_1\,; \tag{4.47a}$$

$$E_b + \bar{E}_b = Q(r_0)\,, \quad r_0 > r_2\,. \tag{4.47b}$$

Thus, $E_a = 1/2Q$ where Q is the charge on the hole and $E_b = 1/2(Q+e)$ where e is the total charge of the sources.

4.5 The Electrodynamics of the Event Horizon

The electrodynamic properties of the event horizon and the space–time near the horizon are determined by looking at sources of Poisson's equations at small lapse function, $\alpha \to 0$ (i.e., $r \to r_+$). Poisson's and Laplace's equation, as described in Sects. 4.3 and 4.4, do not determine the electrodynamic properties of the horizon. As we learned in the study of quasi-stationary equilibria (and lack thereof) near the horizon (see Chap. 3), one must incorporate both Maxwell's equations as well as the momentum equations of the charges that source Maxwell's equation. Similarly, it is the implementation of the momentum equations for the constituent species near the horizon that reveals the electrodynamic nature of the horizon. Since particle motion is not arbitrary near the horizon (see (3.94) and (3.95)), the currents in Poisson's equation are not either. For example, all particles must be corotating with the horizon to $\mathrm{O}(\alpha^2)$, as seen by external observers. Thus, an arbitrary azimuthal particle drift can not be achieved between species, and the azimuthal current is restricted. Thus, we need to explore the four current density and its tetrad components near the horizon.

4.5.1 Electromagnetic Sources of Poisson's Equations Near the Horizon

The source of the spinorial Poisson's equations is the current J_2 in (4.31). This is a first order differential equation in $J_{\bar{m}}$ and J_n that we expand in terms of

Boyer–Lindquist current density $\tilde{J}^\mu$, using (4.13) and (4.17):

$$J_{\bar{m}} = \bar{m}_\mu \tilde{J}^\mu = \left[\sqrt{2}(r - \mathrm{i}a\cos\theta)\right]^{-1} \times \left[-\mathrm{i}a\cos\theta \tilde{J}^t - \rho^2 \tilde{J}^\theta + \mathrm{i}(r^2+a^2)\sin\theta \tilde{J}^\phi\right] , \tag{4.48a}$$

$$J_n = n_\mu \tilde{J}^\mu = \frac{1}{2}\left[\frac{\Delta}{\rho^2}\tilde{J}^t + \tilde{J}^r - \frac{a\Delta}{\rho^2}\sin^2\theta \tilde{J}^\phi\right] . \tag{4.48b}$$

The decomposition of tetrad currents in Boyer–Lindquist components is useful because each species of charge satisfies the asymptotic (space–time near the horizon) boundary conditions (3.95). Expanding the Boyer–Lindquist current density as we did for the ZAMO current density in (3.74) and (3.78) yields

$$\tilde{J}^\mu = n_+(-e)\tilde{u}^\mu_+ + n_-(e)\tilde{u}^\mu_- , \tag{4.49}$$

where n_+ and n_- are number densities in the frames of the fluids moving with four velocities $\tilde{u}^\mu_+$ and $\tilde{u}^\mu_-$, respectively. From (3.95a), the expansion (4.49) implies

$$\lim_{\alpha\to 0} \tilde{J}^t = \alpha^{-2}\left[J_0^t(\theta,\phi) + j_t(r,\theta,\phi)\right] , \tag{4.50a}$$

where

$$\frac{\partial}{\partial r} J_0^t(\theta,\phi) = 0 , \tag{4.50b}$$

$$\lim_{\alpha\to 0} j_t(r,\theta,\phi) = 0 . \tag{4.50c}$$

Similarly, (3.95b) applied to the expansion (4.49) implies

$$\lim_{\alpha\to 0} \tilde{J}^\phi = \frac{a}{r_+^2+a^2}\tilde{J}^t\left[1 + \mathrm{O}\left(\alpha^2\right)\right] + J_0^\phi(\theta,\phi) + j_\phi(r,\theta,\phi) , \tag{4.51a}$$

where

$$\frac{\partial}{\partial r} J_0^\phi(\theta,\phi) = 0 , \tag{4.51b}$$

$$\lim_{\alpha\to 0} j_\phi(r,\theta,\phi) = 0 . \tag{4.51c}$$

The first term represents the bulk motion of charge corotating with the horizon, since

$$\Omega_H = \frac{a}{r_+^2+a^2} . \tag{4.51d}$$

The second and third terms include the effects of particle drifts between species.

The radial current density decomposes similarly with the aid of (3.95c)

$$\lim_{\alpha\to 0}\tilde{J}^r = -\frac{\Delta}{r^2+a^2}\tilde{J}^t\left[1+\mathrm{O}\left(\alpha^2\right)\right] \\ +\alpha^2\left[J_0^r(\theta,\phi)+j_r(r,\theta,\phi)\right], \tag{4.52a}$$

where

$$\frac{\partial}{\partial r}J_0^r(\theta,\phi)=0, \tag{4.52b}$$

$$\lim_{\alpha\to 0}j_r(r,\theta,\phi)=0. \tag{4.52c}$$

Using (3.17c) and (3.95a), $\tilde{J}^\theta$ is asymptotically given by

$$\lim_{\alpha\to 0}\tilde{J}^\theta = J_0^\theta(\theta,\phi)+j_\theta(r,\theta,\phi), \tag{4.53a}$$

where

$$\frac{\partial}{\partial r}J_0^\theta(\theta,\phi)=0, \tag{4.53b}$$

$$\lim_{\alpha\to 0}j_\theta(r,\theta,\phi)=0. \tag{4.53c}$$

Inserting (4.50)–(4.53) into (4.48) reveals how the horizon boundary condition (3.95) on the constituent charges restricts the asymptotic form of the tetrad currents near the horizon. We write, in anticipation of inserting these expressions into the spinorial source term J_2 of (4.31), the asymptotic tetrad current $J_{\bar{m}}$ as

$$\lim_{\alpha\to 0}(r-\mathrm{i}a\cos\theta)J_{\bar{m}} = J_m^0(\theta,\phi)+j_m(r,\theta,\phi), \tag{4.54a}$$

where,

$$\frac{\partial}{\partial r}J_m^0(\theta,\phi)=0, \tag{4.54b}$$

$$\lim_{\alpha\to 0}j_m(r,\theta,\phi)=0, \tag{4.54c}$$

and the tetrad current, J_n, as

$$\frac{\rho^2(r-\mathrm{i}a\cos\theta)}{\Delta}J_n = J_n^0(\theta,\phi)+j_n(r,\theta,\phi), \tag{4.55a}$$

where,

$$\frac{\partial}{\partial r}J_n^0(\theta,\phi)=0, \tag{4.55b}$$

$$\lim_{\alpha\to 0}j_n(r,\theta,\phi)=0, \tag{4.55c}$$

Equations (4.54) and (4.55) can be used to evaluate the asymptotic form of the source of the spinorial Maxwell's equations, J_2, in (4.31). At this point we will differentiate between axisymmetric fields and nonaxisymmetric fields ($m \neq 0$). This is motivated by the discussion in Sect. 3.7.1 that $m \neq 0$ fields, even in the stationary ($\partial/\partial t = 0$) case, appear as waves to observers near the horizon since they must rotate with the horizon. Consequently, the physical interpretation is very different. Thus, we keep the two types of solutions ($m = 0$ and $m \neq 0$) separate for physical clarity.

There is only one complicated step in the substitution of (4.54) and (4.55) into (4.31). Namely, the $(\partial/\partial r) j_n(r,\theta,\phi)$ term. Expand this term as

$$\lim_{r\to r_+} \Delta \frac{\partial}{\partial r} j_n(r,\theta,\phi) = \lim_{r\to r_+} \left(r - r_-\right)\left(r - r_+\right) \left[\lim_{r\to r'} \frac{j_n(r,\theta,\phi) - j_n(r',\theta,\phi)}{r - r'}\right] . \tag{4.56a}$$

Thus ($r' \to r_+$ and $r \to r_+$ with $r > r'$) we have

$$\lim_{r\to r_+} \Delta \frac{\partial}{\partial r} j_n(r,\theta,\phi) = \left(r_+ - r_-\right)\left[j_n(r,\theta,\phi) - j_n(r',\theta,\phi)\right] = \left(r_+ - r_-\right) j_n(r,\theta,\phi) , \tag{4.56b}$$

where the last term is a consequence of (4.55c).

Consider a distribution of sources to Poisson's equations with an ingoing flow front at $r_1 \gtrsim r_+$. Then, from (4.54)–(4.56) applied to (4.31), we get

$$\lim_{\alpha\to 0} J_2 = \left[\sum_{k=0}^{\infty} \alpha^{2k} J_k(\theta,\phi)\right] j_2(r,\theta,\phi)\Theta\left(r - r_1\right) , \tag{4.57a}$$

where

$$\frac{\partial}{\partial r} J_k(\theta,\phi) = 0 , \tag{4.57b}$$

$$\lim_{\alpha\to 0} j_2(r,\theta,\phi) = 0 , \quad m = 0 \tag{4.57c}$$

$$\lim_{\alpha\to 0} j_2(r,\theta,\phi) \sim \alpha^0 , \quad m \neq 0 \tag{4.57d}$$

and Θ is the Heaviside step function.

Similarly, the integrated source term of the radial equation (4.29), the quantity ${}^2J_{lm}(r)$ defined in (4.30), can be expanded near the event horizon in terms of the radial coordinate X of (4.32b) as

$$\lim_{X\to 1} {}^2J_{lm} = \left[\sum_{k=0}^{\infty} J^0_{lm}(\theta,\phi)(X-1)^k\right] j_{lm}(X,\theta,\phi)\Theta\left(X - X_1\right) , \tag{4.58a}$$

where

$$\frac{\partial}{\partial r} J^0_{lm}(\theta,\phi) = \frac{\partial}{\partial X} J^0_{lm}(\theta,\phi) = 0\,, \tag{4.58b}$$

$$\lim_{X\to 1} j_{lm}(X,\theta,\phi) = 0\,, \quad \text{when } m = 0 \tag{4.58c}$$

$$\lim_{X\to 1} j_{lm}(X,\theta,\phi) \sim X^0\,, \quad \text{when } m \neq 0 \tag{4.58d}$$

$$\lim_{\alpha\to 0} X = 1\,. \tag{4.58e}$$

In the axisymmetric case the source term ${}^2J_{lm}$ actually goes to zero near the horizon. When $m \neq 0$ it goes to a constant. However, this is not the whole story since the expansion coefficients for external fields in (4.42), b_{lm}, also depend on the limits of integration in (4.45b), X_1 and X_2. When this effect is incorporated into the analysis, one introduces gravitational redshift into the physics by means of the freezing of the flow near the horizon.

4.5.2 *External Fields From Electromagnetic Sources Near the Horizon*

4.5.2.1 The Freezing of the Flow

Consider a distribution of electromagnetic sources with a finite proper radial thickness that accretes toward a black hole. We can keep track of the innermost and outermost radial coordinates r_1 and r_2, respectively, during accretion. Equation (3.95c) can be integrated at small lapse function to show that at late times

$$\lim_{t\to\infty} \left(r - r_+\right) = \text{constant} \times \mathrm{e}^{-2\kappa t}\,, \tag{4.59}$$

where κ is the surface gravity of the hole defined in (1.35). Thus,

$$\lim_{t\to\infty} (r_2 - r_1) = \text{constant} \times \mathrm{e}^{-2\kappa t}\,, \tag{4.60a}$$

$$\lim_{t\to\infty} (X_2 - X_1) = \text{constant} \times \mathrm{e}^{-2\kappa t}\,. \tag{4.60b}$$

The constant time cross section of space–time constructed by an external observer makes the radial thickness of an electromagnetic source shrink exponentially in time as it nears the horizon. By (4.59) no external observer ever "sees" any particle or source actually reach horizon. This is known as the "freezing of the flow." Often in wave scattering problems one changes coordinates (see, for example, [99] and [68]) to

$$\mathrm{d}r_* = \frac{r^2 + a^2}{\Delta}\mathrm{d}r\,. \tag{4.61}$$

In this “tortoise coordinate” it appears as if the flow is approaching an infinity at $r_* = -\infty$ instead of stagnating at r_+ in corotation with the hole.

4.5.2.2 No Hair Theorem

The result above will help explain the asymptotic form of the expansion coefficients $b_{lm}(t)$ in (4.42) for sources approaching the hole. Since the sources approach the hole exponentially in the time coordinate by (4.59) as $\alpha \to 0$, we can use Poisson’s equations to show the how the fields of an accreting source die off near the hole (i.e., the time derivatives are negligible in Maxwell’s equations near the hole in Boyer–Lindquist coordinates). In order to evaluate (4.45b) for $b_{lm}(t)$, we need the asymptotic form of the radial function. From the asymptotic form of the hypergeometric function ${}^2y^{(I)}_{lm}$ in (4.34a) as given in [98] and (4.32a)

$$\lim_{X\to 1} {}^2R^{(I)}_{lm} = (X-1)\left[R_0 + \tilde{R}(X)\right] , \tag{4.62a}$$

where,

$$\frac{\partial}{\partial X} R_0 = 0 , \tag{4.62b}$$

$$\lim_{X\to 1} \tilde{R}(X) = 0 . \tag{4.62c}$$

Inserting (4.62) and the asymptotic form of the current source ${}^2J_{lm}(X)$ in (4.58) and (4.54b), we find

$$\lim_{X\to 1} b_{lm}(t) = -\frac{4\pi(l+1)!\ \Gamma(l+1)}{(2l+1)!\ \Gamma(2)} \mathcal{J}(X_2)\left[(X_2 - X_1) + \mathrm{O}\left((X_2 - X_1)^2\right)\right] , \tag{4.63a}$$

where

$$\lim_{X_2\to 1} \mathcal{J}(X_2) = 0 , \quad m = 0 \tag{4.63b}$$

$$\lim_{X_2\to 1} \mathcal{J}(X_2) \sim {X_2}^0 . \quad m \neq 0 \tag{4.63c}$$

Using the gravitational redshift effect in (4.60b), we know that $X_2 \to X_1$ for any accreting finite source. Thus, (4.63) shows that an accreting source that is charge neutral (i.e., $E_b = 0$ in (4.42b)) produces fields that die off as $(X_2 - X_1)$ in general and even faster in the astrophysically interesting case of axisymmetry:

$$\lim_{t\to\infty} b_{lm}(t) \sim \mathrm{e}^{-\kappa t} . \tag{4.64}$$

Equation (4.64) captures the essential physics. However, it shown in [100] and [101] that the full time dependent problem in Schwarzschild geometry is more complicated. There is a second order effect due to backscattering of the radiated fields by the curvature potential of space–time that peaks at $r \approx 3M$. The outgoing waves are not free to radiate to infinity, but backscatter off this centrifuge-like barrier. Thus, the multipole moments die off slower than exponentially. In the Kerr space–time we would expect that $b_{lm}(t) \sim t^{-2l+2}$ as in Schwarzschild space–time. However, this extremely complicated analysis provides no further insight into (4.64) so it is not pursued here.

The result (4.64) is essentially a proof of a no hair theorem for charge neutral ($l \neq 0$) perturbations of a Kerr black hole. The accretion of the $l = 0$ moment adds to the charge in the Kerr–Newman solution.

4.5.2.3 Electromagnetic "Bootstrapping"

In this section we show that a charge neutral electromagnetic source near the horizon can not communicate information through electromagnetic characteristics by a "bootstrap" effect. Namely, we explore the question: as a source approaches the horizon, even though its $l \neq 0$ fields are dying off, can the innermost source still affect a source slightly upstream and in turn can this source effect a third source at a slightly larger r coordinate and so on, until a signal escapes to infinity? The answer is no since the $l = 0$ fields die off over smaller and smaller proper distances as the hole is approached.

We will show this result by introducing a third coordinate X_3. As before, X_1 and X_2 represent the inner and outer coordinate of the electromagnetic source. An infalling probe at X_3 falls in behind the source, $X_3 > X_2 > X_1$, and monitors the fields. Thus, X_3 experiences the solutions to Laplace's equations (4.42) as $X_3 \to 1$. Again, using the asymptotic form of the hypergeometric function from [98] and (4.36)

$$\lim_{X_3 \to 1} {}^0R_{lm}^{(II)}(X_3) \sim (X_3 - 1)^{-1} . \tag{4.65a}$$

Similarly, from (4.34b) and (4.32a),

$$\lim_{X_3 \to 1} {}^2R_{lm}^{(II)}(X_3) \sim {X_3}^0 , \tag{4.65b}$$

and using (4.40b),

$$\lim_{X_3 \to 1} \frac{\mathrm{d}}{\mathrm{d}X}\left({}^2y_{lm}^{(II)}\right) \sim \ln(X_3 - 1) . \tag{4.65c}$$

Inserting the asymptotic expressions (4.63) and (4.65) into the solution of Laplace's equations (4.42) for $l \neq 0$ multipole moments of the fields yields

$$\lim_{X_3\to 1} \Phi_0(X_3) \sim (X_2 - X_1)(X_3 - 1)^{-1} \mathcal{J}(X_2) , \tag{4.66a}$$

$$\lim_{X_3\to 1} \Phi_1(X_3) \sim (X_2 - X_1)\ln(X_3 - 1) \mathcal{J}(X_2) , \tag{4.66b}$$

$$\lim_{X_3\to 1} \Phi_2(X_3) \sim (X_2 - X_1) \mathcal{J}(X_2) . \tag{4.66c}$$

The Boyer–Lindquist Maxwell tensor components can be found from the spin coefficients in (4.26) and the expansions of (4.10). Using the asymptotic expressions (4.66), we have

$$\lim_{X_3\to 1} \tilde{F}^{tr}(X_3) \sim (X_2 - X_1)\ln(X_3 - 1) \mathcal{J}(X_2) , \tag{4.67a}$$

$$\lim_{X_3\to 1} \tilde{F}^{t\theta}(X_3) \sim (X_2 - X_1)(X_3 - 1)^{-1} \mathcal{J}(X_2) , \tag{4.67b}$$

$$\lim_{X_3\to 1} \tilde{F}^{t\phi}(X_3) \sim (X_2 - X_1)(X_3 - 1)^{-1} \mathcal{J}(X_2) , \tag{4.67c}$$

$$\lim_{X_3\to 1} \tilde{F}^{r\phi}(X_3) \sim (X_2 - X_1)\ln(X_3 - 1) \mathcal{J}(X_2) , \tag{4.67d}$$

$$\lim_{X_3\to 1} \tilde{F}^{r\theta}(X_3) \sim (X_2 - X_1) \mathcal{J}(X_2) , \tag{4.67e}$$

$$\lim_{X_3\to 1} \tilde{F}^{\theta\phi}(X_3) \sim (X_2 - X_1)(X_3 - 1)^{-1} \mathcal{J}(X_2) . \tag{4.67f}$$

It is not straightforward to interpret the results in (4.67) since the coordinates are not orthonormal, i.e., $\tilde{F}^{tr} \neq \tilde{F}_{tr}$. The physical significance of these quantities is revealed by computing the coordinate independent fluxes.

First consider the electric fluxes defined by the relation

$$\Phi_{E} \equiv \int *\tilde{F}_{\alpha\beta}\, \mathrm{d}X^{\alpha} \wedge \mathrm{d}X^{\beta} . \tag{4.68}$$

The radial electric flux through a surface at X_3 is

$$\Phi_{E}(\tilde{F}^{rt}) = \int_{\phi_1}^{\phi_2} \int_{\theta_1}^{\theta_2} \tilde{F}^{tr} \sqrt{-\tilde{g}}\, \mathrm{d}\theta \wedge \mathrm{d}\phi . \tag{4.69a}$$

The asymptotic expression in (4.67a) inserted into (4.69a) implies

$$\lim_{X_3\to 1} \Phi_{E}(\tilde{F}^{rt}) \sim (X_2 - X_1)\ln(X_3 - 1) \mathcal{J}(X_2) \to 0 ,$$

$$X_3 > X_2 > X_1 > 1 . \tag{4.69b}$$

The amount of flux through an arbitrary surface from the radial electric field, $\Phi_{E}(\tilde{F}^{rt})$, goes to zero as the infalling probe approaches the horizon. Clearly, the fields from the electromagnetic sources are dying off over smaller and smaller proper distances as $X_1 \to 1$.

Now consider the azimuthal electric flux in the region between the infalling source and the infalling probe near the horizon, $\Phi_E(\tilde{F}^{\phi t})$,

$$\Phi_E(\tilde{F}^{\phi t}) = \int_{\theta_1}^{\theta_2} \int_{r_2}^{r_3} \tilde{F}^{\phi t} \sqrt{-\tilde{g}}\, \mathrm{d}r \wedge \mathrm{d}\theta \,. \tag{4.70a}$$

The flux in (4.71a) is evaluated through an arbitrary surface at a constant ϕ coordinate. From (4.67c)

$$\lim_{X_3 \to 1} \Phi_E(\tilde{F}^{\phi t}) \sim (X_2 - X_1) \ln (X_3 - X_2) \mathcal{J}(X_2) \to 0 \,, \quad X_3 > X_2 > X_1 > 1 \,. \tag{4.70b}$$

Since the flux between the source and the probe is decreasing by (4.70b) we conclude that the electromagnetic fields are dying off over smaller and smaller proper distances.

Similarly, we compute the electric flux $\Phi_E(\tilde{F}^{\theta t})$,

$$\Phi_E(\tilde{F}^{\theta t}) = \int_{\phi_1}^{\phi_2} \int_{r_2}^{r_3} \tilde{F}^{\theta t} \sqrt{-\tilde{g}}\, \mathrm{d}r \wedge \mathrm{d}\phi \,. \tag{4.71a}$$

$$\lim_{X_3 \to 1} \Phi_E(\tilde{F}^{\theta t}) \sim (X_2 - X_1) \ln (X_3 - X_2) \mathcal{J}(X_2) \to 0 \,, \quad X_3 > X_2 > X_1 > 1 \,. \tag{4.71b}$$

All of the electric fields in Maxwell's equations are dying off faster and faster as the sources approach the horizon. In general, one can write from (4.64) to (4.71) using (4.60b), for the flux evaluated at X_3, $\Phi_E(X_3)$,

$$\lim_{t \to \infty} \Phi_E(X_3) \sim \text{constant} \times (t - t_0)\, \mathrm{e}^{-\kappa t} \,. \tag{4.72}$$

Now consider the magnetic flux Φ_B defined by

$$\Phi_B \equiv \int \tilde{F}_{\alpha\beta}\, \mathrm{d}X^\alpha \wedge \mathrm{d}X^\beta \,. \tag{4.73}$$

From (4.67e) and (1.24) we have

$$\tilde{F}_{r\theta} \sim (X_2 - X_1)(X_3 - 1)^{-1} \mathcal{J}(X_2) \,. \tag{4.74a}$$

Computing the azimuthal magnetic flux as in (4.70a)

$$\Phi_B(\tilde{F}_{r\theta}) = \int_{\theta_1}^{\theta_2} \int_{r_2}^{r_3} \tilde{F}_{r\theta}\, \mathrm{d}r \wedge \mathrm{d}\theta \,, \tag{4.74b}$$

$$\lim_{X_3 \to 1} \Phi_B(\tilde{F}_{r\theta}) \sim (X_2 - X_1) \ln (X_3 - X_2) \mathcal{J}(X_2) \to 0 \,, \quad X_3 > X_2 > X_1 > 1 \,. \tag{4.74c}$$

In order to compute the poloidal magnetic flux we need the covariant components of the Maxwell field, $\tilde{F}_{\theta\phi}$ and $\tilde{F}_{r\phi}$ as indicated in (4.73). These are linear combinations of contravariant electric and magnetic fields in (4.67). We can greatly facilitate this calculation if we note that (4.10), (4.42) and (4.63) imply

$$\lim_{X_3\to 1} \frac{\tilde{F}^{\phi\theta}(X_3)}{\tilde{F}^{\theta t}(X_3)} \sim \frac{\tilde{F}^{\phi r}(X_3)}{\tilde{F}^{rt}(X_3)} = \Omega_H + \mathrm{O}\left[(X_2 - X_1)\ln(X_3 - 1)\,\mathcal{J}(X_2)\right] . \tag{4.75a}$$

Thus, (1.24) and (4.67) imply

$$\tilde{F}_{\theta\phi} \sim \tilde{F}_{r\phi} \sim (X_2 - X_1)\ln(X_3 - 1)\,\mathcal{J}(X_2) . \tag{4.75b}$$

We can compute the radial magnetic flux in analogy to (4.69a),

$$\Phi_B(\tilde{F}_{\theta\phi}) = \int_{\phi_1}^{\phi_2}\int_{\theta_1}^{\theta_2} \tilde{F}_{\theta\phi}\, \mathrm{d}\theta \wedge \mathrm{d}\phi , \tag{4.76a}$$

$$\lim_{X_3\to 1} \Phi_B(\tilde{F}_{\phi\theta}) \sim (X_2 - X_1)\ln(X_3 - 1)\,\mathcal{J}(X_2) , \quad X_3 > X_2 > X_1 > 1 . \tag{4.76b}$$

Similarly

$$\lim_{X_3\to 1} \Phi_B(\tilde{F}_{r\phi}) \sim (X_2 - X_1)\ln(X_3 - 1)\,\mathcal{J}(X_2) , \quad X_3 > X_2 > X_1 > 1 . \tag{4.77a}$$

It is instructive to incorporate the results (4.74), (4.76) and (4.77) into one expression as we did in (4.72). The magnetic flux at X_3 decays in time as

$$\lim_{t\to\infty} \Phi_B(X_3) \sim \text{constant} \times (t - t_0)\,\mathrm{e}^{-\kappa t} . \tag{4.78}$$

Equations (4.72) and (4.78) imply that $l \neq 0$ moments of the Maxwell field die off over smaller and smaller proper distances as an electromagnetic source approaches the horizon. Thus, even in a continuous flow, sources near the horizon can not initiate a chain reaction upstream (the "bootstrap" effect) to communicate a signal upstream. The sources are causally disconnected from the asymptotic region of space–time at $r \gg r_+$. Equations (4.72) and (4.78) indicate that the event horizon is an asymptotic infinity to charge neutral ($l \neq 0$) electromagnetic sources. This is substantiated by the freezing of the flow condition (4.60) where external observers see sources always approaching but never reaching the horizon. In the language of (4.61), the sources are approaching an infinity at $r_* = -\infty$. As they approach this asymptotic infinity, the fields they produce die off according to (4.72) and (4.73). This is the relevant (causal) horizon electromagnetic boundary condition.

The analysis can be extended to the case of $a = M$. This causes a confluence of singular points as $r_- \to r_+$. This solution is more of mathematical interest than physical interest since maximal rotation, $a = M$, is never attained in practice. The appropriate radial functions and spin components of the Maxwell field are solved in [90] and [102]. The freezing of the flow condition (4.59) changes as well into $(r-r_{+}) \sim 1/(t-t_0)$. However, the physics of the no hair theorem is the same as when $a \neq M$.

4.6 Simple Solutions to Laplace's Equations

4.6.1 The Kerr Newman Solution

The $l=0$ moment of the electromagnetic field is the Kerr–Newman field. In (4.47b), it was shown that the accretion of a source with charge e made the constant $E_b = (1/2)e$ in the vacuum solution (4.42). As the charge at r_1 approaches the horizon for $r > r_1 \gtrsim r_{+}$, by (4.63), (4.64) and (4.42)

$$\lim_{r_1 \to r_+} \Phi_0(r) = 0\,, \qquad r > r_1 \tag{4.79a}$$

$$\lim_{r_1 \to r_+} \Phi_1(r) = \frac{1}{2}\frac{e}{(r - \mathrm{i}a\cos\theta)^2}\,, \qquad r > r_1 \tag{4.79b}$$

$$\lim_{r_1 \to r_+} \Phi_2(r) = 0\,, \qquad r > r_1\,. \tag{4.79c}$$

Inserting (4.79) into (4.10) and using the metric (1.24)

$$\tilde{F}_{\mu\nu} = e\rho^{-4}\left(r^2 - a^2\cos^2\theta\right)\,\mathrm{d}r \wedge \left[\mathrm{d}t - a\sin^2\theta\,\mathrm{d}\phi\right] \\ -2e\rho^{-4} a r\cos\theta\sin\theta\,\mathrm{d}\theta \wedge \left[a\,\mathrm{d}t - \left(r^2+a^2\right)\mathrm{d}\phi\right]\,. \tag{4.80}$$

Using (3.3) we transform the Maxwell tensor components into the orthonormal ZAMO frame

$$E^r = \frac{e\left(r^2+a^2\right)\left(r^2 - a^2\cos^2\theta\right)}{\rho^4\left[\left(r^2+a^2\right)^2 - a^2\Delta\sin^2\theta\right]^{1/2}}\,, \tag{4.81a}$$

$$E^\theta = -\frac{2ea^2 r\Delta^{1/2}\cos\theta\sin\theta}{\rho^4\left[\left(r^2+a^2\right)^2 - a^2\Delta\sin^2\theta\right]^{1/2}}\,, \tag{4.81b}$$

$$E^\phi = 0\,, \tag{4.81c}$$

$$B^r = \frac{2ear\cos\theta\left(r^2+a^2\right)}{\rho^4\left[\left(r^2+a^2\right)^2 - a^2\Delta\sin^2\theta\right]^{1/2}}\,, \tag{4.81d}$$

$$B^\theta = \frac{ea\Delta^{1/2}\left(r^2 - a^2\cos^2\theta\right)\sin\theta}{\rho^4\left[\left(r^2+a^2\right)^2 - a^2\Delta\sin^2\theta\right]^{1/2}}\,, \tag{4.81e}$$

$$B^\phi = 0\,. \tag{4.81f}$$

The lines of force can be found in the ZAMO frames are given in [103] in terms of the ZAMO four velocity, u^μ,

$$\frac{dX^\mu}{d\lambda} = -*F^\mu{}_\nu u^\nu\ , \quad \text{magnetic} \tag{4.82a}$$

$$\frac{dX^\mu}{d\lambda} = F^\mu{}_\nu u^\nu\ , \quad \text{electric.} \tag{4.82b}$$

The lines of force in the ZAMO frame can be found from (4.81) and (4.82). The electric lines of force are given by

$$\left(r^2 - a^2\cos^2\theta\right)\left(r^2 + a^2\right)\, d\theta + 2a^2 r\cos\theta\sin\theta\, dr = 0\ , \tag{4.83a}$$

and the magnetic lines of force are parameterized by

$$2r\cos\theta\left(r^2 + a^2\right)\, d\theta - \left(r^2 - a^2\cos^2\theta\right)\sin\theta\, dr = 0\ . \tag{4.83b}$$

These lines of force are tangent to the local magnetic and electric field at every point and are lines of constant flux [104, 105]. The lines of force were plotted in [106] and are reproduced here in Fig. 4.1. Asymptotically, the electromagnetic field is that of an electric monopole and a magnetic dipole. Even though the vacuum field solutions in (4.42) were derived for a Kerr black hole background, they can be applied to Kerr–Newman backgrounds as well, even for substantial charge Q on the hole. In essence, we can ignore the effects of space–time curvature induced by the stress-energy tensor of the Maxwell field (i.e., ignore the right hand side of Einstein's equations) even for huge astronomical charges. Inspection of the metric in (1.24) shows that for a $10\,M_\odot$ black hole, a polar magnetic field strength of 10^{15} G at the horizon of a rapidly rotating black hole, $a^2 \lesssim M^2$, produces changes to the Kerr metric on the order of $Q^2/M^2 \approx 10^{-6}$. Thus, the results of Sect. 4.5 are unchanged for astrophysically reasonable charges on the hole and the fields in (4.42) differ from the Kerr case in that $E_b = (Q+e)/2$, i.e., only the Kerr–Newman ($l = 0$) moment of the fields differs from the Kerr case.

4.6.2 The Wald Solution

The next simplest solution is the $l = 0$, $m = 0$ solution found originally by Wald [107]. This field is uniform, magnetic and aligned with the rotation axis of the hole at infinity. The electromagnetic field in the ZAMO frames is

$$B^r = \frac{B_0}{A^{\frac{1}{2}}\rho^4}\left\{\left(r^2+a^2\right)\left[\left(r^2-a^2\right)\left(r^2-a^2\cos^2\theta\right)\right.\right.$$
$$\left.\left. +2a^2 r(r-M)\left(1+\cos^2\theta\right)\right] - a^2\Delta\rho^2\sin^2\theta\right\}\cos\theta\ , \tag{4.84a}$$

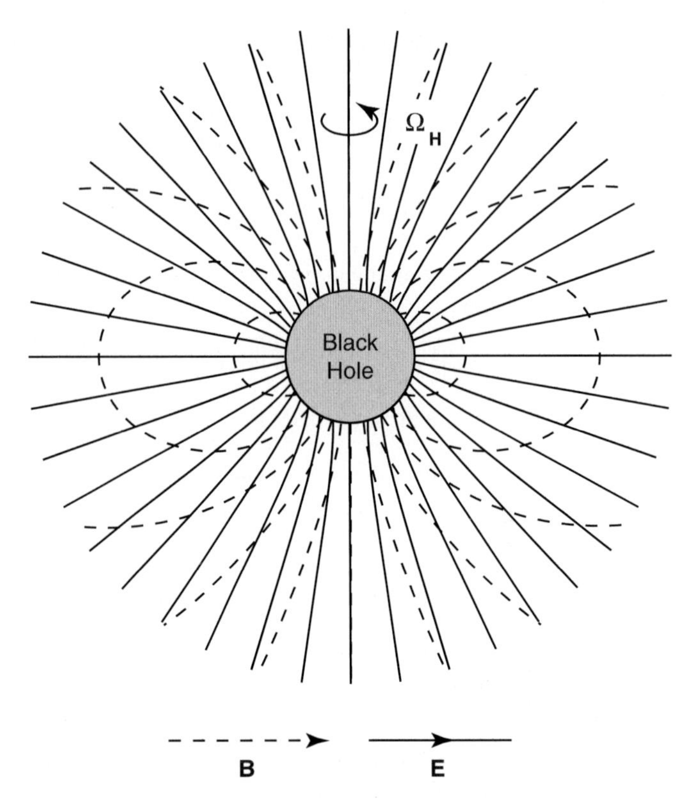

Fig. 4.1 The electric and magnetic field lines of a Kerr–Newman black hole adapted from [104], equivalently this is the $l=0,\ m=0$ moment of the electromagnetic field in the Kerr space–time. Asymptotically, the field is that of an electric monopole and a magnetic dipole. The polar magnetic field lines are obscured by the overlapping polar electric field lines. Note the ubiquitous condition $*F^{\mu\nu}F_{\mu\nu}\neq 0$ that is found in vacuum solutions to Maxwell's equations in the Kerr space–time

$$B^{\theta} = -\frac{B_0\Delta^{\frac{1}{2}}}{A^{\frac{1}{2}}\rho^4}\left\{a^2\left[2r\left(r^2-a^2\right)\cos^2\theta \right.\right.$$
$$\left.-(r-M)\left(r^2-a^2\cos^2\theta\right)\left(1+\cos^2\theta\right)\right]$$
$$\left.+\left(r^2+a^2\right)\rho^2 r\right\}\sin\theta\ , \tag{4.84b}$$

$$E^{r} = -\frac{B_0 a}{A^{\frac{1}{2}}\rho^4}\left\{\left(r^2+a^2\right)\left[2r\left(r^2-a^2\right)\cos^2\theta \right.\right.$$
$$\left.-(r-M)\left(r^2-a^2\cos^2\theta\right)\left(1+\cos^2\theta\right)\right]$$
$$\left.+r\Delta\rho^2\sin^2\theta\right\}\ , \tag{4.84c}$$

$$E^{\theta} = -\frac{B_0 a\Delta^{\frac{1}{2}}}{A^{\frac{1}{2}}\rho^4}\left[\left(r^2-a^2\right)\left(r^2-a^2\cos^2\theta\right)\right.$$
$$\left.+2a^2 r(r-M)\left(1+\cos^2\theta\right)-\left(r^2+a^2\right)\rho^2\right]\cos\theta\sin\theta\ , \tag{4.84d}$$

$$B^{\phi} = E^{\phi} = 0 \,, \tag{4.84e}$$

$$A = \left(r^2 + a^2\right)^2 - \Delta a^2 \sin^2\theta \,, \tag{4.84f}$$

where B_0 is the strength of the uniform magnetic field at infinity.

The solution bears a superficial resemblance to the vacuum electromagnetic field of a conductive sphere rotating in a uniform magnetic field [108]. Firstly, $*F^{\mu\nu}F_{\mu\nu} \neq 0$, i.e., there is a component of the electric field parallel to the magnetic field. Secondly, there is a voltage drop across the magnetic field lines. The voltage drop is a global concept so this needs to be evaluated in a global coordinate system. In the stationary frames (Boyer–Lindquist coordinates) this voltage drop at the horizon is manifested by the relation

$$\lim_{r\to r_+} \tilde{F}_{t\theta} = \frac{\Omega_H}{c} \lim_{r\to r_+} \tilde{F}_{\phi\theta} \,. \tag{4.85}$$

The voltage drop across the magnetic field at the horizon, ΔV_+, is given in terms of the magnetic flux at the horizon, Φ_+, as defined by (4.73) and (4.85),

$$\Delta V_+ (\theta_2, \theta_1) = \frac{\Omega_H}{c} \left[\Phi_+ (\theta_2) - \Phi_+ (\theta_1)\right] \,. \tag{4.86}$$

The voltage drop between θ_2 and θ_1 depends on the amount of flux threading the horizon between θ_2 and θ_1, $\Phi_+ (\theta_2) - \Phi_+ (\theta_1)$. In the same spirit, Thorne, Price and Macdonald [80] make the analogy of the event horizon with a rotating conductor by suggesting that one could terminate the radial electric field at the horizon with a fictitious quadrupolar surface charge density on the horizon, σ_+. (Clearly, there can be no physical charges on the horizon which is a null hypersurface).

$$\sigma_+ = \frac{B_0 a r_+ \left(r_+ - M\right)}{4\pi} \; \frac{r_+ \sin^4\theta - 2M\cos^2\theta \left(1 + \cos^2\theta\right)}{\left(r_+ + a^2\cos^2\theta\right)^2} \,. \tag{4.87}$$

The Wald electric and magnetic field lines in Boyer–Lindquist coordinates are indicated diagrammatically in Fig. 4.2. From (4.84c), the electric field dies off with radius as $1/r^2$ instead of $1/r^4$ as in a flat space quadrupole:

$$\lim_{r\to\infty} E^r = -\frac{B_0(Ma)}{r^2} \left(3\cos^2\theta - 1\right) \,. \tag{4.88}$$

This indicates that there is a fundamental difference between the induced (by frame dragging of the magnetic field) electric field of the Wald solution and the electrostatically sourced quadrupole field from a rotating conductor.

Before leading too much credence to the analogy of a conductor and an event horizon, we critique this idea in the next section. Clearly, a conductor boundary condition is generally inconsistent with the vacuum infinity aspect of the horizon in the no hair theorem.

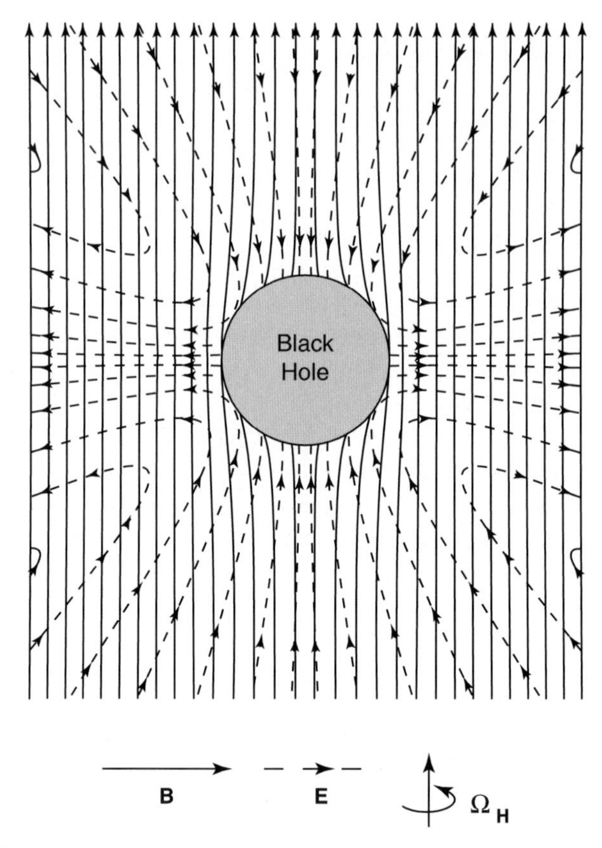

Fig. 4.2 The electromagnetic field associated with the magnetic flux threading the event horizon of a Kerr black hole, in the $\ell = 1$, $m = 0$ moment of the electromagnetic field. The field lines are indicated in Boyer–Lindquist coordinates, *dash* (*solid*) *lines* represent the electric (magnetic) field. The black hole rotates about the vertical axis [112]

One should note that the Wald field is the unique stationary axisymmetric solution of Maxwell's equations that is well behaved on both the horizon and asymptotic infinity [96]. From (4.21), (4.34) and (4.36) we have the following asymptotic forms of the radial functions

$$\lim_{X\to 1} {}^{2}R^{(I)}_{lm} \sim (X-1)\left[1-\frac{1}{X}\right]^{\mathrm{i}Z_m}, \tag{4.89a}$$

$$\lim_{X\to 1} {}^{0}R^{(I)}_{lm} \sim (X-1)^{-1}\left[1-\frac{1}{X}\right]^{\mathrm{i}Z_m}, \qquad m \neq 0 \tag{4.89b}$$

$$\lim_{X\to 1} {}^{0}R^{(I)}_{lm} \sim \text{constant}, \qquad m = 0 \tag{4.89c}$$

$$\lim_{X\to 1} {}^{2}R^{(II)}_{lm} \sim \text{constant}, \tag{4.89d}$$

$$\lim_{X\to 1} {}^0R^{(II)}_{lm} \sim (X-1)^{-1}\left[1-\frac{1}{X}\right]^{iZ_m}, \tag{4.89e}$$

$$\lim_{X\to\infty} {}^2R^{(I)}_{lm} \sim X^{l+1}, \tag{4.89f}$$

$$\lim_{X\to\infty} {}^0R^{(I)}_{lm} \sim X^{l-1}, \tag{4.89g}$$

$$\lim_{X\to\infty} {}^2R^{(II)}_{lm} \sim X^{-l}, \tag{4.89h}$$

$$\lim_{X\to\infty} {}^0R^{(II)}_{lm} \sim X^{-l-2}. \tag{4.89i}$$

To keep the magnetic flux at the horizon finite (see (4.73)), we require $\Phi_1 \sim (X-1)$ and $\Phi_0 \sim$ (constant) by (4.10f) and the metric (1.24). Thus, only the solutions $R^{(I)}_{lm}$ are allowable at the horizon. The only asymptotic multipole of this solution at $X \to \infty$ that does not diverge is $l = 1$ in (4.89g), the Wald field. Consequently, this regularity condition implies that external axisymmetric sources create predominately the $l = 1$ moment of the electromagnetic field near the horizon [96].

4.6.3 Axisymmetric Time Stationary Fields

The $m = 0$ approximation creates some simplification to the solutions of Laplace's equation. Using $m = 0$ in (4.33), which defines the radial functions used in the solutions to Laplace's equation, (4.42), yields

$$\phi_0 = -\frac{2}{\Delta(r - ia\cos\theta)^2}\phi_2, \quad m = 0. \tag{4.90a}$$

Inserting this result into the Boyer–Lindquist field expansion (4.10) implies

$$\tilde{F}^{t\phi} = \tilde{F}^{r\theta} = 0, \quad m = 0. \tag{4.90b}$$

Thus, there can be no poloidal Poynting flux in the field. Therefore, axisymmetric vacuum solutions to Laplace's equation do not extract energy from a rotating black hole.

One can also compute the magnetic flux through a hemisphere of the event horizon using (4.73). We noted in the last section that setting $l = 1$ and $m = 0$ will yield the dominant field component at the horizon. Performing the integral using (4.84a) for B^r of the Wald field gives

$$\Phi_B(\text{hemisphere}) = \pi r_+^2 B_0 \left[1 - \left(a/r_+\right)^4\right]. \tag{4.90c}$$

Rapidly rotating black holes exclude axisymmetric magnetic flux. The same is not true of nonaxisymmetric magnetic flux [102].

4.7 The Horizon Electromagnetic Boundary Condition

4.7.1 Displacement Currents at the Ingoing Flow Front

Consider the ingoing wave and flow boundary conditions near the horizon as given by (3.94) and (3.95). Combining this result with the "freezing of flow" conditions (4.59), we conclude that all fields and particles appear to be approaching the horizon exponentially slowly as if it were an asymptotic infinity for all external observers. Thus, there is always a flow front or wave front just outside of the horizon as viewed globally. There is a change in field strengths across the inward propagating flow front, hence there is displacement current at the head of the inflow.

The magnetic field is well behaved in the frame of the ingoing particles, which requires that (3.90) hold (i.e., the ingoing wave condition). Using (3.90) in Ampere's law (3.53b) and (3.60b)

$$\frac{\partial}{\partial X^r}\left(\alpha\sqrt{g_{\phi\phi}}F^{\theta r}\right) \approx -\frac{\partial}{\partial X^0}\left(\alpha\sqrt{g_{\phi\phi}}F^{\theta 0}\right), \tag{4.91a}$$

$$\frac{\partial}{\partial X^r}\left(\alpha\sqrt{g_{\theta\theta}}F^{\phi r}\right) \approx -\frac{\partial}{\partial X^0}\left(\alpha\sqrt{g_{\theta\theta}}F^{\phi 0}\right). \tag{4.91b}$$

Therefore, the magnetic field near the horizon in the ZAMO frames is created by displacement currents, not physical currents, as a result of the inward ultrarelativistic boundary condition (3.94).

We expand (4.91) as we did for the semi-infinite Faraday wheel terminated transmission line in Sect. 2.9.4 in terms of step functions and the functions RB^ϕ, RE^θ, RB^θ and RE^ϕ that are independent of the coordinates X^0 and X^r to $\mathrm{O}\left(\alpha^2\right)$:

$$\alpha\sqrt{g_{\phi\phi}}F^{\theta r} \equiv RB^\phi\ \Theta\left[VX^0 - X^r\right], \tag{4.92a}$$

$$\alpha\sqrt{g_{\phi\phi}}F^{\theta 0} \equiv RE^\theta\ \Theta\left[VX^0 - X^r\right], \tag{4.92b}$$

$$\alpha\sqrt{g_{\theta\theta}}F^{\phi r} \equiv RB^\theta\ \Theta\left[VX^0 - X^r\right], \tag{4.92c}$$

$$\alpha\sqrt{g_{\theta\theta}}F^{\phi 0} \equiv RE^\phi\ \Theta\left[VX^0 - X^r\right], \tag{4.92d}$$

where $V \equiv \mathrm{d}X^r/\mathrm{d}X^0 = \omega^r/\omega^0$ as a result of (3.94b). We evaluate (4.92) as the flow front passes by the ZAMO. At an instant later, we can define accurately for small displacements

$$X^r \equiv \int \mathrm{d}X^r \equiv \int \omega^r, \tag{4.93a}$$

$$X^0 \equiv \int \mathrm{d}X^0 \equiv \int \omega^0. \tag{4.93b}$$

Equations (4.93ab) represent a local coordinate system. The basis vectors of the coordinate system are aligned with the legs of the ZAMO tetrad at a point near the flow front. The origin of time is chosen so that the flow front passes this point at $X^0 = 0$. We examine the flow an instant after it passes this point. Clearly, for small enough times, X^0, the flow front will be contained in an open set in which the legs of this coordinate system can be considered orthonormal to an excellent approximation. Denote differential operators in the coordinate system by the same symbols used in (3.25). One has the coordinate vector field condition

$$\left[\frac{\partial}{\partial X^i}, \frac{\partial}{\partial X^j}\right] = 0\,, \quad \forall i, j\,, \tag{4.94}$$

since the coefficients of $\partial/\partial r$, $\partial/\partial\phi$, $\partial/\partial\theta$ and $\partial/\partial t$ in (3.25) are chosen to be constants on the open set. By choosing X^0 small enough, any desired accuracy in Ampere's law, (3.53) and (3.60), can be achieved. With the mathematical rigor now established, we insert the expression (4.92) into (4.91) to find the displacement current at the flow front

$$J_D^\theta = -\frac{c}{4\pi}F^{\theta r}\delta\left(VX^0 - X^r\right)\,, \tag{4.95a}$$

$$J_D^\phi = -\frac{c}{4\pi}F^{\phi r}\delta\left(VX^0 - X^r\right)\,. \tag{4.95b}$$

This is the source of the poloidal magnetic field in the flow front.

As in (2.106a) we define a surface displacement current by integrating across the flow front

$$\mathcal{J}_D^\mu = \lim_{\varepsilon\to 0}\int_{VX^0+\varepsilon}^{VX^0-\varepsilon} J_D^\mu \,\mathrm{d}X^r\,. \tag{4.96}$$

Then, using (3.90) and (4.95a) inserted into (4.96), we obtain

$$\frac{4\pi}{c}\mathcal{J}_D^i = E^i\,. \tag{4.97}$$

Thus, we can associate with the vacuum a surface impedance for displacement currents,

$$Z_D = \frac{4\pi}{c}\,, \tag{4.98}$$

the same value as in (2.106c). Since a vacuum space–time has the same electrodynamic structure regardless of location, it is not surprising that the equivalence principle yields the same result here as it does in the distant vacuum of the semi-infinite transmission line discussed in Sect. 2.9.4. Displacement current at an asymptotic flow front is not a causative agent for global physics and is not of much physical significance in a dynamical sense. One could use IZ_D^2 power to describe a vacuum infinity as a resistor. However, this just represents Poynting flux radiation to infinity by a relativistic flow or light wave using Ampere's law. Representing a nonconductive vacuum space–time infinity as an imperfect conductor is not very useful and is in fact misleading. Many authors ([80, 109, 110]) have tried to describe the horizon as an imperfect conductor with a surface impedance of $Z = 4\pi/c$. However, this is clearly inconsistent with the no hair theorem calculated in Sect. 4.5 which states that the horizon is formally a vacuum infinity to charge neutral electromagnetic sources. The horizon can not radiate $l \neq 0$ moments of the Maxwell field as can a true conductor. The Faraday wheel at the end of a semi-infinite transmission line was discussed in detail in Sect. 2.9.4 precisely to make this contrast.

4.7.2 The Horizon as a Circuit Element

In this section we strengthen the analogy of the horizon to an infinity in a transmission line circuit. Consider a generalized battery as shown in Fig. 4.3. The battery pumps charges toward the horizon. The charges are allowed to free fall inward. To simplify matters, one can choose an electron–positron plasma in this gedanken experiment. Given a high enough plasma density in a positronic plasma, a four current can be generated by particle drifts where the difference in injected constants

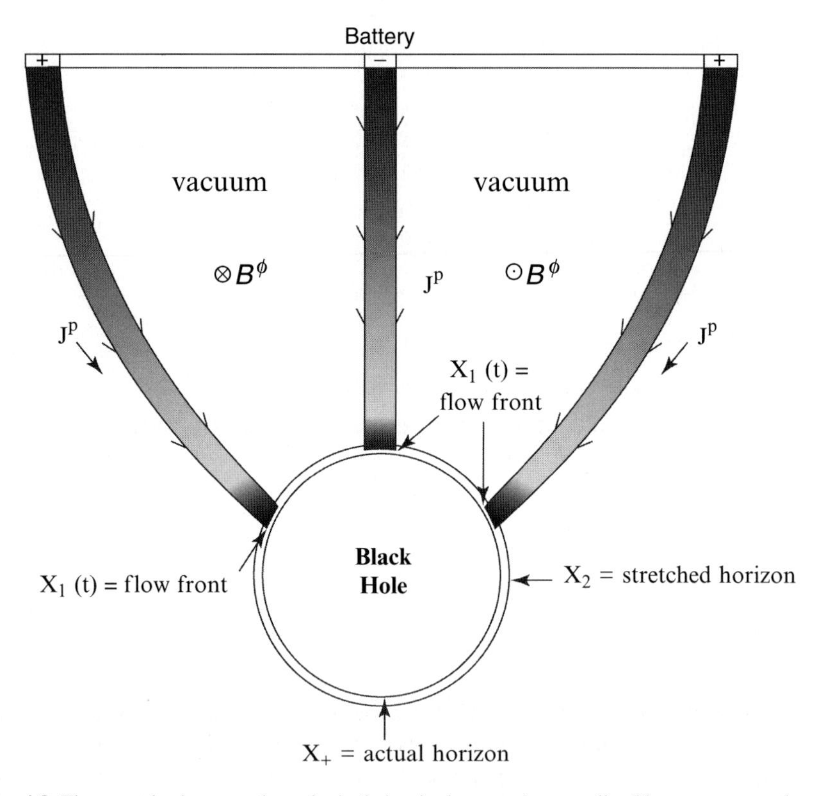

Fig. 4.3 The event horizon as a hypothetical circuit element. A generalized battery pumps electromagnetic sources from large "r" coordinate towards the black hole. External observers never see the flows pumped by the battery ever reach the event horizon. The horizon is physically disconnected from any electrical contact that one tries to make. The horizon is also causally disconnected and is electrodynamically equivalent to an infinity in a transmission line in the hypothetical circuit

of motion (see Sect. 3.3) is virtually the same for each species. Thus, the inflow can maintain an essentially charge neutral character as the two species spiral inward along virtually identical trajectories (as in Figs. 3.1 and 3.2).

The battery has two terminals. The central terminal pumps charges toward the pole. The outer terminal is cylindrical and these charges spiral inward as in Figs. 3.1 and 3.2. They approach the horizon with a small spread in latitude just above the equator. The outer cylindrical current path is seen in cross section in Fig. 4.3. The generalized battery concept is quite versatile. For example, by choosing an alternating array of positron and electron guns, a drift entirely in the azimuthal direction can be created within the plasma, (i.e., $J^{\phi} \neq 0$; all other components equal zero; $J^{\mu} = 0$ for $\mu \neq \phi$). This current ring can be allowed to accrete intact as the quantum numbers of the flow can be chosen to be virtually the same for both species for large number densities.

Consider the simplest experiment as shown in Fig. 4.3. Two oppositely directed equal magnitude poloidal currents, J^P, flow from the battery terminals. The density decreases along the inflows as gravity accelerates the particles. This is indicated by the shading intensity decreasing inward of the battery. Near the horizon, the shading intensity increases due to the freezing of the flow. The radial currents create a B^{ϕ} in the vacuum all the way to the flow front at $X_1(t)$ where the displacement current switches off B^{ϕ} (see 4.95a). In the "membrane paradigm" [80], the displacement current is replaced by a fictitious surface current on a spacelike stretched horizon, $X_2 > X_1(t)$. The vacuum infinity approach and the conductive membrane approach agree when the horizon is passively accepting information.

It is interesting to note that by the no hair theorem, B^{ϕ} is sourced near the horizon by inward and sideways directed vacuum electromagnetic characteristics from the current source J^P, not by the displacement current which carries only ingoing information.

It is also instructive to realize that the battery deposits one sign of charge near the pole and the other near the equator. The Kerr–Newman $l = 0$ moment of the field manifests itself the same whether it is accreted at the pole or $\theta \neq 0$. Thus, between $X_1(t)$ and X_+ there is a radial electrical field. Outside of $X_1(t)$, however, the Kerr–Newman pieces from the two electrodes cancel.

4.7.3 The Horizon is not a Conductor

A superficial resemblance to a rotating conductor in a uniform magnetic field was noted in connection with the fictitious quadrupolar horizon surface charge density on the horizon in (4.87) for the Wald field. We show that this is a poor analogy in this section.

Consider a battery attached to a spherical laboratory conductor with axisymmetric electrodes at θ_1 and θ_2. A current J_{θ} flows between θ_1 and θ_2 on the conductor. This current J_{θ} is a source for B^{ϕ} by the Biot–Savart law. In the black hole case consider a large amount of J^θ deposited inside of a stretched horizon. No matter how large we make J^θ, it is not a source for an external B^ϕ by the no hair theorem.

Of more significance astrophysically, consider a spherical shell of magnetized gas that accretes onto a star. The shell contains an azimuthal current J^ϕ that supports the magnetic field. The stellar surface is a good conductor (particle drifts in the solar corona can persist for the lifetime of the Sun based solely on its high conductivity and ignoring plasma instabilities). When the shell reaches the star, the magnetic flux does not die off as long as the particle drifts exist between species on the stellar surface.

Now consider an azimuthal current source created by the generalized battery in Fig. 4.3 as described in the last section. Note that we only need one terminal of the battery to accomplish this. After a finite time the battery is turned off.

At late times, there will always be a particle drift between species in the plasma, u^ϕ_D, as a consequence of (3.94ac),

$$u^\phi_D = u^\phi_+ - u^\phi_- \sim \alpha\,. \tag{4.99}$$

However, there is no detectable magnetic field upstream as a result of the no hair theorem as given by (4.79). Thus, the horizon does not behave like a good conductor.

Ascribing a surface impedance of $4\pi/c$ to the horizon does not tell one if it is a good conductor or a poor conductor. We contrast the horizon with a poor conductor. In this instance particle collisions quickly absorb any differences in momenta between the species, killing off particle drifts in the accreting plasma. This property of a poor conductor contradicts (4.99).

The astrophysically interesting property of a conductor is that it supports $l \neq 0$ moments of the electromagnetic field. This is precisely what the horizon can not do, so the conductor analogy is misleading in cases of actual interest.

4.7.4 The Absence of Unipolar Induction Near the Horizon

Astrophysically, the most interesting electrodynamic question of the space–time near the event horizon is whether or not it can behave as a unipolar inductor of Faraday wheel. The authors of [80] have tried to make such a comparison in an effort to physically justify the Blandford–Znajek [66] mechanism. Clearly, by the no hair theorem there is no such analogy to a unipolar inductor. Yet, it is instructive to understand the flaws in the reasoning in [80] in order to elucidate the role of black hole GHM near the horizon. The long discussion of the Faraday wheel at the end of a transmission line or plasma filled waveguide in Sect. 2.9.4 was prepared to explicitly contrast the horizon with a rotating conductor.

Consider a rotating black hole immersed in the Wald field of Sect. 4.6.2. There is a voltage drop across the magnetic field lines as given by (4.86). In [80], this is called a "battery-like EMF." Even though no physical electrical lead can actually touch the horizon, the claim is made in [80] that if no current flows in the leads, a spark gap

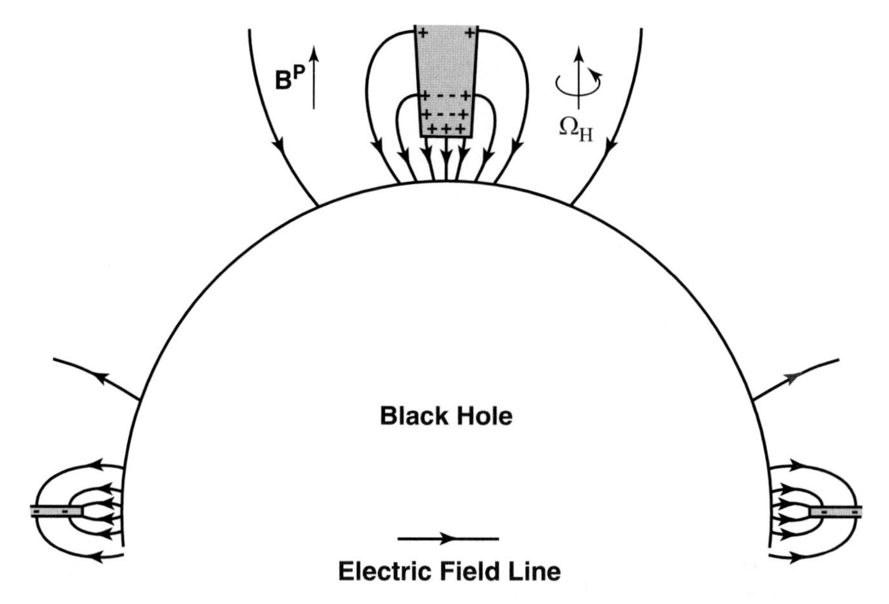

Fig. 4.4 The electrostatic equilibrium of conductive leads (*shaded*) that connect to a hypothetical circuit near the event horizon of a black hole that is rotating in the Wald magnetic field [112]

forms as in the Faraday wheel slightly detached from the end of a transmission line or plasma filled waveguide (Figs. 2.6 and 2.7). The voltage near the horizon in (4.86) is dropped across the gap, driving a global circuit. The mistake in [80] is that they ignore the horizon boundary condition on the flow (3.94) and (3.95) for the charges that create the electromagnetic fields. The inaccuracy of utilizing only Maxwell's equations or only the momentum equations in studying equilibria near the horizon was stressed in Chap. 3. When the ends of the leads are no longer held stationary arbitrarily close to the horizon, but flow inward and rotate according to (3.94) and (3.95), the voltage drop across the magnetic field lines becomes associated with an ingoing wave front by Faraday's law ($B^{\phi} \approx -E^{\theta}$) and the voltage drop in the gap remains small as for the detached rotating disk at the end of a transmission line with a large gap. There is a voltage drop across the magnetic field, but no current flows as in the open circuited transmission line. In terms of the no hair theorem, as far as the electromagnetic characteristics are concerned, the horizon is always "far away" in a global context. As with the open circuited transmission line, we expect an electrostatic equilibrium to be achieved at the ends of the leads (Fig. 4.4). We justify these statements by direct calculation in the remainder of this section.

4.7.4.1 Attaching the Leads

It is no trivial matter to connect perfectly conductive leads to the horizon from large distances as posited in the gedanken experiments of [80, 109]. This expedience assumes away all of the relevant physics. In fact we will show in the remaining chapters that significant black hole GHM interaction must occur within the leads.

Noting that the leads will interact with the magnetic field requires a very careful statement of the problem in order to isolate any putative unipolar properties of the near horizon space–time and the unipolar properties of the leads themselves. Clearly, an equatorial conductive disk as used in [80] is the most difficult choice since the plasma must cross the Wald field to reach the hole. A simpler choice is to just slide two azimuthally symmetric plasma flows down the Wald magnetic field lines in Fig. 4.2, so the flows resemble those in the circuit of Fig. 4.3. Near the hole, the leads are shown in Fig. 4.5. It is unreasonable to assume that the flow can proceed down the field lines at $\theta > 0$ from infinity to near the equator of the horizon and satisfy perfect MHD. We do not require such a circumstance in our analysis. The voltage drop and dissipation in the external circuit is arbitrary. We only require that the tips of the leads obey perfect MHD.

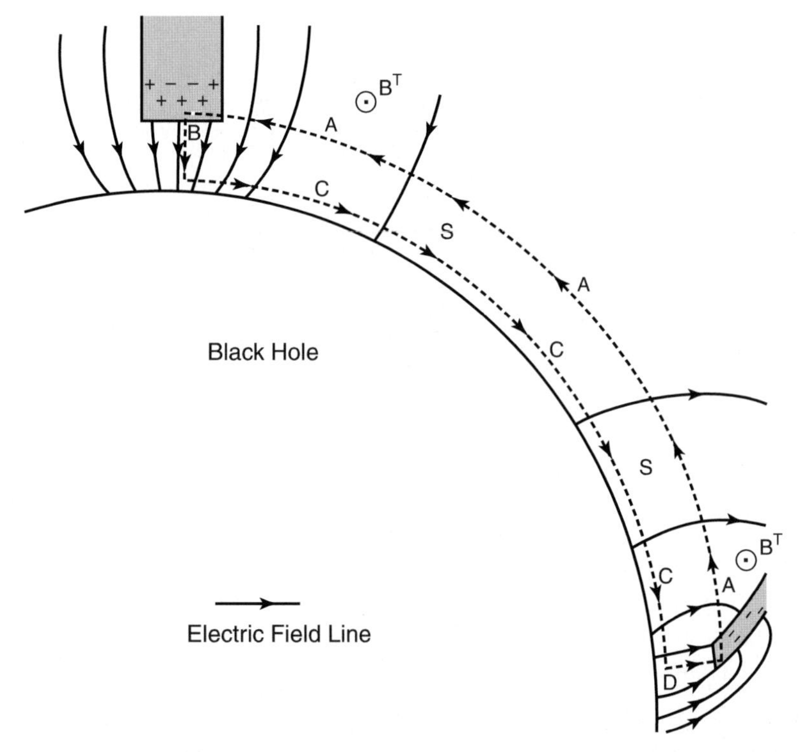

Fig. 4.5 The contour of integration ∂S that bounds the surface S used in the analysis of Faraday's law near the event horizon. The *curve* is broken into four pieces. The *curves B* and *D* are integral curves of $\partial/\partial r$ and the *curves A* and *C* are integral curves of $\partial/\partial\theta$ [112]

4.7.4.2 Faraday's Law in the 3 + 1 Splitting of Space–Time

Hypersurface orthogonality of the ZAMO four velocity was used in [80] to rewrite Maxwell's equations in a 3 + 1 splitting of space–time. The advantage of this is that Maxwell's equations, (3.53)–(3.60), can be written in integral form with ZAMO evaluated fields. Faraday's law is

$$\oint_{\partial S} \alpha \left[\boldsymbol{E} + \frac{\boldsymbol{v}}{c} \times \boldsymbol{B} \right] \cdot \mathrm{d}\boldsymbol{l} = -\frac{1}{c}\frac{\mathrm{d}}{\mathrm{d}t}\Phi_{B} = -\frac{1}{c}\frac{\mathrm{d}}{\mathrm{d}t}\int_{S} \boldsymbol{B} \cdot \mathrm{d}\Sigma , \tag{4.100}$$

where $\boldsymbol{v}$ is the velocity of the boundary of integration as viewed by a ZAMO. The magnetic flux is evaluated through a surface, S, with surface area element, $\mathrm{d}\Sigma$. We consider the contour of integration, ∂S, in the following to consist of four pieces: A, B, C, and D as shown in Fig. 4.5. We pick a contour of integration that is stationary with respect to infinity. Thus, by (3.42),

$$\boldsymbol{v} = -\frac{\Omega}{\alpha}\sqrt{g_{\phi\phi}}\,\hat{\boldsymbol{e}}_{\phi} \, . \tag{4.101}$$

The curve C in Fig. 4.5 is close enough to the horizon that the fields associated with the leads have yet to be propagated to such a small lapse function. Thus, the fields along C can be described by the background Wald field. The curves B and D just barely penetrate the tips of the perfectly conductive leads. The curves B and D are integral curves of the vector field $\partial/\partial r$, and the curves C and D are integral curves of the vector field $\partial/\partial\theta$. This contour of integration is used in the next two sections.

4.7.4.3 Proof by Induction that the Near Horizon is not a Unipolar Generator

Construct a pair of perfect MHD conductive leads near the horizon as indicated in the cross sectional image in Fig. 4.5. These leads are attached to some global circuit that possibly does not obey perfect MHD everywhere. The proof by induction that there is no unipolar induction in the space–time near the horizon hinges on the fact that if the tips of the leads (i.e., the portion of the leads in the asymptotic region of space–time near the horizon which obey perfect MHD by assumption) are introduced with zero current flow, $I = 0$, then no large electric fields are created by the near horizon space–time to disturb electrostatic equilibrium and drive currents in the ends of the leads. Thus, I remains zero and perfect MHD is maintained in the tips of the leads.

The Wald electric field $E^{r} \sim \alpha^{0}$ in (4.84c). Thus, a well behaved finite surface charge can shield the Wald electric field at the tips of the leads, $\sigma_{tip} \sim \alpha^{0}$. The surface charges on the leads greatly modifies the electric field from the background Wald field. The resulting E^{θ} has a contribution from the Wald field that scales as α^{1} (see 4.84d) and another contribution from the surface charges at the tip, σ_{tip}. Thus, a finite amount of electric flux is terminated on the sides of the leads when $I = 0$ and $\alpha \to 0$. The surface charge density that shields the sides of the leads is therefore well behaved, $\sigma_{sides} \sim \alpha^{0}$.

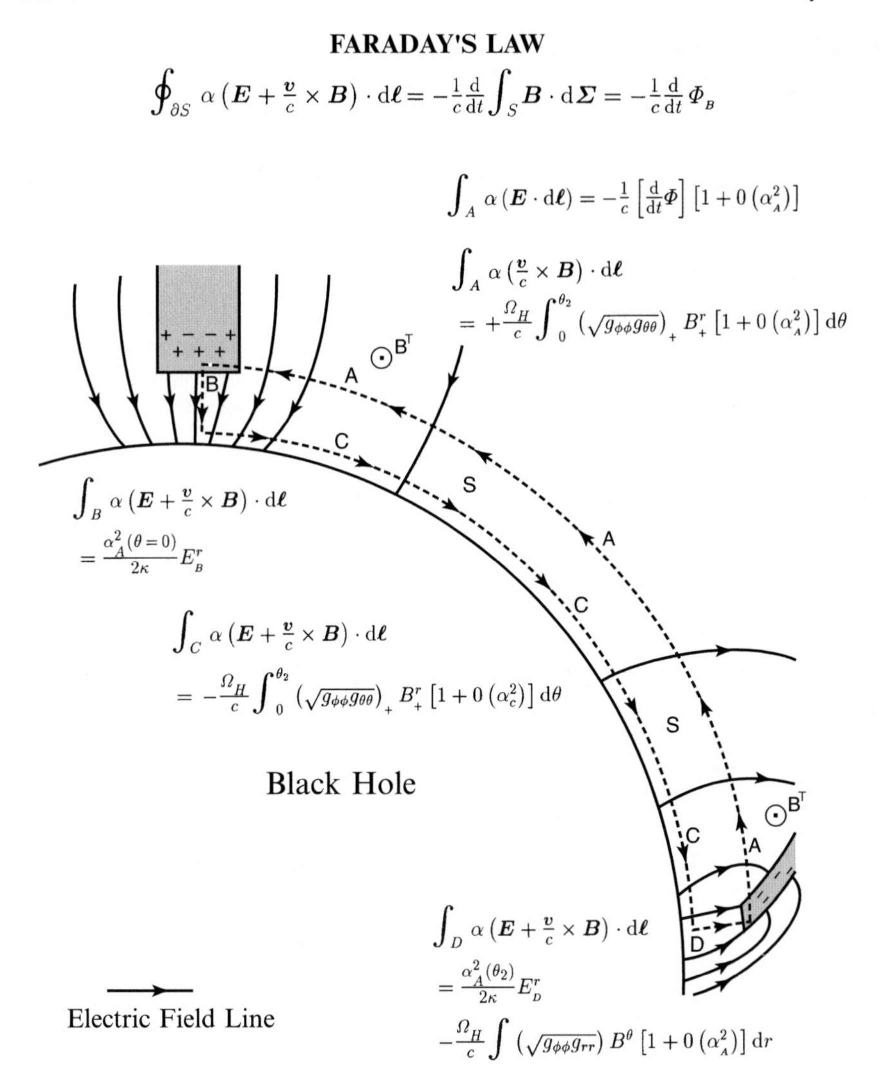

Fig. 4.6 The contributions to Faraday's law from each piece of the contour of integration in Fig. 4.5. The dynamical nature of space–time assures that any voltage drop across the magnetic field lines at the tips of the leads is a consequence of toroidal magnetic flux advected with the inflow in Faraday's law

We now compute the integrals in Faraday's law for the surface, S, drawn in Fig. 4.5. The contributions to each piece of the curve are indicated in Fig. 4.6. As the surface charge moves inward according to the horizon boundary condition (3.94) and (3.95), a toroidal magnetic field, $F^{r\theta}$, is created by Ampere's law (3.56b). Since $\sigma \sim \alpha^0$, $J^r \sim \alpha^0$ in the vacuum between the leads, and

$$F^{r\theta} \approx F^{\theta 0} \sim \alpha^0 . \tag{4.102}$$

Equation (4.102) follows from the axisymmetric Ampere's law, (3.56b) and (3.53b), and represents an ingoing wave condition similar to (3.90). Expanding out (4.100) gives the following terms:

$$\int_B \alpha\left(\boldsymbol{E}+\frac{\boldsymbol{v}}{c}\times\boldsymbol{B}\right)\cdot \mathrm{d}\boldsymbol{l} = \int_B \alpha\boldsymbol{E}\cdot\mathrm{d}\boldsymbol{l} = -\frac{\alpha_{_A}^2(\theta=0)}{2\kappa}E^r_{_B}\,, \tag{4.103a}$$

where $E^r_{_B}$ is the electric field strength in the gap and κ is the surface gravity. Similarly,

$$\int_D \alpha\left(\boldsymbol{E}+\frac{\boldsymbol{v}}{c}\times\boldsymbol{B}\right)\cdot \mathrm{d}\boldsymbol{l} = \frac{\alpha_{_A}^2(\theta=\theta_2)}{2\kappa}E^r_{_D}$$
$$-\frac{\Omega_H}{c}\int_D \sqrt{g_{\phi\phi}g_{rr}}B^\theta\left[1+\mathrm{O}\left(\alpha_{_A}^2\right)\right]\mathrm{d}r\,, \tag{4.103b}$$

where θ_2 is the latitude at which the tip of the lead approaches the horizon and $E^r_{_D}$ is the electric field in the gap. Along curve C one has

$$\int_C \alpha\left(\boldsymbol{E}+\frac{\boldsymbol{v}}{c}\times\boldsymbol{B}\right)\cdot \mathrm{d}\boldsymbol{l} = \int_C \alpha\left(\frac{\boldsymbol{v}}{c}\times\boldsymbol{B}\right)\cdot \mathrm{d}\boldsymbol{l} =$$
$$-\frac{\Omega_H}{c}\int_0^{\theta_2}\left(\sqrt{g_{\phi\phi}g_{\theta\theta}}\right)_+ B^r_{_+}\left[1+\mathrm{O}\left(\alpha_{_C}^2\right)\right]\mathrm{d}\theta\,, \tag{4.103c}$$

where "+" means to evaluate at the horizon. The E^θ term was ignored because $E^\theta \sim \alpha$ in the background Wald field ahead of the electromagnetic wave front carried by the leads. Finally,

$$\int_A \alpha\left(\boldsymbol{E}+\frac{\boldsymbol{v}}{c}\times\boldsymbol{B}\right)\cdot \mathrm{d}\boldsymbol{l} = -\frac{1}{c}\left[\frac{\partial}{\partial t}\Phi_{_B}\right]\left[1+\mathrm{O}\left(\alpha_{_A}^2\right)\right]$$
$$+\frac{\Omega_H}{c}\int_0^{\theta_2}\left(\sqrt{g_{\phi\phi}g_{\theta\theta}}\right)_+ B^r_{_+}\left[1+\mathrm{O}\left(\alpha_{_A}^2\right)\right]\mathrm{d}\theta\,. \tag{4.103d}$$

The line integral $\int_A \alpha\,(\boldsymbol{E}\cdot\mathrm{d}\boldsymbol{l})$ is equal to the rate that flux is deposited into S as the leads propagate inward as indicated in (4.104d). As the surface charge moves inward, the poloidal current it makes deposits more toroidal flux close to the horizon. The ingoing wave condition (4.102) assures cancelation in (4.100) to $\mathrm{O}\left(\alpha_{_A}^2\right)$. From the divergence equation (3.55b) for an axisymmetric field in the gap D and the regularity of the poloidal flux near the horizon (4.73), we have

$$B^\theta_{_D} \sim \alpha\,. \tag{4.104}$$

Thus, an electrostatic equilibrium is established in the tips of the leads as a consequence of (4.103) substituted into Faraday's law (4.100):

$$E^r_{_D} \sim E^r_{_B} \sim \alpha^0\,. \tag{4.105}$$

These are well behaved fields in the gap and there is no reason why a conductor could not be shielded from these fields. If the tips of the leads were in a perfect MHD state at $t=0$ they should remain so. Furthermore, if $I=0$ initially, then $I=0$ at later times. There is no current generation or unipolar induction near the horizon, just a surface charge density that is virtually static in global time t. □

Let us contrast this with the open circuited transmission line in Sect. 2.9.4. In electrostatics $B^{\phi}=0$, so the voltage drop across $\boldsymbol{B}$ in the first term of (4.103d) can not be balanced by a change in flux on the right hand side of Faraday's law (4.100). Thus, the entire voltage drop is in the gaps B and D, so we have $E^{r}\sim d^{-1}$ in the disconnected transmission line. Similarly, [80] ignore the influx of toroidal magnetic field by holding the leads fixed just outside of the horizon in contradiction to the horizon boundary condition (3.94) and (3.95). Thus, there is no magnetic flux change on the right hand side of Faraday's law (4.100) to balance the voltage drop across the magnetic field lines. Consequently, they find $E^{r}_{D}\sim E^{r}_{B}\sim \alpha_{A}^{-2}$ and charges are ripped out of the leads driving a current. This is the assumed physics of the membrane paradigm that simulates a unipolar inductor in the Blandford–Znajek [66] mechanism. However, the analysis does not consider the equations of motion of the charges in the leads, only Maxwell's equations, hence, it gives the wrong answer. One concludes that the near horizon voltage drop does not drive currents. The horizon and near horizon space–time is like the space in the gap between the rotating disk and the open circuited transmission line of Sect. 2.9.4 as depicted in the top views of Figs. 2.6 and 2.7. The fictitious horizon surface currents (displacement currents at the ingoing flow front) are not equivalent to the physical currents in a Faraday wheel that terminates a semi-infinite transmission line.

4.7.4.4 The Near Horizon Passively Accepts All Voltage Drops

Assume as in the last section that a pair of perfect MHD leads exist near the horizon and they are attached to some global circuit that need not have infinite conductivity. We further assume that an arbitrary current $I=I_0$ flows in the leads. We show that independent of the value of I_0, the space–time near the horizon never reacts back on the leads to alter the current, consistent with the no hair theorem. By Ampere's law (3.62b), the toroidal magnetic field in the vacuum region between the leads in Fig. 4.5 is a function of the current I, $F^{r\theta}=F^{r\theta}(I)$. The law of current conservation in the ZAMO frames $J^{\mu}{}_{;\mu}=0$ is derived in the same manner as the mass flux conservation law (3.66b):

$$\alpha\frac{\partial}{\partial X^{0}}\left(J^{0}\right)=\nabla^{(3)}\cdot(\alpha\boldsymbol{J})\ . \tag{4.106}$$

Thus, a finite current $I=I_0$ in the leads corresponds to an asymptotic poloidal current density that scales as

$$J^{P}\sim\alpha^{-1}\ . \tag{4.107}$$

The scaling (4.107) also follows from the horizon boundary condition on the individual species (3.94ab) that $u^r \sim \alpha^{-1}$. Combining (4.107) with Ampere's law (3.56b) and the regularity of the magnetic field in a surface layer of the infalling conductive lead (proper frame) reproduces the ingoing wave condition (3.90a). By (3.4) and the definition of the voltage drop in the stationary frames,

$$\tilde{F}_{t\theta} = -\alpha\rho F_{\theta 0} + \frac{\Omega\sqrt{g_{\phi\phi}}}{c}\rho F_{\theta\phi} \approx \alpha\rho F_{r\theta} + \frac{\Omega\sqrt{g_{\phi\phi}}}{c}\rho F_{\theta\phi}\,, \quad (4.108a)$$

$$\Delta V = \int_{\theta_1}^{\theta_2} \tilde{F}_{t\theta}\mathrm{d}\theta\,. \quad (4.108b)$$

Near the horizon (4.108) implies that $\Delta V = \Delta V\,(I = I_0)$ which is a function of the current as well (for a given poloidal magnetic flux). Consider an imposed value of I_0 (and therefore $\Delta V\,(I_0)$) in the leads near the horizon. At the tips of the leads, only the Wald radial electric field, $E^r \sim \alpha^0$, needs to be shielded as was the case in the last section with $I = 0$, $\sigma_{\text{tip}} \sim \alpha^0$. However, by (3.90a), $E^\theta \sim \alpha^{-1}$, thus $\sigma_{\text{side}} \sim \alpha^{-1}$ in order to shield the conductor from E^θ. Even though the surface charge on the sides of the conductor scales as α^{-1}, it is well behaved in the proper frame and the α^{-1} scaling is from a Lorentz contraction as observed in the ZAMO frame, $u^r \sim \alpha^{-1}$. This surface charge density is the analog of the Goldreich–Julian charge density from pulsar physics discussed in the plasma filled waveguide analysis of Sect. 2.9.4. The new question in this scenario is whether the charges needed to supply $\sigma_{\text{side}}(V)$ are equivalent to a unipolar current? The value of the voltage drop between the leads $V\,(I_0)$ is a constant in the region where the leads obey perfect MHD as a consequence of Ampere's law (4.100) with no $\mathrm{d}\Phi_B/\mathrm{d}t$ term. However, if one applies (4.100) to the contour of integration in Fig. 4.6, the only difference from the last section is that the ingoing wave condition (4.102) is replaced by the ingoing wave condition (3.90a). Thus, all of the calculation is identical to (4.103) and we again find that

$$E^r_D \sim E^r_B \sim \alpha^0\,. \quad (4.109)$$

This implies that the near horizon space–time never reacts back to change $\Delta V\,(I_0)$ near the ends of the leads by creating a large E^r to drive current. This result is independent of I_0. The near horizon passively accepts any value of ΔV or I. By contrast, a unipolar inductor imposes ΔV and I in a transmission line circuit (see (2.107) and (2.108)) or in a plasma filled waveguide circuit (see (2.120) and (2.121)). The causal nature of $\sigma_{\text{side}}(V)$ is clear. This is the charge necessary to support the ingoing wave. The currents, charges and the waves they create are generated upstream and radiated toward the hole. The amount of charge in the asymptotic region near infinity (the near horizon space–time) grows in time as long as the source of plasma keeps radiating inward.

4.7.5 The Horizon is an Electrodynamic Infinity

As the flow front of infalling leads (or any charge neutral electrodynamic source) approaches the horizon, the poloidal voltage drop between the flow front and asymptotic infinity (the horizon), ΔV_r dies off by (4.103ab), (4.105) and (4.109):

$$\lim_{\alpha \to 0} \Delta V_r \sim \alpha^2 . \tag{4.110a}$$

This compares to an outgoing electrodynamic flow propagating toward infinity:

$$\lim_{r \to \infty} \Delta V_r \sim \frac{1}{r} . \tag{4.110b}$$

This contrasts to a rotating conductor detached from the end of a transmission line or plasma filled waveguide. As the Faraday wheel approaches the end of a transmission line or waveguide, $\Delta V =$ constant (see Figs. 2.6 and 2.7). The near horizon space–time passively accepts any voltage drop across the magnetic field lines and any poloidal currents as does asymptotic infinity at $r \to \infty$. A unipolar inductor, by contrast, imposes a voltage drop and poloidal current in a circuit. Electrostatic equilibrium is achieved at a flow front as it approaches the horizon as occurs for an outgoing flow at $r \to \infty$. By contrast, electrostatic equilibrium breaks down at the end of a transmission line or plasma filled waveguide as they approach a Faraday wheel (see Sects. 2.9.4 and bottom of Figs. 2.6 and 2.7). Clearly, the space–time near the event horizon behaves as an asymptotic infinity to charge neutral flows and not like a conductor. The imperfect conductor interpretation is valid when the horizon passively accepts information and the impedance of the vacuum of $4\pi/c$ is assigned. Treating the vacuum like a conductor is not a particularly useful physical construct and is quite misleading.

4.8 The Charge of a Rotating Black Hole

Consider the nonvanishing electromagnetic invariant, $\boldsymbol{E} \cdot \boldsymbol{B}$, of the Wald field (4.84):

$$\begin{aligned} *F^{\mu\nu}F_{\mu\nu} = \frac{B_0^2 a \cos\theta}{\rho^4} & \{\Delta r \rho^2 \sin^2\theta + \left[\left(r^2 - a^2\right)\left(r^2 - a^2\cos^2\theta\right)\right. \\ & \left. + 2a^2 r(r-M)\left(1+\cos^2\theta\right)\right]\left[2r\left(r^2-a^2\right)\cos^2\theta\right. \\ & \left. -(r-M)\left(r^2 - a^2\cos^2\theta\right)\left(1+\cos^2\theta\right)\right]\} . \end{aligned} \tag{4.111}$$

By (4.87), the component of $\boldsymbol{E}$ parallel to $\boldsymbol{B}$, $E_{\|}$, changes sign with latitude on the horizon (see Fig. 4.2). For $\alpha \approx 0$ the sign changes at $\theta \approx 35°$ and when $\alpha \approx M$ it changes at $\theta \approx 29°$ [96]. The change of sign in $E_{\|}$ led Phinney [111] to suggest that these vacuum electric fields can drive a charge-starved and therefore charge separated current system as in the early models of pulsars [59]. However, this does not work for black holes for the same reason that is does not work for pulsars: the "Goldreich–Julian conundrum" [67]. Inspection of the global field in Fig. 4.2 shows no problem with attracting the positrons in a tenuous pair plasma inward along the magnetic field lines at the poles. The problem occurs at lower latitudes where $E_{\parallel}$ changes signs and the putative return current flows. In these flux tubes, even though $E_{\parallel} < 0$ near the horizon, $E_{\parallel} > 0$ a few black hole radii away. Thus, $E_{\parallel}$ can not drive a charge separated tenuous plasma. The plasma being tenuous must flow along the magnetic field lines, yet in a charge separated plasma, currents can only be made by the bulk motion of charge, so $J_{\parallel}$ and $E_{\parallel}$ have the same sign in a semi-vacuum magnetosphere. Without a unipolar inductor (like the neutron star in a pulsar) the currents can only be driven by $E_{\parallel}$ in the semi-vacuum magnetosphere. Consequently, since $E_{\parallel}$ switches signs in the putative return current path, the magnetosphere is open circuited. However, charges can still be accreted along the polar field lines and the result is not a current system, but a net charge on the hole. Wald [107] pointed out that the change in electromagnetic energy ε for a charge accreting down the polar field lines is found in terms of the vector potential (see (1.30b)):

$$\varepsilon = e\left(\tilde{A}_{t|_{\text{horizon}}} - \tilde{A}_{t|_{\infty}}\right) . \tag{4.112}$$

Because of the dragging of interial frames, the $\boldsymbol{E}$ and $\boldsymbol{B}$ components are mixed in the Kerr background space–time. Thus, the minimum electrostatic energy a particle can experience occurs when some of this electric field is canceled by the $l = 0$ field component. Anticipating this consequence, Pettersen [93] realized that the minimum energy state of a plasma filled magnetosphere required charge on the hole and charge neutrality would therefore require an equal but opposite charge in the axisymmetric magnetosphere. Thus, he considered a charged current ring as the source of the Wald field, with current I and charge q at a radial coordinate r_0 from the hole:

$$\tilde{J}^t = \frac{q}{2\pi r_0^2}\,\delta\left(r - r_0\right)\,\delta\left(\cos\theta\right) , \tag{4.113a}$$

$$\tilde{J}^r = \tilde{J}^\theta = 0 , \tag{4.113b}$$

$$\tilde{J}^\phi = \frac{I}{2\pi r_0^2}\,\delta\left(r - r_0\right)\,\delta\left(\cos\theta\right) . \tag{4.113c}$$

It is shown in [93] that the change in electrostatic energy for polar accretion in the presence of the charged current ring (4.112) is

$$\varepsilon = e\left[-\frac{Q}{2M} + \frac{aI\left(r_0^2 + a^2\right) - qa^2}{2r_0\left(r_0^2 + a^2\right)}\right] . \tag{4.114}$$

Since a conduction path only exists along the polar field lines, charges will accrete at the pole until the change in electrostatic energy vanishes in (4.114). Thus, setting $\varepsilon = 0$ and $Q = -q$ in steady state gives

$$Q = aMI \left[\frac{(r_0^2 + a^2)}{r_0 \left(r_0^2 + a^2\right) - a^2 M} \right] . \tag{4.115}$$

A real magnetosphere is clearly more complicated than a charged current ring and a tenuous plasma. However, the mixing of $\boldsymbol{E}$ and $\boldsymbol{B}$ occurs and a minimum energy state would almost certainly require a net charge on the hole. For $a \approx M$, the poloidal magnetic field from the $l = 0$ moment (the Kerr–Newman field) can be comparable to the Wald magnetic field. For reasonable magnetic field values $(Q^2/M^2) \ll 1$ in (4.115) and the metric is essentially the Kerr metric. Therefore, the Kerr–Newman electromagnetic field is essentially a test field. We acknowledge that the hole can be a significant source of magnetic flux in general, but we will not discuss the Kerr–Newman field explicitly except in Chapter 11. In most of the book, the $l = 0$ moment is just considered a second order source of poloidal magnetic flux around the black hole.

4.9 The Example of Axisymmetric Current Loops

The details of the no hair theorem can be illustrated by the behavior of the fields of uncharged axisymmetric current loops in the equatorial plane near the event horizon. The contents of this section are computer generated solutions of Laplace's equations that were created by Tomas Ledvinka expressly for the purposes of this book.

4.9.1 Magnetic Flux Exclusion From Rapid Rotators

The first thing that we illustrate is the exclusion of magnetic flux as the rotation rate of a black hole increases. Figure 4.7 shows the magnetic field of an axisymmetric current loop at $r = 1.5r_+$ near a black hole with $a/M = 0.9$. Figure 4.8 is the same current loop at $r = 1.5r_+$ near a very rapidly spinning black hole with $a/M = 0.995$. Notice how there is less flux threading the horizon when $a/M = 0.995$ as pointed out in (4.90c).

4.9.2 The No Hair Theorem

Next we illustrate the no hair theorem through a current loop that is adiabatically contracted onto the black hole. We take the current loop from Fig. 4.7 near a black

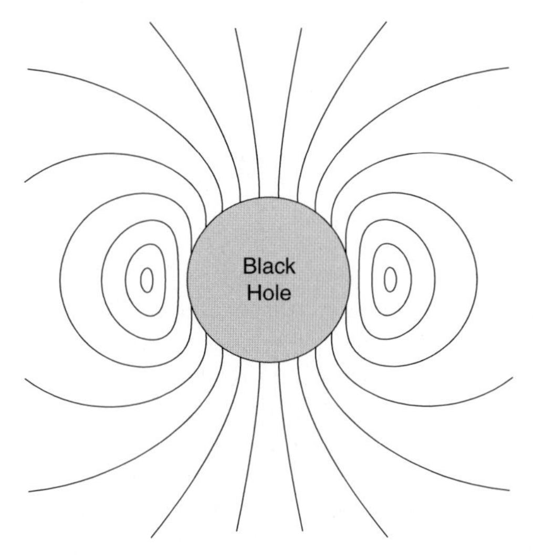

Fig. 4.7 The magnetic field of an axisymmetric current loop with a radius of $r = 1.5r_+$, $\alpha = 0.43$, centered about a black hole with $a/M = 0.9$

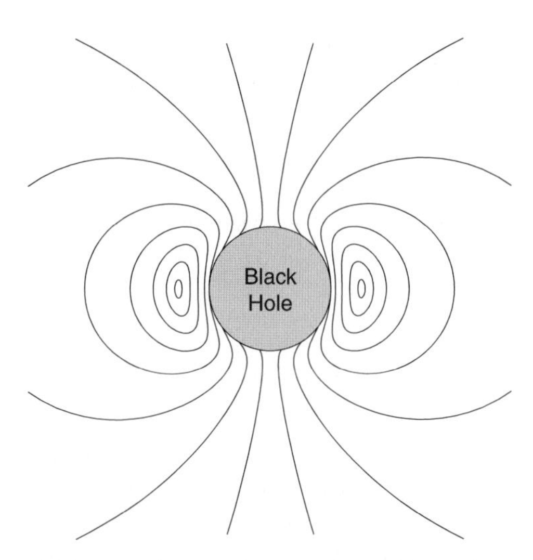

Fig. 4.8 The magnetic field of an axisymmetric current loop with a radius of $r = 1.5r_+$, $\alpha = 0.29$, centered about a black hole with $a/M = 0.995$

hole with $a/M = 0.9$ and contract it to a radius $r = 1.05r_+$ and then to $r = 1.001r_+$ in Figs. 4.9 and 4.10, respectively. Comparison of Figs. 4.7, 4.9 and 4.10 shows the large scale magnetic flux dying off as the loop contracts toward the horizon.

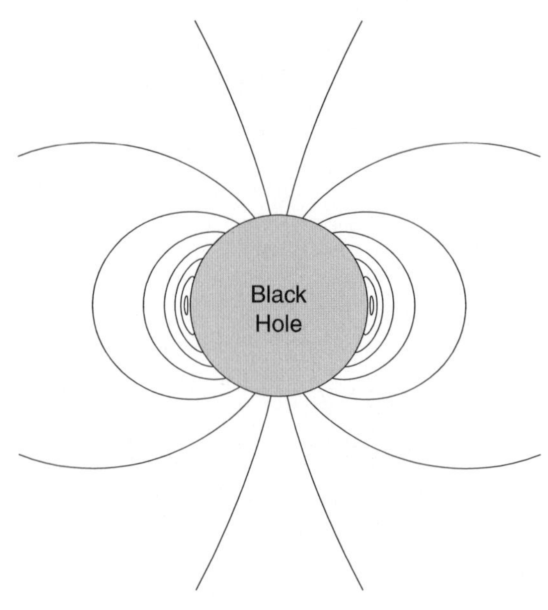

Fig. 4.9 The magnetic field of an axisymmetric current loop with a radius of $r = 1.05r_+$, $\alpha = 0.124$, centered about a black hole with $a/M = 0.9$

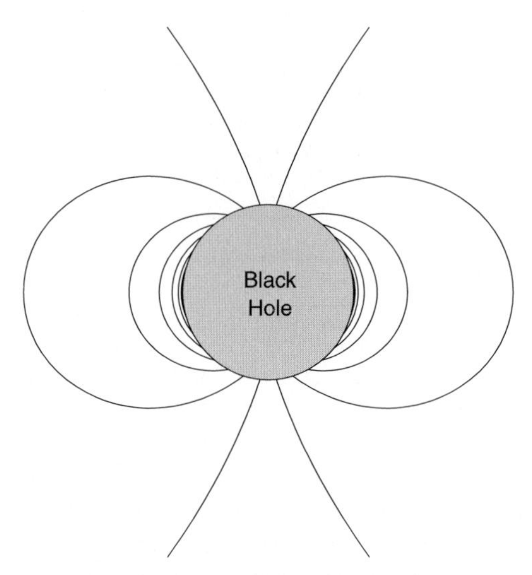

Fig. 4.10 The magnetic field of an axisymmetric current loop with a radius of $r = 1.001r_+$, $\alpha = 0.017$, centered about a black hole with $a/M = 0.9$

4.9.3 Magnetic Field Line Reconnection Near the Event Horizon

The computer calculation depicted in Fig. 4.11 is of two axisymmetric concentric current loops carrying the same total azimuthal current (as measured in the local

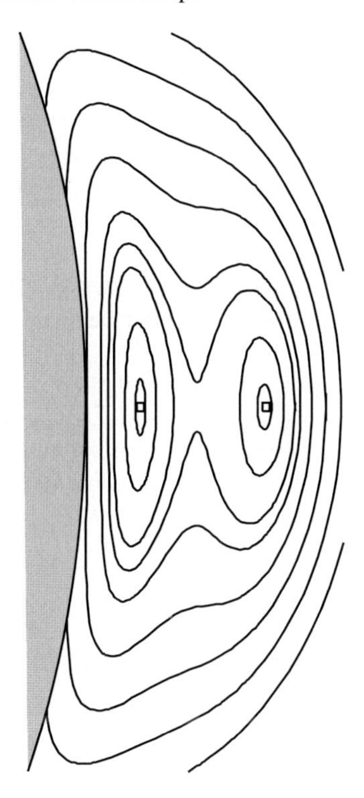

Fig. 4.11 The magnetic field of two concentric axisymmetric current loops centered about a black hole with $a/M = 0.9$. The inner loop is at a radial coordinate $r = 1.05r_+$, $\alpha = 0.124$, and the outer loop is a proper distance $0.7M$ farther out, at a lapse function of $\alpha = 0.24$. The image is a high resolution depiction of the near field region. Notice the X-point forming between the two current rings

ZAMO frames) near the event horizon of a black hole with $a/M = 0.9$. The inner loop in Fig. 4.11 is at $r = 1.05r_+$ as in Fig. 4.9 and the outer loop is located a proper distance $0.7M$ farther out. Fig. 4.11 illustrates the "no electromagnetic bootstrapping" effect that was discussed in Sect. 4.5.2. The magnetic field of the inner loop is dying off much faster than the magnetic field of the outer loop as the pair of current loops approach the event horizon. The net effect is the formation of an X-point in the near field structure in the high resolution image of Fig. 4.11. The result is of great importance in bridging the gap between vacuum solutions and a continuous plasma flow. This insight is crucial for the correct implementation of the horizon boundary condition on magnetic flux accretion which is essential for the analysis of Chap. 8.

4.9.4 The Physical Interpretation of the Results

The computer calculations described in this section do not capture the entire physical foundation of the no hair theorem. These results represent the effects of the redshifting of vacuum electromagnetic characteristics in the curved space–time for sources near the event horizon. There are also frame dragging effects associated with the no hair theorem that arise from the horizon boundary conditions (3.94c) and (3.95b). These are particularly significant for charge neutral current loops near the event horizon. More specifically, all inflowing sources of Maxwell's equations must corotate with the event horizon to $O(\alpha^2)$ as seen globally. Consequently, the azimuthal drift between particle species dies off in current loops near the event horizon. Alternatively stated, the event horizon boundary condition makes the azimuthal current in a current loop turn off as $O(\alpha^2)$ near the event horizon, as seen globally. For more details see Sect. 8.1 and (8.3).

Thus, our choice of keeping the current in the loop a constant as the horizon is approached in the computer simulations is artificial. One expects that current loops in the interval $1.001r_+ < r < 1.05r_+$ are almost certainly in the asymptotic zone of space–time and the azimuthal ZAMO evaluated current in the loops should be scaling (dying off) with lapse function as in (8.3a). With these frame dragging effects taken into account, one would expect that the ZAMO evaluated current in the loop depicted in Fig. 4.10 should equal $0.017/0.124 = 0.137$ of the ZAMO evaluated current in the current loop in Fig. 4.9. Hence, frame dragging contributions to the no hair theorem make the magnetic field of a contracting current loop die off much faster near the event horizon than is depicted in Figs. 4.9 and 4.10. Similarly, frame dragging effects make the formation of the X-point in the field topology of Fig. 4.11 more pronounced and occur farther from the event horizon.

4.10 The Implications of Vacuum Electrodynamics to GHM

The most important aspect of vacuum electrodynamics to black hole GHM is that there is no meaningful electrodynamic boundary condition associated with the event horizon. Unfortunately, the popular notion that the space–time near the event horizon is electrodynamically equivalent to a unipolar inductor has masked the actual physics of black hole GHM. The space–time near the horizon passively accepts any electrodynamic parameters imposed on it. The event horizon is an asymptotic infinity to charge neutral accretion flows. The horizon boundary condition is that the space–time near the horizon has no relevance electrodynamically in any global plasma flow. The only importance of the horizon is that it is a sink for inflowing plasma. Consequently, one needs to look outside of the asymptotic space–time to find a significant ergospheric GHM interaction. We showed that axisymmetric vacuum fields can not extract energy from a black hole. In the remainder of the book, we explore plasma filled magnetospheres and GHM in the ergosphere. Two other interesting results are that the hole is likely to have a net charge and that magnetic flux is excluded from rapidly rotating black holes. Most of this chapter is from [112]. Some additional analysis of idealized vacuum electrodynamic problems can be found in that reference.

Chapter 5
Magnetically Dominated Time Stationary Perfect MHD Winds

In Chap. 4, the solutions of Maxwell's equations demonstrate that no axisymmetric vacuum electromagnetic field can extract energy from a rotating black hole (see 4.90b). Yet, based on Lens–Thirring torques we expect an external source of magnetic flux (as well as the field of the hole itself which has $m = 0$) to be axisymmetric to first order (as discussed in the introduction to Chap. 4). Therefore, plasma must exist in a black hole magnetosphere if an effective energy extraction method exists. Furthermore, a charge starved magnetosphere is open circuited by the Goldreich–Julian conundrum of axisymmetric pulsar magnetospheres: the vacuum electric field switches sign in the magnetic flux tubes that are potential return current paths. Since there is no unipolar inductor associated with the horizon, this is a fatal flaw of the charge starved scenario as there is nothing to drive the return current (see Sect. 4.8). Thus, we expect that enough plasma is present in any viable energy extracting scheme so that any memory of the background vacuum electric field is erased from the magnetosphere. Locally, this condition is given by the Goldreich–Julian charge density [81]. In a black hole magnetosphere, we crudely estimate the Goldreich–Julian charge density for a $10^9\,M_\odot$ black hole,

$$\rho_{G-J} \sim \frac{\Omega_H B}{2\pi ce} \sim 1 \left(\frac{B}{10^4\,\mathrm{G}}\right) \mathrm{cm}^{-3} . \tag{5.1}$$

Where the value of $B \sim 10^4$ G will provide enough magnetic pressure to stop accretion onto a $10^9\,M_\odot$ black hole and therefore a field larger than this at the horizon cannot be supported by external sources (see Chap. 10). Clearly by (5.1), for magnetospheres that contain strong magnetic fields, even a very tenuous plasma can support $n_e > \rho_{G-J}$ and short out the vacuum electric field. The simplest description of flows from a highly magnetized magnetosphere are time stationary perfect axisymmetric MHD winds in the magnetically dominated limit. The language of relativistically wind theory will permeate the remainder of the text and we introduce the nomenclature in this chapter.

The theory of magnetized winds began with the study of the solar wind [62]. Relativistic pulsar winds in the context of a split monopole geometry were discussed

B. Punsly, *Black Hole Gravitohydromagnetics, 2nd. ed.*,
Astrophysics and Space Science Library 355, doi: 10/1007/978-3-540-76957-6_5,

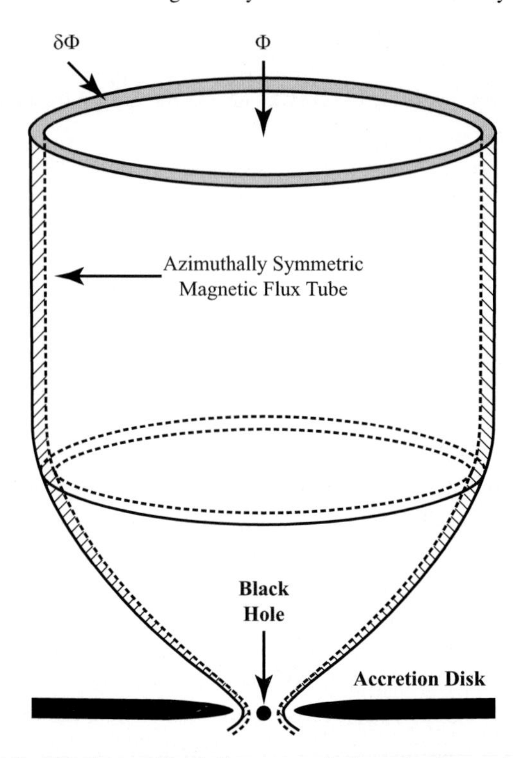

Fig. 5.1 The hatched region is an azimuthally symmetric flux tube that encircles a magnetic flux Φ and contains a magnetic flux $\delta\Phi$

in [113]. The relativistic wind theory of [113] has been extended to more general magnetic field configurations in [114, 115]. By the frozen-in condition and the definition of the rotated ZAMO basis in (3.8), there is flow only along the magnetic field lines,

$$u^P = u^1, \quad u^2 = 0, \quad F^{\mu\nu}u_\nu = \forall\mu \,. \tag{5.2}$$

Thus, each axisymmetric flux tube can be considered to have its own MHD flow (see Fig. 5.1). The formalism of relativistic winds was introduced into black hole magnetospheres in [111, 116].

5.1 The Perfect MHD Wind Equations

The perfect MHD wind equations are the momentum equations (3.39) with no dissipation,

$$n\frac{\mathrm{d}}{\mathrm{d}\tau}(\mu u^\alpha) + n\mu\left[\Gamma^\alpha{}_{\mu\nu}u^\mu u^\nu\right] + \frac{\partial}{\partial x^\alpha}(P) = \frac{F^{\mu\nu}J_\nu}{c}\,, \tag{5.3}$$

Maxwell's equations of Sect. 3.6, with the subsidiary condition (5.2) and the law of mass conservation (3.36b). The combined system of equations essentially contain four pieces of information. The first two pieces of information are simple conservation laws. The time component of the momentum equations corresponds to the law of energy conservation in an azimuthally symmetric magnetic flux tube. The azimuthal momentum equation similarly represents angular momentum flux conservation in each azimuthally symmetric magnetic flux tube.

A third component of the wind solution is the field aligned poloidal momentum equation for the plasma in the $\hat{\mathbf{e}}_1$ direction (see 3.8a and 3.40), known as the wind equation. The wind equation can also be found algebraically using the normalization condition of the plasma four velocity $u \cdot u = -1$ and the conservation of energy and angular momentum (remember $u^2 = 0$ in perfect MHD) relations in an azimuthally symmetric flux tube (this is done in 5.24). If one specifies the constants of motion of the flow within an azimuthally symmetric flux tube, then the solutions of the wind equation reveal which regions of spacetime can be linked by a perfect MHD flow with those particular constants of the flow. The constants of motion are injected into the flux tube and the wind equation gives no information as to what are physically reasonable constants of motion. There is a large history of the wind equations being used to determine the constants of motion. However, constants of motion by definition are not created within the perfect MHD regime of the flow. Remember that in the Faraday wheel terminated plasma-filled waveguide of Sect. 2.9.4, that the Faraday wheel injected the constants of motion into the wind zone. Similarly, the neutron star injects the constants of motion into an MHD pulsar wind and the sun into the solar wind.

The seven constants of motion are: Φ = the amount of magnetic flux encapsulated by an azimuthal magnetic flux tube (see Fig. 5.1), k = the mass flux per unit magnetic flux in a flux tube, $\delta\Phi$ = the amount of magnetic flux within the flux tube (see Fig. 5.1), ΔV =the voltage drop across the magnetic field in the flux tube, ke = the energy flux in the flux tube, $k\ell$ = the angular momentum flux in the flux tube, and $\mathbb{S}$ = the specific entropy flux within the magnetic flux tube.

The fourth equation is the poloidal momentum equation in the trans-field ($\hat{\boldsymbol{e}}_2$ direction). This is known as the Grad–Shafranov equation. It determines the magnetic flux distribution in the wind. It is the most complex equation of the four and typically can be solved only in the force free limit. It is discussed briefly in Sect. 5.7.

5.2 Constants of Motion within a Flux Tube

In this section, we elaborate on the seven constants of motion. Firstly, since the frozen-in condition (5.2) implies that plasma does not cross the magnetic field lines ($u^2 = 0$), the amount of flux contained within cylindrical radii less than that of the flux tube, Φ, is conserved. Sometimes this flux is labeled by the stream function, Ψ, in wind theory, where Ψ is a constant along each stream line (flux tube),

$$\Phi = \Phi(\Psi). \tag{5.4}$$

Since the Maxwell tensor derives from the vector potential,

$$F_{\mu\nu} = A_{\mu;\nu} - A_{\nu;\mu}, \tag{5.5}$$

$$B^P = F^{2\phi} = \frac{1}{\sqrt{g_{\phi\phi}}} \frac{\partial}{\partial X^2} \left(\sqrt{g_{\phi\phi}} A^\phi\right), \tag{5.6}$$

where we used (3.32d) to write (5.6) in the rotated ZAMO basis. By definition, $B^2 = F^{1\phi} = 0$ (see 3.7), therefore (3.32c) and (5.5) imply that

$$\frac{\partial}{\partial X^1} \left(\sqrt{g_{\phi\phi}} A^\phi\right) = 0 . \tag{5.7}$$

Thus, the stream function is

$$\frac{\Psi}{2\pi} = \sqrt{g_{\phi\phi}} A_\phi = \tilde{A}_\phi . \tag{5.8}$$

Similarly, we find that the flux contained within the magnetic flux tube is a constant, $\delta\Phi$,

$$\delta\Phi = \int F_{2\phi}\, \mathrm{d}X^2 \wedge \mathrm{d}X^\phi . \tag{5.9}$$

The quantity $\mathrm{d}X^2 \wedge \mathrm{d}X^\phi = \omega^2 \wedge \omega^\phi$ by (3.8b), and it is the cross sectional area element in an azimuthally symmetric flux tube, $\mathrm{d}A_\perp$,

$$\mathrm{d}A_\perp = \mathrm{d}X^2 \wedge \mathrm{d}X^\phi . \tag{5.10}$$

The third constant of motion is found by integrating the mass conservation law (3.36b) with the source term $S = 0$, and $u^P = u^1$:

$$\int n\alpha u^P \mathrm{d}X^2 \wedge \mathrm{d}X^\phi = \text{constant} . \tag{5.11}$$

Taking the ratio of (5.11) and (5.9) in the limit $A_\perp \to 0$ yields the third constant of motion, the mass flux per unit magnetic flux, k,

$$k = \frac{n\alpha u^P}{B^P}, \quad \frac{\partial}{\partial X^1}(k) = 0 \tag{5.12}$$

It is instructive to define a frame that corotates with the magnetic field at a fixed radial coordinate "r." The "corotating frame" has a four velocity $\overline{u}_\mu$. The frozen-in condition in this frame becomes

$$F^{20}\overline{u}_0 + F^{2\phi}\overline{u}_\phi = 0 . \tag{5.13a}$$

The azimuthal velocity of the field as viewed by a ZAMO is

$$\beta_F^\phi = \frac{\overline{u}^\phi}{\overline{u}^0}, \tag{5.13b}$$

$$F^{20} = \beta_F^{\phi} B^P . \tag{5.13c}$$

One can define a field line angular velocity, Ω_F, as viewed from asymptotic infinity implicitly from (3.42)

$$\beta_F^{\phi} = \left(\frac{\Omega_F - \Omega}{c\alpha}\right)\sqrt{g_{\phi\phi}} . \tag{5.14}$$

Since $\Omega \to 0$ as $r \to \infty$ and $\Omega \to \Omega_H$ as $r \to r_+$ in a flux tube, $\beta_F^{\phi} \to +\infty$ as $r \to \infty$ and $\beta_F^{\phi} \to -\infty$ as $r \to r_+$, if $0 < \Omega_F < \Omega_H$.

We see from (5.13c) that β_F^{ϕ} is associated with the cross-field electric potential. Using the time stationary Faraday's law in the 3 + 1 splitting of spacetime (4.100), the voltage drop across the magnetic field is given by (5.13c) and (5.14) as

$$\Delta V = -\int \mathbf{E} \cdot dl = -\int \frac{\Omega_F \sqrt{g_{\phi\phi}}}{c} B^P \mathrm{d}X^2 . \tag{5.15a}$$

Since $F^{10} = 0$, by the frozen-in condition, (5.15a) implies $\Delta V =$ constant in a flux tube. Thus it follows from (5.9) that

$$\Delta V = -\frac{\Omega_F}{2\pi c}\,\delta\Phi = \text{constant} . \tag{5.15b}$$

Thus, Ω_F is a constant in a flux tube that is related to the electrostatic potential.

In the rotated ZAMO basis, the frozen-in condition becomes one relation for the toroidal magnetic field, $F^{12} = F^{r\theta}$:

$$F^{20}u_0 + F^{21}u_1 + F^{2\phi}u_{\phi} = 0 , \tag{5.16a}$$

$$F^{12} = \frac{(\beta^{\phi} - \beta_F^{\phi})}{\beta^P} B^P , \tag{5.16b}$$

where β^{ϕ} and β^P are the azimuthal and poloidal three velocities of the plasma in the ZAMO frame, respectively. In terms of conservation laws, it is more desirable to define the Boyer–Lindquist toroidal magnetic field density, B^T

$$B^T = \sqrt{-\tilde{g}}\,\tilde{F}^{r\theta} = \alpha\sqrt{g_{\phi\phi}}\,F^{12} . \tag{5.17}$$

Boyer–Lindquist coordinates are suitable for global conservation laws because the metric is independent of the coordinates "t" and "ϕ" ($\partial/\partial t$ and $\partial/\partial\phi$ are Killing vectors). Thus, the "t" and "ϕ" components of the stress-energy tensor are associated with conserved quantities. In the absence of plasma inertia, the stress-energy tensor is given by the electromagnetic term (3.23). The conserved quantities are called the poloidal redshifted Poynting and angular momentum fluxes [117]. The total redshifted Poynting flux in a magnetic flux tube is found from (3.23) in Boyer–Lindquist coordinates using (5.13), (5.14) and (5.17) to be

$$\int \alpha S^P \mathrm{d}A_{\perp} = -\frac{\Omega_F}{4\pi}\int B^T F_{2\phi}\,\mathrm{d}X^2 \wedge \mathrm{d}X^{\phi} . \tag{5.18}$$

Similarly, the poloidal angular momentum flux, S_L^P, is given by $\tilde{T}_{\phi P}$ in (3.23) and the total angular momentum flux about the symmetry axis of the black hole in a flux tube is

$$\int \alpha S_L^P \mathrm{d}A_\perp = -\frac{c}{4\pi} \int B^T F_{2\phi}\, \mathrm{d}X^2 \wedge \mathrm{d}X^\phi\,. \tag{5.19}$$

The fifth and sixth constants of motion are the energy and angular momentum fluxes, ke and $k\ell$, respectively. Including the plasma stress-energy tensor in (3.21) and the conserved mass flux (5.12), the conserved specific energy can be found from (5.18) and (3.7) to be

$$e = \omega - \frac{\Omega_F B^T}{4\pi k}\,. \tag{5.20}$$

Similarly, the conserved specific angular momentum is found from (3.21) and (5.19),

$$\ell = m - \frac{cB^T}{4\pi k}\,. \tag{5.21}$$

This result is also shown in the derivation of (5.48). The specific mechanical energy and angular momentum (ω and m, respectively) were defined in (1.36). Note that this "m" is not the same quantity as the multipole moment of the electromagnetic field discussed in Chap. 4.

The seventh constant of motion is the entropy per unit magnetic flux $\mathbb{S}$. For warm plasmas that are not relativistically hot, the entropy does not strongly affect the dynamics of a magnetically dominated wind.

5.3 The Wind Equations

The wind equations can be described as a set of algebraic relations for the four velocity. From (3.14), (5.16), (5.20), and (5.21) one finds,

$$u^\phi = \frac{\dfrac{\alpha\ell}{\sqrt{g_{\phi\phi}}}\left[M^2 + (\beta_F^\phi)^2\right] - \beta_F^\phi\left(e - \dfrac{\Omega\ell}{c}\right)}{\alpha\mu\left[M^2 + \left(\beta_F^\phi\right)^2 - 1\right]}\,, \tag{5.22a}$$

$$u^0 = \frac{(M^2-1)\left(e - \dfrac{\Omega\ell}{c}\right) + \dfrac{\alpha\beta_F^\phi\ell}{\sqrt{g_{\phi\phi}}}}{\alpha\mu\left[M^2 + \left(\beta_F^\phi\right)^2 - 1\right]}\,, \tag{5.22b}$$

$$F^{12} = \frac{4\pi ncu^P \left[\beta_F^\phi \left(e - \frac{\Omega \ell}{c}\right) - \frac{\alpha \ell}{\sqrt{g_{\phi\phi}}}\right]}{\alpha B^P \left[M^2 + \left(\beta_F^\phi\right)^2 - 1\right]} . \tag{5.22c}$$

In (5.22) we used the pure Mach number defined in terms of the pure Alfvén speed

$$M^2 \equiv \frac{\left(u^P\right)^2}{U_A^2 c^2} , \tag{5.23a}$$

$$U_A^2 = \frac{\left(B^P\right)^2}{4\pi n \mu c^2} . \tag{5.23b}$$

Combining (5.22a,b) with the normalization condition on the bulk four velocity, $u \cdot u = -1$, yields the wind equation:

$$\left(u^P\right)^2 + 1 = \frac{M^4 \left[e - \frac{\Omega_{\min}}{c}\ell\right]\left[e - \frac{\Omega_{\max}}{c}\ell\right] - \left[2M^2 - 1 + \left(\beta_F^\phi\right)^2\right]\left[e - \frac{\Omega_F \ell}{c}\right]}{\alpha^2 \mu^2 \left[M^2 - 1 + \left(\beta_F^\phi\right)^2\right]^2} , \tag{5.24}$$

where $\Omega_{\min}$ and $\Omega_{\max}$ are defined in (3.43).

5.4 The Critical Surfaces

When the plasma bulk velocity u^P exceeds the group velocity of a plasma wave propagating in a flux tube along the poloidal field lines, then no plasma waves of this type can be propagated in the antiflow direction as seen globally. This condition is satisfied at and beyond a critical surface for each wave mode. There are potentially three critical wave surfaces, corresponding to the slow, Alfvén, and fast waves. At the critical surface, a wave radiated opposite to the poloidal flow direction stagnates. This condition is given by the vanishing of the poloidal component of the group four velocity for that mode, $u_g^P = 0$, as given by (2.51). Expression (2.51) is not trivial since u_g is a function of θ. The critical surface condition $u_g^P = 0$ is defined by the value of u^P in the flow through (2.51) and in this section we solve for u^P at the critical surface.

Since the antidirected wave stagnates at the critical surface, the poloidal component of the wave vector becomes infinite (i.e., the wavelength of oscillation goes to zero):

$$|k|^P \to \infty , \qquad \frac{k^P}{|k|} \approx \pm 1 . \tag{5.25}$$

The positive (negative) sign in (5.25) corresponds to outgoing (ingoing) waves in an ingoing (outgoing) wind. Inserting the condition (5.25) into (2.51) and setting $u_g^P = 0$ yields the expected result

$$u^P = |u_g(\theta)| \,. \tag{5.26a}$$

Similarly, (5.25) applied to (2.40) yields the phase velocity at the critical surface,

$$u_\varphi^2 \approx \frac{(u^P)^2 (k^P)^2}{(k^P)^2} = (u^P)^2 \,. \tag{5.26b}$$

Consequently we must determine $u_g(\theta)$ in order to find the poloidal flow velocity at the critical surface.

First we consider the four velocity of an axisymmetric flux tube. From (5.13) we write the four velocity of the corotating frame as

$$F^\mu = \frac{\partial}{\partial X^0} + \beta_F^\phi \frac{\partial}{\partial X^\phi} \,. \tag{5.27}$$

We do not bother with normalizing (5.27) since F^μ is not necessarily timelike as $|\beta_F^\phi| > 1$ is an allowed value as discussed below (5.14).

The magnetic four vector can be simplified in terms of F^μ [111]. The magnetic four vector, B^μ, is of interest because we need it to find the angle of propagation relative to the magnetic field in the expression for $u_g(\theta)$ that is required to evaluate (5.26). Expanding B^μ in the orthonormal rotated ZAMO basis,

$$B^\mu = {*F^{\mu\nu}} u_\nu = \left(u_0\; u_1\; u_2\; u_\phi\right) \begin{pmatrix} 0 & -B^P & 0 & -B^\phi \\ B^P & 0 & 0 & -E^2 \\ 0 & 0 & 0 & 0 \\ B^\phi & E^2 & 0 & 0 \end{pmatrix} . \tag{5.28}$$

We find from (5.27) that

$$B^\mu = \frac{B^P}{u^P} \left[-F^\mu - (F^\mu u_\mu) u^\mu\right] . \tag{5.29}$$

Noting that

$$u_\mu F^\mu = -u^0 \left[1 - \beta^\phi \beta_F^\phi\right] , \tag{5.30}$$

(5.29) becomes

$$B^\mu = \frac{B^P}{u^P} \left[-F^\mu + u^0 u^\mu \left(1 - \beta^\phi \beta_F^\phi\right)\right] . \tag{5.31}$$

Substituting (5.29) and (5.31) into expression (2.34) for $\cos\theta$ and applying the definition of phase velocity in (2.39) we have

$$\frac{1}{2}F^{\mu\nu}F_{\mu\nu}\cos^2\theta = \left(\frac{B^P}{u^P}\right)^2 \left[-\frac{F^{\mu}k_{\mu}}{|\mathbf{k}|} + v_{\varphi}u^0\left(1-\beta^{\phi}\beta^{\phi}_{F}\right)\right]^2 , \tag{5.32a}$$

where $\boldsymbol{k}$ is defined in (2.35). Using (5.25) and (5.27)

$$\lim_{|k^P|\to\infty} \frac{F^{\mu}k_{\mu}}{|\mathbf{k}|} = 0 . \tag{5.32b}$$

At the critical surface $\left(v_{\varphi}u^0\right)^2 = \left(u^P\right)^2$ by (5.26). Thus, (5.32) reduces to

$$\frac{1}{2}F^{\mu\nu}F_{\mu\nu}\cos^2\theta = \left(1-\beta^{\phi}\beta^{\phi}_{F}\right)^2\left(B^P\right)^2 . \tag{5.33}$$

In order to find u^P at the Alfvén critical surface, we rewrite (2.41) as

$$u_{P}^2 = U_{I}^2 = \frac{F^{\mu\nu}F_{\mu\nu}}{8\pi n\mu}\left[\left(\frac{U_{I}^2}{c^2}+1\right)\cos^2\theta - \frac{U_{I}^2}{c^2}\right] . \tag{5.34}$$

In the rotated ZAMO basis, the proper magnetic field is very simple using the frozen-in condition (5.13c) and (5.16)

$$\frac{1}{2}F^{\mu\nu}F_{\mu\nu} = \left[1-\left(\beta^{\phi}_{F}\right)^2\right]\left(B^P\right)^2 + \left(F_{12}\right)^2 = \left[1-\left(\beta^{\phi}_{F}\right)^2 + \frac{\left(\beta^{\phi}-\beta^{\phi}_{F}\right)^2}{\left(\beta^{P}\right)^2}\right]\left(B^P\right)^2 . \tag{5.35}$$

Using expressions (5.33) and (5.35) in (5.34) at the Alfvén critical surface, the poloidal four velocity satisfies

$$\left(u^P\right)^2 = U_{I}^2 = \left[1-\left(\beta^{\phi}_{F}\right)^2\right]\frac{\left(B^P\right)^2}{4\pi n\mu} . \tag{5.36}$$

Similarly to (5.34), we can write the magnetic-acoustic four speed dispersion relation (2.44) as

$$U_{F,SL}^4 - U_{F,SL}^2\left[U_{S}^2 + \frac{F^{\mu\nu}F_{\mu\nu}}{8\pi n\mu}\right] + \left[\left(c^2+U_{F,SL}^2\right)\cos^2\theta - U_{F,SL}^2\right]\frac{U_{S}^2F^{\mu\nu}F_{\mu\nu}}{8\pi n\mu c^2} = 0 . \tag{5.37}$$

Again, we substitute (5.35) and (5.33) to reduce the last term in (5.37) to get the magneto-acoustic critical surface conditions with $U_{F,SL} = |u^P|$:

$$U^4_{F,SL} - U^2_{F,SL}\left[U^2_S + \frac{F^{\mu\nu}F_{\mu\nu}}{8\pi n\mu}\right] + \frac{\left[1-\left(\beta^{\phi}_F\right)^2\right]}{4\pi n\mu}\left(B^P\right)^2 U^2_S = 0\,. \tag{5.38}$$

The critical surfaces play a role in the poloidal momentum equation, or wind equation (5.3). This has been explored through the algebraic wind equation (5.24). From (5.24), Beskin [118] computed $\partial/\partial x^a\,(M\alpha)$ where $a = r,\theta$ and $b = r,\theta$ in the expressions to follow. First, he notes that

$$\begin{aligned}\frac{\partial}{\partial X^\alpha}\mu = {} & \frac{c_S^2}{c^2-c_S^2}\mu\left[2\frac{\frac{\partial}{\partial X^\alpha}k}{k} - \frac{\frac{\partial}{\partial X^\alpha}(M\alpha)^2}{M^2\alpha^2}\right] \\ & + \frac{c^2}{c^2-c_S^2}\left[\frac{1}{nm_p}\left(\frac{\partial P}{\partial \mathbb{S}}\right)_n + T\right]\frac{\partial}{\partial X^\alpha}\mathbb{S}\,,\end{aligned} \tag{5.39}$$

where $\mathbb{S}$ is the entropy. Then he finds from (5.24) that one can write

$$\frac{\partial}{\partial X^a}\left(M^2\alpha^2\right) = \frac{N_a}{D}\,. \tag{5.40a}$$

Note that (5.40a) is the differential part of the wind equation (5.3). Using the stream function Ψ of (5.8)

$$N_a = \frac{\alpha^2\left[1-\left(\beta^2_F\right)^2 - M^2\right]}{(\nabla\Psi)\cdot(\nabla\Psi)}\left[\nabla^b\Psi\nabla_a\nabla_b\Psi + \frac{1}{2}\nabla'_a\left[(\nabla\Psi)\cdot(\nabla\Psi)\right]\right], \tag{5.40b}$$

where ∇'_a acts on all quantities except M^2 and

$$D = \frac{\left[1-\left(\beta^{\phi}_F\right)^2 - M^2\right]}{M^2} + \frac{\left(F^{12}\right)^2}{M^2\left(B^P\right)^2} - \frac{c^2}{\left(u^P\right)^2}\frac{\left[1-\left(\beta^{\phi}_F\right)^2 - M^2\right]}{M^2}\frac{c_S^2}{c^2 - nc_S^2}\,. \tag{5.40c}$$

The denominator, D, vanishes when (5.38) is satisfied with $|u^P| = U_{F,SL}$. Thus, the magneto-acoustic critical surfaces are singular points of the poloidal momentum equation.

Similarly in [119], this result is generated by taking $\partial/\partial X^1\left[\ln\left(u^P\right)\right]$, where u^P is given by the algebraic relation (5.24). However, the dispersion relation they find involves unnatural definitions of magnetic field components as a result of using Boyer–Lindquist (nonorthonormal) coordinates. Thus, it is difficult to interpret the wave speeds. The important consequence of computing the plasma wave speeds in Chap. 2 is that it is not at all obvious that (5.36) and (5.38) represent the appropriate plasma wave speeds at the critical surfaces. The plasma wave speeds depend on the angle of propagation and the value of $\cos\theta$ from (5.33) and (5.35) is not at all intuitive,

$$\cos^2\theta = \frac{\left(1-\beta^\phi\beta_F^\phi\right)^2\left(\beta^P\right)^2}{\left[1-\left(\beta_F^\phi\right)^2\right]\left(\beta^P\right)^2+\left(\beta^\phi-\beta_F^\phi\right)^2}. \tag{5.41}$$

Note that the algebraic relations (5.22) and the algebraic wind equation (5.24) all have a singularity at the intermediate speed, by (5.36), since the denominator vanishes. The only way that the flow can proceed smoothly through the Alfvén critical surface is if the numerators in (5.22) and (5.24) vanish as well. This reduces to a single constraint

$$\left(\beta_F^\phi\right)_A\left(e-\frac{\Omega\ell}{c}\right)_A = \frac{\alpha_A\ell}{\left(\sqrt{g_{\phi\phi}}\right)_A}, \tag{5.42a}$$

or alternatively written,

$$e-\frac{\Omega_F\ell}{c} = \left[1-\left(\beta_F^\phi\right)^2\right]_A\left(e-\frac{\Omega\ell}{c}\right)_A. \tag{5.42b}$$

5.5 The Topology of the Outgoing MHD Wind Solution Space

Kennel et al. [113] published the first treatment of MHD wind theory that successfully clarified the critical point structure of outgoing magnetized winds. The critical points are mathematical singularities in the combined set of differential equations comprised of Maxwell's equations and the momentum equations of the wind plasma. Physically, the critical points represent points at which the flow exceeds one of the MHD plasma wave speeds (slow, intermediate, and fast speeds or waves). Once beyond the critical point, the corresponding wave cannot be antidirected to the bulk flow. Wind solutions pass successively through the slow and intermediate critical points if they propagate beyond the light cylinder (i.e., the surface at which the magnetic field lines rotate at the speed of light) as depicted in Fig. 5.2. There is a solution that also passes through the fast critical point on its way towards asymptotic infinity, the critical solution or minimum torque solution. For every value of Ω_F there is one value of conserved energy, E_{CR}, that defines this solution, where Ω_F is the field line angular velocity. Even though the analysis was carried out in a monopolar magnetic field geometry, the results are qualitatively correct for winds in axisymmetric flux tubes [113]. This is adequate for our purposes as the winds considered in this book are really nested sets of axisymmetric winds (i.e., a unique wind for each flux tube).

We consider the solution space topology through Figs. 6, 7, and 10 of [113]. We have a particular emphasis on the finite temperature wind solution space of their Fig. 10. The results are modified slightly for our situation in Fig. 5.2. Notice that there are two critical solutions (one physical and one unphysical) defined by E (the conserved energy) equal to E_{CR} (the critical energy) for a given value of Ω_F.

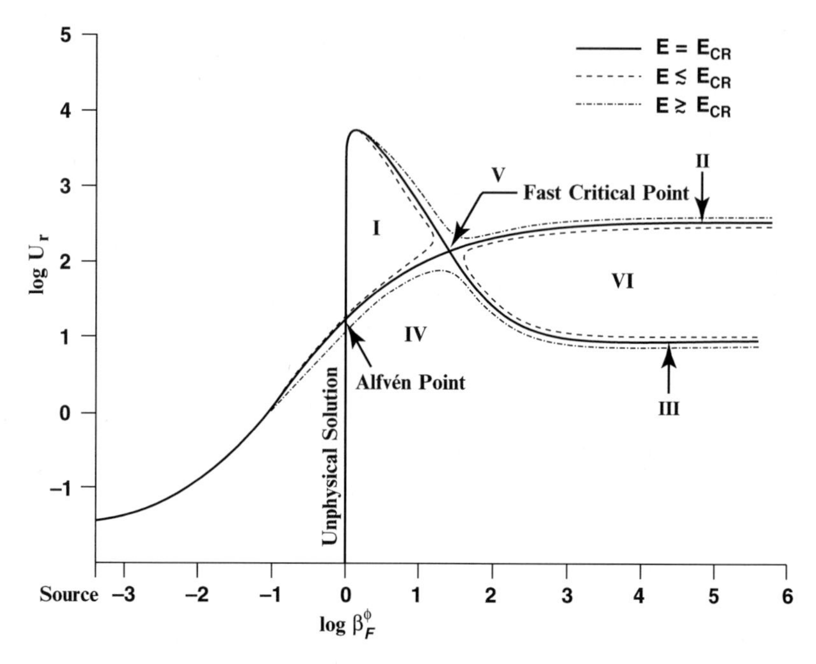

Fig. 5.2 The topology of the outgoing MHD wind solution space as derived in [111]. The vertical axis is the logarithm of the poloidal four velocity of a sample wind divided by the speed of light, $\log U_r$. The displacement from the source is measured in terms of the logarithm of the field line rotational three velocity divided by the speed of light, $\log \beta_F^\phi$. There are six classes of solutions considered in the text. Regions I and VI of the solution space are occupied by subcritical solutions. Curves II (physical) and III (unphysical) are the minimum torque solutions for the given value of the field line angular velocity, Ω_F. Regions IV (physical) and V (unphysical) contain supercritical wind solutions

There are six disjoint regions of solution spaces labeled I–VI in the figure; these are described below.

I There is a two-dimensional subset of solution space that is comprised of subcritical solutions, $E < E_{CR}$. A subcritical solution extends just beyond the light cylinder then turns back towards the source, even though the poloidal four velocity, U_r, is positive. These solutions are clearly unphysical beyond the turnaround point. The subcritical solutions join onto the vertical degenerate solution, commonly called the unphysical branch of the minimum torque solution (i.e., the nearly vertical solid line in Fig. 5.2).

II There is a singular (i.e., one-dimensional set in solution space), physical minimum torque solution that connects the source to asymptotic infinity with $E = E_{CR}$. The solution accelerates all the way out and asymptotes to a maximum flow speed that exceeds the fast magnetosonic speed.

III The unphysical branch of the minimum torque solution, $E = E_{CR}$, does not connect to the source. The solution is singular (i.e., a one-dimensional set in solution space) and extends inward only to vertical solution at the light cylinder. For cold winds, the unphysical branch asymptotes to the physical branch of the minimum torque solution at asymptotic infinity. However, for the case of more interest with finite temperature, the two branches of the minimum torque solution cross at the fast point. It is the self-intersecting nature of the one-dimensional set of minimum torque solutions that creates the nontrivial topology of the MHD solution space.

IV A supercritical subset of solution space is defined by $E > E_{CR}$. These solutions connect the source to asymptotic infinity. The solutions are accelerated to just beyond the light cylinder where they begin to slow down to an asymptotic speed. The flow speed is always less than the fast speed.

V There is also a set of supercritical solutions, $E > E_{CR}$, that are unphysical. These solutions connect asymptotic infinity to the degenerate vertical solution at the light cylinder.

VI In the finite temperature case, there exists a set of subcritical solutions that connect asymptotic infinity to itself with a turnaround point outside the light cylinder.

One should note that there is a breeze solution that emanates from the source and never crosses the light cylinder. However, it does not appear in Fig. 5.2 because the range of coordinates on the axes that were chosen do not provide adequate resolution.

Figure 5.2 shows that the fast critical point is an X-type critical point in the wind equation since solutions cross at this point. The slow critical point is of X-type as well, but does not occur when the winds injected above the slow wave group velocity.

5.6 The Minimum Torque Solution

The critical solution in Fig. 5.2 is of particular physical interest because it is the solution that extracts the minimum energy and angular momentum from the central engine (i.e., a neutron star or black hole). We are interested in the magnetically dominated limit of this solution, i.e., $U_A^2 \gg 1$. Thus, (2.44) implies that $U_F^2 \gg 1$, and the outgoing wind is relativistic $\beta^P \approx +1$.

Michel [120] showed that this wind obeys the relation (recall the definition of conserved quantities in 5.20 and 5.21),

$$e \approx m_e c^2 + \frac{\Omega_F \ell}{c}, \tag{5.43}$$

where the use of m_e for the plasma mass implies a positronic wind. The frozen-in condition (5.16b) can be written in terms of B^T, using (5.17)

$$B^T = \frac{(\Omega_p - \Omega_F)}{c\beta^P} B^P g_{\phi\phi}, \tag{5.44}$$

where $\Omega_p = d\phi/dt$ is the angular velocity of the plasma as viewed from asymptotic infinity. Integrating, the time stationary version of Ampere's law (3.62b) and integrating the time stationary current conservation law (4.106) as was done in (5.11) yields

$$B^T = \frac{2}{c}\int \alpha J^1 dX^2 \wedge dX^\phi \equiv \frac{2I^1}{c}\,, \tag{5.45}$$

where I^1 is the total field aligned current at cylindrical radii inside the azimuthal flux tube. Inspection of the azimuthal momentum equation,

$$n\left[u^P\frac{\partial}{\partial X^1}(\mu u^\phi) + \mu u^P u^\phi \Gamma^1{}_{\phi\phi}\right] = -\frac{J^2 B^P}{c}\,, \tag{5.46}$$

points out that in the magnetically dominated limit J^2 is negligibly small in a wind. The cross-field current J^2 is known as the inertial current as it is driven by the inertial loading of the field lines and breaks the force-free condition. Applying Ampere's law (3.61b) and the expansion of the connection (3.32c) to the azimuthal momentum equation yields

$$\frac{n\alpha u^P}{B^P}\frac{1}{\sqrt{g_{\phi\phi}}}\frac{\partial}{\partial X^1}\left(\mu u^\phi\sqrt{g_{\phi\phi}}\right) = \frac{c}{4\pi\sqrt{g_{\phi\phi}}}\frac{\partial}{\partial X^1}\left(B^T\right)\,. \tag{5.47}$$

Then by (5.12)

$$\frac{\partial}{\partial X^1}\left(k\mu u^\phi g_{\phi\phi} - \frac{cB^T}{4\pi}\right) = 0\,, \tag{5.48}$$

which is just the conservation of angular momentum equation (5.21). Equations (5.46), (5.48) and (5.45) show that $B^T \approx$ constant and $I^1 \approx$ constant in a magnetically dominated wind. The wind is almost pure Poynting flux with very little inertial loading, thus the minimum torque condition (5.43) is satisfied by (5.20) and (5.21) in this limit.

As $r \to \infty$, $\beta^\phi < 1$ and we must therefore have

$$\lim_{r\to\infty} \Omega_p = 0\,. \tag{5.49}$$

Consequently, for the relativistic minimum torque wind we have an approximately constant toroidal magnetic field density given by the asymptotic form of the frozen-in condition (5.44), B^T_∞

$$B^T \approx B^T_\infty \approx -\frac{\Omega_F B^P}{c g_{\phi\phi}}\,, \tag{5.50a}$$

$$B^T_\infty \equiv -\frac{\Omega_F \Phi}{c k_F}\,. \tag{5.50b}$$

where k_F is a geometrical factor. If asymptotically, the wind collimates as in Cygnus A and is uniform as well, $k_F \approx \pi$. Note that the Poynting flux is also approximately constant in the flux tube by (5.20) and (5.21).

$$\int S^P \mathrm{d}A_\perp = 4\pi \frac{\Omega_F^2}{c^2} k \frac{\Phi}{k_F} \delta\Phi \,. \tag{5.51}$$

In general from (5.20), (5.21) and the asymptotic frozen-in condition (5.44) with the constraint as $r \to \infty$ of (5.49), one has

$$\int S^P_\infty \mathrm{d}A_\perp = \frac{\Omega_F}{c} \int \left(S^P_L\right)_\infty \mathrm{d}A_\perp = \frac{4\pi}{c} \frac{\Omega_F^2}{c^2} \frac{\Phi}{k_F} k \frac{\delta\Phi}{\beta^P} \,. \tag{5.52}$$

Thus, by (5.52), for five given constants of motion in a flux tube, $\Phi, \delta\Phi, k, \Omega_F$ and $\mathbb{S}$, the conserved energy and angular momentum fluxes in a magnetically dominated wind are minimized as $\beta^P \to 1$. Hence, the critical solution is the minimum energy and minimum torque solution for all possible winds in a flux tube with $\Phi, \delta\Phi, k, \Omega_F$ and $\mathbb{S}$ given as initial conditions (see Fig. 5.2). Note that this solution implies that two of the constants, k and $\mathbb{S}$, are completely negligible.

5.7 The Grad–Shafranov Equation

So far we have analyzed the flow in isolated azimuthally symmetric flux tubes. In order to study a global wind problem, we need to study the shape and distribution of the flux tubes. This is determined by the trans-field poloidal momentum equation. Since the seven constants of motion are unique in each flux tube and the flux tubes can be labeled by the value of the stream function (Ψ = a constant in each flux tube), the constants in the global wind are not functions of coordinate, but functions of Ψ. When the trans-field poloidal momentum equation is rewritten as an equation of the constants of motion that are differentiated with respect to Ψ, it is known as the Grad–Shafranov equation. We have avoided discussion of this so far because the equation is intractable in practice, even within the perfect MHD assumption.

Nitta et al. [121] have derived the Grad–Shafranov equation in the cold plasma limit in Boyer–Lindquist coordinates. In Boyer–Lindquist coordinates the trans-field momentum equation is given by

$$\frac{\tilde{F}^A{}_\phi}{\tilde{F}_{B\phi}\tilde{F}^B{}_\phi} \left(\mu n \tilde{u}^\beta \tilde{u}_{A;\beta} - \tilde{F}_{A\beta} \tilde{J}^\beta\right) = 0\,, \quad A,B = r,\theta \,. \tag{5.53}$$

Defining

$$\tilde{B}_P^2 = \tilde{F}_{B\phi}\tilde{F}^B{}_\phi\,, \quad A,B = r,\theta\,, \tag{5.54}$$

the transfield momentum equation (5.53) can be rewritten using $u \cdot u = -1$ as

$$\frac{\tilde{F}^{A}{}_{\phi}}{\tilde{B}_{p}^{2}}\left\{\mu n\left[-\tilde{u}^{B}\left(\tilde{u}_{B;A}-\tilde{u}_{A;B}\right)-\tilde{u}^{t}\partial_{A}\tilde{u}_{t}-\tilde{u}^{\phi}\partial_{A}\tilde{u}_{\phi}\right]-\tilde{F}_{AB}\tilde{J}^{B}-\tilde{F}_{A\phi}\tilde{J}^{\phi}-\tilde{F}_{At}\tilde{J}^{t}\right\}=0\,. \tag{5.55}$$

We will rewrite (5.55) in terms of $\tilde{J}^{\phi}$ since we have a second equation in $\tilde{J}^{\phi}$ from Maxwell's equations. After elimination of $\tilde{J}^{\phi}$ from the two equations, we can obtain the Grad–Shafranov equation. Using the frozen-in condition and (3.64a) for Ampere's law

$$\begin{aligned}\tilde{J}^{\phi}=-\frac{1}{4\pi\sqrt{-\tilde{g}}}&\left[\partial_{r}\left[\frac{\sqrt{-\tilde{g}}}{g_{rr}}\left(\frac{g_{tt}+g_{t\phi}\Omega_{F}}{\alpha^{2}g_{\phi\phi}}\right)\partial_{r}\Psi\right]\right.\\&\left.+\partial_{\theta}\left[\frac{\sqrt{-\tilde{g}}}{g_{\phi\phi}}\left(\frac{g_{tt}+g_{t\phi}\Omega_{F}}{\alpha^{2}g_{\phi\phi}}\right)(\partial_{\theta}\Psi)\right]\right]=0\,.\end{aligned} \tag{5.56}$$

Nitta et al. [121] introduce the direction derivative along a field line

$$\partial_{\Psi}=\left(\frac{\tilde{F}^{A}{}_{\phi}}{\tilde{B}_{p}^{2}}\right)\partial_{A} \tag{5.57}$$

Using this definition, they rewrite the transfield momentum equation, (5.55), in four pieces as follows:

$$-\frac{\mu n\tilde{F}^{A}{}_{\phi}\tilde{u}^{B}}{\tilde{B}_{P}^{2}}\left(\tilde{u}_{B;A}-\tilde{u}_{A;B}\right)=\frac{\alpha^{2}M^{2}}{g_{tt}+g_{t\phi}\,\Omega_{F}}\tilde{J}^{\phi}+\frac{\mu kB_{P}^{2}}{g_{tt}+g_{t\phi}\,\Omega_{F}}\partial_{\Psi}\left[\frac{k}{(g_{tt}+g_{t\phi}\,\Omega_{F})\,n}\right], \tag{5.58a}$$

$$-\frac{\mu n}{\tilde{B}_{P}^{2}}\tilde{F}^{A}{}_{\phi}\left(\tilde{u}^{t}\partial_{A}\tilde{u}_{t}+\tilde{u}^{\phi}\partial_{A}\tilde{u}_{\phi}\right)=-\mu n\left(\tilde{u}^{t}\partial_{\Psi}\tilde{u}_{t}+\tilde{u}^{\phi}\partial_{\Psi}\tilde{u}_{\phi}\right), \tag{5.58b}$$

$$-\frac{\tilde{F}^{A}{}_{\phi}}{\tilde{B}_{P}^{2}}\left(\tilde{F}_{AB}\tilde{J}^{B}+\tilde{F}_{A\phi}\tilde{J}^{\phi}\right)=-\tilde{J}^{\phi}-\frac{\tilde{F}_{r\theta}}{4\pi\sqrt{-\tilde{g}}}\partial_{\Psi}\left(\sqrt{-\tilde{g}}\tilde{F}^{r\theta}\right), \tag{5.58c}$$

$$-\frac{\tilde{F}^{A}{}_{\phi}}{\tilde{B}_{P}^{2}}\tilde{F}_{At}\tilde{J}^{t}=-\Omega_{F}\frac{g_{t\phi}+g_{\phi\phi}\,\Omega_{F}}{g_{tt}+g_{t\phi}\,\Omega_{F}}\tilde{J}^{\phi}-\frac{\Omega_{F}}{4\pi}\frac{B_{P}^{2}}{g_{tt}+g_{t\phi}\,\Omega_{F}}\partial_{\Psi}\left[\frac{g_{t\phi}+g_{\phi\phi}\,\Omega_{F}}{g_{tt}+g_{t\phi}\,\Omega_{F}}\right], \tag{5.58d}$$

where the following definition was used:

$$B_P^2 \equiv -\frac{\left(g_{tt}+g_{t\phi}\,\Omega_F\right)^2}{\alpha^2 g_{\phi\phi}}\left[g^{rr}\left(\partial_r\Psi\right)^2+g^{\theta\theta}\left(\partial_\theta\Psi\right)^2\right]$$

$$= -\frac{\left(g_{tt}+g_{t\phi}\,\Omega_F\right)^2}{\alpha^2 g_{\phi\phi}}\left(\nabla\Psi\cdot\nabla\Psi\right)\,. \tag{5.58e}$$

Inserting the expansion (5.58) back into the transfield poloidal momentum equation (5.55) yields

$$\begin{aligned}\frac{\alpha^2\left[1-\left(\beta_F^\phi\right)^2-M^2\right]}{g_{tt}+g_{t\phi}\,\Omega_F}\tilde{J}^\phi &= -\mu n\left(\tilde{u}^t\partial_\Psi\tilde{u}_t+\tilde{u}^\phi\partial_\Psi\tilde{u}_\phi\right)-\frac{\tilde{F}_{r\theta}}{4\pi\sqrt{-\tilde{g}}}\partial_\Psi\left(\sqrt{-\tilde{g}}\tilde{F}^{r\theta}\right)\\ &+\frac{\mu k B_P^2}{g_{tt}+g_{t\phi}\Omega_F}\partial_\Psi\left[\frac{k}{\left(g_{tt}+g_{t\phi}\Omega_F\right)n}\right]\\ &-\frac{\Omega_F}{4\pi}\frac{B_P^2}{g_{tt}+g_{t\phi}\Omega_F}\partial_\Psi\left[\frac{g_{t\phi}+g_{\phi\phi}\Omega_F}{g_{tt}+g_{t\phi}\Omega_F}\right]\end{aligned} \tag{5.59}$$

We can express the first two terms in (5.59) as derivatives of constants of motion with respect to Ψ by using (5.20) and (5.21),

$$\partial_\Psi\left(e-\frac{\Omega_F\,\ell}{c}\right)=\mu\left(\partial_\Psi\tilde{u}_t+\Omega_F\,\partial_\Psi\tilde{u}_\phi+\tilde{u}_\phi\partial_\Psi\Omega_F\right)\,, \tag{5.60}$$

$$\partial_\Psi\ell=-\mu\partial_\Psi\tilde{u}_\phi+\frac{1}{4\pi k}\partial_\Psi\left(\sqrt{-\tilde{g}}\tilde{F}^{r\theta}\right)+\frac{\sqrt{-\tilde{g}}}{4\pi}\tilde{F}^{r\theta}\partial_\Psi\left(\frac{1}{k}\right)\,. \tag{5.61}$$

Combining (5.59), (5.60) and (5.61) yields

$$\begin{aligned}\frac{\alpha^2\left[1-\left(\beta_F^\phi\right)-M^2\right]}{g_{tt}+g_{t\phi}\,\Omega_F}\tilde{J}^\phi &= -n\tilde{u}^t\,\partial_\Psi\left(e-\frac{\Omega_F\,\ell}{c}\right)+\mu n\tilde{u}^t\,\tilde{u}_\phi\,\partial_\Psi\Omega_F\\ &+\frac{kB^T}{\alpha^2 g_{\phi\phi}}\partial_\Psi\ell+\frac{k\left(B^T\right)^2}{4\pi\alpha^2 g_{\phi\phi}}\partial_\Psi\left(\frac{1}{k}\right)\\ &+\frac{\mu k B_P^2}{g_{tt}+g_{t\phi}\Omega_F}\partial_\Psi\left[\frac{k}{\left(g_{tt}+g_{t\phi}\Omega_F\right)n}\right]\\ &-\frac{\Omega_F}{4\pi}\frac{B_P^2}{g_{tt}+g_{t\phi}\Omega_F}\partial_\Psi\left[\frac{g_{t\phi}+g_{\phi\phi}\Omega_F}{g_{tt}+g_{t\phi}\Omega_F}\right]\,.\end{aligned} \tag{5.62}$$

Eliminating $\tilde{J}^\phi$ from (5.62) using Ampere's law (5.56) yields the cold plasma limit of the Grad–Shafranov equation in the Kerr spacetime:

$$
-\frac{\alpha^4\left[1-(\beta_F^\phi)^2-M^2\right]}{4\pi\sqrt{-\tilde{g}}}g_{\phi\phi}\left[\partial_r\left[\frac{\sqrt{-\tilde{g}}\left(1-(\beta_F^\phi)^2-M^2\right)\partial_r\Psi}{g_{\phi\phi}\quad g_{rr}}\right]\right.
$$

$$
\left.+\partial_\theta\left[\frac{\sqrt{-\tilde{g}}\left[1-(\beta_F^\phi)^2-M^2\right]\partial_\theta\Psi}{g_{\phi\phi}\quad g_{\theta\theta}}\right]\right]
$$

$$
=-2\pi\left[\frac{g_{\phi\phi}}{M^2}\left[k^2\left(e-\frac{\Omega_F\ell}{c}\right)^2\right]'+g_{tt}\left(k^2\ell^2\right)'+2g_{t\phi}\left(k^2e\ell\right)'+g_{\phi\phi}\left(k^2e^2\right)'\right]
$$

$$
+\frac{4\pi\mu^2\alpha^2g_{\phi\phi}\left[1-(\beta_F^\phi)^2-M^2\right]}{M^2}kk'+\frac{4\pi\mu^2g_{\phi\phi}}{M^4}\left(g_{t\phi}+g_{\phi\phi}\Omega_F\right)k^2
$$

$$
\times\left[\left(\frac{e-\frac{\Omega_F\ell}{c}}{\mu}\right)^2-\alpha^2\left[1-(\beta_F^\phi)^2-M^2\right]\right]\Omega_F', \tag{5.63}
$$

where the prime denotes derivative with respect to Ψ.

Beskin [118] gives the Grad–Shafranov equation in the warm plasma limit in more compact form as

$$
\frac{1}{\alpha}\nabla_k\left\{\frac{\alpha}{g_{\phi\phi}}\left[1-(\beta_F^\phi)^2-M^2\right]\nabla^k\Psi+\frac{1}{\alpha\sqrt{g_{\phi\phi}}}(\beta_F^\phi)(\nabla\Psi\cdot\nabla\Psi)\frac{\partial\Omega_F}{\partial\Psi}\right\}
$$

$$
+\frac{64\pi^4}{\alpha^4g_{\phi\phi}M^2}\frac{\partial}{\partial\Psi}\left[\frac{g_{\phi\phi}\left[\left(e-\frac{\Omega_F\ell}{c}\right)^2+\left(e-\frac{\Omega_{\min}\ell}{c}\right)\left(e-\frac{\Omega_{\max}\ell}{c}\right)\right]}{1-\left(\beta_F^\phi\right)^2-M^2}\right]
$$

$$
-16\pi^3\mu n\frac{1}{k}\frac{\partial k}{\partial\Psi}-16\pi^3nT\frac{d\mathbb{S}}{\partial\Psi}=0\,. \tag{5.64}
$$

Beskin [118] shows that the Grad–Shafranov equation has singular points at the fast, Alfvén and slow critical surfaces, as does the wind equation. Nitta et al. [121] show that the regularity conditions at the Alfvén surface reduce to those of the wind equation in (5.42).

The only somewhat tractable version of the Grad–Shafranov equation is in the cold force-free limit of (5.64):

$$
\frac{1}{\alpha}\nabla_k\left\{\frac{\alpha}{g_{\phi\phi}}\left[1-(\beta_F^\phi)^2\right]\nabla^k\Psi\right\}+\frac{\beta_F^\phi}{\alpha\sqrt{g_{\phi\phi}}}(\nabla\Psi\cdot\nabla\Psi)\frac{\mathrm{d}\Omega_F}{\mathrm{d}\Psi}+\frac{4\pi^2}{\alpha^2g_{\phi\phi}}B^T\frac{\mathrm{d}B^T}{\mathrm{d}\Psi}=0\,. \tag{5.65}
$$

In the limit of the flat space and rigid rotation with a star, $g_{\phi\phi}$ becomes the cylindrical radius and $d\Omega_F/d\Psi=0$, $\Omega_F=\Omega_{\rm star}$ and (5.65) becomes the "pulsar" equation [122].

Equation (5.65) is not very useful in the ergosphere since the boundary condition at the horizon given by Sect. 3.7 is the opposite of force-free, the plasma is inertially dominated at the horizon. Consequently, we will never explicitly solve for the distribution of flux in the magnetosphere of a black hole as the Grad–Shafranov equation in a realistic circumstance is far too complicated. Furthermore, it will be discussed in Chap. 9 that dissipative $(F^{\mu\nu}u_{\nu} \neq 0)$ regions are likely to bound the perfect MHD portion of the magnetosphere. The solutions of the Grad–Shafranov equation are very sensitive to boundary conditions. These dissipative regions do not obey the perfect MHD Grad–Shafranov equation. Thus, it is difficult to know what boundary conditions exist on the perfect MHD distribution of magnetic flux in the magnetosphere. Unfortunately, physical reality renders the discussion of the Grad–Shafranov equation probably no more than a mere pedantic exercise. The distribution of magnetic flux near the black hole is still an unsolved problem in realistic plasma-filled magnetospheres.

Chapter 6
Perfect MHD Winds and Waves in the Ergosphere

In this chapter, we discuss the nature of ingoing perfect MHD winds in the ergosphere. It is illustrated in Sect. 6.3 that the spacetime near the event horizon is an asymptotic infinity to MHD winds and waves similar to the infinity of the semi-infinite plasma filled waveguide of Sect. 2.9.4. As we found in the study of electrodynamics of black holes in Chap. 4, the spacetime near the event horizon has no significance for the determination of a global flow (e.g., the constants of motion in a magnetic flux tube), except that it is a sink for mass influx. Most of this chapter is concerned with the MHD causal structure of black hole GHM. We continue to emphasize the differences between oblique Alfvén waves and fast waves beyond just the wave speeds.

Section 6.4 is a very long calculation of the structure of fast waves that propagate outward from the inner ergosphere. This analysis has important implications that mandate the application of the spacetime near the horizon as a causal MHD boundary as physically inappropriate (Sect. 6.5). Although this is evident by the fact that fast waves do not carry field aligned currents or charge (see 2.58), it is possible as in a dipolar magnetic field that modal characteristics of the Alfvén wave become mixed with those of the fast wave [123] due to the curvature of the magnetic field lines. The curvature of the dipolar field lines induces the most significant changes in long wavelength modes. Recall that the derivations of MHD wave properties in Chap. 2 were predicated on the assumption that the waves varied on length scales much shorter than those of the background field variations. These long wavelength hybrid modes represent deviations from the short wavelength approximation and were not addressed in Chap. 2. The situation in the black hole magnetosphere is ostensibly exacerbated since there are curvature effects in both the field and the background spacetime. This raises the question as to whether the hybrid modes in the inner region of the black hole magnetosphere can significantly alter the wave properties investigated in Chap. 2 and therefore the causal structure of the MHD wind system. In order to resolve this issue, we quantify the effects of hybridization of outgoing fast waves near the horizon due to curvature effects in Sect. 6.4. This laborious calculation is a necessary technical difficulty that is required to complete the picture of the causal structure of black hole GHM developed in this book.

B. Punsly, *Black Hole Gravitohydromagnetics, 2nd. ed.*,
Astrophysics and Space Science Library 355, doi: 10/1007/978-3-540-76957-6_6,

Section 6.4 is a detailed computation of the structure of globally outgoing fast waves in the inner ergosphere in both the long and short wavelength limits. It is found that the deviations from the local analysis of fast waves are negligible in this context; we show that these fast waves carry predominantly cross-field poloidal currents and they can not change the electrostatic potential. The second of these two wave properties is a consequence of the magnetically dominated condition of $B^T \approx$ constant. We actually show that outgoing fast waves in the inner ergosphere can not affect significant global changes in B^T in a magnetic flux tube. From the frozen-in condition for B^T, and the horizon boundary conditions $\Omega_p \approx \Omega_H$ and $\beta^P \approx -1$ (see 3.95 and 3.94, respectively), the inability to change B^T is equivalent to the ineffectiveness of these waves to alter the global value of Ω_F or equivalently, by (5.15b), the cross-field potential that is set by the Goldreich–Julian charge density. This result adds to the depth of our knowledge of the causal structure of black hole GHM, but it is learned at great calculational expense; we need to describe the electromagnetic field structure of the MHD wave equations in the full time dependent version of the Newman–Penrose formalism for Maxwell's equation.

The calculation in Sect. 6.4 is very complex and yields a null result. Namely, outgoing fast waves near the horizon in the long wavelength limit are no different from other fast waves in the magnetosphere in the sense that they can not significantly alter the causal structure of MHD winds. The causality of MHD winds in a black hole magnetosphere is determined primarily by Alfvén waves as inferred by the results of Chap. 2. For this reason, some readers might prefer to skip Sect. 6.4 (without much loss of content for understanding black hole GHM) the first time through the book.

6.1 Paired MHD Winds

Consider a magnetic flux tube that threads the ergosphere. In the ergospheric region the flow will be a magnetized accretion flow and if there is an outward energy flux to infinity, an outgoing wind of magnetized plasma will exist at large r coordinate. Thus, there are two new aspects to this theory that do not occur in relativistic stellar wind theory. First, the flow must divide into an accretion flow and an outgoing wind. Thus, there is a source of mass flux in the magnetic flux tube itself, with some of the mass accreting toward the hole and some driven to infinity. The mass flux constant, k, in (5.12), is not conserved because the source function, S, in the mass conservation law (3.36b) is significant in some region of the flux tube. The second new aspect for the accretion flow is that the mass flux is anti-directed to the energy flux. Since the equations of perfect MHD were written in the ZAMO frames in Chap. 5, they are equally valid near the hole or at asymptotic infinity. Consequently, the same formalism can be used to describe the perfect MHD regime of the accretion flow and the outgoing wind. We will therefore refer to this as a paired wind system. The ingoing wind is the accretion flow and the outgoing wind is of astrophysical interest.

The ingoing wind has interesting black hole GHM properties in the ergosphere and is potentially capable of extracting energy from the hole.

The details of the plasma injection process has been discussed by various authors. Charged starved vacuum gap models were discussed in [66] and [124] in analogy to models of pulsars. However, unlike pulsars there is likely to be a significant luminosity of background γ-rays from equatorial accretion disks and coronae in AGN. As will be shown below, $\gamma + \gamma \rightarrow e^+ + e^-$ scattering will almost certainly short out the vacuum electric fields making any semi-vacuum pair creation process irrelevant. A crucial distinction from pulsars is that the scale lengths are at $\sim 10^8$ larger for astrophysical black holes. Even though the voltage drops along vacuum field lines are sufficient for vacuum or semi-vacuum pair creation, the Goldreich–Julian charge density is only $\sim$1 cm^{-3} for a $10^9 M_\odot$ black hole magnetosphere, since it is proportional to the magnetic field strength (which is $\sim 10^{-8}$ of B^P of a pulsar) and the field line angular velocity (which is $\sim 10^{-5}$ that of a pulsar). Thus, if a pair creation process greatly exceeds the Goldreich–Julian density in a stationary state then the electric fields will vanish locally and perfect MHD will be established by local dynamics, and the memory of the vacuum fields will be erased. This concept was first noted by Phinney [125]. He crudely estimated the stationary pair density by balancing the infall (free-fall) rate with the pair creation rate yielding

$$n \sim \left(\frac{m_p}{m_e}\right)\left(\frac{L_C}{L_{Edd}}\right)^2 10^{13} M_8^{-1}\,\mathrm{cm}^{-3}\,, \tag{6.1}$$

where L_C is the luminosity of γ-rays > 1 MeV from the accretion disk and corona, and M_8 is the mass of the black hole in units of $10^8\,\mathrm{M}_\odot$. For a $10^9 M_\odot$ central black hole, (6.1) implies that

$$n > 10\rho_{G-J}\,, \quad L_c > 10^{41}\,\mathrm{ergs/sec}\,. \tag{6.2}$$

EGRET measurements of γ-rays from AGN can only detect $L_C > 10^{46}$ ergs/sec for the nearest radio loud quasars and $L_C > 10^{45}$ ergs/sec for the nearest radio galaxies [126]. Thus, it is still likely that AGN accretion disks have a value of $L_C > 10^{41}$ ergs/sec which is sufficient to short out the strongest magnetospheres of black holes. Taking a nominal value of $L_C = 10^{43}$ ergs/sec in a radio loud quasar (from the accretion flow alone, not including jet emission which is far away and beamed away from the black hole) would yield stationary pair densities of

$$n \sim 10^5 - 10^6\,\mathrm{cm}^{-3}\,. \tag{6.3}$$

Phinney [125] also introduced the useful concept of a particle creation zone. This construct is based on the assumption that the flux tube is long enough so that one can consider most of the particles to have been created in a finite segment of the flux tube. Outside of this region, the effects of particle creation are negligible. In terms of the mass conservation law (3.36b), we note that $S \ll (nu^P/r)$ in the wind zones because n has been built up in a long length of pair creation in the flux tube upstream.

The flow beyond the particle creation zone can be considered to emanate as perfect MHD with conserved values of k, ℓ and e (see Fig. 6.1). The outflow is initiated by thermal expansion and centrifugal forces. In the inflow, the effective gravity of (3.41) is inward directed since the centrifugal force cannot balance radial gravity. We expect that a geometry similar to Fig. 6.1 can feed magnetized pair plasma into the ergosphere.

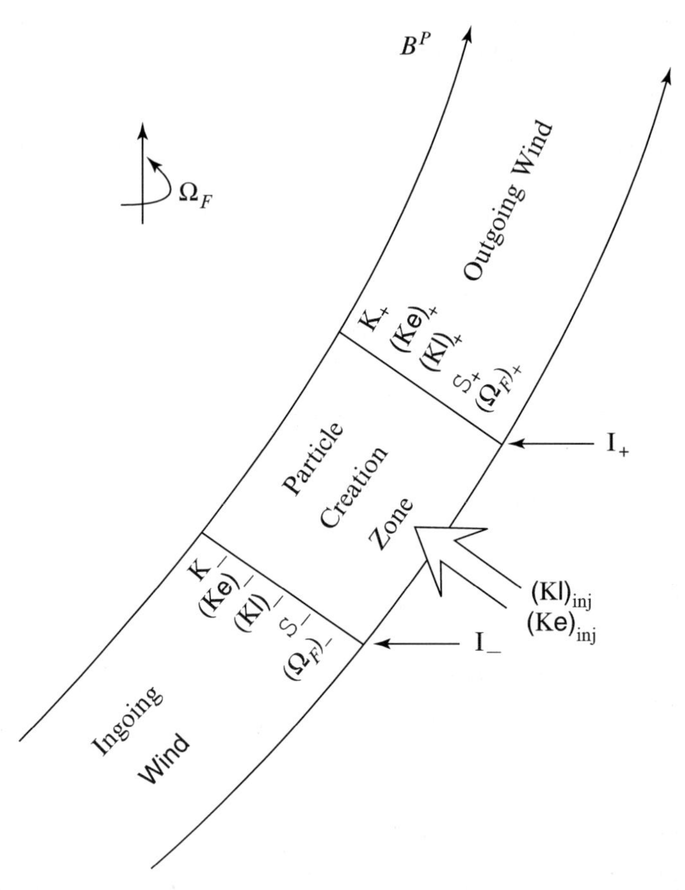

Fig. 6.1 Plasma can be created on large scale magnetic field lines that thread the ergosphere. Pair creation in the γ-ray field of an AGN is the most likely source of plasma. Far from the black hole, centrifugal forces and thermal expansion can initiate an outgoing wind that can be magnetically slung to infinity. Closer to the black hole centrifugal forces can not balance gravity and thermal expansion resulting in an ingoing wind (a magnetized accretion flow). The bulk of the particle creation can be considered to exist in a finite section of magnetic flux tube, the particle creation zone. The process of pair creation injects energy and angular momentum flux into the axisymmetric magnetic flux tube, $(ke)_{inj}$ and $(k\ell)_{inj}$, respectively. The ingoing and outgoing winds can be considered to result from the integration of the MHD wave equations away from the boundary surfaces, I_- and I_+, respectively. The constants of motion for each wind are given by the initial data on these surfaces

6.2 Ingoing Perfect MHD Ergospheric Winds

We specialize the wind analysis of Chap. 5 that was formulated in the ZAMO frames to the ingoing wind. The critical surfaces defined by (5.36) and (5.38) are expressed in terms of rest frame evaluated quantities. Since the asymptotic horizon boundary condition is $u^P \sim \alpha^{-1}$ (see 3.94), any MHD wind must pass sequentially through the slow, Alfvén and fast critical surfaces. This is a trivial result since light waves can not escape the gravitational pull at the horizon, so no subluminal wave such as a plasma wave can escape the horizon either.

By (5.36) for either the ingoing or outgoing wind, the flow goes super Alfvénic before reaching the light cylinder. Consider the interesting case of $0 < \Omega_F < \Omega_H$ in which the magnetosphere transports an outward directed angular momentum flux (see 6.8 below). The light cylinder of the ingoing wind is a result of the dragging of inertial frames, $\beta_F^\phi = -1$. This is in contrast to an outgoing wind in which $\beta_F^\phi = +1$. For the ingoing wind,

$$\Omega_F = \Omega_{min}\left[r_{L.C.}(\theta)\right] , \tag{6.4}$$

and for the outgoing wind,

$$\Omega_F = \Omega_{max}\left[r_{L.C.}(\theta)\right] , \tag{6.5}$$

where $r_{L.C.}$ is the radial coordinate parameterization of the speed of light surface, i.e., the "light cylinder."

We are particularly interested in magnetically dominated winds in AGN. From (6.3) we expect values of the pure Alfvén speed of $U_A \sim 10^3 - 10^4$ to be typical in the ergosphere. Thus, the Alfvén point occurs just before the inner light cylinder. By (5.36),

$$(\beta_F^\phi)_A \gtrsim -1 . \tag{6.6}$$

Since the ZAMOs are a physical frame, the energy of the plasma $E > 0$ (see 3.14),

$$E = \frac{e - \frac{\Omega \ell}{c}}{\alpha} > 0 . \tag{6.7}$$

Combining (6.7) and (6.6) with the Alfvén point condition (5.42a) yields a constraint on the angular momentum if a flow reaches the Alfvén point,

$$\ell < 0 , \quad 0 < \Omega_F < \Omega_H . \tag{6.8a}$$

Similarly, the Alfvén point condition combined with (6.6) and (6.7) yields the following conditions for inner Alfvén point accessibility with perfect MHD:

$$e \approx \frac{\Omega_F \ell}{c} < 0 , \qquad 0 < \Omega_F < \Omega_H , \tag{6.8b}$$

$$e - \frac{\Omega_F \ell}{c} > 0 . \tag{6.8c}$$

For inflow, $k < 0$, thus the energy and angular momentum fluxes, ke and $k\ell$, are outward directed even though the flow is inward directed. The fact that the specific energy and angular momentum are negative is reminiscent of the energy extracting processes discussed in Sect. 3.5. This is not coincidental and the fact that a connection between field line angular velocity and negative energy states is established as far out as the Alfvén surface is not coincidental either. Remember, in a magnetically dominated wind or paired wind system the current flowing parallel to the field and the field line angular velocity are essentially constant from the source to asymptotic infinity. The field line angular velocity is equivalent to the electrostatic potential drop across the magnetic field by (5.15b) which is supported by the Goldreich–Julian charge density by Gauss' law (3.59). Field aligned poloidal currents and a charge separation (as is needed to support ΔV) can only be transported by Alfvén waves, not fast waves as shown in (2.56) and (2.58), respectively. The nature of the time dependent waves is imprinted into the elliptic time stationary wind equations through these Alfvén point relations and their global causal significance.

Note that for $0 < \Omega_F \ll \Omega_H$, the Alfvén point is just inside of the stationary limit. For $\Omega_F \sim (1/2)\Omega_H$, $\alpha_A \sim 0.1$ and as $\Omega_F \to \Omega_H$ the Alfvén point goes toward the event horizon.

If one writes the fast critical surface condition (5.38) and divides through by $u_1^2 = U_F^2$, one has

$$U_F^2 = \left(\frac{F^{\mu\nu}F_{\mu\nu}}{8\pi n\mu} + U_S^2\right) + \frac{U_S^2 n\alpha^2\left[1-(\beta_F^\phi)^2\right]}{k^2\mu} . \tag{6.9}$$

In the magnetically dominated case the sound speed, U_S^2, is just a second order correction to U_F^2. When $\Omega_F \sim \Omega_H$, the fast speed at the critical surface is $\sim U_A^2$ by (5.35) and (6.9). In this case the fast surface is near the horizon. By contrast, when $0 < \Omega_F \ll \Omega_H$, the toroidal magnetic field near the inner light cylinder is small $|F^{12}| \ll B^P$, if the outgoing wind satisfies the minimum torque condition (5.50). Then by (5.35) and (6.9), $U_F^2 \gtrsim U_I^2$ and the fast critical surface is near the light cylinder.

6.3 The Horizon is an Asymptotic Infinity to MHD Winds

We elucidate the nature of the asymptotic spacetime near the horizon by studying an ingoing perfect MHD wind front at late times. By (4.59) and (3.95c) the wind front is always approaching, but never reaches the event horizon as viewed by external observers. Alternatively, by (4.61), the wind is approaching $r_* = -\infty$.

Introduce the local coordinate system of (4.94) in an arbitrarily small open set about the ingoing wind front. Take $\partial/\partial X^2$ of Ampere's law (3.61b) and combine with $\partial/\partial X^1$ of Ampere's law (3.62b) to obtain the current conservation law on the open set:

$$\frac{\partial}{\partial X^2}\left[\alpha\sqrt{g_{\phi\phi}}\frac{\partial}{\partial X^0}F^{20}\right] = \frac{\partial}{\partial X^2}\left[\frac{4\pi J^2\alpha\sqrt{g_{\phi\phi}}}{c}\right] + \frac{\partial}{\partial X^1}\left[\frac{4\pi J^1\alpha\sqrt{g_{\phi\phi}}}{c}\right]. \quad (6.10)$$

Next, consider the azimuthal momentum equation (5.46) and note that dynamical quantities on the left hand side can be expressed in terms of Heaviside step functions near the wind front. For example, if $\bar{X}_1$ is the position of the flow front in local coordinates, then $\beta^{^P} = \mathrm{d}\bar{X}_1/\mathrm{d}X^0$ and

$$\mu u^\phi = \mu(r,\theta)\,u^\phi(r,\theta)\,\Theta\left[\bar{X}_1 - \int_0^{X^0}\beta^{^P}\,\mathrm{d}X^0\right]. \quad (6.11)$$

Then writing the left hand side of the azimuthal momentum equation as in (5.47),

$$J^2 = -\left\{\left[\frac{ncu^1}{B^{^P}\sqrt{g_{\phi\phi}}}\frac{\partial}{\partial X^1}\left(\sqrt{g_{\phi\phi}}\,\mu u^\phi\right)\right](r,\theta)\right\}\Theta\left[\bar{X}_1 - c\int_0^{X^0}\beta^{^P}\mathrm{d}X^0\right]. \quad (6.12)$$

Noting that $\partial k/\partial X^1 = 0$ in (5.12), we can rewrite (6.12) as

$$J^2 = -\left\{\left[\frac{c}{\alpha\sqrt{g_{\phi\phi}}}\frac{\partial}{\partial X^1}\left(\alpha\sqrt{g_{\phi\phi}}\frac{n\mu u^\phi u^1}{B^P}\right)\right](r,\theta)\right\}\Theta\left[\bar{X}_1 - c\int_0^{X^0}\beta^{^P}\mathrm{d}X^0\right]. \quad (6.13)$$

Similarly, near the flow front, the frozen-in condition (5.13) implies

$$F^{20} = \beta_F^\phi B^P\Theta\left[\bar{X}_1 - c\int_0^{X^0}\beta^{^P}\mathrm{d}X^0\right]. \quad (6.14)$$

This approximation notes that tangential fringing fields near the flow front will be Lorentz contracted into a very thin layer $\sim\alpha^{-1}$ as viewed in a local ZAMO basis.

Combining Maxwell's equation (6.10) which is essentially the law of current conservation with the expansions (6.13) and (6.14), the field aligned current that sinks at the flow front is

$$J^1 = -\left\{\frac{1}{\alpha\sqrt{g_{\phi\phi}}}\frac{\partial}{\partial X^2}\left[\left(\alpha\sqrt{g_{\phi\phi}}\frac{\beta^{^P}\beta_F^\phi cB^P}{4\pi}\right) + \frac{n\mu u^1 u^\phi c}{B^P}\right](r,\theta)\right\}$$
$$\times\Theta\left[\bar{X}_1 - c\int_0^{X^0}\beta^{^P}\mathrm{d}X^0\right]. \quad (6.15)$$

Inserting (6.15) into Ampere's law (3.62b), one finds that

$$F^{12} = -\left\{\left[\frac{\beta^{^P}\beta_F^\phi cB^P}{4\pi} + \frac{n\mu u^1 u^\phi c}{B^P}\right](r,\theta)\right\}\Theta\left[\bar{X}_1 - c\int_0^{X^0}\beta^{^P}\mathrm{d}X^0\right]. \quad (6.16)$$

Comparing this expression to the frozen-in condition (5.16b) yields β^ϕ in the wind,

$$\beta^\phi = \left\{\left[\frac{\left[1-(\beta^{P})^2\right]\beta^\phi_F}{1+M^2}\right](r,\theta)\right\}\Theta\left[\bar{X}_1 - c\int_0^{X^0}\beta^{P}\,\mathrm{d}X^0\right] , \tag{6.17}$$

where M is the pure Alfvénic Mach number defined in (5.23a). Equation (6.17) can be used to write an expression for u^ϕ as $\alpha(\bar{X}_1)\to 0$. Combining (6.17), the horizon boundary condition, (3.95), and $u\cdot u = -1$ gives

$$u_\phi = \left\{\left[\frac{\beta^\phi_F}{u^0(1+M^2)}\right](r,\theta)\right\}\Theta\left[\bar{X}_1 - c\int_0^{X^0}\beta^{P}\,\mathrm{d}X^0\right]\left[1+\mathrm{O}(\alpha^2)\right] . \tag{6.18}$$

Thus, from (6.18), (6.15) and (6.10), current closure at the ingoing wind front is accomplished primarily with displacement current. The expression (6.18) is the same as that found from pulsar winds [67, 113]. Substituting (6.18) into the expression for the field aligned current (6.15) yields

$$J^1 = \left\{\frac{1}{\alpha\sqrt{g_{\phi\phi}}}\frac{\partial}{\partial X^2}\left[\left(\frac{\alpha\sqrt{g_{\phi\phi}}cB^P\beta^{P}\beta^\phi_F}{4\pi}\left[\frac{1+M^2(\beta^{P})^{-2}}{1+M^2}\right]\right)\right](r,\theta)\right\}$$
$$\times\Theta\left[\bar{X}_1 - c\int_0^{X^0}\beta^{P}\,\mathrm{d}X^0\right] . \tag{6.19}$$

We can show that J^1 is almost purely electrodynamic in nature. From the axisymmetric version of the divergence equation (3.55b), we have the limiting form,

$$\lim_{\alpha\to 0} B^P = B^r\left[1+\mathrm{O}(\alpha^2)\right] , \tag{6.20a}$$

as well as,

$$\sqrt{g_{\phi\phi}} \approx \left(\frac{r^2+a^2}{\rho^2}\right)\left[1-\frac{\Delta a^2\sin^2\theta}{2(r^2+a^2)}\right]\sin\theta . \tag{6.20b}$$

Since $\beta^{P}\approx -1$ (it is constant to $\mathrm{O}(\alpha^2)$), (6.19) can be written at small lapse function as

$$J^1 \approx \left\{\left[\frac{1}{\beta^{P}\sin\theta}\frac{\partial}{\partial\theta}\left(\frac{\sin\theta\,\beta^\phi_F B^P}{4\pi}\right)\right](r,\theta)\right\}\Theta\left[\bar{X}_1 - c\int_0^{X^0}\beta^{P}\,\mathrm{d}X^0\right] . \tag{6.21}$$

Gauss' law (3.59b) can be approximated near the event horizon to find the Goldreich–Julian charge density, ρ_{G-J},

$$\rho_{G-J} \equiv J^0 \approx \frac{c}{4\pi}\frac{1}{\sqrt{g_{\phi\phi}g_{rr}}}\frac{\partial}{\partial X^0}\left[\sqrt{g_{\phi\phi}g_{rr}}F^{00}\right] . \tag{6.22}$$

Comparing (6.22) to (6.21) we find, as for asymptotic pulsar winds [67],

$$J^1 \approx \frac{c\rho_{G-J}}{\beta^P} \,. \tag{6.23}$$

The current is purely electrodynamic up to a correction equal to $\left[1-(\beta^P)^2\right]J^1$. Note that this expression is the same as the asymptotic field aligned current in a plasma filled waveguide terminated by a Faraday wheel (see 2.119). It is also the same relation found to be characteristic of Alfvén waves in a cylindrical plasma. We conclude that the ingoing perfect MHD wave front is an Alfvén wave advected with the bulk plasma flow velocity and it must be radiated inward by something equivalent to a Faraday wheel upstream. The event horizon seems analogous to the vacuum infinity of a plasma filled waveguide.

Next we explore the radial electric field in the gap between the horizon and the wind front as we did for the idealized circuits in Sect. 4.7.4. We use the same contour of integration that was used in Figs. 4.5 and 4.6 except it now barely penetrates the wind zone as opposed to the leads (see Fig. 6.2). From the frozen-in condition (5.16b) and (5.13c), the ingoing wave condition of (3.90a) holds. The various pieces of the contour integration of Faraday's law in (4.100) are identical to those in Fig. 4.6 and (4.103). There is a voltage drop across the magnetic field lines near the horizon in the wind, yet this is canceled to $O(\alpha^2)$ in Faraday's law by the inflow of toroidal magnetic flux through S, and again we find $E^r \sim \alpha^0$ in the gap between the wind and the horizon. The voltage drop across the magnetic field is not a "battery-like" EMF.

This result is independent of the value of Ω_F. The spacetime near the event horizon passively accepts any voltage drop (or Ω_F as given by (5.15b)) or current from an MHD wind. It never reacts back on the incoming flow by creating a large E^r to break the electrostatic equilibrium at the flow front and therefore alter the current flow in the wind. The MHD wind plasma is easily shielded from the electric field in the gap by a well behaved charge density on the surface of the wind front. Again, as in Chap. 4, we find that the voltage drop between the wind front and the horizon, ΔV_r, obeys

$$\lim_{\alpha\to 0} \Delta V_r \sim \alpha^2 \,, \tag{6.24}$$

as would be expected for an asymptotic infinity. The spacetime near the event horizon behaves like an electrically disconnected Faraday wheel at the end of a plasma-filled waveguide as shown in the top view of Fig. 2.7. The contrast to a laboratory conductor that approaches the end of a plasma filled waveguide, is that the effective length for electromagnetic characteristics becomes infinitely long near the horizon at an ever increasing rate as the horizon is approached as shown in Sect. 4.5.2.

The spacetime near the event horizon has no relevance to MHD flows in the ergosphere and is completely characterized as a vacuum infinity to perfect MHD flows.

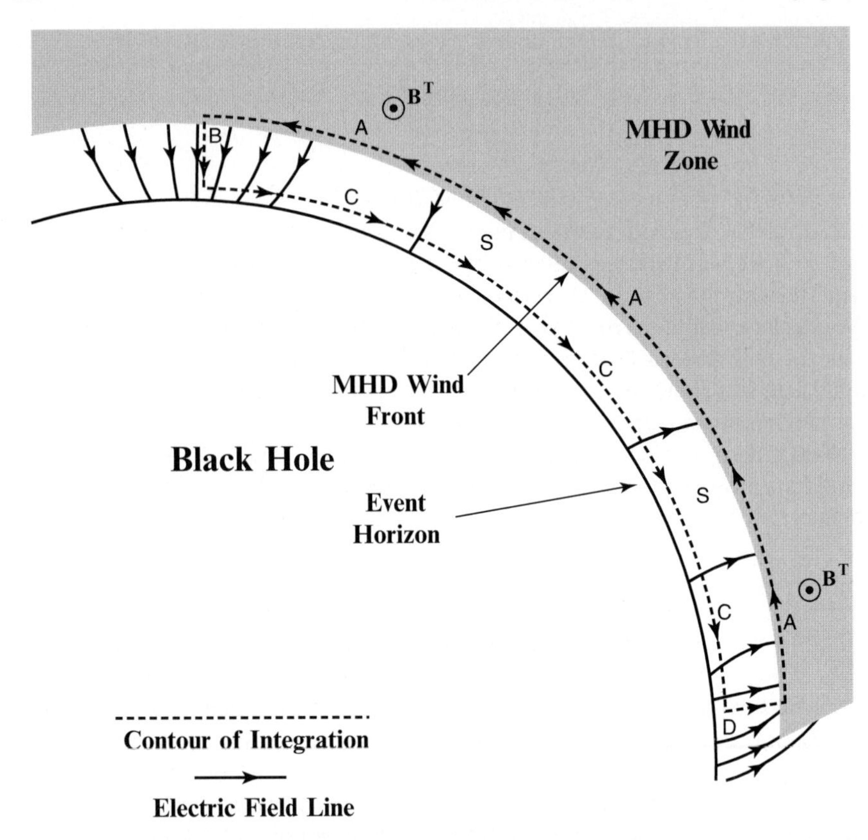

Fig. 6.2 Faraday's law can be applied at an ingoing perfect MHD wind front in order to show that the radial electric field in the gap between the wind zone and the event horizon is well behaved. The dashed contour, ABCD, is used to evaluate Faraday's law in (4.100). The piecewise contributions to the line integrals are identical to those indicated in Fig. 4.6. The contour ABCD bound a surface, S, that contains a toroidal magnetic flux that increases in time with the advance of the ingoing MHD wind front. As in vacuum electrodynamics (Chap. 4), this cancels the voltage drop across the magnetic field lines near the event horizon to O(α^2) in Faraday's law. The MHD wind zone is indicated by the shaded region in the figure and it is threaded by a large scale poloidal magnetic flux

6.4 Outgoing Fast Waves Near the Horizon

In this book, we have emphasized the Alfvén waves ability to transport electric charge and field aligned currents and the lack of such an ability for fast waves to do the same. Similarly, outgoing and ingoing wind fronts of strongly magnetized MHD winds have been shown to be step Alfvén waves. However, motivated by the concerns expressed in the introductory comments to this chapter, we need to address the issue of hybrid modes resulting from curvature effects to complete our GHM understanding of the causal structure of black hole magnetospheres. In this section, we explore the possibility that such effects can increase the relevance of fast waves as participants in the causal structure of the paired wind systems.

We noted in Sect. 6.2 that if $\Omega_F \sim \Omega_H$ then the inner fast point is near the horizon. In this section, we look at the electrodynamic structure of globally outgoing fast waves emitted from ergospheric plasma near the horizon when $\Omega_F \sim \Omega_H$. We will demonstrate that such waves are very inefficient modes for propagating torsional magnetic stresses, i.e., the perturbed toroidal magnetic field, $\delta B^T \sim \alpha_0^2$, where α_0 is the lapse function at the point of emission. We will also confirm the result of (2.58) that the poloidal current in a fast wave is orthogonal to the magnetic field.

There are two classes of fast waves that appear outgoing globally near the horizon. Firstly, most of the momentum space of such waves have very large energies as measured in the proper frame of the plasma and can be treated in the short wavelength approximation as was done in [129]. For these waves, hybrid effects are negligible since the wavelengths are much less than the radii of curvature of both spacetime and the magnetic field. Thus, the assumptions and results of the locally covariant analysis in Chap. 2 are valid. However, outgoing waves emitted from near the fast point appear to stagnate (i.e., they propagate globally outward very slowly), even in the short wavelength limit. This phenomenon is unique to a magnetosphere and therefore requires a new comprehensive analysis. This long calculation is performed in Sects. 6.4.6–6.4.8.

Secondly, there are long wavelength outgoing fast waves (i.e., wavelengths comparable to the radii of curvature of spacetime and the magnetic field) near the event horizon that need to be explored. The existence of such waves requires a fine tuning of the wave parameters since most waves that are capable of overcoming the intense gravitational redshift near the event horizon are very high energy (short wavelength) modes. These long wavelength solutions require a full relativistic treatment using the Newman–Penrose formalism introduced in Chap. 4. The calculation is very long and is broken up into many subsections, beginning with background material in Sects. 6.4.1–6.4.4 and culminating with the main calculation in Sect. 6.4.5. The results of all of these calculations are summarized in tabular form in Sect. 6.4.10.

6.4.1 The Vacuum Electrodyanmic Equations

In this chapter we are interested in causality and wave phenomena, so we need the time dependent solutions to Maxwell's equations as given by the set of equations (4.23). Following Chandrasekhar [92] we redefine the Newman–Penrose field components as

$$\tilde{\Phi}_0 \equiv \phi_0 \,, \tag{6.25a}$$

$$\tilde{\Phi}_1 \equiv -\frac{\phi_1}{\sqrt{2}\tilde{\rho}} \,, \tag{6.25b}$$

$$\tilde{\Phi}_2 \equiv \frac{\phi_2}{2\tilde{\rho}^2} \,. \tag{6.25c}$$

Chandrasekhar [127] finds the separated solutions of (4.23) to be

$$\tilde{\Phi}_0(r,\theta,\phi,t) = \mathrm{e}^{-\mathrm{i}\omega t}\mathrm{e}^{-im\phi}\, R_{+1}(r)\,{}_{+1}S_k{}^m(\theta)\,, \tag{6.26a}$$

$$\tilde{\Phi}_2(r,\theta,\phi,t) = \mathrm{e}^{-\mathrm{i}\omega t}\mathrm{e}^{-im\phi}\, B\bar{\rho}^2 R_{-1}(r)\,{}_{-1}S_k{}^m(\theta)\,, \tag{6.26b}$$

$$B^2 = (A+\omega^2 a^2 + 2a\omega m)^2 - 4a^2\omega^2 - 4a\omega m\,. \tag{6.26c}$$

Power series solutions to the vacuum Maxwell's equations given by (4.23) with $G_{\pm 1}(r) = 0$, were found in [128] near the horizon. There are two independent solutions for each radial equation of the form,

$$\overset{+}{R}(r-r_{+}) = |r-r_{+}|^{\lambda_{+}} \sum_{n=0}^{\infty} a_n (r-r_{+})^n\,, \tag{6.27a}$$

$$\overset{-}{R}(r-r_{+}) = |r-r_{+}|^{\lambda_{-}} \sum_{n=0}^{\infty} a_n (r-r_{+})^n\,. \tag{6.27b}$$

Solving the indicial equations for λ by the method of Frobenius yields

$$(\lambda_{+1})_{\pm} = -\frac{1}{4} \mp \frac{1}{4} \mp \frac{\mathrm{i}P_{+}}{2\sqrt{M^2-a^2}}\,, \tag{6.28a}$$

$$(\lambda_{-1})_{\pm} = \frac{1}{4} \pm \frac{1}{4} \mp \frac{\mathrm{i}P_{+}}{2\sqrt{M^2-a^2}}\,, \tag{6.28b}$$

where P_{+} is P of (4.23g) evaluated at the horizon. For ingoing vacuum electromagnetic waves near the horizon (6.27) and (6.28) imply that one has the simplification that $R_{+1}/R_{-1} \sim \alpha^{-2}$, so the R_{-1} solution is negligible. For the outgoing modes of interest $R_{+1}/R_{-1} \sim \alpha^0$, and by (4.10), both Newman–Penrose components are needed in general to describe the field near the horizon.

Solutions that are very accurate near the event horizon are obtained in [128],

$$R_{+1} \underset{r\to r_{+}}{\simeq} \exp\left[\mathrm{i}\int^{r}\left[\frac{P}{\Delta}+c_{+1}\right]\mathrm{d}r'\left[1+\mathrm{O}(\alpha^2)\right]\right]\,, \tag{6.29a}$$

$$R_{-1} \underset{r\to r_{+}}{\simeq} \exp\left[\mathrm{i}\int^{r}\left[\frac{P}{\Delta}+c_{-1}\right]\mathrm{d}r'\left[1+\mathrm{O}(\alpha^2)\right]\right]\,, \tag{6.29b}$$

where $c_{\pm 1}$ are constants in the free field solutions. When we study fast waves $c_{\pm 1}$ are functions in general.

Substituting (6.29) into (4.23) yields in the source free case,

$$c_{+1} = \frac{P_{+}k_{+} - 12\omega r_{+}\sqrt{M^2-a^2} - \mathrm{i}\left(6\omega r_{+}P_{+} + 2k_{+}\sqrt{M^2-a^2}\right)}{2P_{+}{}^2 + 8\left(M^2-a^2\right)}\,,$$

$$c_{-1} = -\frac{(k_{-}+2\mathrm{i}\omega r)}{P_{+}}\,. \tag{6.30a}$$

One can write (6.29) as

$$R_{\pm 1}(r) \underset{r \to r_+}{\sim} \exp\left[\mathrm{i}\int^r k_r(r')\,\mathrm{d}r'\right], \tag{6.31}$$

where the difference in the small corrections c_{+1} and c_{-1} can be factored out as a slowly varying complex amplitude. This phase function, $k_r(r)$, appears like a wave momentum in the WKB approximation.

6.4.2 Current Sources Near the Horizon

We now look at the restriction of the sources of Maxwell's equations near the horizon resulting from the horizon boundary condition (3.95) as we did for Poisson's equations in Sect. 4.5.1. The sources are more complicated than (4.31) because we now must include time derivatives in (4.19) and (4.21). These are the currents that support the electromagnetic fields in outgoing fast waves:

$$\begin{aligned} J_0 = &\left\{\frac{-\tilde{\rho}^*}{\sqrt{2}}\left[\mathrm{i}a\sin\theta\left(\frac{\partial}{\partial t}\right)+\left(\frac{\partial}{\partial\theta}\right)+\frac{1}{\sin\theta}\left(\frac{\partial}{\partial\phi}\right)\right]+\frac{\sqrt{2}\mathrm{i}a}{\rho^2}\sin\theta\right\}J_l \\ &-\left\{\left[\frac{r^2+a^2}{\Delta}\left(\frac{\partial}{\partial t}\right)+\left(\frac{\partial}{\partial r}\right)+\frac{a}{\Delta}\left(\frac{\partial}{\partial\phi}\right)\right]-2\tilde{\rho}-\tilde{\rho}^*\right\}J_m\,, \end{aligned} \tag{6.32a}$$

$$\begin{aligned} J_2 = &\frac{\Delta\tilde{\rho}^2}{2\sqrt{2}\rho^2}\left\{\sqrt{2}\left[\frac{r^2+a^2}{\Delta}\left(\frac{\partial}{\partial t}\right)-\left(\frac{\partial}{\partial r}\right)+\frac{a}{\Delta}\left(\frac{\partial}{\partial\phi}\right)-\tilde{\rho}\right]\tilde{\rho}^{-2}J_{\bar{m}}\right. \\ &\left.+2\left[-\mathrm{i}a\sin\theta\left(\frac{\partial}{\partial t}\right)+\left(\frac{\partial}{\partial\theta}\right)-\frac{\mathrm{i}}{\sin\theta}\left(\frac{\partial}{\partial\phi}\right)-\mathrm{i}a\tilde{\rho}\sin\theta\right]\frac{\rho^2}{\tilde{\rho}\Delta}J_n\right\}. \end{aligned} \tag{6.32b}$$

In order to use the asymptotic form of the Boyer–Lindquist current density from Sect. 4.5.1 in the Newman–Penrose expression for sources of Maxwell's equations (6.32), we write the perturbed current density as in (4.25a) using the general expression in (6.31),

$$\delta\tilde{J}^\mu = \int \mathrm{d}\omega \sum_{s,k,m} \tilde{J}^\mu_0 \mathrm{e}^{-\mathrm{i}\omega t}\mathrm{e}^{-\mathrm{i}m\phi}\mathrm{e}^{\mathrm{i}\int k_r(r)\,\mathrm{d}r}\,{}_sS^m_k(\theta)\,, \tag{6.33}$$

where $\tilde{J}^\mu_0$ is a function given by the expansions in Sect. 4.5.1. From (4.15), (4.50)–(4.53) and (6.33) we find the asymptotic form of the sources of Maxwell's equations as restricted by the horizon boundary condition (3.95):

$$\delta J_0 \underset{\alpha\to 0}{\simeq} \int \mathrm{d}\omega \sum_{k,m} \Bigg\{ (J_{+1})_0 + J_{+1}(r,\theta) - \mathrm{i}\left[(J_m)_0 + J_m(r,\theta)\right] \\ \times \left[\frac{P}{\Delta} - k_r(r)\right]\Bigg\} \mathrm{e}^{-\mathrm{i}\omega t} \mathrm{e}^{-\mathrm{i}m\phi} \mathrm{e}^{\mathrm{i}\int k_r(r)\,\mathrm{d}r} {}_{+1}S_k^m(\theta), \quad (6.34\mathrm{a})$$

$$\delta J_2 \underset{\alpha\to 0}{\simeq} \int \mathrm{d}\omega \sum_{k,m} \Bigg\{ (J_{-1})_0 + J_{-1}(r,\theta) + \mathrm{i}\left[(J_{\bar{m}})_0 + J_{\bar{m}}(r,\theta)\right] \\ \times \left[\frac{P + \Delta k_r(r)}{2\rho^2}\right]\Bigg\} \mathrm{e}^{-\mathrm{i}\omega t} \mathrm{e}^{-\mathrm{i}m\phi} \mathrm{e}^{\mathrm{i}\int k_r(r)\,\mathrm{d}r} {}_{-1}S_k^m(\theta)\,, \quad (6.34\mathrm{b})$$

where

$$\frac{\partial}{\partial r}(J_{+1})_0 = \frac{\partial}{\partial r}(J_m)_0 = \frac{\partial}{\partial r}(J_{-1})_0 = \frac{\partial}{\partial r}(J_{\bar{m}})_0 = 0\,, \quad (6.34\mathrm{c})$$

and

$$\lim_{\alpha\to 0} J_{+1}(r,\theta) = \lim_{\alpha\to 0} J_m(r,\theta) = \lim_{\alpha\to 0} J_{-1}(r,\theta) = \lim_{\alpha\to 0} J_{\bar{m}}(r,\theta) = 0\,. \quad (6.34\mathrm{d})$$

We note that by (6.29) and (6.31), the second term in (6.34a) acts as a "feedback" term and can be very significant for outgoing fast waves near the event horizon.

The asymptotic expression for the current in (6.34) can be used in (4.23) and (4.25) with the expansions (6.29)–(6.31) that define $c_{\pm 1}$ to yield the asymptotic sources to the radial equations in (4.23):

$$\lim_{r\to r_+} G_{+1}(r) = -(G_{+1})_0 - \bar{G}_{+1}(r,\theta) - \mathrm{i}c_{+1}\left[(\mathcal{J}_{+1})_0 + \bar{\mathcal{J}}_{+1}(r,\theta)\right] \\ \equiv -G_{+1} - \mathrm{i}\mathcal{J}_{+1}c_{+1}\,, \quad (6.35\mathrm{a})$$

$$\lim_{r\to r_+} G_{-1}(r) = -(G_{-1})_0 - \bar{G}_{-1}(r,\theta) + \mathrm{i}\Delta\, c_{-1}\left[(\mathcal{J}_{-1})_0 + \bar{\mathcal{J}}_{-1}(r,\theta)\right] \\ \equiv -G_{-1} + \mathrm{i}\Delta\, \mathcal{J}_{-1}c_{-1}\,, \quad (6.35\mathrm{b})$$

where

$$\frac{\partial}{\partial r}(G_{+1})_0 = \frac{\partial}{\partial r}(\mathcal{J}_{+1})_0 = \frac{\partial}{\partial r}(\mathcal{J}_{-1})_0 = \frac{\partial}{\partial r}(G_{-1})_0 = 0\,, \quad (6.35\mathrm{c})$$

and

$$\lim_{\alpha\to 0} \bar{G}_{+1}(r,\theta) = \lim_{\alpha\to 0} \bar{\mathcal{J}}_{+1}(r,\theta) = \lim_{\alpha\to 0} \bar{\mathcal{J}}_{-1}(r,\theta) = \lim_{\alpha\to 0} \bar{G}_{-1}(r,\theta) = 0\,. \quad (6.35\mathrm{d})$$

Equation (6.35) represents the relevant form of the sources of the inhomogeneous Maxwell's equations near the horizon as restricted by the horizon boundary condition (3.95). In the next subsection we look at solutions near the horizon, but outside of the inner fast critical surface, $r_F(\theta)$.

6.4.3 Solutions of the Inhomogeneous Maxwell's Equations Near the Horizon

The reason for writing the vacuum solutions as in (6.29) was in anticipation of the more general wave structure in which $c_{\pm1}$ are functions. We are interested in the wave equation at radial coordinate $r > r_F \gtrsim r_+$, where globally outgoing fast waves can be generated. Substitution of (6.35) for the sources and (6.29) for the general solution into the radial equations (4.23cd) yield equations for the correction terms $c_{\pm1}$. The resulting pair of equations have analytic coefficients when $r > r_F$. Thus, solutions exist for the Ricatti equations in (6.36) if $r > r_F$,

$$\mathrm{i}\frac{\mathrm{d}}{\mathrm{d}r}c_{+1} - \frac{2P - 4\mathrm{i}(r-M)}{\Delta}c_{+1} + \frac{(6\mathrm{i}\omega r - k_+)}{\Delta} - c_{+1}^2 = \frac{G_{+1}}{\Delta} + \frac{\mathrm{i}c_{+1}\mathcal{J}_{+1}}{\Delta}\,, \tag{6.36a}$$

$$\mathrm{i}\frac{\mathrm{d}}{\mathrm{d}r}c_{-1} - \frac{2P}{\Delta}c_{-1} - \frac{(2\mathrm{i}\omega r + k_-)}{\Delta} - c_{-1}^2 = \frac{G_{-1}}{\Delta} - \mathrm{i}c_{-1}\mathcal{J}_{-1}\,. \tag{6.36b}$$

If one makes the substitution $c \equiv [\mathrm{i}y^{-1}(\mathrm{d}y/\mathrm{d}r)]$ then y satisfies a second order ordinary differential equation with regular singular points at r_+ and r_-. Thus, solutions for $c_{\pm1}$ can be found in terms of power series. In equation (6.36) a third regular singular point occurs at r_F from the source term $G_{\pm1}$ (see (6.94)).

When $c_{\pm1}$ is a small correction to (6.29) we have from (6.36) a solution that is accurate to $\mathrm{O}(\alpha^4)$:

$$c_{+1} \approx \frac{6\mathrm{i}\omega r - k_+ - G_{+1}}{2P - 4\mathrm{i}\sqrt{M^2 - a^2} + \mathrm{i}\mathcal{J}_{+1}}\,, \quad \omega \not\approx m\frac{\Omega_H}{c}\,, \tag{6.37a}$$

$$c_{-1} \approx \frac{-2\mathrm{i}\omega r - k_- - G_{-1}}{2P}\,, \quad \omega \not\approx m\frac{\Omega_H}{c}\,. \tag{6.37b}$$

For example these solutions are valid when $1 \gg \alpha^2 - \alpha_F^2 \gg \alpha_F^2$, as occurs for $\Omega_F \sim \Omega_H$ in a magnetically dominated wind with $U_A^2 \ggg 1$. The fast point can be driven arbitrarily close to the horizon for the approximately force free magnetospheres considered in [66, 80]. Thus, as long as the fast wave radiated outward from deep in the ergosphere does not originate near the fast point, (6.37) and (6.29) are accurate expressions for the radial wave function. In this regime the oscillatory nature of the short wavelength fast waves (and ingoing waves for that matter) are determined primarily by the spacetime geometry. The current terms only modify the second order vacuum term (compare (6.30) and (6.37)). Near the fast point, the singular nature of $G_{\pm1}$ dominates the dynamics and $c_{\pm1}$ is no longer a small correction.

6.4.4 Outgoing Fast Waves Near the Fast Point

The solutions given by (6.37) and (6.29) are valid as long as $1 \gg \alpha^2 - \alpha_F^2 \gg \alpha_F^2$. However, for $r - r_F$ small enough, i.e., $r - r_F \sim r_F - r_+$, this condition is not true.

Thus, we extend our discussion to globally outgoing waves originating from very close to the fast point. We will find two types of solutions: those in which wave variations are manifested on scales small compared to the radius of curvature in the proper frame of the plasma and long wavelength solutions in which curvature terms in the local plasma wave equations cannot be dropped (they do not resemble plane waves locally). The first types of solutions are essentially more accurate versions of the short wavelength modes discussed in [129]. The long wavelength modes require the tight constraint $\omega \approx m\Omega_{_H}/c$ and are referred to as singular solutions due to the approximate three dimensional support for these solutions in four dimensional momentum space (as a consequence of the aforementioned relation between ω and m).

6.4.4.1 Short Wavelength Modes Near the Fast Point

If we have short wavelength solutions then the radial component of the momentum vector is well defined in Boyer–Lindquist coordinates by (6.31) as $\tilde{k}_r \equiv k_r(r)$. We find the dispersion relation for fast waves by transforming the fast wave dispersion relation (5.34) in the proper frame by a relativistic boost to the rotated ZAMO frame. Then the basis transformation (3.3) - (3.8) can be used to write the dispersion relation in the stationary frames.

The transformation from the rotated ZAMO basis, $\hat{\mathbf{e}}_\alpha$, to the proper basis, $\bar{\mathbf{e}}_\alpha$, is given by [72]

$$\begin{bmatrix} \bar{\mathbf{e}}_u \\ \bar{\mathbf{e}}_\phi \\ \bar{\mathbf{e}}_\rho \end{bmatrix} = \begin{bmatrix} u^0 & u_\phi & u^1 \\ \dfrac{u^\phi}{\sqrt{1+u_1^2}} & \dfrac{u^0}{\sqrt{1+u_1^2}} & 0 \\ \dfrac{u^0 u^1}{\sqrt{1+u_1^2}} & \dfrac{u^\phi u^1}{u^0\sqrt{1+u_1^2}} & \sqrt{1+u_1^2} \end{bmatrix} \begin{bmatrix} \hat{\mathbf{e}}_0 \\ \hat{\mathbf{e}}_\phi \\ \hat{\mathbf{e}}_1 \end{bmatrix}, \tag{6.38a}$$

and the inverse transformation is

$$\begin{bmatrix} \hat{\mathbf{e}}_0 \\ \hat{\mathbf{e}}_\phi \\ \hat{\mathbf{e}}_1 \end{bmatrix} = \begin{bmatrix} u^0 & -\dfrac{u^\phi}{\sqrt{1+u_1^2}} & -\dfrac{u^0 u^1}{\sqrt{1+u_1^2}} \\ -u_\phi & \dfrac{u^0}{\sqrt{1+u_1^2}} & \dfrac{u^\phi u^1}{u^0\sqrt{1+u_1^2}} \\ -u^1 & 0 & \sqrt{1+u_1^2} \end{bmatrix} \begin{bmatrix} \bar{\mathbf{e}}_u \\ \bar{\mathbf{e}}_\phi \\ \bar{\mathbf{e}}_\rho \end{bmatrix}. \tag{6.38b}$$

The unit vector, $\bar{\mathbf{e}}_u$, is the bulk four velocity of the plasma. As noted in Sect. 5.4, the fast critical surface is defined in terms of purely poloidally propagating waves. Thus, we write the fast speed in the proper frame as

$$v_F = \frac{\bar{k}^u}{\bar{k}^\rho} . \tag{6.39}$$

Using the horizon boundary conditions (3.94) and the transformation from the rotated ZAMO basis (3.4)-(3.8) to Boyer–Lindquist coordinates, (6.38) and (6.39) imply (with the definition $\beta_F = v_F/c$):

$$\frac{\tilde{k}_r}{\tilde{k}^t} = \frac{\rho^2}{r^2+a^2}\left\{\frac{\left[\beta_F - \sqrt{1+(u^P)^{-2}}\right]\left[1+\mathrm{O}(\alpha^2)\right]}{1-\beta_F\sqrt{1+(u^P)^{-2}}+\mathrm{O}(\alpha^2)\mathrm{O}(\bar{k}^\phi/\bar{k}_\rho)}\right\} . \tag{6.40}$$

Expanding (6.40) about the fast point using the asymptotic expressions from (3.94) and (3.4) (see 6.85a, for example), one finds from (3.11a) an expression that eliminates $\tilde{k}^t$ in terms of ω and m in (6.40),

$$\tilde{k}_r = \frac{\Delta_F + \Delta}{\Delta - \Delta_F}\left(\frac{r^2+a^2}{\Delta}\right)\left[\omega - \frac{m\Omega_H}{c} + \mathrm{O}(\alpha^2)\right] , \quad \omega \not\approx \frac{m\Omega_H}{c} , \tag{6.41}$$

where the subscript "F" means to evaluate at the fast point.

The transformation (6.38) combined with the asymptotic bulk flow poloidal velocity from the horizon boundary condition, (3.94), and the fast speed in (6.39) yields a blueshift effect in the proper frame for outgoing fast waves near the critical surface,

$$\bar{P}^u \underset{r\to r_F}{\sim} \frac{\beta_F}{\alpha^2 - \alpha_F^2}\left(\omega - \frac{\Omega}{c}m\right) , \tag{6.42a}$$

$$\bar{P}^\rho \underset{r\to r_F}{\sim} \frac{1}{\alpha^2 - \alpha_F^2}\left(\omega - \frac{\Omega}{c}m\right) . \tag{6.42b}$$

The blueshift effect in (6.42) justifies the short wavelength approximation if $\omega \not\approx m(\Omega/c)$. Compare (6.42) to the similar expression for light waves in (3.19). Note that $\tilde{k}_r \to \infty$ at the fast point.

6.4.5 The Singular Set of Long Wavelength Solutions

Both the solution (6.37) and the dispersion relation (6.41) depend on the fact that $\alpha/[\omega - m(\Omega/c)]$ is a small parameter. When $\omega \approx m(\Omega/c)$ this approximation breaks down and one has long wavelength modes and these must be discussed in terms of the Newman–Penrose formalism. If $w \gg \Omega_H$ then the gradient in Ω causes $\alpha/[\omega - m(\Omega/c)]$ to transition to a small parameter over very short proper distances deep in the ergosphere. Thus, the long wavelength solution transitions to a short wavelength solution almost immediately. However, for $\omega \sim \Omega_H$ one can have a legitimate long wavelength solution deep in the ergosphere in the limit of small values of the parameter $\alpha/[\omega - m(\Omega/c)]$.

Consider the situation in which the high frequency, short wavelength condition, (6.42), does not hold in the proper frame near the fast point. These local long wavelength solutions in the proper frame are characterized by the conditions,

$$\tilde{k}^{\rho} \underset{r \to r_F}{\sim} \alpha^0 \,, \tag{6.43a}$$

$$\tilde{k}^{u} \underset{r \to r_F}{\sim} \alpha^0 \,. \tag{6.43b}$$

In order to understand the global behavior of these modes requires the evaluation of the Boyer–Lindquist components of the propagation vector. To obtain this information, first note that (6.42) implies the long wavelength condition near the fast point for an outgoing mode emitted form a point, r_0, just outside the fast critical surface, $r_0 \gtrsim r_F$,

$$\left(\omega - m\frac{\Omega}{c}\right) \underset{r_0 \to r_F}{\sim} r - r_F \,, \quad r - r_0 \ll r_0 - r_+ \,. \tag{6.44a}$$

With the aid of (3.11a), or equivalently the inverse metric in (6.90e), (6.44a) can be transformed to the Boyer–Lindquist contravariant wave vector scaling near the fast point,

$$\tilde{k}^{t} \underset{r_0 \to r_{(F)+}}{\sim} \frac{r - r_F}{r - r_+} \,, \quad r - r_0 \ll r_0 - r_+ \,. \tag{6.44b}$$

Similarly, one can consider the more general case of a long wavelength mode emitted from a point, r_0, satisfying the condition, $1 \gg \alpha_0^2$ (i.e., it is radiated from deep in the ergosphere, but not necessarily "near" the fast point). In this instance, the wave function (6.29) implies the following parameter constraint on long wavelength outgoing fast modes deep within the ergosphere at the point of emission:

$$\tilde{k}^{t}\left(r_0\right) = \frac{\omega - m\dfrac{\Omega_0}{c}}{\alpha_0^2} = \varepsilon \gtrsim 0 \,, \tag{6.44c}$$

where the subscript "0" means that the corresponding quantity is evaluated at r_0 and ε is a very small positive constant. Note that even with the small ε constraint that a wave emitted from very deep in the ergosphere can eventually transition to a short wavelength solution farther out at a lapse function, $1 \gg \alpha^2 \gg \alpha_0^2$. We consider a long wavelength solution to be one that can propagate outward a proper distance that is a significant fraction of the radius of curvature of either spacetime ($\sim M$) or the magnetic field before the solution transitions to a short wavelength mode. Thus, we have the long wavelength condition,

$$\varepsilon \ll \omega \,, \quad \text{if } \omega \sim \Omega_H \,. \tag{6.44d}$$

Recall that $\omega \approx m(\Omega_H/c)$ is not allowed for massive particles deep in the ergosphere, which are dispersive waves, since $\beta^\phi \sim \alpha^{-1}$ (see (3.16)). However, by contrast, fast waves are nondispersive and this restriction yields a well behaved group velocity.

Note that (6.44d) implies that the quantity P of (4.23g) is negative at r_0,

$$P_0 = \frac{\rho_0^2 (\varepsilon - \omega)}{2Mr_0} \Delta(r_0) , \tag{6.45a}$$

and this circumstance has profound implications for the wave functions in (6.29). Secondly, expanding P about P_0,

$$P = P_0 + \omega (r^2 - r_0^2) , \tag{6.45b}$$

one can see that P vanishes at a coordinate value of r given by

$$r^2 = r_0^2 + \frac{\Delta(r_0) [\omega - \varepsilon] \rho_0^2}{2Mr_0 \omega} , \quad P = 0 . \tag{6.45c}$$

Combining (6.45b) and (6.45c) we get the highly restrictive bound on P in the ergosphere,

$$|P| < \alpha^2 r_+ \frac{\omega}{\Omega_H} , \quad r > r_0 , \quad \alpha^2 \ll 1 . \tag{6.45d}$$

Thus, we have the small dimensionless parameter in the ergosphere,

$$\frac{|P|}{r_+ - r_-} \ll 1 , \quad \forall \alpha^2 \ll 1 . \tag{6.45e}$$

The fact that $P \approx 0$ near r_+ implies that c_{-1} can not be considered a small correction to (6.29) as was the case in (6.37b). In the Ricatti equation (6.36b), the coefficient of c_{-1} is small near r_+, thus the solution of (6.37b) is no longer justified. We can solve (6.36b) at small lapse function in terms of our small parameter in (6.45) using the asymptotic form of the current sources in (6.35),

$$c_{-1} = \frac{-\mathrm{i} - \dfrac{2P}{r_+ - r_-}}{r - r_+} + \mathrm{i}\frac{\mathcal{J}_{-1}}{2} - \frac{\mathrm{i}}{2(r_+ - r_-)} \left[2\mathrm{i}\omega r + k_- + G_{-1}\right] + \mathrm{O}\left(\frac{P}{r_+ - r_-}\right) ,$$
$$\alpha_0^2 \ll 1 , \quad \omega - m\frac{\Omega_0}{c} \approx 0 , \quad \omega \sim \Omega_H . \tag{6.46a}$$

It was implicitly assumed in the derivation of (6.46a) that the current sources of (6.35) do not have a singularity in the ergosphere. This is demonstrated explicitly for the long wavelength modes in Sect. 6.4.8 (see 6.92). Inserting (6.46a) into (6.29b) yields an expression for $\tilde{\Phi}_2$,

$$\tilde{\Phi}_2(r) \underset{r_0 \to r_+}{\sim} r - r_+ , \quad \omega - m\frac{\Omega_0}{c} \approx 0 , \quad \omega \sim \Omega_H . \tag{6.46b}$$

There exists another disjoint class of singular solution that is also of the long wavelength that is axisymmetric, $m = 0$. However, a long wavelength solution in this instance requires that ω is very small,

$$\omega \sim \alpha_0^2 \Omega_H \,. \tag{6.47a}$$

For a central black hole in an AGN this corresponds to frequencies less than $10^{-7}\mathrm{sec}^{-1}$! Note that in this branch of the singular solution space, $\varepsilon \gg \omega$, in contrast to the other long wavelength solutions that obey the opposite constraint, (6.44d). The quantity P never vanishes for these waves, but is very small at the point of emission,

$$P_0 \sim \alpha_0^2 r_+ \ll r_+ - r_- \,. \tag{6.47b}$$

Thus, (6.45b) and (6.47ab) imply the global ergospheric parameter constraint,

$$\frac{P}{r_+ - r_-} \ll 1 \,, \forall\, \alpha^2 \ll 1 \,. \tag{6.47c}$$

Acknowledging this small parameter, we can solve (6.36b) at small lapse function, finding the solution (6.46a) again,

$$c_{-1} = \frac{-\mathrm{i} - \dfrac{2P}{r_+ - r_-}}{r - r_+} + \mathrm{i}\frac{\mathcal{J}_{-1}}{2} - \frac{\mathrm{i}}{2\left(r_+ - r_-\right)}\left[2\mathrm{i}\omega r + k_- + G_{-1}\right] + \mathrm{O}\left(\frac{P}{r_+ - r_-}\right), \quad m = 0\,, \quad \omega \gtrsim 0\,. \tag{6.47d}$$

Inserting the wave solution of (6.47d) into the general wave form of (6.29b) for $\tilde{\Phi}_2$ we find,

$$\tilde{\Phi}_2 \underset{r \to r_+}{\simeq} \alpha^2 \,, \quad m = 0\,, \quad \omega \gtrsim 0\,. \tag{6.47e}$$

Not surprisingly, this result is similar to the vacuum electrodynamic results found in Sect. 4.5 that were used to prove the no-hair theorem. The small ω waves are in essence a DC limit to the general fast wave phenomenon. Note that such waves are not governed by the notions of fast speed and fast critical surfaces which are derived in the short wavelength limit. The electromagnetic properties of these hybrid waves were derived near the event horizon without regard to any properties associated with fast waves.

By contrast, notice that ϕ_0 is well behaved and does not scale with lapse function as $r_0 \to r_+$ because $P \sim \alpha_F^2$ is just a small correction to the denominator of c_{+1} in (6.37a). Consequently, one can combine the wave function information from (6.37a), (6.46) and (6.47) to find the asymptotic amplitudes of the spin components of the electromagnetic field carried by outgoing long wavelength fast waves in the ergosphere. In particular, using these amplitudes with (4.10) and (5.17), we can find the toroidal magnetic field density transported by outgoing long wavelength solutions emitted near the event horizon, at a value of lapse function, $\alpha_0 \equiv \alpha\left(r_0\right)$,

$$\delta B^T \underset{r_0 \to r_+}{\sim} \frac{\tilde{\rho}^2 \Delta}{2} \phi_0 + \phi_2 \underset{r_0 \to r_+}{\sim} \alpha_0^2 , \quad \omega - m\frac{\Omega_0}{c} \approx 0 . \tag{6.48}$$

In the next subsection we find that this result is independent of wavelength.

6.4.6 The Linearized Perturbation Equations for Short Wavelength Modes

In this subsection, we study the complement of the singular solutions in the outgoing fast wave parameter space, $\omega \not\approx m(\Omega_H/c)$. The existence of solutions to the Ricatti equations (6.36a,b) and the general form of the solutions (6.29) allow us to express perturbed Boyer–Lindquist field components $\delta\tilde{F}^{\mu\nu}(r,\theta,\phi,t)$ in terms of complex amplitudes, $\delta\tilde{F}^{\mu\nu}$, as

$$\delta\tilde{F}^{\mu\nu}(r,\theta,\phi,t) = \int \mathrm{d}\omega \sum_{s,k,m} \delta\tilde{F}^{\mu\nu} \mathrm{e}^{-\mathrm{i}\omega t} \mathrm{e}^{-im\phi} \mathrm{e}^{\mathrm{i}\int k_r(r)\mathrm{d}r} {}_sS_k{}^m(\theta). \tag{6.49}$$

The dispersion relation (6.41) implies that we need to know more than the asymptotic currents near the horizon, we also need the linearized momentum equations of the fluid in order to understand behavior near the fast point. In contrast, if $1 \gg \alpha^2 - \alpha_F^2 \gg \alpha_F^2$, the dispersion relation (6.41) yields the result of (6.29) and (6.37) that the wave function is determined by the spacetime metric,

$$\tilde{k}_r \approx \frac{P}{\Delta}\left[\mathrm{O}(1+\alpha^2)\right] . \tag{6.50}$$

Consequently, there is clearly some type of singular behavior of the current sources near the fast point. The nature of this singular behavior and its implications are elucidated in the calculations of the next three subsections.

At this juncture it is worth commenting on the use of “short wavelength” in the title to this subsection in the context of the general form of the wave expansion in (6.49). According to the dispersion relation (6.41), once the condition, $\omega \approx m\Omega_H/c$, is violated (as is the assumption of the analysis to follow), the radial wavelength becomes very small. Even though the variations orthogonal to the radial direction can take place on large distance scales (the waves are linear combinations of spin weighted spheroidal harmonics), these are short wavelength modes in the radial direction.

In this section, we perform a full linearized perturbation in Boyer–Lindquist coordinates following [129] in order to find the wave behavior for $\alpha^2 - \alpha_F^2 \sim \alpha_F^2$. It is worth contrasting the more general present analysis to the exponential “plane wave” type short wavelength solutions of [129]. Firstly, (6.49) implies that the waves are not merely simple exponential functions of imaginary argument in the θ coordinate. Furthermore, by (6.41), near the fast point, the radial wavelength is rapidly varying. This is not like plane wave behavior. The wavefronts are compressed together as the wave propagates away from the spacetime near the inner fast point and the wavelengths get more and more stretched out the farther that the wave propagates away from the horizon and inner fast critical surface. The WKB type radial wave functions of (6.31) that are implemented in the wave expansions of (6.49) are accurate descriptions of this type of wavelength variation phenomena.

The calculations of this subsection incorporate the effects of magnetic field line curvature, since the current does not vanish in the black hole magnetosphere (zero current flow is assumed in the calculations of Chap. 2). Secondly, the curvature of spacetime is reflected in the use of spin weighted spheroidal harmonics and the radial wave functions in (6.49). Strictly speaking, the calculations of this section are the most accurate in the simple exponential, "plane wave" limit of [129]. The solutions derived in this subsection are only approximate when there is additional wave behavior associated with θ variations. The integrity of the approximate calculations can be verified by comparison with the last subsection. All of the relevant results are identical to the limiting case of the longest wavelength solutions (the singular solutions), implying that no physical content has been lost by our mathematical expediences.

6.4.6.1 The Perturbed Electromagnetic Field

Consider the linearized Maxwell's equations (3.64b) using (6.49) for a wave with quantum numbers m, ω, and k:

$$-\mathrm{i}\,\delta\tilde{F}_{tr,\theta}-\omega\,\delta\tilde{F}_{r\theta}+k_r\,\delta\tilde{F}_{\theta t}=0\,, \tag{6.51a}$$

$$m\,\delta\tilde{F}_{r\theta}+k_r\,\delta\tilde{F}_{r\phi}-\mathrm{i}\,\delta\tilde{F}_{\phi r,\theta}=0\,, \tag{6.51b}$$

$$m\,\delta\tilde{F}_{tr}-\omega\,\delta\tilde{F}_{r\phi}+k_r\,\delta\tilde{F}_{\phi t}=0\,, \tag{6.51c}$$

$$m\,\delta\tilde{F}_{t\theta}-\omega\,\delta\tilde{F}_{\theta\phi}-\mathrm{i}\,\delta\tilde{F}_{\phi t,\theta}=0\,. \tag{6.51d}$$

One can define perturbed poloidal field components as in [129],

$$\varepsilon^{\parallel}=\frac{1}{\tilde{u}^P}\left[\tilde{u}^r\,\delta\tilde{F}_{rt}+\tilde{u}^\theta\,\delta\tilde{F}_{\theta t}\right]\,, \tag{6.52a}$$

$$\varepsilon^{\perp}=\frac{1}{\tilde{u}^P}\frac{1}{\sqrt{g_{rr}g_{\theta\theta}}}\left[\tilde{u}_\theta\,\delta\tilde{F}_{rt}-\tilde{u}_r\,\delta\tilde{F}_{\theta t}\right]\,, \tag{6.52b}$$

$$b^{\parallel}=\frac{1}{\tilde{u}^P}\left[\tilde{u}_r\,\delta\tilde{F}_{\theta\phi}-\tilde{u}_\theta\,\delta\tilde{F}_{r\phi}\right]\,, \tag{6.52c}$$

$$b^{\perp}=\frac{1}{\tilde{u}^P}\sqrt{g_{rr}g_{\theta\theta}}\left[\tilde{u}^\theta\,\delta\tilde{F}_{\theta\phi}+\tilde{u}^r\,\delta\tilde{F}_{r\phi}\right]\,, \tag{6.52d}$$

$$\left(\tilde{u}^P\right)^2=\left[g_{rr}\,(\tilde{u}^r)^2+g_{\theta\theta}\left(\tilde{u}^\theta\right)^2\right]\,. \tag{6.52e}$$

The perturbed frozen-in condition is

$$\delta\tilde{F}_{\phi t}\tilde{u}^t + \tilde{F}_{\phi r}\,\delta\tilde{u}^r + \tilde{F}_{\phi\theta}\,\delta\tilde{u}^\theta - b^\perp\alpha\sqrt{g_{\phi\phi}}\,\tilde{u}^P = 0\,, \tag{6.53a}$$

$$\tilde{F}_{rt}\,\delta\tilde{u}^t + \delta\tilde{F}_{rt}\tilde{u}^t + \tilde{F}_{r\phi}\,\delta\tilde{u}^\phi + \delta\tilde{F}_{r\phi}\tilde{u}^\phi + \tilde{F}_{r\theta}\,\delta\tilde{u}^\theta + \delta\tilde{F}_{r\theta}\tilde{u}^\theta = 0\,, \tag{6.53b}$$

$$\tilde{F}_{\theta t}\,\delta\tilde{u}^t + \delta\tilde{F}_{\theta t}\tilde{u}^t + \tilde{F}_{\theta\phi}\,\delta\tilde{u}^\phi + \delta\tilde{F}_{\theta\phi}\tilde{u}^\phi + \tilde{F}_{\theta r}\,\delta\tilde{u}^r + \delta\tilde{F}_{\theta r}\tilde{u}^r = 0\,, \tag{6.53c}$$

$$\delta\tilde{F}_{t\phi}\tilde{u}^\phi + \tilde{F}_{tr}\,\delta\tilde{u}^r + \tilde{F}_{t\theta}\,\delta\tilde{u}^\theta + \varepsilon^\parallel\tilde{u}^P = 0\,, \tag{6.53d}$$

$$\Omega_F = -\frac{\tilde{F}_{r\phi}}{\tilde{F}_{rt}} = -\frac{\tilde{F}_{\theta\phi}}{\tilde{F}_{\theta t}}\,. \tag{6.53e}$$

Next we express $\delta\tilde{F}^{\alpha\beta}$ in terms of $\delta\tilde{F}_{\phi t}$ and $\delta\tilde{F}^{r\theta}$. We define

$$b_T \equiv \delta\tilde{F}^{r\theta}\,, \tag{6.54a}$$

and

$$ik_\theta \equiv \frac{\delta\tilde{F}_{\phi t,\theta}}{\delta\tilde{F}_{\phi t}}\,, \tag{6.54b}$$

where k_θ is not a component of a propagation vector since $\delta\tilde{F}^{\mu\nu}$ in (6.49) is not a plane wave, but a spin weighted spheroidal harmonic. Multiply $\tilde{u}^r$ by (6.51c) and $\tilde{u}^\theta$ by (6.51d); add then substitute into (6.53a), (6.53d) and (6.53e) to yield

$$\varepsilon^\parallel = \frac{\left(\omega - \dfrac{m\Omega_F}{c}\right)\tilde{u}^\phi + \Omega_F\tilde{u}^\mu k_\mu}{\left(\omega - \dfrac{m\Omega_F}{c}\right)\tilde{u}^P}\,\delta\tilde{F}_{t\phi}\,. \tag{6.55a}$$

Similarly, multiply $\tilde{u}^r$ by (6.51c) and $\tilde{u}^\theta$ by (6.51d); add then substitute into (6.53a) and (6.53c) to get

$$b^\perp = \frac{\left(\omega - \dfrac{m\Omega_F}{c}\right)\tilde{u}^t + \tilde{u}^\mu k_\mu}{\left(\omega - \dfrac{m\Omega_F}{c}\right)\alpha\sqrt{g_{\phi\phi}}\,\tilde{u}^P}\,\delta\tilde{F}_{t\phi}\,. \tag{6.55b}$$

We can decompose the propagation vector into components parallel and perpendicular to the field,

$$k_\parallel = \frac{\tilde{k}_r\tilde{u}^r + \tilde{k}_\theta\tilde{u}^\theta}{\tilde{u}^P}\,, \tag{6.56a}$$

$$k_\perp = \frac{1}{\sqrt{g_{rr}g_{\theta\theta}}\,\tilde{u}^P}\left[\tilde{k}_r\tilde{u}_\theta - \tilde{k}_\theta\tilde{u}_r\right]\,. \tag{6.56b}$$

As $\alpha \to 0$, we have from (6.41),

$$\lim_{\alpha\to 0} k_\parallel \approx \frac{\tilde{k}_r\tilde{u}^r}{\tilde{u}^P}\left[1 + \mathrm{O}(\alpha^2)\right]\,, \tag{6.57a}$$

and

$$\lim_{\alpha \to 0} k_{\perp} \sim \alpha k_{\parallel} \; . \tag{6.57b}$$

We can crudely approximate $k_{\perp}$ when we don't have a plane wave by setting k_θ of (6.54b) equal to $\tilde{k}_\theta$ in (6.56b), or ignoring all θ variations except when $\alpha^2 \gtrsim \alpha_F^2$, i.e., set $\tilde{k}_\theta = 0$ in (6.56b). Either will give an inaccurate expression for $k_{\perp}$, but by (6.57b) this will not affect the final results which depend primarily on $k_{\parallel}$.

Setting $k_\theta = \tilde{k}_\theta$ in (6.56), these expressions become

$$\tilde{u}^P k_{\parallel} = \left[\tilde{k}_r \tilde{u}^r - \frac{\mathrm{i}\,\delta \tilde{F}_{\phi t,\theta}}{\delta \tilde{F}_{\phi t}} \tilde{u}^\theta \right] , \tag{6.58a}$$

$$\tilde{u}^P k_{\perp} = \left[\tilde{k}_r \tilde{u}_\theta + \frac{\mathrm{i}\,\delta \tilde{F}_{\phi t,\theta}}{\delta \tilde{F}_{\phi t}} \tilde{u}_r \right] . \tag{6.58b}$$

Inverting these expressions yields

$$-\frac{\mathrm{i}\,\delta \tilde{F}_{\phi t,\theta}}{\delta \tilde{F}_{\phi t}} = \frac{\left[k_{\parallel} \tilde{u}_\theta - \sqrt{g_{rr} g_{\theta\theta}}\, \tilde{u}^r k_{\perp} \right]}{\tilde{u}^P} , \tag{6.59a}$$

$$\tilde{k}_r = \frac{\left[k_{\parallel} \tilde{u}_r + \sqrt{g_{rr} g_{\theta\theta}}\, k_{\perp} \tilde{u}^\theta \right]}{\tilde{u}^P} . \tag{6.59b}$$

Similarly, the perturbed electric field decomposes as

$$\delta \tilde{F}_{\theta t} = \frac{\left[\varepsilon^{\parallel} \tilde{u}_\theta - \sqrt{g_{rr} g_{\theta\theta}}\, \tilde{u}^r \varepsilon^{\perp} \right]}{\tilde{u}^P} , \tag{6.60a}$$

$$\delta \tilde{F}_{rt} = \frac{\left[\varepsilon^{\parallel} \tilde{u}_r + \sqrt{g_{rr} g_{\theta\theta}}\, \varepsilon^{\perp} \tilde{u}^\theta \right]}{\tilde{u}^P} . \tag{6.60b}$$

Define

$$\Delta_0 \equiv k_\theta\, \delta \tilde{F}_{rt} - \mathrm{i}\, \delta \tilde{F}_{rt,\theta} , \tag{6.61a}$$

then (6.59), (6.60) and (6.54a) substituted into (6.51a) yields

$$\varepsilon^{\perp} = \frac{\left[\omega \left(\tilde{u}^\phi - \frac{\Omega_F}{c} \tilde{u}^t \right) + \Omega_F \tilde{u}^P k_{\parallel} \right] k_{\perp}\, \delta \tilde{F}_{t\phi}}{\left(\omega - \frac{m \Omega_F}{c} \right) \tilde{u}_P k_{\parallel}} - \frac{\omega \sqrt{g_{rr} g_{\theta\theta}}\, b_T + \Delta_0}{k_{\parallel}} . \tag{6.61b}$$

The quantity Δ_0 represents a deviation from plane wave behavior by the general wave form in (6.49). Multiplying $\tilde{u}_\theta$ by (6.51c) and subtracting the product of $\tilde{u}_r$ and (6.51d) yields

$$-m \varepsilon^{\perp} + \omega \alpha \sqrt{g_{\phi\phi}}\, b^{\parallel} = k_{\perp}\, \delta \tilde{F}_{t\phi} . \tag{6.62}$$

Inserting (6.62) into (6.61) gives

$$b^{\parallel} = \frac{1}{\alpha\sqrt{g_{\phi\phi}}}\left\{\left[m\left(\tilde{u}^{\phi} - \frac{\Omega_F}{c}\tilde{u}^{t}\right) + \tilde{u}_P k_{\parallel}\right] k_{\perp}\,\delta\tilde{F}_{t\phi} - \frac{m\sqrt{g_{rr}g_{\theta\theta}}\,b_T + \frac{m}{\omega}\Delta_0}{k_{\parallel}}\right\}. \tag{6.63}$$

6.4.6.2 The Perturbed Fluid Equations

Next we express $\delta\tilde{u}_{\phi}$ and $\delta\tilde{u}_t$ in terms of b_T and $\delta\tilde{F}_{t\phi}$. We perturb the azimuthal momentum equation in Boyer–Lindquist coordinates in the cold plasma limit,

$$n\mu\left[\delta\left(\frac{\mathrm{d}}{\mathrm{d}t}\right)\tilde{u}_{\phi} + \frac{\mathrm{d}}{\mathrm{d}\tau}\delta\tilde{u}_{\phi}\right] = \frac{\delta\tilde{F}_{\phi\alpha}\tilde{J}^{\alpha}}{c} + \frac{\tilde{F}_{\phi\alpha}\,\delta\tilde{J}^{\alpha}}{c}. \tag{6.64}$$

Inserting (6.33) into Ampere's law (3.64a), one can eliminate the perturbed currents in (6.64) in favor of perturbed field strengths. Then note that $\tilde{k}_{\beta}\,\delta\tilde{F}^{\beta\alpha}/(\partial\delta\tilde{F}^{\theta\alpha}/\partial\theta) \sim \mathrm{O}(1/\tilde{k}_r M)$ and (by 6.41), $\tilde{k}_r M \gg 1$, to create an approximate perturbed azimuthal momentum equation:

$$n\mu\tilde{k}_{\mu}\tilde{u}^{\mu}\,\delta\tilde{u}_{\phi} \approx \frac{\mathrm{i}\tilde{k}^{\beta}\left[\tilde{F}_{\phi}{}^{r}\,\delta\tilde{F}_{\beta r} + \tilde{F}_{\phi}{}^{\theta}\,\delta\tilde{F}_{\beta\theta}\right]}{c} + \frac{\delta\tilde{F}_{\phi t}\tilde{J}^{t}}{c}. \tag{6.65}$$

Similarly, the time component of the perturbed momentum equation is

$$n\mu\tilde{k}_{\mu}\tilde{u}^{\mu}\,\delta\tilde{u}_{t} \approx \frac{\mathrm{i}\tilde{k}^{\beta}\left[\tilde{F}_{t}{}^{r}\,\delta\tilde{F}_{\beta r} + \tilde{F}_{t}{}^{\theta}\,\delta\tilde{F}_{\beta\theta}\right] + \delta\tilde{F}_{t\phi}\tilde{J}^{\phi}}{c}. \tag{6.66}$$

Combining (6.65) and (6.66) with the frozen-in equation (6.53f) yields

$$\delta\tilde{u}_{t} + \frac{\Omega_F}{c}\,\delta\tilde{u}_{\phi} \approx \frac{\frac{\delta\tilde{F}_{\phi t}}{c}\left[\tilde{J}^{\phi} - \frac{\Omega_F}{c}\tilde{J}^{t}\right]}{n\mu\tilde{k}_{\mu}\tilde{u}^{\mu}} \approx \mathrm{O}(\alpha^2)\,\delta\tilde{u}_{\phi}. \tag{6.67a}$$

This leads directly to the very simple approximate relation

$$\delta\tilde{u}_{t} + \frac{\Omega_F}{c}\,\delta\tilde{u}_{\phi} \approx 0. \tag{6.67b}$$

Expanding out (6.65) yields

$$\delta\tilde{u}_{\phi} \approx \frac{1}{4\pi\mu k}\frac{\alpha\sqrt{g_{\phi\phi}}}{\tilde{u}^{\mu}\tilde{k}_{\mu}k_{\parallel}}\left[\frac{\left(\tilde{u}^{\phi} - \frac{\Omega_F}{c}\tilde{u}^{t}\right)\left(k_{\mu}k^{\mu} + k_{\perp}^2\right)k_{\parallel}\,f}{\left(\omega - \frac{m\Omega_F}{c}\right)} - k_{\perp}\,\delta\tilde{F}_{t\phi} + \left(k^{\mu}k_{\mu} + k_{\perp}^2\right)\tilde{u}^{P}\left(\sqrt{g_{rr}g_{\theta\theta}}\,b_T + \frac{\Delta_0}{\omega}\right)\right], \tag{6.68a}$$

$$f \equiv \tilde{u}^{P}\left(\tilde{k}^{\phi} - \frac{\Omega_F}{c}\tilde{k}^{t}\right) + k_{\parallel}\left(\tilde{u}^{\phi} - \frac{\Omega_F}{c}\tilde{u}^{t}\right). \tag{6.68b}$$

6.4.6.3 Eliminating $\delta\tilde{F}_{t\phi}$

We eliminate $\delta\tilde{F}_{t\phi}$ in the expressions above by finding a second relation for $\delta\tilde{u}_{\phi}$ using $u \cdot u = -1$. Perturbing the normalization condition gives

$$\tilde{u}^{\mu}\,\delta\tilde{u}_{\mu} = -\tilde{u}^{P}\,\delta\tilde{u}^{\parallel} + \tilde{u}^{t}\,\delta\tilde{u}_{t} + \tilde{u}^{\phi}\,\delta\tilde{u}_{\phi} = 0\,, \tag{6.69}$$

where the parallel component of the poloidal velocity is given by

$$\delta\tilde{u}^{\parallel} = \frac{\delta\tilde{u}_{r}\,\tilde{u}^{r} + \delta\tilde{u}_{\theta}\,\tilde{u}^{\theta}}{\tilde{u}^{P}} = \frac{\delta\tilde{u}^{r}\,\tilde{u}_{r} + \delta\tilde{u}^{\theta}\,\tilde{u}_{\theta}}{\tilde{u}^{P}}\,. \tag{6.70}$$

Multiplying $\tilde{u}_{\theta}$ by (6.53b) and subtracting $\tilde{u}_{r}$ times (6.53c) yields

$$\begin{aligned}\delta u^{\parallel}\tilde{F}_{r\theta} = {} & -\varepsilon^{\perp}\sqrt{g_{rr}g_{\theta\theta}}\,\tilde{u}_{t} + \tilde{B}^{P}\left[\delta\tilde{u}^{\phi} - \frac{\Omega_F}{c}\,\delta\tilde{u}^{t}\right] \\ & + \tilde{u}^{P}\,\delta\tilde{F}_{r\theta} + b^{\parallel}\tilde{u}^{\phi}\,,\end{aligned} \tag{6.71}$$

where

$$\left(\tilde{B}^{P}\right)^{2} \equiv g_{rr}\left[\frac{\tilde{F}_{\theta\phi}}{\sqrt{-\tilde{g}}}\right]^{2} + g_{\theta\theta}\left[\frac{\tilde{F}_{\phi r}}{\sqrt{-\tilde{g}}}\right]^{2}. \tag{6.72}$$

Note that (the components of the inverse metric can be found in 6.90e, but are not explicitly required at this point)

$$\delta\tilde{u}^{\phi} - \frac{\Omega_F}{c}\,\delta\tilde{u}^{t} = \left[g^{\phi\phi} - 2\frac{\Omega_F}{c}g^{\phi t} + \frac{\Omega_F^{2}}{c^{2}}g^{tt}\right]\delta\tilde{u}_{\phi}\,. \tag{6.73}$$

Equations (6.67b) and (6.70) imply that

$$\delta u^{\parallel} \approx -\delta\tilde{u}_{\phi}\frac{\left[\tilde{u}^{\phi} - \frac{\Omega_F}{c}\,\delta\tilde{u}^{t}\right]}{\tilde{u}^{P}}\,. \tag{6.74}$$

Combining (6.71)–(6.74) and noting that

$$\frac{\tilde{F}^{\mu\nu}\tilde{F}_{\mu\nu}}{\left(\tilde{B}^{P}\right)^{2}} = g^{rr}g^{\theta\theta}\left[g^{\phi\phi} - 2\frac{\Omega_F}{c}g^{\phi t} + \frac{\Omega_F^{2}}{c^{2}}g^{tt} + \frac{\left(\tilde{u}^{\phi} - \frac{\Omega_F}{c}\tilde{u}^{t}\right)^{2}}{\left(\tilde{u}^{P}\right)^{2}}\right], \tag{6.75}$$

we find using expression (5.38) for the fast speed in the cold plasma limit,

$$\delta\tilde{u}_\phi = \frac{1}{4\pi\mu k}\frac{\alpha\sqrt{g_{\phi\phi}}\tilde{u}^\mu\tilde{k}_\mu}{U_F^2 k_\parallel}\left[\frac{\tilde{u}^\phi - \dfrac{\Omega_F}{c}\tilde{u}^t}{\omega - \dfrac{m\Omega_F}{c}}k_\perp\,\delta\tilde{F}_{t\phi} + \tilde{u}^P\sqrt{g_{rr}g_{\theta\theta}}\left(b_T + \frac{\Delta_0}{\omega}\right)\right]. \tag{6.76}$$

Combining (6.76) and (6.68) we find,

$$\left\{\left(\tilde{u}^\phi - \frac{\Omega_F}{c}\tilde{u}^t\right)\left[(k^\mu u_\mu)^2 + U_F^2\left(k^\mu k_\mu + k_\perp^2\right)\right] + U_F^2 k_\parallel f\right\}k_\perp\,\delta\tilde{F}_{t\phi}$$
$$= -\left(\omega - \frac{\Omega_F}{c}\right)\left[(k^\mu u_\mu)^2 + U_F^2\left(k^\mu k_\mu + k_\perp^2\right)\right]\left[\tilde{u}^P\sqrt{g_{rr}g_{\theta\theta}}\left(b_T + \frac{\Delta_0}{\omega}\right)\right]. \tag{6.77}$$

Using the covariant expression for the propagation vector for fast waves (2.40) in (6.77) gives

$$\delta\tilde{F}_{t\phi} \approx -\frac{\tilde{u}^P\left(\sqrt{g_{rr}g_{\theta\theta}}\,b_T + \dfrac{\Delta_0}{\omega}\right)\left(\omega - \dfrac{m\Omega_F}{c}\right)k_\perp}{\left[k_\perp^2\left(\tilde{u}^\phi - \dfrac{\Omega_F}{c}\tilde{u}^t\right) + k_\parallel f\right]}. \tag{6.78}$$

6.4.6.4 Solving For δb_T Near the Fast Point

Combining (6.74), (6.76) and (6.78) gives an expression for poloidal velocity variation,

$$\frac{\delta u^\parallel}{\tilde{u}^P} = \frac{\alpha^2 g_{\phi\phi}\tilde{u}^\mu\tilde{k}_\mu\left(\tilde{u}^\phi - \dfrac{\Omega_F}{c}\tilde{u}^t\right)^2 f}{U_F^2 M_A^2\tilde{u}^P\left[\left(\tilde{u}^\phi - \dfrac{\Omega_F}{c}\tilde{u}^t\right)k_\perp^2 + k_\parallel f\right]}\frac{\left(b_T + \dfrac{\Delta_0}{\omega}\sqrt{g_{rr}g_{\theta\theta}}\right)}{\tilde{F}^{r\theta}}, \tag{6.79}$$

where we used the frozen-in condition to write

$$k = -\frac{\sqrt{-\tilde{g}}\,n\left(\tilde{u}^\phi - \dfrac{\Omega_F}{c}\tilde{u}^t\right)}{\tilde{F}_{r\theta}}, \tag{6.80}$$

and pure Alfvén Mach number was expressed as

$$M_A^2 \equiv \frac{4\pi\mu k^2}{n}. \tag{6.81}$$

Using the dispersion relation (6.41) and the near the horizon approximation, $\alpha^2 \ll 1$, and a long algebraic manipulation, we find

$$\lim_{\alpha \to 0} \frac{\delta u^{\|}}{\tilde{u}^P} \approx \frac{\delta \tilde{u}^r}{\tilde{u}^r} \approx \frac{2\Delta^2 (\tilde{u}^t)^2 \sin^2\theta \left[\Omega_H - \Omega_F\right]^2}{U_F^2 M_A^2 (\Delta_F + \Delta)} \frac{b_T}{\tilde{F}^{r\theta}} . \tag{6.82}$$

By (3.95) $\tilde{u}^t \sim \alpha^{-2}$, and $U_F^2 \sim \alpha^0$ since it is a proper frame evaluated quantity and $M_A^2 \sim \alpha^0$ as a result of (6.81). Consolidating this information, we rewrite (6.82) as

$$\frac{\delta u^r}{u_0^r} = F(\theta, \Omega_F, \omega, m) \left(\alpha_F^2 + \alpha^2\right)^{-1} \frac{\delta B^T}{B_0^T} , \tag{6.83}$$

where the subscript "0" represents unperturbed quantities. F is a well behaved function that scales as α^0 near the horizon, and u^r is a ZAMO evaluated velocity as a consequence of the defining relation (6.52e) and (3.3) in the limit as $\alpha \to 0$. Mathematically we have

$$u^r \equiv u_0^r + \delta u^r , \quad u^r < 0 , \tag{6.84a}$$

$$B^T \equiv B_0^T + \delta B^T , \quad B^T < 0 , \quad B_0^T < 0 . \tag{6.84b}$$

The condition $B^T < 0$ and $\delta B^T < 0$ is the necessary condition to extract angular momentum from the hole (see (5.21)) when $B^P > 0$ in the northern hemisphere. From (3.94), we have the asymptotic form of the four velocity in terms of the constant, U,

$$u_0^r \underset{r \to r_+}{\simeq} -\alpha^{-1} U , \quad U > 0 , \tag{6.85a}$$

$$U < \alpha U_F , \tag{6.85b}$$

where (6.85b) is the subfast condition. For a wave to be emitted upstream we require $-u^r < U_F$, or from (6.84) and (6.85),

$$\frac{\delta u^r}{u_0^r} < \frac{\alpha U_F - U}{U} . \tag{6.86}$$

Combining (6.86) with (6.83) gives

$$\frac{\delta B^T}{B_0^T} < \frac{\alpha_F^2 + \alpha^2}{F(\theta, \Omega_F, \omega, m)} \frac{\alpha U_F - U}{U} . \tag{6.87}$$

Thus, δB^T is bounded by a constant times $\alpha_F^2 + \alpha^2$ for an outgoing fast wave. We find the same result in the parameter space, $\omega \not\approx m(\Omega_H/c)$, that we found in the long wavelength singular solution, (6.48):

$$\delta B^T \underset{r \to r_+}{\simeq} \alpha_F^2 + \alpha^2 . \tag{6.88}$$

6.4.7 *Outgoing Magnetic Stresses Carried Fast Waves Near the Horizon*

We verify that the $\delta B^T \sim \alpha^2$ condition is equivalent globally to weak magnetic stresses in the wave. Consider a closed surface, S_0, just outside the horizon with $\alpha_0(\theta,\phi) \ll 1$ and a second closed surface at infinity S_∞. Compute the integral law of angular momentum conservation in the $3+1$ split of [80] of a fast wave emitted outward from just inside S_0. Computing time averaged fluxes and letting the boundary surface be stationary with respect to infinity gives

$$\int_{S_0} \alpha \mathbf{S}_{\mathrm{L}} \cdot \mathrm{d}\mathbf{A} = \mathrm{Re}\left[\int_{S_0} \frac{\sqrt{-\tilde{g}}}{4\pi}\left(\delta\tilde{F}^{r\theta}\delta\tilde{F}^*_{\theta\phi} + \delta\tilde{F}^{rt}\delta\tilde{F}^*_{t\phi}\right)\mathrm{d}\theta\mathrm{d}\phi\right] =$$
$$\int_{S_\infty} \alpha \mathbf{S}_{\mathrm{L}} \cdot \mathrm{d}\mathbf{A} = \mathrm{Re}\left[\int_{S_\infty} \frac{\sqrt{-\tilde{g}}}{4\pi}\left(\delta\tilde{F}^{r\theta}\delta\tilde{F}^*_{\theta\phi} + \delta\tilde{F}^{rt}\delta\tilde{F}^*_{t\phi}\right)\mathrm{d}\theta\mathrm{d}\phi\right]. \quad (6.89)$$

We list the relevant scalings near S_0 of perturbed quantities in (6.89) below. From (6.88), (6.78), the definition of f in (6.68b) and the horizon boundary conditions (3.94),

$$\delta\tilde{F}^{r\theta} \underset{r\to r_F}{\simeq} \alpha_0^2 + \alpha_F^2 , \quad (6.90a)$$

$$\delta\tilde{F}_{t\phi} \underset{r\to r_F}{\simeq} \alpha_0^2\left(\alpha_0^2 + \alpha_F^2\right) . \quad (6.90b)$$

Equations (6.55b) and (6.63) can be applied to (6.52c) and (6.52d) to get

$$\delta\tilde{F}_{\theta\phi} \underset{r\to r_F}{\simeq} \alpha_0^2 + \alpha_F^2 , \quad (6.90c)$$

$$\delta\tilde{F}_{r\phi} \underset{r\to r_F}{\simeq} \alpha_0^2\left(\alpha_0^2 + \alpha_F^2\right) . \quad (6.90d)$$

Equations (6.55a), (6.60b), (6.61) and (6.90d) combined with the inverse metric in (6.90e),

$$g^{\mu\nu} = -\frac{1}{\Delta\rho^2}\left[\left(r^2+a^2\right)\frac{\partial}{\partial t} + a\frac{\partial}{\partial\phi}\right]^2 + \frac{1}{\rho^2\sin^2\theta}\left[\frac{\partial}{\partial\phi} + a\sin^2\theta\frac{\partial}{\partial t}\right]^2$$
$$+\frac{\Delta}{\rho^2}\left(\frac{\partial}{\partial r}\right)^2 + \frac{1}{\rho^2}\left(\frac{\partial}{\partial\theta}\right)^2 , \quad (6.90e)$$

yields the final perturbed field scaling from the perturbed frozen-in condition (6.53),

$$\delta\tilde{F}^{rt} = -\frac{1}{\rho^4}\left[\frac{\left(r^2+a^2\right)^2 - \Delta a^2\sin^2\theta}{c}\Omega_F - 2Mra\right]\delta\tilde{F}_{r\phi} , \quad (6.90f)$$

$$\delta\tilde{F}^{rt} \underset{r\to r_F}{\simeq} \alpha_0\left(\alpha_0^2 + \alpha_F^2\right) . \quad (6.90g)$$

Inserting the scalings of (6.90) into the angular momentum conservation law of an outgoing fast wave emitted from near the horizon, (6.89), yields

$$\int_{S_\infty} \alpha \mathbf{S}_{\mathrm{L}} \cdot \mathrm{d}\mathbf{A} \approx \mathrm{Re}\left[\int_{S_\infty} \frac{\sqrt{-\tilde{g}}}{4\pi}\left(\delta \tilde{F}^{r\theta} \delta \tilde{F}^{*}_{\theta\phi}\right) \mathrm{d}\theta \mathrm{d}\phi\right] \underset{r \to r_F}{\sim} \alpha_{_0}^4 + \alpha_{_F}^4 \,. \tag{6.91}$$

Consequently, (6.91) shows that fast waves emitted at small values of lapse function in either the long or short wavelength regimes transport negligible torsional electromagnetic stresses outward to infinity. This results from the vanishing of the $\tilde{\Phi}_2$ component of the electromagnetic field in the short wavelength limit (see 6.95j). Note that this is also the case for the outgoing long wavelength solutions emitted from small values of lapse function (see 6.46b and 6.47e). The net result, at a physical level, is that fast waves become ever more feeble at propagating poloidal angular momentum flux outward as $\alpha \to 0$. The effective redshifting of the poloidal angular momentum flux carried by fast modes is consistent with the interpretation that the spacetime near the event horizon is an asymptotic infinity for MHD waves.

6.4.8 The Singular Point Structure of the Wave Equation Near the Fast Critical Surface

In order to see the singular point structure of the radial wave equation in (6.36), we elucidate the singular structure of $\delta \tilde{J}^\theta$ near the fast point. Ampere's law (3.64a) implies that

$$\delta \tilde{J}^\theta = \frac{\mathrm{i}}{4\pi}\left[-\omega \delta \tilde{F}^{\theta t} + m \delta \tilde{F}^{\theta\phi} + \tilde{k}_r \delta \tilde{F}^{\theta r}\right] . \tag{6.92a}$$

From the dispersion relation (6.41), near the fast point, and (5.17) we have an approximate version of the perturbed Ampere's law above,

$$\delta \tilde{J}^\theta \approx -\frac{\mathrm{i}}{4\pi} \tilde{k}_r \frac{\delta B^T}{\sqrt{-\tilde{g}}} \,. \tag{6.92b}$$

From (6.88) and (6.41), near the fast point, (6.92b) scales like

$$\delta \tilde{J}^\theta \underset{r \to r_F}{\sim} \frac{(\Delta_{_F} + \Delta)^2}{\Delta(\Delta - \Delta_{_F})} \underset{r \to r_F}{\sim} \frac{\Delta}{\Delta - \Delta_{_F}} \underset{r \to r_F}{\sim} \frac{\alpha^2}{\alpha^2 - \alpha_{_F}^2} \,. \tag{6.93a}$$

Therefore, in the somewhat extended region $1 \gg \alpha^2 - \alpha_{_F}^2$, (6.92a) can be expressed asymptotically as in (6.90) in the form

$$\delta \tilde{J}^\theta \underset{\alpha_0 \to 0}{\sim} J \frac{\alpha_{_0}^2}{\alpha_{_0}^2 - \alpha_{_F}^2} + V \frac{\alpha_{_0}^2}{\alpha_{_0}^2 + \alpha_{_F}^2} \,, \tag{6.93b}$$

where V and J are constants. It follows from (6.92b) that there is no singular point in the current sources associated with the long wavelength modes that were discussed in Sect. 6.4.5, since $\tilde{k}_r$ is well behaved in these solutions by definition.

Equation (6.93a) implies that there is a singular point at r_F in (6.32), the defining relations for J_0 and J_2, since they are functions of $\delta\tilde{J}^\theta$. The singularity in the spinorial current sources translates into a singular point in the radial equations (4.23c,d). In order to see the nature of the singular point, note that there is both an imaginary and real component of the source expansion for Maxwell's equations in (6.35) at the fast point. Inserting the expansion for $\delta\tilde{J}^\theta$, (6.93b), into (6.32a) and (6.35), the real part of the source term in (4.25a) is given primarily by the following expansion near the fast point:

$$\bar{G}_{+1} \underset{r\to r_F}{\sim} \frac{A\alpha^2}{(r-r_F)^2}\,, \tag{6.94a}$$

where A is a constant. Note that in accordance with the expansions in (6.35), (6.94a) has the limiting form near the horizon,

$$\lim_{\alpha\to 0} \bar{G}_{+1} = 0\,, \tag{6.94b}$$

as required by the horizon boundary conditions (3.94) and (3.95). The imaginary part of the singular current at $r = r_F$ is given primarily by the $\mathrm{i}c_{+1}\mathcal{J}_{+1}$ term in (6.35a). The singular nature is partially absorbed in the functional dependence of c_{+1} that results from the dispersion relation, (6.41), applied to the wave function, (6.29):

$$\lim_{r\to r_F} c_{+1} = \frac{2\Delta_F P}{\Delta(\Delta-\Delta_F)} + \beta_{+1}\frac{\Delta+\Delta_F}{\Delta-\Delta_F} + \mathrm{i}\gamma_{+1}\,, \tag{6.95a}$$

$$\lim_{r\to r_F} \beta_{+1} \ll \frac{|P|}{\Delta_F}\,. \tag{6.95b}$$

The functions β and γ are well behaved and their general properties are described in [130]. The function β transitions the singular wave function near the fast point in (6.95a) to the small correction term, (6.37a), upstream. Substituting (6.95a) into the Ricatti equation (6.36a), yields the imaginary part of the spinorial current source according to the decomposition of (6.35a):

$$(\mathcal{J}_{+1})_0 = -2\sqrt{M^2-a^2} + 4\mathrm{i}P_+\,, \tag{6.95c}$$

$$\bar{\mathcal{J}}_{+1}(r,\,\theta) \underset{r\to r_F}{\sim} \frac{C\alpha^2}{(r-r_F)} + \mathrm{O}\left[(r-r_F)^0\right], \tag{6.95d}$$

where C is a constant and the complete expression for $\bar{\mathcal{J}}_{+1}(r,\,\theta)$ is very long and can be found in [130]. The current source $\bar{\mathcal{J}}_{+1}(r,\,\theta)$ vanishes at the event horizon as required by the horizon boundary conditions as expressed through (6.35). Furthermore, note that this functional form is consistent with the expansion for $\delta\tilde{J}^\theta$ near the fast point, (6.93b), inserted into (6.32a). Substitution of $c_{+1} \equiv \mathrm{i}y^{-1}(\mathrm{d}y/\mathrm{d}r)$ into the Ricatti equation (6.36a) along with the functional forms of the current source in (6.94) and (6.95b,c) yields a second order ordinary differential equation with variable coefficients that has a regular singular point at r_F.

It is instructive to perform the same analysis on the negative spin wave function in the short wavelength limit, near the fast point, that was performed above for the positive spin wave function. Note, by (6.32), the current source of the negative spin wave has the same radial derivative of $\delta \tilde{J}^{\theta}$ as the positive spin current source. Thus, the current source has the same singular point structure that was described in (6.94) and (6.95b,c),

$$\bar{G}_{-1} \underset{r \to r_F}{\simeq} \frac{A\alpha^4}{(r-r_F)^2}\,, \tag{6.95e}$$

$$\bar{\mathcal{J}}_{-1}(r,\theta) \underset{r \to r_F}{\simeq} \frac{C\alpha^2}{(r-r_F)} + \mathrm{O}\left[(r-r_F)^0\right]. \tag{6.95f}$$

There is an additional multiplicative factor of α^2 in (6.32b) that damps both the "feedback effect" as well as the current source G_{-1} for the negative spin component of the electromagnetic field transported by outgoing fast waves. Note that the lapse function scaling was absorbed in our definition of $\mathcal{J}_{-1}$ in (6.35) that is transcribed over into the Ricatti equation, (6.36b). The scaling of the "feedback current" with α^2 and G_{-1} with α^4 significantly modifies the solution from that found in (6.95a). In order to see this, first note that the solution (6.95a) is valid near the fast point because the leading order divergence in the imaginary part of (6.36a) is $\mathrm{O}[(r-r_F)^{-2}]$ and one has the asymptotic equality,

$$\mathrm{Im}\left[(r-r_F)^2 \mathrm{i}\frac{\mathrm{d}}{\mathrm{d}r}c_{+1}\right] = \left[(r-r_F)^2 \frac{c_{+1}\mathrm{Re}(\mathcal{J}_{+1}) + \mathrm{Im}(G_{+1})}{\Delta}\right] + \mathrm{O}(r-r_F)\,. \tag{6.95g}$$

Similarly, the leading order divergence at the fast point of the real part of the wave equation is satisfied by the balancing of the terms,

$$\mathrm{Re}\left[(r-r_F)^2 c_{+1}^2\right] = \left[(r-r_F)^2 \frac{c_{+1}\mathrm{Im}(\mathcal{J}_{+1}) - \mathrm{Re}(G_{+1})}{\Delta}\right] + \mathrm{O}(r-r_F)\,. \tag{6.95h}$$

The extra α^2 multiplicative factor on the negative spin current source in (6.36b) that appears in (6.95e,f) does not allow an asymptotic solution of the leading order real and imaginary divergences near the singular point as in (6.95g,h). In view of the asymptotic form of the currents in (6.95e,f), the only solution to (6.36b) near the fast point is,

$$\lim_{r \to r_F} c_{-1} = \frac{-\mathrm{i}}{r-r_F}\left[1 - \mathcal{J}_{-1}(r-r_F) + \frac{G_{-1}}{\Delta_F}(r-r_F)^2 + \mathrm{O}(\alpha^4)\right] + \frac{\mathrm{i}}{3}\left[\frac{P_F}{\Delta_F}\right]^2 (r-r_F) - \frac{P_F}{\Delta_F} + \mathrm{O}\left(\frac{r-r_F}{r-r_+}\right). \tag{6.95i}$$

This expression is significantly different from the wave function in (6.95a) due to the suppressed current source. The wave solution is long wavelength with a damped amplitude as opposed to the short wavelength undamped oscillations of the positive spin component wave function in (6.95a). The ergospheric geometry presents a birefringent medium for high frequency outgoing fast waves. Note the similarity of (6.95i) to the long wavelength solutions for the negative spin component that were derived in (6.46a) and (6.47d). The low frequency solutions (6.46a) and (6.47d) vanish at the event horizon. The solutions in (6.95i) vanish at the fast point, since it is the relevant singular point when (6.36b) is evaluated at high frequencies. The wave function in (6.95i), inserted into (6.29) yields the following asymptotic form of the negative spin component of the Maxwell field for outgoing "short wavelength" fast modes near the inner fast point,

$$\tilde{\Phi}_2 \underset{r \to r_F}{\sim} \alpha^2 - \alpha_F^2 , \quad \omega - m\frac{\Omega}{c} \not\approx 0 . \tag{6.95j}$$

Note that (6.95j) implies that one could combine the wave functions given by (6.95a,i) with (4.10) and (5.17) to derive the asymptotic scaling, (6.88), of the perturbed toroidal magnetic field density for outgoing fast waves emitted from just beyond the inner fast point.

6.4.9 Comparison to the Locally Covariant Calculation of Chapter 2

In this subsection we compare and contrast the wave properties of the globally outgoing fast waves in the inner ergosphere derived in this chapter with the locally covariant (special relativistic) wave properties found in Chap. 2. To begin with, we compute the ratio of cross-field poloidal current to field aligned poloidal current in the outgoing fast waves at small lapse function.

From Ampere's law in Boyer–Lindquist coordinates, (3.64a), and the field strength scalings in (6.90), we have

$$\delta \tilde{J}^r = \frac{\mathrm{i}}{4\pi} \tilde{k}_\mu \delta \tilde{F}^{r\mu} \underset{r \to r_F}{\sim} \alpha_0 \left(\alpha_0^2 + \alpha_F^2 \right) . \tag{6.96}$$

From the transformation to the rotated ZAMO basis, (3.3)-(3.8), as $\alpha \to 0$ and the asymptotic expressions for the Boyer–Lindquist current densities, (6.93) and (6.96), the asymptotic ratio of ZAMO current densities in a fast wave scales as

$$\lim_{\alpha_0 \to 0} \frac{J^2}{J^1} \sim \frac{1}{\alpha_0^2 - \alpha_F^2} . \tag{6.97a}$$

As was discussed in relation to (5.45), the total field aligned current that appears in the global current conservation law near the event horizon is

$$\lim_{\alpha_0 \to 0} I^1 \sim \alpha_0 J^1 . \tag{6.97b}$$

Combining (6.97a) and (6.97b), we can find the asymptotic behavior of the ratio of the total cross-field poloidal current to the total field aligned poloidal current in outgoing fast waves radiated from near the event horizon

$$\lim_{\alpha_0 \to 0} \frac{I^2}{I^1} \sim \frac{1}{\alpha_0 \left(\alpha_0^2 - \alpha_F^2\right)} . \tag{6.97c}$$

Thus, the poloidal current is predominantly in the cross-field direction in outgoing fast waves that are emitted from small values of lapse function. This is consistent with the results of the locally Lorentz covariant calculations given in (2.58) that fast waves can only transport field aligned currents. Equation (6.97) includes the effects of the curvature of the magnetic field lines deep in the ergosphere that can, in principle, couple the properties of the oblique Alfvén and fast modes as occurs in dipolar geometry [123]. The long calculation of this section shows that the mixing of these characteristics is negligible deep in the ergosphere.

In order to understand the electrostatic polarization properties of fast waves emitted from deep in the ergosphere in a perfect MHD plasma requires a global perspective if we are concerned with hybrid wave properties resulting from curvature effects. Thus, we discuss these fast waves in the context of the paired wind system of Fig. 6.1. First, consider the constraints on δB^T of (6.48) and (6.88) carried by outgoing fast waves as $\alpha_0 \to 0$. The frozen-in condition reduces the functional form of the toroidal magnetic field near the event horizon as a consequence of the plasma boundary conditions (3.94) and (3.95),

$$\lim_{\alpha \to 0} B^T = \frac{\Omega_F - \Omega_H}{c} B^P g_{\phi\phi} . \tag{6.98}$$

Since $B^T \approx$ constant in a paired magnetically dominated perfect MHD wind system and Ω_F is a constant in a magnetic flux tube, (6.98) implies that the aforementioned inability of outgoing fast modes radiated from deep in the ergosphere to significantly change B^T in the paired wind system is equivalent to the fast waves inability to affect a significant change in the global value of Ω_F. Thus, by (5.15b), fast waves radiated outward from deep in the ergosphere can not alter the global cross-field electrostatic potential in the magnetic flux tubes (i.e.,they can not alter the Goldreich–Julian charge density in any significant way). Thus, to an excellent approximation this agrees with the local covariant result of (2.61) that fast waves do not transport electric charge.

Including curvature effects in both the magnetic field and spacetime near the event horizon does not introduce hybrid wave characteristics that allow fast waves from deep in the ergosphere to be a causative agent in determining the global wind constants of a paired MHD wind system.

6.4.10 Summary of Results

In this section, we have computed the properties of outgoing fast waves deep in the ergosphere under a variety of physical circumstances. In Table 6.1, we summarize these results and indicate where the reader can find the relevant expressions. In order to compactify the notation in the table, we describe the parameter space in terms of the ZAMO evaluated propagation vector. In particular,

$$\alpha k^0 = \omega - m\frac{\Omega}{c} , \tag{6.99}$$

where αk^0_+ means to evaluate near the event horizon at the point of wave emission. Column (1) of Table 6.1 indicates the region of momentum space that is occupied by the particular class of outgoing fast mode deep in the ergosphere in terms of the notation of (6.99). The second column is the location of the point of emission in terms of lapse function. The next two columns of the table show where in the text that the wave function for the two spin components are found. The value in parenthesis is the equation of relevance and the subsection in which the discussion appears is located directly to the right (no parenthesis). Note that (6.29) was not included in order to save space since it is the common general expression for all wave functions. The last column indicates the equation that expresses the toroidal magnetic field scaling transported by the outgoing wave. The last two regions of parameter space represent short wavelength modes that are not near a singular point. Thus, the curvature of the magnetic field and the background spacetime are not sensed by these modes and the locally covariant calculations of Chap. 2 should be very accurate. This is reflected in the entry of (2.61b) in the last column. In order to understand the reference to

Table 6.1 Summary of fast wave properties deep in the ergosphere

Wave parameters	Location	$\tilde{\Phi}_0$	$\tilde{\Phi}_2$	δB^T
$\alpha k^0_+ \approx 0\,,\ \omega \sim \Omega_H$	$1 \gg \alpha^2 \gtrsim \alpha_F^2$	(6.37a) 6.4.5	(6.46) 6.4.5	(6.48)
$m = 0\,,\ \omega \ll \Omega_H$	$1 \gg \alpha^2 \gtrsim \alpha_F^2$	(6.37a) 6.4.5	(6.47) 6.4.5	(6.48)
$\alpha k^0_+ \approx 0\,,\ \omega \gg \Omega_H$	$1 \gg \alpha^2 \gtrsim \alpha_F^2$	(6.95a) 6.4.5	(6.95i) 6.4.5	(6.88)
$\alpha k^0_+ \not\approx 0$	$1 \gg \alpha^2 \gtrsim \alpha_F^2$	(6.95a) 6.4.8	(6.95i) 6.4.8	(6.88)
$\alpha k^0_+ \approx 0\,,\ \omega \sim \Omega_H$	$1 \gg \alpha^2 \gg \alpha_F^2$	(6.37a) 6.4.5	(6.46) 6.4.5	(6.48)
$m = 0\,,\ \omega \ll \Omega_H$	$1 \gg \alpha^2 \gg \alpha_F^2$	(6.37a) 6.4.5	(6.47) 6.4.5	(6.48)
$\alpha k^0_+ \approx 0\,,\ \omega \gg \Omega_H$	$1 \gg \alpha^2 \gg \alpha_F^2$	(6.37a) 6.4.3	(6.37b) 6.4.3	(2.61b)
$\alpha k^0_+ \not\approx 0$	$1 \gg \alpha^2 \gg \alpha_F^2$	(6.37a) 6.4.3	(6.37b) 6.4.3	(2.61b)

(2.61b), one needs to recall the discussion of (6.98) concerning the fact that the electrostatic potential and the toroidal magnetic field contain redundant information in a perfect MHD magnetically dominated magnetosphere.

The most important aspect of Table 6.1 is that regardless of the momentum space description of the outgoing fast wave, the transport of toroidal magnetic field (magnetic stresses) is negligible for MHD waves originating deep within the ergosphere.

Finally, it is useful to comment on the coupling of the ergospheric geometry to the Maxwell spin components that is common to all of the wave solutions in Table 6.1. There is a feedback term in (6.34a) that allows the outgoing fast wave to modify the pure geometrically induced properties of the current source at small lapse function (as discussed in the derivation of the no-hair theorem in Chap. 4). This feedback current source manifests itself in the Ricatti equation, (6.36a), in the term, $ic_{+1}\mathcal{J}_{+1}/\Delta$. In the language of the discussion of Sect. 6.4.8, this feedback effect allows J_0 through its dependence on $\delta\tilde{J}^{\theta}$ to be modified form its geometrically induced asymptotic form as a consequence of the back reaction of the wave function. In particular, this feedback is most pronounced at small lapse function near the fast point due to the singular nature of the current source as described in Sect. 6.4.8. This allows the sources for $\tilde{\Phi}_0$ to be significant near the fast point. By contrast, the feedback term is very small in (6.34b) and (6.36b). Thus, the electromagnetic fields carried by fast waves do not have the ability to induce the currents that are required to support the $\tilde{\Phi}_2$ component of the field near the fast point. This is in contrast to short wavelength outgoing pure electromagnetic waves which are capable of transporting a significant field component, $\tilde{\Phi}_2$, from the inner ergosphere. Similarly, we found that outgoing long wavelength solutions were incapable of supporting the currents required to create a substantial $\tilde{\Phi}_2$ near the event horizon. There appears to be an angular momentum coupling of the wave to the geometry that inhibits the outward propagation of the negative spin current sources in plasma waves near the event horizon.

In essence the calculation of toroidal magnetic fields carried by waves emitted from the vicinity of the horizon mirrors the calculation of the no-hair theorem in Chap. 4. The same redshifting of electromagnetic characteristics is fundamental to both. In the fast wave calculation this information is captured in the restricted form of the current sources near the event horizon in (6.35). The lack of a "bootstrap effect" near the horizon was computed in Sect. 4.5. This is the underlying physics behind the inability of a continuous plasma near the horizon to modify the electrodynamic infinity of the spacetime near the event horizon. It truly is an infinity.

6.5 Causality and the Blandford–Znajek Horizon Boundary Condition

Early treatments of black hole magnetospheres [66, 117] determined the rate of global energy extraction from a black hole by a paired wind system by invoking the force-free condition at the event horizon, (6.98). The same boundary condition was obtained in [111] using the frozen-in condition at the event horizon. The boundary condition, (6.98), allowed them to determine Ω_F and in the magnetically dominated limit that they assume, this quantity also determines the energy and angular momentum fluxes (see 5.20 and 5.21), if one also assumes a minimum torque outgoing wind (see Sect. 5.6). As was discussed in Sect. 5.6, such a magnetically dominated wind system has a virtually constant field aligned poloidal current, I^1, and toroidal magnetic field density, B^T, in each azimuthally symmetric magnetic flux tube.

The use of the force-free condition at the event horizon is the opposite of the actual horizon boundary condition proved in Chap. 3: all plasma flows are inertially dominated, in a global sense, near the event horizon (i.e., gravity overpowers all other astrophysical forces near the event horizon). The use of the force-free boundary condition on the magnetically dominated paired wind system allows the event horizon to behave implicitly like a unipolar inductor since field aligned poloidal currents appear to emanate from the event horizon and there is no significant source of poloidal current anywhere within the wind zone. Similarly, the frozen-in condition that is implemented at the horizon in order to determine Ω_F in [111] makes the spacetime near the event horizon appear as the causative agent for the charge separation necessary to support the global potential (or equivalently, Ω_F as discussed in 5.15b), as well as the poloidal current, I^1. This conflicts at the most fundamental level with the demonstration in Sect. 6.3 that the spacetime near the event horizon is an asymptotic infinity for MHD winds that passively accepts any externally imposed value of Ω_F and I^1.

In an attempt to justify the causal nature of the horizon boundary condition and the field aligned poloidal current system in these models [131] noted that for $\Omega_F \sim \Omega_H$, the fast critical surface is very close to the event horizon. The claim was that this surface can behave as the effective causal boundary surface for the wind system. Plasma near the inner fast point therefore rotates with an angular velocity, $\mathrm{d}\phi/\mathrm{d}t \approx \Omega_H$, and it was conjectured that this plasma can radiate information upstream through the fast mode that determines B^T (or effectively, I^1) and Ω_F (or equivalently, the charge separation). However, this argument ignores the properties of the fast wave other than the wave speed. It was shown in (2.58) that the short wavelength fast wave does not carry field aligned current or charge. In (2.146), it wa shown that the fast step wave does not transport electric charge. Thus, the fast mode seems incapable of transporting the appropriate information along characteristics of the MHD wave equation from the putative "causal boundary" that can establish the constants of motion in the paired wind system. However, it still needed to be checked if very long wavelength, higher dimensional fast waves could possibly transport charge and field aligned current changes. Thus, a very long calculation was developed in this chapter. In Sect. 6.4, it was shown, in the curved spacetime of the black hole magnetosphere, that fast waves radiated outward from near the putative "causal boundary" carry very weak poloidal angular momentum fluxes $\sim \alpha_0^4 + \alpha_F^4$ (see 6.91), and predominantly cross-field poloidal currents (see 6.97).

This discussion reaffirms that the spacetime near the event horizon is nothing more than an asymptotic infinity to the paired wind system. The wind solution of [66] is a valid solution to the time stationary, elliptic MHD wind equations.

However, such solutions need not be physical and do not necessarily arise from causal time dependent MHD evolution. The spacetime near the event horizon is unstable to any perturbation of Ω_F made upstream since it can not react back on the flow (i.e., it does not have Alfvén radiation at its disposal). Thus, in general, a paired wind system does not have the physical capabilities to evolve to the value of Ω_F given by [66]. In particular in an astrophysical environment, strong MHD waves will be created by accreting gas and high temperature coronal gas. These waves can in principle perturb the black hole magnetosphere. The black hole magnetosphere will not be driven back to the [66] wind parameters by the putative event horizon unipolar inductor. Instead, it simply adapts to the newly imposed parameters that are injected by active the boundary surfaces. The spacetime near the event horizon passively accepts these changes to the black hole magnetosphere. The 3-D numerical simulations that are discussed in detail in Sect. 11.4 clearly show that this is indeed the case. The 3-D simulations indicate that highly nonstationary flaring phenomena near the boundary of the black hole magnetosphere seems to be the norm, not the exception in an astrophysical environment. Thus, these causal comments are particularly relevant and not just a pedantic argument.

It should be stressed that the acauslity of the method of solution in [66, 111] does not invalidate the central concept of electrodynamic energy extraction if $0 < \Omega_F < \Omega_H$. Although, in a highly turbulent environment, it is not clear that a stationary value of Ω_F has much physical significance. The most general electrodynamic energy extraction method is likely a mix of the large scale field concept in [66] and the wave concept of super-radiant scattering.

As a point of clarification, if one assumes

- Magnetically dominated perfect MHD everywhere with a relativistic outflow
- or equivalently a force-free magnetosphere,
- and boundaries that are passive mathematical sources in Maxwell's equations for the electromagnetic field

then there is only one solution, the [66, 111] solution, by definition. However, this is a highly restrictive class of solutions by astrophysical standards. Even if a perfect MHD, magnetically dominated magnetosphere were established near a black hole, the physical nature of the boundaries that create the poloidal flux and carry the poloidal current source would unlikely be trivial. It was assumed in [66,80,111] that the electrodynamic properties of the horizon would induce the poloidal current flow and the surface charge on the boundary surfaces. But, in reality, these boundary surfaces are likely to be turbulent dense accretion flows. The dynamics of these flows are determined by inertial forces that drive large scale current systems. The numerical simulations in Sect. 11.4 show that these inertially driven currents persist right up to the boundary between the accreting gas and the black hole magnetosphere. Hence, these time dependent boundary surfaces are the initial data for the time evolution of the system of perfect MHD partial differential equations in the black hole magnetosphere. In Sect. 11.4, the numerical simulations show that these perturbations to the [66, 111] solution can range anywhere form same order of magnitude changes to the wind solution, to completely erasing any imprint of the [66, 111] wind parameters.

In summary, there is no meaningful boundary condition that can be imposed near the event horizon except that the black hole is a sink for mass flux. Therefore, there is no unique electrodynamic solution. Besides the [111] solution, there are the solutions found in the simulations of Chap. 11 that are driven by the lateral boundaries of the event horizon magnetosphere. Even if the lateral boundaries are passive, there are perfect MHD solutions in which the outgoing wind is subrelativistic, i.e., it is a supercritical wind as discussed in Sect. 5.5 and Fig. 5.2 [136]. In general, one must actually deal with the inertial physics of the bounding plasma and GHM in the black hole ergosphere in order to understand the electrodynamic energy extracting mechanism that can realize the Christodoulou/Ruffini energy extracting process described in Sect. 1.4.

Chapter 7
Ergosphere Driven Winds

This is by far the shortest chapter in the book, yet it is the most important as it synthesizes the fundamental physics of all ergospheric dynamos that create and sustain a toroidal magnetic flux (and the currents that support it). The content is not more than a typical section of the book, but is broken off into its own chapter for emphasis.

In this chapter, we describe an internally self-consistent physical process that allows magnetically dominated winds to be driven by the gravitational field of a black hole. It will be demonstrated that it is the coupling of the gravitational field to the plasma in the ergosphere through the dragging of inertial frames (see Sect. 3.5) that is responsible for driving the global poloidal current system (i.e., the dynamo for the toroidal magnetic field). Since there is no boundary condition in which to sink the ergospheric physics (see Sect. 6.5), we need to understand the microscopic forces in the ergosphere that can drive the macroscopic currents.

7.1 Analogy to the Physics of the Faraday Wheel

In Sect. 2.9.4, we learned that a unipolar inductor drives current because the rotationally induced EMF is unbalanced by the electrostatic force in a Faraday wheel. This rotational inertia is converted to Poynting flux through $F^{\mu\nu}J_{\nu}$ forces. The key piece of physics is to isolate the unbalanced source of EMF in the black hole magnetosphere. In analogy to the Faraday wheel, we find that a dynamo for the toroidal magnetic flux results from a rotationally induced EMF that is created by the dragging of inertial frames. This process converts the relativistic rotational inertia imparted to the plasma by the gravitational field into Poynting flux.

B. Punsly, *Black Hole Gravitohydromagnetics, 2nd. ed.*,
Astrophysics and Space Science Library 355, doi: 10/1007/978-3-540-76957-6_7,

7.2 Causal Determination of the Constants of Motion

Consider a paired wind system as depicted in cross section in Fig. 6.1. The constants of motion $\delta\Phi$ and Φ (see Fig. 5.1) are considered to be made primarily by external sources and Φ can be altered by azimuthal currents in the flux tubes through the Grad–Shafranov equation (see Sect. 5.7). The mass flux constants in the outgoing and ingoing winds, k_+ and k_-, respectively, are created by the plasma injection mechanism. It is assumed that the plasma is not relativistically hot, so we set $\mathbb{S}_+ = \mathbb{S}_- = 0$, without loss of physical content outside of the particle creation region. We assume that a tenuous pair plasma exists in the strong magnetic field of a black hole magnetosphere. Since the plasma inertia is relatively small, the dissipation associated with the pair injection mechanism is therefore negligible (a full theoretical treatment can be found in Chap. 9). Thus, $(ke)_+ \approx (ke)_-$, $(k\ell)_+ \approx (k\ell)_-$ and $(\Omega_F)_+ \approx (\Omega_F)_-$. Consequently, the paired wind solution in this limit requires the designation of three constants of motion as given by the discussion of Sect. 5.2, ke, $k\ell$ and Ω_F. From (5.20) and (5.21), the dynamo for B^T determines ke and $k\ell$ for a given value of Ω_F. The physics that determines B^T and $k\ell$ is a manifestation of a torsional "tug of war" between ingoing ergospheric plasma and the outgoing magnetically dominated wind. In Chap. 8, we will consider an example where Ω_F is determined by the ergospheric dynamo. By contrast, in Chaps. 9 and 10 we will look at models in which Ω_F is determined by dynamics external to the dynamo and the analogy to the Faraday wheel is somewhat weaker than for the models discussed in Chap. 8.

In order to uncover the ergospheric physics of the dynamo and understand the causal structure of the global "tug of war," we keep an open mind and list all the potential sources of dynamo-like dynamics. The dynamo might be the result of one of the following agents:

1. Axisymmetric vacuum electromagnetic fields,
2. The gravitational field,
3. Light waves,
4. Perfect MHD waves,
5. MHD waves as modified by small dissipation,
6. Waves in a highly dissipative medium.

7.2.1 Axisymmetric Vacuum Electromagnetic Fields

We showed in Sect. 4.6.3 that axisymmetric vacuum electromagnetic fields do not torque Kerr black holes (see 4.90b), nor do they extract energy. However, the dragging of inertial frames can induce an electric field with a component parallel electric field that can potentially attract charges toward the black hole in a semivacuum magnetosphere. The resulting poloidal current system superficially appears to be sustainable and would create a B^T by Ampere's law. However, recall the discussion of Sect. 4.8 that demonstrated that such a magnetosphere is open circuited in the low latitude magnetic flux tubes that carry any putative return current, since $\boldsymbol{E} \times \boldsymbol{B}$ changes sign in these flux tubes and currents can only flow as a result of bulk motion of charge in this scenario. This is the well known Goldreich–Julian conundrum of charge-nstarved pulsar magnetospheres. The pulsar is still viable because the unipolar induction of the neutron star can drive MHD currents that do not depend on the vacuum field properties. By contrast, the semi-vacuum ergosphere has no unipolar induction near the horizon, so the physics of current generation still requires an explanation, if it exists.

7.2.2 The Gravitational Field

The problem with invoking the gravitational field to drive a global current system is that gravity couples to mass-energy not electric charge. Thus, gravity cannot drive currents directly nor establish the charge separation in the magnetosphere associated with the global potential and Ω_F. Trivially, the gravitational field alters plasma trajectories through the dependence of the momentum of the flow on the metric. However, there needs to be an intermediate step that couples to the electric charge and can also be communicated upstream to the plasma source and the current system in the outgoing wind. It is the physics of this facet of the interaction that determines the causal structure of the paired wind system.

The solution of [66] relies on the gravitational field to determine the wind constants. As such, the strong inertial effects ("the intermediate step," mentioned above) are hidden within the boundary surfaces near the equatorial plane. This claim is justified by the 3-D simulations that are discussed in Sect. 11.4. In those simulations, a black hole magnetosphere similar to the [66, 111] solution is set up by strong inertially dominated transients (GHM transients) that create and maintain the currents that are the source for Φ and $k\ell$. If these strong inertial forces in the bounding plasma mysteriously shutoff then indeed a [66, 111] type solution can be sustained in a metastable equilibrium. This equilibrium is essentially maintained by the gravitational field. However, more realistically, the inertial forces in the bounding plasma do not die off mysteriously and persist as in the simulations of Sect. 11.4. The values of ke and $k\ell$ are constantly changing in response to inertial effects in the bounding plasma.

7.2.3 Light Waves and Waves in a Highly Dissipative Medium

The results of Sect. 2.7 demonstrate that in a highly resistive (dissipative) medium that the slow and Alfvén do not propagate and the fast mode becomes the sonic mode [75]. There exist a second class of propagating plasma waves in a hot resistive that are not manifested in cool plasmas. These electrostatic modes are the generalization of the simple MHD longitudinal Langmuir modes. As such they can couple to electric charge and are possibly integral to the causal structure of a dynamo. The myriad of potential plasma modes extant at high temperature are a consequence of a panoply of kinetic effects such as temperature differentials between the constituent species and plasma inhomogeneities. As such they tend to propagate at the thermal speed which is very similar to the sonic speed associated with the fast mode in this limit. The most interesting example of these modes for the present treatment are the electrostatic waves in a magnetic field (Bernstein modes) which we will need to describe the microphysics of the resistive dissipation of the dynamo discussed in Chap. 9. The existence of the magnetic field makes the analysis of the Bernstein modes far more complicated than the treatment of Langmuir waves (for detailed discussions of the variety of high temperature wave phenomena see [77] which is devoted almost entirely to this subject). In summary, the only propagating modes in a highly dissipative medium that can couple to electric charge are light waves, a variety of longitudinal electrostatic modes and the fast mode that all propagate on the order of the sonic speed.

Both electrostatic waves and light waves have a dispersion relation with a cutoff frequency at the electron plasma frequency, ω_{pe} [73,77]. For an astrophysical black hole with a mass on the order of $10^9\,\mathrm{M}_\odot$ one expects that the magnetospheric plasma in the ergosphere to be characterized by $10^5\,\mathrm{sec}^{-1} < \omega_{pe} < 10^{12}\,\mathrm{sec}^{-1}$. By contrast, we also know from (5.18) and (5.44) that an energy extracting paired wind system is characterized by $\Omega_F < \Omega_H \sim 10^{-4}\,\mathrm{sec}^{-1}$. This is significant since the relevant light waves and electrostatic waves that could possibly influence the determination of global potential must have frequencies somewhere in the neighborhood of Ω_F. Therefore, the causally relevant waves are characterized by $\omega \ll \omega_{pe}$. These waves are deep in the reactive range of the dispersion relation and are highly attenuated. Furthermore, the plasma inflow speed is most likely supersonic in the ergosphere, rendering a highly dissipative plasma incapable of radiating fast waves upstream to the plasma source and outgoing wind. We conclude that neither light waves nor plasma waves in a highly dissipative medium can propagate a global value of the electric potential or affect the transport angular momentum and energy fluxes from the black hole magnetosphere.

7.2.4 *Perfect MHD Waves and Mildly Dissipative MHD Waves*

From the discussions of the previous subsections, we deduce that if there is any dynamo for toroidal magnetic flux then it is causally determined by MHD waves with at most mild dissipation. These waves establish Ω_F, ΔV, B^T and I^1 in ergosphere driven paired wind systems. For magnetically dominated winds we know that Alfvén wave radiation is determinant for establishing the global charge separation and field aligned poloidal current systems that constitute the paired winds (see 2.52, 2.56, 2.58 and 2.61 as well as Sect. 6.4). This is reminiscent of the Alfvén wave radiation from the Faraday wheel at the end of a semi-infinite, plasma-filled waveguide of Sect. 2.9.4 that establishes Ω_F, ΔV, B^T and I^1 in the waveguide. These MHD waves constitute the missing intermediate step that was inferred in the discussion of Sect. 7.2.2 concerning the role of the gravitational field in the dynamo dynamics.

7.3 The Causal Structure of the Dynamo

From the analysis of Sect. 7.2.4, we expect that the ergospheric dynamo occupies a region of spacetime upstream of the fast critical surface of the ingoing wind. Furthermore, the field aligned current system and Goldreich – Julian charge density is radiated at or upstream of the Alfvén critical surface. Thus, the outer boundary surface of the dynamo region resides upstream or coincident with the inner Alfvén critical surface. This establishes a causal relationship through Alfvén waves between the dynamo and the plasma source and outgoing wind. By Ampere's law, within the dynamo region, strong cross-field poloidal currents must flow to support the toroidal magnetic field upstream. These must be causally established within the dynamo region by MHD waves. Note that this does not preclude a contribution from standing fast waves and evanescent Alfvén modes in the spacetime between the fast and Alfvén critical surfaces (see Chap. 9).

The microphysics of the interaction within the dynamo occurs through the following causal set of steps:

1. Gravity couples to mass-energy (not electric charge).
2. Plasma is dragged across the magnetic field lines by black hole gravity. This breaks the perfect MHD condition since $u^{\perp} = 0$ by (5.2) in perfect MHD.
3. Each species of charge experiences an opposite $q\boldsymbol{v} \times \boldsymbol{B}$ Lorentz force.
4. This creates particle drifts between the species and therefore a current density $\boldsymbol{J}$.
5. The final stage of the coupling of the black hole to the magnetosphere is through $\boldsymbol{J} \times \boldsymbol{B}$ forces on the plasma and the fields that $\boldsymbol{J}$ produces as a consequence of Ampere's law. This is essentially a plasma wave interaction.

In the following subsections we look at potential gravitational forces that can produce step "2" above, i.e., drag plasma across a strong magnetic field. Then we relate the underlying force to the causal structure of MHD waves. This is the microscopic force that produces an unbalanced EMF, as in the laboratory unipolar inductor, that can drive a global current system from the ergosphere.

We note that the black hole gravitational field can be described by two types of "potentials" [92]. In [80] these are designated as "lapse" and "shift." In any decomposition that one uses, one effect is radial gravity and the other is the dragging of inertial frames. Frame dragging is the dominant physical effect just inside of the stationary limit (see 3.46). Radial gravity dominates the dynamics at $\alpha \ll 1$ as given by the horizon boundary condition (3.94) that equates to the inertial dominance of the flow.

7.3.1 Radial Gravity

Firstly, we look at the relevance that radial gravity has in the putative microphysics of current generation in the ergosphere through its ability to drag plasma across magnetic field lines, step "2" above. Radial gravity induces a $\boldsymbol{v}^P$, so plasma would be dragged across the magnetic field lines in the poloidal direction. This can create $q\boldsymbol{v}^P \times \boldsymbol{B}^P$ Lorentz forces that can generate a J^ϕ. However, this current relates to the poloidal magnetic flux in the Grad–Shafranov equation (see Sect. 5.7) and is not directly related to the dynamo for B^T.

To first order, $\boldsymbol{v}^P \times (B^T \hat{\boldsymbol{e}}_\phi) = 0$ in an axisymmetric vacuum magnetosphere into which the tenuous plasma has been injected (see 4.90b). However, consider a circumstance in which some other physical process has seeded the magnetosphere with a small B^T. It is of interest to investigate the possibility that radial gravity can create a perturbation δB^T by dragging plasma across the seed B^T. If this δB^T were to grow unstably and can be radiated upstream into the outgoing wind then this would be a possible ergospheric dynamo for B^T.

Choose $B^P > 0$ in the northern hemisphere, as usual, then by (5.20) and (5.21) the angular momentum and energy extracting conditions require $\delta B^T < 0$. Radial gravity can pull plasma inward creating a v^r across the seed B^T. The $qv^r B^T \hat{\boldsymbol{e}}_r \times \hat{\boldsymbol{e}}_\phi$ Lorentz force induced by this dynamic (step "3" above) creates a J^θ directed poleward in step "4" above. By Ampere's law, J^θ switches off δB^T upstream in step "5" above. Thus, the induced current damps the instability. We conclude that there is no dynamical mechanism in which $qv^r B^T \hat{\boldsymbol{e}}_r \times \hat{\boldsymbol{e}}_\phi$ forces can drive currents to create a growing B^T upstream in the outgoing wind.

Radial gravity cannot be the force creating an unbalanced EMF in the ergospheric dynamo for B^T. One should also note that by the time that the inflow is dominated by the force of radial gravity, it is probably already propagating inward supermagnetosonically and is therefore out of causal contact with the outgoing wind.

7.3.2 The Dragging of Inertial Frames

Frame dragging can pull plasma across the poloidal magnetic field in the $+\hat{\boldsymbol{e}}_\phi$ direction relative to black hole rotation. The azimuthal velocity of the plasma relative to the magnetic field lines created by this component of the gravitational force produces a $qv^\phi B^P \hat{\boldsymbol{e}}_\phi \times \hat{\boldsymbol{e}}_1$ Lorentz force in step "2" above. This in turn generates a cross-field poloidal current, $J^\perp = J^2$, directed equatorward. By Ampere's law, if the flow is submagnetosonic, this creates a $B^T < 0$ upstream as required for outgoing energy and angular momentum fluxes from the ergosphere (see 5.20 and 5.21). This is the same microphysics that occurs in the Faraday wheel, i.e., there is an unbalanced $v^\phi \hat{\boldsymbol{e}}_\phi \times \boldsymbol{B}^P$ EMF.

Current generation can occur in the outer ergosphere where frame dragging is strong and the flow is yet to go supermagnetosonic. Note that the generation of

$J^{\perp}$ is crucial to the dynamics and there is no purely field aligned poloidal current generation mechanism associated with the microphysics of the ergospheric plasma.

The fact that this behavior can exist in the outer ergosphere is significant. We know that the constants of motion are propagated into the outgoing wind primarily by Alfvén waves, since they can create charge separation and transport poloidal field aligned currents. This is critical because Goldreich – Julian charge is advected away with the outgoing wind and must be replenished by the dynamo. This is a manifestation of the fact that the asymptotic outgoing wind zone is essentially an Alfvén wave front (see Sect. 6.3). As we saw in (6.6), the Alfvén critical surface of an ingoing magnetically dominated wind is determined by the dragging of inertial frames that creates the inner speed of light surface. Thus, the Alfvén critical surface is always within the ergosphere, even when $0 < \Omega_F \ll \Omega_H$. We will find in Chaps. 8 and 9 that the field line azimuthal velocity condition at the inner Alfvén critical surface, $(\beta_F^{\phi})_A \gtrsim -1$, is fundamental as it connects the outgoing Alfvén mode to the negative energy states of plasma through (3.50).

7.4 The Torsional Tug of War

Consider a large scale magnetic flux tube that threads the ergosphere and also extends to large distances from the black hole. Assume that there is a tenuous plasma frozen onto the magnetic field lines in gyro-orbits both in the ergosphere and far from the hole as depicted in Fig. 7.1. Far from the hole, by (3.45) the plasma rotates with an angular velocity

$$\lim_{r\to\infty} \Omega_p \approx 0. \tag{7.1}$$

Similarly, in the ergosphere by (3.44) and (3.46) the angular velocity of the plasma is bounded from below:

$$\Omega_p > \Omega_{min} > 0, \tag{7.2a}$$

$$\lim_{r\to r_+} \Omega_{min} = \Omega_H. \tag{7.2b}$$

As discussed in the text below (5.15b), the value of Ω_F is a constant in a perfect MHD flux tube and as a consequence of (7.1) and (7.2), the plasma cannot corotate globally with the magnetic flux tube no matter what value of Ω_F is chosen.

The only global resolution is that there must exist a toroidal magnetic field, B^T, so that the plasma can slide azimuthally with respect to the corotating frame of the magnetic field and remain frozen-in globally. The existence of B^T, as plasma is introduced globally on vacuum axisymmetric poloidal field lines is a result of a torsional tug of war between plasma at $r \to \infty$ and plasma in the ergosphere. The plasma at $r \to \infty$ sends torsional Alfvén waves inward that communicate condition (7.1), i.e., it tells plasma and the field near the black hole not to rotate so fast. Similarly, plasma in the ergosphere sends torsional Alfvén upstream telling the plasma

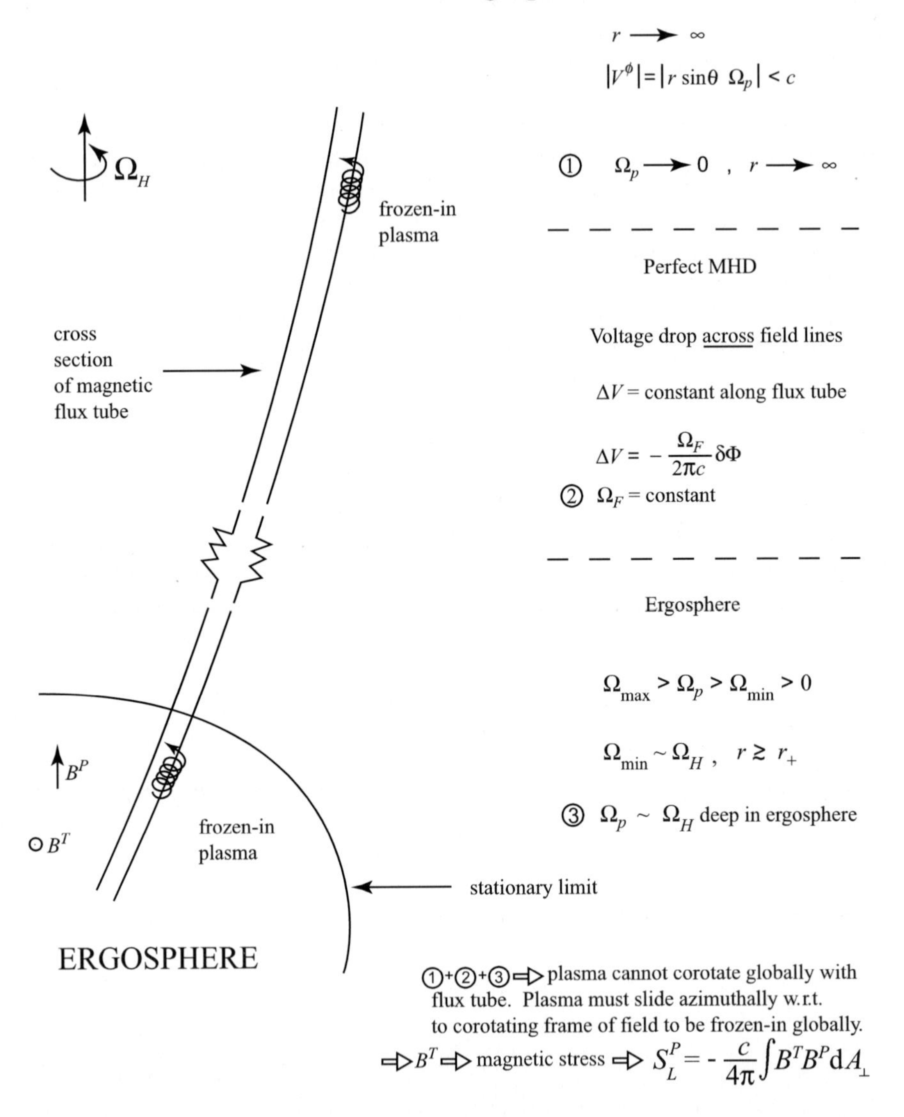

Fig. 7.1 The source of large scale torque in a magnetic flux tube that threads the ergosphere and extends to large distance from the hole. Due to the dragging of inertial frames, plasma cannot corotate global with the field lines and they are consequently twisted by a torsional tug of war

and the field at $r \to \infty$ to rotate in the positive sense as reflected in (7.2a). Plasma "near" the event horizon plays no role in the torsional tug of war since it is out of Alfvén wave communication with asymptotic infinity. The only component that can yield in this scenario is the poloidal magnetic field which gets bent azimuthally.

Implicitly, this dynamic requires a region in the flux tube in which plasma inertia actually accomplishes this bending of the magnetic field. At first glance this seems paradoxical for a paired magnetically dominated system (i.e., a strong magnetic field and a tenuous plasma). However, we know that one cannot impose the magnetically dominated constraint globally on the ingoing wind because of the horizon boundary condition of inertial dominance given by (3.94) and (3.95). So, it follows that if the tenuous plasma is to attain a large inertia from relativistic effects (as required to bend the magnetic field lines azimuthally) then this must occur in the ingoing wind. The natural place to look is where the rotational effects of frame dragging can bend the field lines, namely approximate corotation of the plasma with the magnetic field near the inner light cylinder. Not only does this keep the putative dynamo plasma within Alfvén wave communication with plasma at $r \to \infty$ by (6.6), but it also prepares plasma on negative energy trajectories through (3.50), a necessary occurrence in any physical realization of the Christodoulou/Ruffini energy extraction mechanism (see Sect. 1.4). The frame dragging force is very large near the inner light cylinder if the plasma is close to corotating with the magnetic field. In order for the magnetic field to keep the plasma in corotation near the inner light cylinder it must essentially battle the entire rotational inertia of the black hole as evidenced by the divergently large values of ω found in (3.50) under such a circumstance. Thus, the black hole always wins the torsional tug of war in the ergosphere. The force grows rapidly as the inner Alfvén critical surface is approached, eventually ripping plasma off the field lines in the positive azimuthal direction. This physical interaction is consistent with dynamo behavior near the inner light cylinder with the causal structure deduced from general principles in Sect. 7.3.2.

The ergospheric dynamo represents a transition from a magnetically dominated inflow to an inertial dominated inflow. The frame dragging effects of black hole gravity in the ergosphere eventually overwhelm the electromagnetic forces. This abrupt change cannot happen gracefully. One should note from the discussion of Sect. 2.10 that cross-field poloidal current flow is greatly impeded by the anisotropic electrical conductivity of the tenuous plasma-filled magnetosphere. Thus, the ergospheric dynamo is comprised of highly dissipative cross-field currents(see 2.136). In Chaps. 8 and 9, we will see that this Ohmic heating creates a relativistically hot flow downstream of the dynamo. This torsional tug of war is a very intense interaction between two powerful forces, perhaps the most extreme in the known Universe (as indicated by the incredible radiative losses in radio loud AGN, this is quantified in Chap. 10). The plasma is a mere pawn in this violent struggle as evidenced by its exotic nature: negative energy, relativistically hot, and rotating backwards at the speed of light.

An important aspect of the discussion of Fig. 7.1 is that a GHM driven jet does not require ordered, time stationary poloidal magnetic fields. Any local poloidal field structure that connects the inner ergosphere to a region of tenuous plasma more than

1 M away should causally connect (via MHD plasma waves) two regions of plasma with a large differential in angular velocity, $d\phi/dt$. Thus, large magnetic stresses will be transported between the regions via MHD waves: the GHM interaction. The general implication is that any poloidal field structure, >1 M, that is strong enough to impede the inflow of ergospheric plasma, allows the ergospheric plasma "enough time" to twist up the poloidal field in this semi-suspended state, thereby driving an outward Poynting flux (even in a transient state). In a general turbulent astrophysical environment, ordered field structures are more likely the exception rather than the rule. The numerical simulations in Chap. 11 show that the high efficiency of the GHM dynamo in disordered, transient poloidal magnetic fields renders it a likely mechanism for jet production in astrophysical environments.

Chapter 8
Ergospheric Disk Dynamos

Since there is no unipolar induction associated with the event horizon, in order to elucidate the details of the ergospheric dynamo one can look at the GHM interaction of flux tubes that thread the equatorial plane of the ergosphere and therefore not the horizon. In this chapter we look at a highly idealized model in which a disk of ergospheric plasma is the dynamo (see Fig. 8.1). This has two simplifying features. Firstly, the equatorial disk acts as an MHD piston and thus it is essentially a boundary surface for the paired MHD wind system. Secondly, there is a direct analogy between the ergospheric disk and the Faraday wheel. The model has the same causal structure that was outlined in Chap. 7. This example clearly illustrates how the rotational energy of the hole powers the outgoing wind. The solution emphasizes the crucial role of the Alfvén wave radiation from the dynamo and strong dissipative cross-field currents in the dynamo region. It must be stressed that this is an over-idealize analytical model that is designed to show GHM in its purest and most powerful form.

8.1 Fate of Accreted Magnetic Flux

Before we describe the global flow structure, we derive another horizon boundary condition that is relevant to equatorial flows in the ergosphere. In essence it is a restatement of the no hair theorem for charge neutral sources in the limit of a continuous inflow. Accreting charge neutral plasma cannot approach the event horizon and stay frozen onto large scale poloidal magnetic flux. The black hole "wants" the plasma and it "does not want" the magnetic field. The plasma must disconnect from the large scale poloidal flux before reaching the horizon. We show how this occurs through reconnection of the magnetic field lines at small lapse function. It should be pointed out that this boundary condition does not hold if the accreting plasma attains a charge. However, in a realistic resistive plasma it is likely that the reconnection will occur before the charge separation accumulates. The horizon boundary condition posited is not completely general and it is an assumption of this chapter.

B. Punsly, *Black Hole Gravitohydromagnetics, 2nd. ed.*,
Astrophysics and Space Science Library 355, doi: 10/1007/978-3-540-76957-6_8,

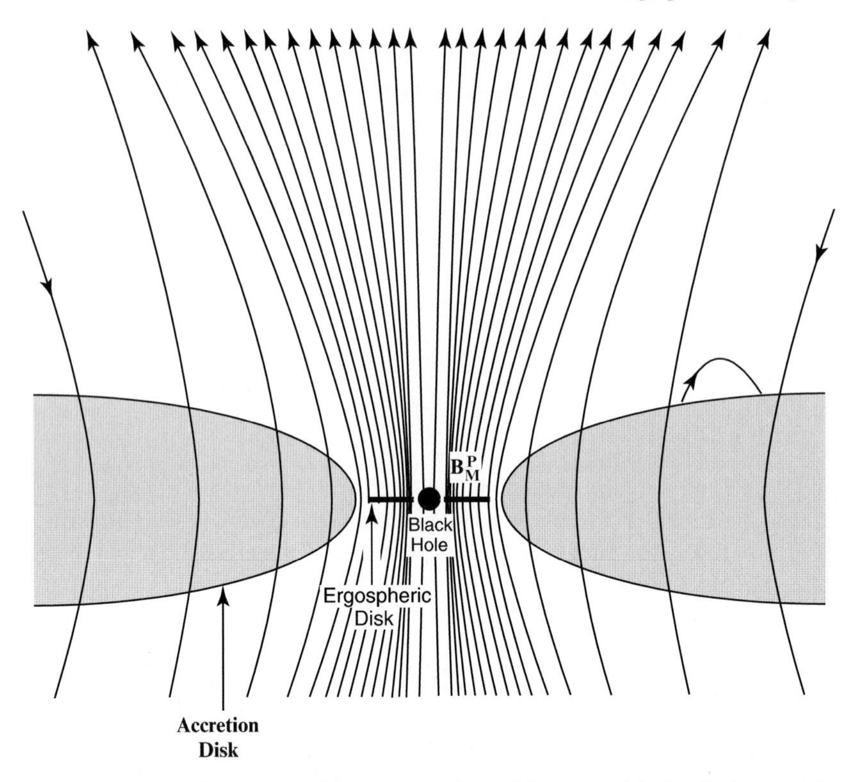

Fig. 8.1 The northern hemisphere of the magnetosphere of the equatorial plane in the ergosphere. A large poloidal magnetic flux, B_M^P, can in principle exist between the accretion disk and the event horizon, since neither is a sink for magnetic flux. The accretion disk is a pathway for flux accreting toward the hole and the event horizon cannot accept magnetic flux advected with charge neutral plasma. Thus, an ergospheric disk can be highly magnetized with trapped flux [153]

We can understand this phenomenon by breaking up the accretion flow and its electromagnetic sources into a series of axisymmetric current rings (see Fig. 8.2). The azimuthal current rings are the discrete elements that approximate the continuous current flow. The dynamics of a contracting current ring is well understood from the calculations of Chap. 4. The most important aspect of that analysis is found in Sect. 4.5.2. In that section, it was shown that the large scale fields from a charge neutral source (such as the current ring which is explored in more detail in Sect. 4.9) not only die off as the source nears the horizon, but die off over smaller and smaller proper distances. This is the physics underlying the reconnection of the field.

In order to understand the global field topology, we describe the field of a current ring as it contracts towards the horizon. We need a frame independent way of quantifying "the size" of the current ring as it contracts towards the hole. For example, the total azimuthal current depends on the frame of reference and scales with lapse function, except in the proper frame. The most natural measure is the total number

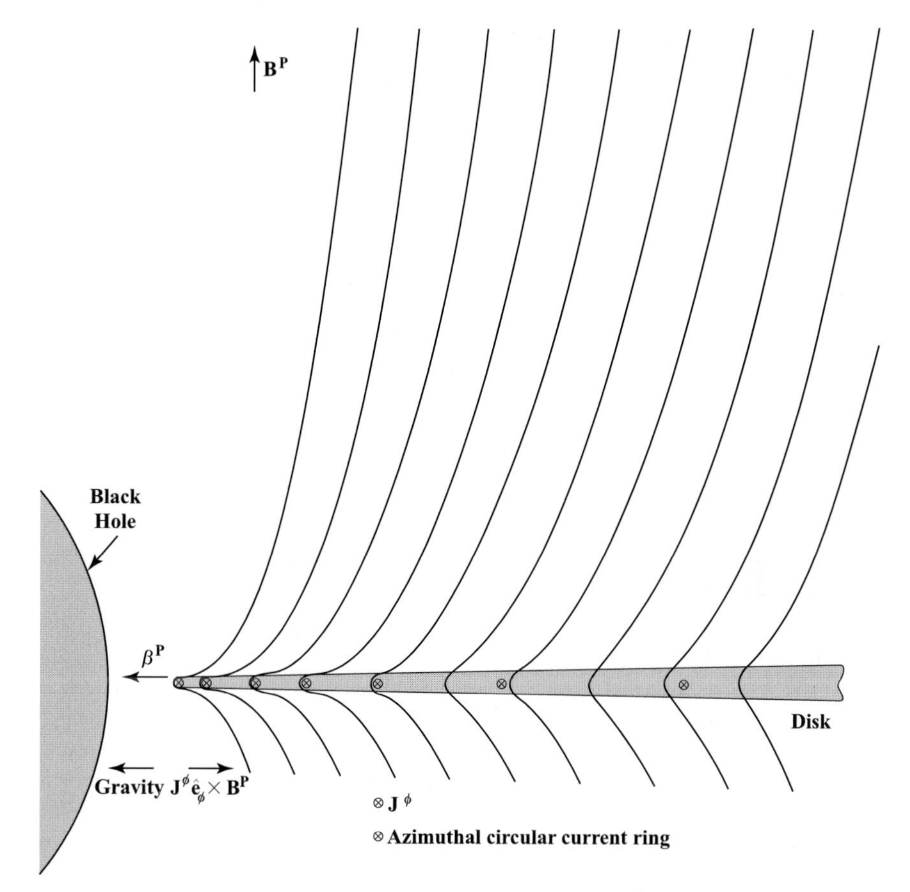

Fig. 8.2 The azimuthal current sheet in the disk supports a radial magnetic field above the disk. There is an approximate balance from the $J^{\phi}\hat{\mathbf{e}}_{\phi} \times \boldsymbol{B}^{P}$ force that is associated with the current and radial gravity. The plasma at the inner edge of the disk slowly advances toward the hole. The current sheet is approximated by a series of current-carrying, concentric circular wires that are centered about the rotation axis of the black hole [132]

of particles, N, in the loop which is frame independent. We describe this in terms of a line density, $\tilde{\rho}$,

$$N = 2\pi\tilde{\rho}\sqrt{g_{\phi\phi}}\,. \tag{8.1}$$

The advantage of the line density description of N is that the azimuthal current takes on the simple form for a charge neutral ring,

$$I_{\phi} = \frac{\tilde{\rho}}{2}\left(v^{\phi}_{+} - v^{\phi}_{-}\right)\,, \tag{8.2}$$

where v^{ϕ}_{+} and v^{ϕ}_{-} are the three velocities of the two species of charge.

The asymptotic form of I_ϕ is easily found from (8.2) and the horizon boundary conditions. From the boundary condition (3.94c), the azimuthal current in the ZAMO frame asymptotically scales as

$$I_\phi \underset{\alpha \to 0}{\simeq} \alpha \,. \tag{8.3a}$$

In the stationary frames by (3.95b) we see that globally the current appears to be turning off as $\alpha \to 0$,

$$\tilde{I}_\phi \underset{\alpha \to 0}{\simeq} \alpha^2 \,. \tag{8.3b}$$

Note that by (6.38) and (8.3a), the time dilation seen in the ZAMO frame implies that I_ϕ is approximately constant near the horizon in the proper frame of the infalling ring as expected.

We are not just interested in the horizon boundary condition on an isolated current ring, but a continuous equatorial flow. Since the poloidal magnetic field of the ring dies off as the horizon is reached (see 4.77 and 4.78; note that 8.3b gives the leading order behavior in 4.77), the current ring that was once connected to the global flow eventually is causally disconnected. It is of physical interest to understand how this process is manifested on global scales.

First consider the quasi-equilibrium that is indicated in Fig. 8.2. Radial magnetic curvature stresses push outward and are in approximate balance with radial gravity. The radial stress is a consequence of the curved poloidal magnetic field near the equator from the component of B^r that is created as radial gravity slowly pulls plasma inward bending the frozen-in field in the process. This is represented by a $J^\phi \hat{\mathbf{e}}_\phi \times \boldsymbol{B}^{\boldsymbol{P}}$ force directed outward. We know from Chap. 3 that this quasi-equilibrium is impossible as $\alpha \to 0$. The magnetic flux accretion process depends not only on Maxwell's equations, but on the momentum equation of the current rings as well. We show how the breakdown of the quasi-equilibrium as $\alpha \to 0$ is manifested in the magnetic flux accretion process.

We explore an intermediate zone where the ring is at small lapse function, yet far enough from the hole that the boundary conditions (8.3) have not been established. This region is the transition from the equilibrium in Fig. 8.2 to the $\alpha \to 0$ boundary condition.

In this region (3.94) and (3.95) have not been established and by definition of a quasi-equilibrium $|v^r| \ll c$ in the ZAMO frame. In order to obtain an approximate balance in the radial momentum equation, the azimuthal current cannot die off. If the ZAMO azimuthal current does not die off, $v^\phi \sim \alpha^0$ in this intermediate zone. By (3.14) this implies a constraint:

$$\omega - \frac{\Omega_H}{c} m \sim \alpha m \sqrt{g_{\phi\phi}} \,. \tag{8.4}$$

In the radial momentum equation we now have $P^r \sim \alpha^0$ and $u^0 \sim \alpha^0$. Radial gravity and the magnetic stresses scale from (3.65) as

$$\left(u^0\right)^2 \Gamma^r{}_{00} \sim \alpha^{-1} \,, \quad a < M \,, \tag{8.5a}$$

$$\left(J^\phi \hat{\mathbf{e}}_\phi \times \boldsymbol{B}^{\boldsymbol{P}}\right) \cdot \hat{\mathbf{e}}_{\mathbf{r}} \sim \alpha^0 \,, \quad a < M \,. \tag{8.5b}$$

By (8.5), the $\left(J^{\phi}\hat{\mathbf{e}}_{\phi}\times\boldsymbol{B^{P}}\right)$ forces will eventually be overwhelmed by gravity and v^{ϕ} must decrease as α. The case $a=M$ is also considered in [132].

This intermediate zone is represented by the innermost current loop in Fig. 8.2. It is also indicated by the third to fifth current loops in Fig. 8.3, that is a higher resolution depiction of the flow at a time later than that of Fig. 8.2. The innermost current loop in Fig. 8.3 is in the asymptotic zone, $v^{\phi}\sim\alpha^{+1}, u^{0}\sim\alpha^{-1}$, and $u^{r}\sim\alpha^{-1}$. In this region, the forces in the poloidal momentum equation (8.5) scale differently,

$$\left(u^{0}\right)^{2}\Gamma^{r}{}_{00}\sim\alpha^{-3}\,, \tag{8.6a}$$

$$\left(J^{\phi}\hat{\mathbf{e}}_{\phi}\times\boldsymbol{B^{P}}\right)\cdot\hat{\mathbf{e}}_{\mathbf{r}}\sim\alpha^{0}\,. \tag{8.6b}$$

Equation (8.6) implies that the electromagnetic forces are totally negligible compared to radial gravity and the current ring is essentially in free fall. Thus, by (8.3) I_{ϕ} is dying off very quickly. Note the X-point in Fig. 8.3 that is forming between the innermost and second current loops. This is a consequence of (4.78) that shows that the magnetic flux from the innermost loop seems to be dying off exponentially in time as viewed in the frame of the second loop. Reconnection of the magnetic field takes place at the X-point [133].

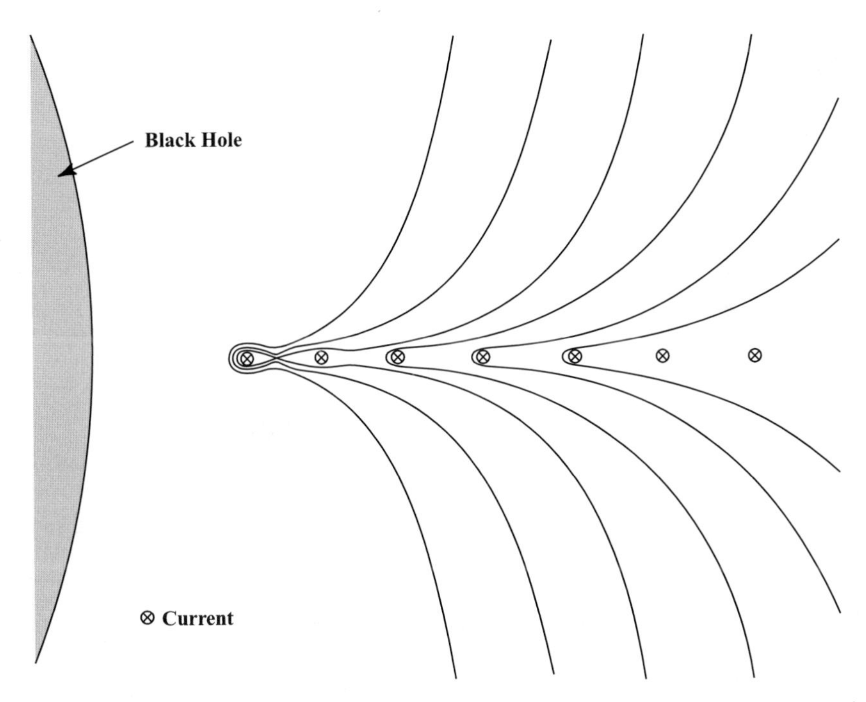

Fig. 8.3 As the inner edge of the current sheet slowly advances toward the hole, the sources in the innermost "circular wire" are redshifted as seen by observers who are farther out in the current sheet. It appears that the contribution of this wire to the global poloidal magnetic field is diminishing. The redshift effect causes the global field, including the contribution from this innermost wire, to develop an X-point between the first and second current rings. Reconnection takes place at the X-point [132]

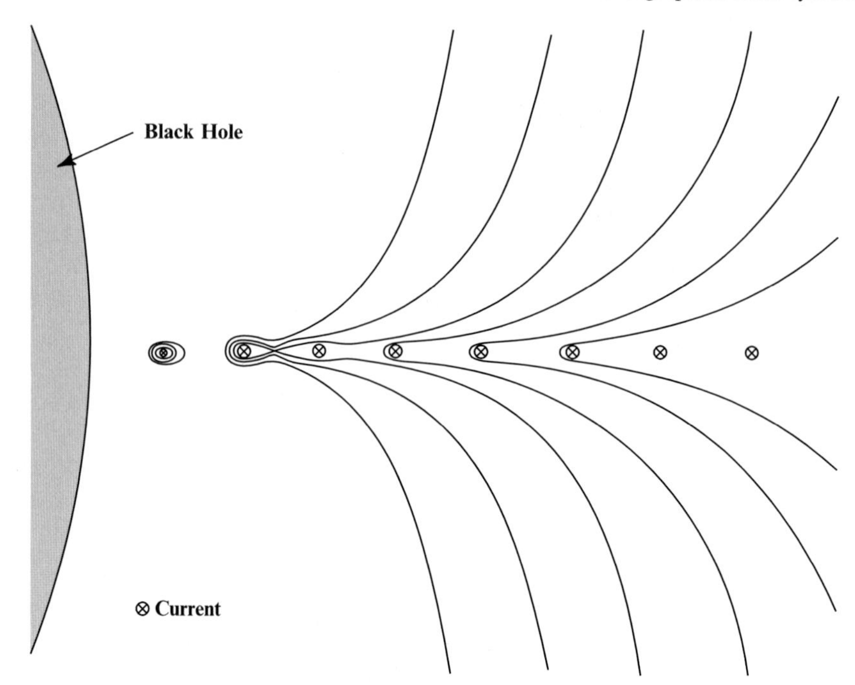

Fig. 8.4 At a time that is later than that of Fig. 8.3, the field from the innermost wire has been disconnected from the large-scale magnetic field. The wire is now a set of O-points, where magnetic flux can be destroyed. The magnetic field loops from this wire contract back into the wire as the current ring is attracted toward the hole by gravity. It is as if the current in this innermost wire is gradually being turned off as viewed globally [132]

At a slightly later time, the field from the innermost current ring has disconnected from the large scale poloidal magnetic field as depicted in Fig. 8.4. The ring is now in the asymptotic zone where B^P dies off faster than α^2 (see 4.77) and the field topology becomes a collection of O-points where magnetic flux can be destroyed. As the innermost current ring approaches the horizon, it eventually encounters new large scale magnetic field lines. The plasma plows through these field lines (since it is inertially dominant by 8.6), by pulling and stretching them toward the horizon until the lines of force reconnect. Reconnection proceeds very quickly since B^P from the ring dies off ever more rapidly (since α is smaller, see 4.77), and the ring keeps contracting toward the horizon. Notice the continuity of the process in Fig. 8.4 as an X-point has developed between the second and third loops of the current sheet. The continual reconnection process allows the hole to accrete plasma without acquiring the poloidal magnetic flux.

The preceding analysis allows one to understand the time evolution of a magnetic flux tube frozen into a magnetic plasma ring that accretes toward the hole. The physics that is captured in Figs. 8.2–8.4 is illustrated for a single flux tube in Fig. 8.5. Reconnection occurs between frames "c" and "d" of Fig. 8.5. The plasma is stripped from the flux tube in frame "d" at which time the flux tube becomes buoyant (i.e.,

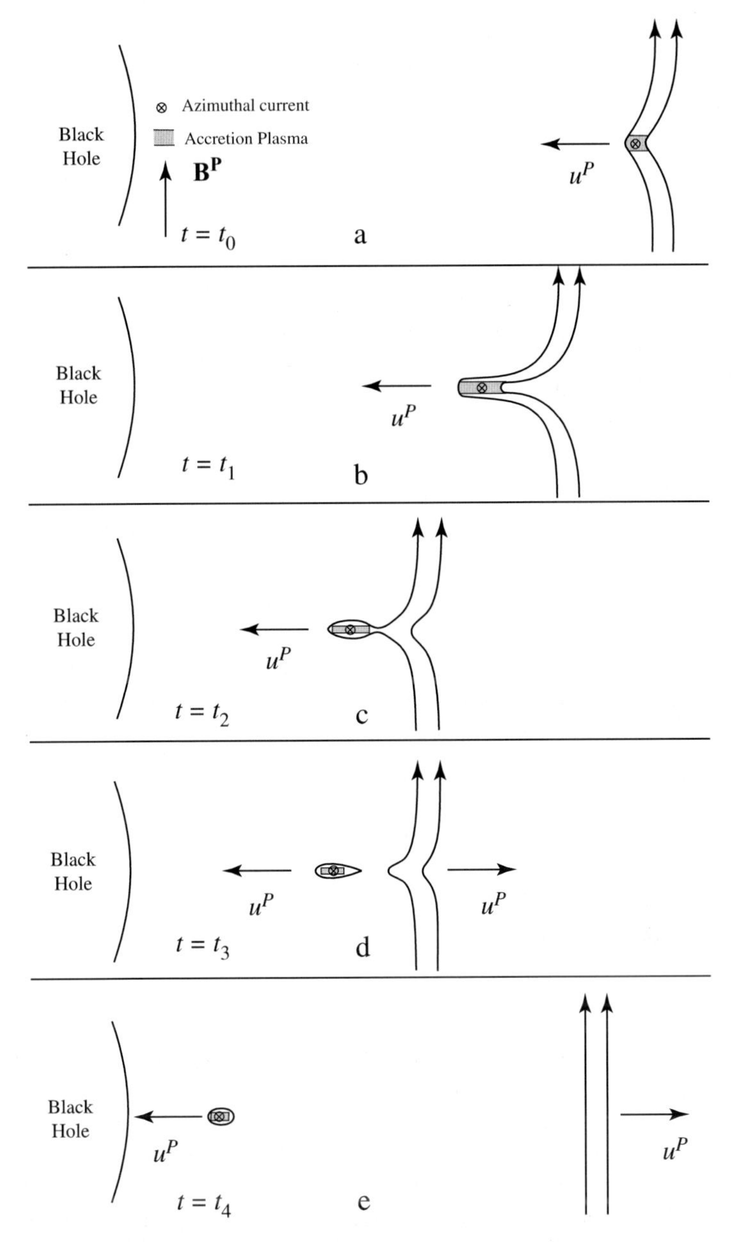

Fig. 8.5 The accretion history of an axisymmetric magnetized plasma ring. A poloidal magnetic field is supported by the azimuthal current in the ring. (**a**) The ring is nearing the hole with poloidal component, u^P, of the four velocity. The shaded region indicates where most of the accreting plasma is located. The time sequence shows the reconnection process. (**c**) A circular ring (seen in cross section) of X-type reconnection sites is about to form. (**e**) The large-scale flux is buoyant and moves outward, completely decoupled from the plasma ring. The accreting plasma ring becomes a circular set of O-points (seen in cross section) at which magnetic loops are destroyed before the plasma reaches the horizon [153]

there are no inertial forces to balance pressure). The buoyant flux tube is no longer connected to the accretion flow and is driven outward by magnetic pressure in the magnetosphere. The physics depicted in Figs. 8.2–8.5 describes the relevant horizon boundary condition for the accretion of magnetic flux in the equatorial plane.

8.2 The Global Structure of the Flow

The horizon boundary condition on the accretion of magnetic flux naturally produces buoyant flux tube that will be temporally trapped by surface currents on the edge of the highly conductive accretion disk. This is a complicated time-dependent scenario, but it does suggest that the existence of large scale magnetic flux inside the accretion disk and outside the horizon is a likely consequence of a magnetized accretion flow in the equatorial plane (see Sect. 11.5.7 for 3-D numerical results that support this posited evolution of accreted flux). In this section, we are interested only in the ergospheric dynamo behavior on these flux tubes. Thus, for simplicity we choose as initial configuration a background field that is sourced by azimuthal currents at the inner edge of the disk that are constant in time (see Fig. 8.1). For conceptual clarity, we do not consider the disk to be accreting but only a source of magnetic flux. We will discuss the effects of accretion and the relevance of this treatment in Chap. 10. Since no accretion is not physically likely, the model discussed in this section is more of a gedanken experiment.

We consider the source of plasma to be pair creation in a γ-ray field as discussed in Sect. 6.1. We then expect a paired wind system as in Fig. 6.1 in both hemispheres. Note that the existence of a γ-ray field is hard to reconcile with no accretion.

8.2.1 Poynting Flux and Disk Formation

From the discussion of Sect. 7.4, large scale magnetic flux tubes anchored in a disk of ergospheric plasma must have an outgoing angular momentum flux due to the torsional "tug of war." Since the field lines are frozen into the conductive ergospheric disk plasma, by (3.46), the field lines must rotate with $\Omega_F > 0$. Therefore (5.20) and (5.21) imply that there is an outgoing Poynting flux in the flux tube.

The field line rotation is imposed by the structure of the ergospheric disk that must form in the equatorial plane, since there is no sink for the inflow of the created pairs as occurs on flux tubes that thread the horizon. The flux tubes have a component along the radial direction. Radial gravity, $\Gamma^r{}_{00}\left(u^0\right)^2$, initiates the ingoing wind and pulls plasma toward the hole. In the Kerr geometry, there is also a component of gravity orthogonal to this given by $\Gamma^\theta{}_{00}\left(u^0\right)^2$ that pulls the flow toward the equatorial plane. Thus, there are two flows toward the equatorial plane along a flux tube, one from each hemisphere. If the flux tube is symmetric about the equatorial plane, the magnetic field is vertical at the equator (orthogonal to the radial direction). Since vertical gravity, $\Gamma^{\theta}{}_{00}\left(u^{0}\right)^{2}$ vanishes at the equator by (3.29b), the antidirected inflows from the two hemispheres should establish a final stationary state of a disk condensate. Initially, the plasma oscillates about the equator with a restoring force that is proportional to the displacement from the equator, $R^{\theta}{}_{0\theta0}\,\delta X^{\theta}$ (where $R^{\theta}{}_{0\theta0}$ is a component of the Riemann curvature tensor). The oscillation will eventually damp via plasma instabilities, or perhaps through radiation damping.

Whatever the damping mechanism turns out to be, the final state should be a highly conductive disk and atmosphere. The high conductivity of the disk implies that $\Omega_{p}\approx\Omega_{F}$ at the base of the flux tube at the equator. Since $\Omega_{p}>\Omega_{\min}>0$, the field must rotate and a Poynting flux must propagate to asymptotic infinity. The ergospheric disk condensate and its relation to Ω_{F} is analogous to the Faraday wheel in a waveguide circuit that was discussed in Sect. 2.9.4.

8.2.2 The Slow Shock and Disk Atmosphere

The function of the atmosphere is to slow the incoming wind to the small poloidal velocity of the accreting disk plasma, in spite of the gravitational acceleration toward the disk. The disk acts like a piston that drives MHD waves upstream to signal the incoming wind that it is approaching a boundary. As discussed in Chap. 2, the two MHD modes that can decelerate the flow are the compressive fast and slow waves. In order to determine which of these waves is dominant, recall the discussion of 2.65 that differentiates the two magneto-acoustic modes in terms of the change in the transverse magnetic field across the wave. First note that $B^{T}<0$ in the northern hemisphere and $B^{T}>0$ in the southern hemisphere; thus $B^{T}=0$ at the equator. This is consistent with the interpretation that the ergospheric disk is the source of Poynting flux in the paired wind system. Thus, B^{T} is switched off in the atmosphere and upper regions of the disk. The transverse magnetic field increases across a fast compression wave and decreases across a slow compression wave. This implies that the signal that brakes the incoming flow is predominantly a compressive slow wave.

The disk is distinguishable from the incoming wind and the atmosphere by its larger rest frame particle number density. Therefore the annihilation rate in the disk is larger as well. The steady state disk must balance the rate at which matter accretes onto the surface with the rate that matter annihilates in the interior. If the disk is to maintain its number density and height, the accreting pair plasma must slowly drift toward the center to replenish the annihilating plasma. The poloidal velocity of the disk plasma, β^{P}_{D}, has a nonvanishing vertical velocity until the flow reaches the middle of the disk (see Fig. 8.6).

The disk plasma is highly conductive, so the frozen-in condition should hold in the middle of the disk near the equator. The dynamics of the atmosphere and the upper regions of the disk are the following: A perfect MHD flow comes in on one side (the ingoing wind) passes through a region where B^{T} is switched off and a perfect

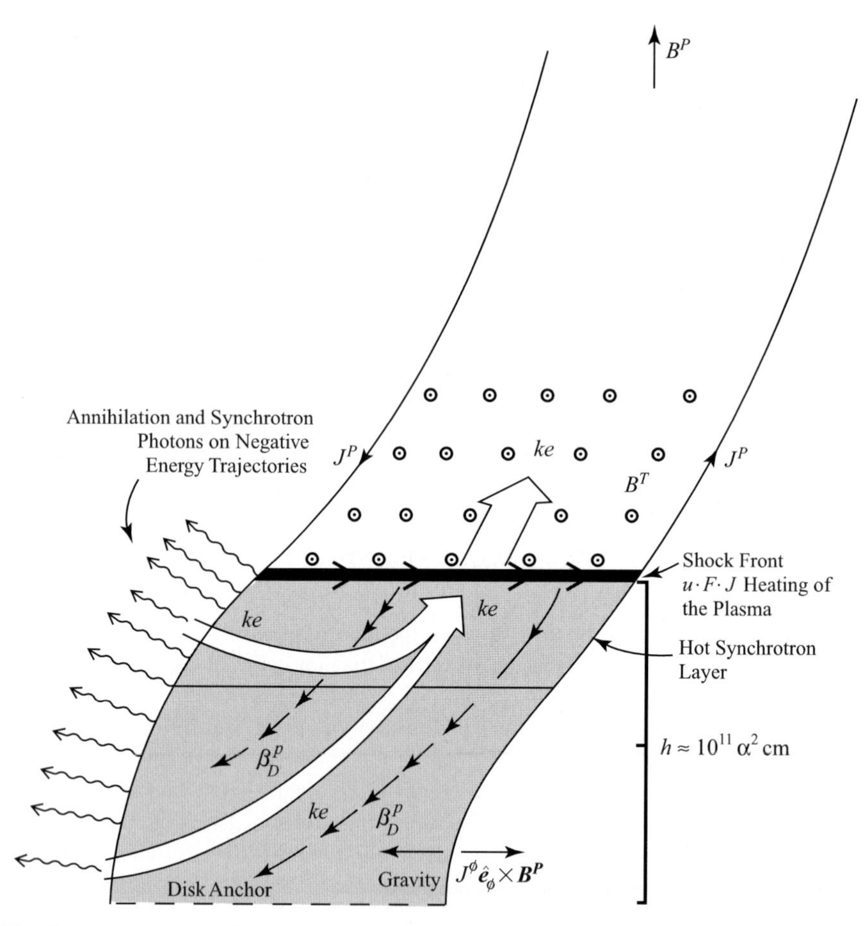

Fig. 8.6 The global energetics of an ergospheric disk dynamo in an azimuthal flux tube. A thin disk (*shaded*) is bounded from above by a slow switch-off shock front at which the energy flux, *ke*, is transformed from mechanical form to Poynting flux radiated from the disk surface. Plasma is dragged azimuthally across the poloidal magnetic field lines in the shock by the dragging of inertial frames. This is the force driving the cross-field dynamo current in the shock front. The enhanced current flow in the shock layer is indicated by the thick arrowed line. Plasma is resistively heated by $u \cdot F \cdot J$ dissipation in the shock. A relativistically hot flow exits the shock and settles into the disk on negative energy trajectories (plasma is rotating at nearly the negative speed of light in the ZAMO frames). The strong "headlight effect" associated with this ultarelativistic motion beams the synchrotron and annihilation radiation from the plasma onto negative energy trajectories. These superradiant photons spin down the black hole. The influx of these negative energy photons can be considered the outflow of energy flux, *ke*, that allows the black hole to power the dynamo [132]

MHD plasma flows out toward the center of the disk. In MHD, these flow conditions are satisfied by a switch off slow shock. The slow waves responsible for the change in B^T must coalesce into a thin shock front as opposed to being distributed throughout the disk because on the downstream side of a slow compressive wave the slow wave speed is increased and all slow waves downstream will overtake the leading wave front (see [75,78]). Thus, in steady state the switch off region is always a thin shock front in MHD. Furthermore, a slow shock is a switch off shock if, and only if, the incoming flow reaches the shock front at the intermediate speed [134]. Thus, the ergospheric disk can communicate Alfvén waves to the plasma source and outgoing wind as required by the causality considerations of Chap. 7.

8.2.3 Some General Disk Structure

Since the plasma in the disk has an enhanced value of $(n\mu c)$, the force of radial gravity will be strong in the equatorial plane which causes the field lines to bend (Fig. 8.6) so that the $\left(J_{\phi}B^{\theta}\right)\hat{\mathbf{e}}_{\mathbf{r}}$ force will approximate balance gravity. The poloidal velocity of the disk plasma will in general have a component in the radial direction, $\beta_{D}^{P}\cdot\hat{e}_{r}\neq 0$. However, the approximate force balance ensures that the plasma radial infall time is much longer than the annihilation time scale. The plasma, for the most part, annihilates in the disk and never enters the hole. The plasma exists long enough in the disk so that the disk can actually be considered to be anchoring the magnetic field like a neutron star in an MHD pulsar model.

There is a wide range of values of Ω_{F} for the flux tubes that thread the equator in the ergosphere. Ω_{F} varies from approximately Ω_{H} for the disk lapse functions, α_{D}, that are mush less than unity to approximately zero near the stationary limit. Most of the analysis in this chapter is confined to $\alpha_{D}\sim 0.1$ where most of the proper surface area of the disk resides. In this region $\Omega_{F}\sim\Omega_{H}$.

8.3 The Rankine–Hugoniot Relations

This section describes the flow at the downstream side of the shock at the top of the ergospheric disk. The paired outgoing wind is the minimum torque solution of Sect. 5.6. On the downstream side of the switch-off shock $F^{12}=0$ and all of the energy and angular momentum are in mechanical form. Thus, in the ZAMO frames on the downstream side of the shock one has

$$\mu u^{0}=\frac{e-\dfrac{\Omega\ell}{c}}{\alpha}\,. \tag{8.7}$$

Combining this with (5.22b) for a frozen-in plasma

$$\beta_{F}^{\phi}\left(e-\frac{\Omega\ell}{c}\right)=\frac{\alpha\ell}{\sqrt{g_{\phi\phi}}}\,. \tag{8.8}$$

Note that (8.8) is the Alfvén point condition (5.24a).

8.3.1 The Field Line Angular Velocity

Balancing the angular momentum flux of the minimum torque wind (5.50) with the mechanical redshifted angular momentum flux downstream of the shock

$$mk_- = \frac{\Omega_F \Phi}{4\pi k_F} = -\frac{c}{4\pi} B_\infty^T , \tag{8.9}$$

where k_- is the mass flux parameter for the ingoing wind and k_+ is the same for the outgoing wind (see Fig. 6.1). Integrating the mass conservation law (3.36b) across the thin shock yields

$$(n\alpha u^P)_u = (n\alpha u^P)_d , \tag{8.10a}$$

$$(k_-)_u = (k_-)_d , \tag{8.10b}$$

where the subscript "u" refers to upstream quantities and "d" refers to downstream quantities. Equation (8.10b) follows from (8.10a) since B^P is constant through the thin shock layer.

Just downstream of the shock $F^{12} = 0$, so by the frozen-in condition (5.16b)

$$\beta^\phi = \beta_F^\phi , \tag{8.11}$$

the magnetic field is anchored into the disk plasma. Combining (8.8), (8.9) and (8.11) yields the following useful relation in terms of downstream evaluated quantities

$$1 - \left(\beta_F^\phi\right)^2 = \frac{4\pi k_- \left(e - \dfrac{\Omega_F \ell}{c}\right) \beta_F^\phi k_F}{\alpha \Phi \Omega_F} \sqrt{g_{\phi\phi}} \tag{8.12}$$

A switch-off shock exists if and only if the incoming flow is at the Alfvén speed. From (5.36) we can rewrite the Alfvén point condition as

$$1 - \left(\beta_F^\phi\right)^2 = \frac{(k_-)^2 \left(B^P\right)^2}{c^2 (n_u)^2 (U_A)^2 \alpha^2} , \tag{8.13}$$

where the pure Alfvén speed, U_A, is evaluated upstream. Combining (8.13) and (8.12) we find two new relationships

$$\frac{k_-}{\alpha} = \left[\frac{4\pi k_F \beta_F^\phi \left(e - \dfrac{\Omega_F \ell}{c}\right)}{z \Phi \Omega_H}\right] \left[\frac{c^2 n_u^2 U_A^2}{(B^P)^2}\right] \sqrt{g_{\phi\phi}} , \tag{8.14}$$

$$1-\left(\beta_F^\phi\right)^2 = \frac{c^2 k_F^2 \left(\beta_F^\phi\right)^2}{z^2 U_A^2 k_S^2 \Omega_H^2 g_{\phi\phi}} \left(\frac{e - \frac{\Omega_F \ell}{c}}{m_e c^2} \right) , \tag{8.15}$$

where the following definitions are made:

$$z \equiv \frac{\Omega_F}{\Omega_H} , \tag{8.16}$$

and the analog of k_F at the shock front is

$$k_S \equiv \frac{\Phi}{\left(g_{\phi\phi}\right)_S B_S^P} , \tag{8.17}$$

where the subscript "S" means to evaluate at the shock front.

The expression (8.15) can be used to estimate how close β_F^ϕ is to -1. The quantity $e - \Omega_F \ell / c$ for a plasma that has not been heated to ultrarelativistic temperatures, as in the warm incoming wind, is on the order of $m_e c^2$ from (5.20) and (5.21). For a magnetically dominated flow $U_A \gg 1$. Furthermore, by (8.11) $\Omega_F > \Omega_{\text{min}}$ in the disk, thus for $\alpha_D \sim 0.1, \Omega_F \sim \Omega_H$ and $z \sim 1$ in (8.16). For a cylindrical asymptotic field geometry $k_F \sim \pi$. Applying all of these estimates to (8.15) yields the desired result at the shock

$$1-\left(\beta_F^\phi\right)^2 \ll 1 . \tag{8.18}$$

From (8.11), (3.50) and (8.18) we know that the ergosphere disk plasma has a large relativistic inertia and is of negative energy as viewed from asymptotic infinity. Furthermore, by (8.18) and (3.43) $\beta_F^\phi \approx -1$ yields the field line angular velocity on the disk

$$\Omega_F(r) \gtrsim \Omega_{\text{min}}(r) . \tag{8.19}$$

Note that the creation of the negative energy plasma in the ergospheric disk by the slow switch-off shock is exactly the physical process necessary for extracting the rotational energy of the hole as deduced in Sect. 1.4.

Combining (8.15) with the general expression for the intermediate speed of the ingoing wind (5.36), at the shock front

$$U_I^2 \approx \frac{c^2 k_F^2}{z^2 k_S^2 \Omega_H^2 g_{\phi\phi}} \approx \frac{c^2 k_F^2}{k_S^2 \left(\Omega_{\text{min}}\right)^2 g_{\phi\phi}} . \tag{8.20}$$

8.3.2 The Specific Enthalpy of the Post Shock Gas

The second Rankine–Hugoniot relation is energy flux conservation along the flux tube $\left(T^{0P}\right)_u = \left(T^{0P}\right)_d$, where $T^{\mu\nu}$ is the total stress-energy tensor obtained from combining the fluid and electromagnetic contributions of (3.21) and (3.23)

$$\left(n\mu c u^P u^0\right)_d + \frac{c}{4\pi}\left(\boldsymbol{E}\times\boldsymbol{B}\right)_d^P = \left(n\mu c u^P u^0\right)_u + \frac{c}{4\pi}\left(\boldsymbol{E}\times\boldsymbol{B}\right)_u^P . \tag{8.21a}$$

Using the frozen-in condition (5.13) and (8.10), (8.21a) can be expressed as

$$\frac{k\_B^P\left(\mu c^2\right)_d \gamma_d}{\alpha} - \frac{c}{4\pi}\beta_F^\phi B^P F_d^{12} = \frac{k\_B^P\left(\mu c^2\right)_u \gamma_u}{\alpha} - \frac{c}{4\pi}\beta_F^\phi B^P F_u^{12} . \tag{8.21b}$$

For a magnetically dominated flow into a switch-off shock, (8.21b) implies that the Lorentz γ-factor downstream (with $F^{12} \equiv F_u^{12}$) is

$$\gamma_d \approx -\frac{\alpha\beta_F^\phi F^{12}}{4\pi k_-\left(\mu c\right)_d} \approx \frac{\alpha F^{12}}{4\pi k_-\left(\mu c\right)_d} \tag{8.21c}$$

We can use (8.21c) to make an estimate of the thermal inertia $\left(\mu c^2\right)_d$. The key point is that we can approximate $\gamma_d^{-2} \approx 1-\left(\beta_F^\phi\right)^2$. In order to demonstrate this, we note that the outflow downstream of a slow shock must emerge subslowly, otherwise slow waves could not propagate upstream, steepen and then coalesce to form the shock front. This is shown mathematically in [75]. Thus we have

$$\left(u^P\right)_d^2 < U_S^2 . \tag{8.22a}$$

Dropping the subscript "d," (8.22a) expands out as

$$\left(\beta^P\right)^2\left[1-\left(\beta^\phi\right)^2-\left(\beta^P\right)^2\right]^{-1} < U_S^2 . \tag{8.22b}$$

Therefore, we have the inequality

$$\left(\beta^P\right)^2\left[1-\left(\beta^\phi\right)^2\right]^{-1} < \left(\beta^P\right)^2\left[1-\left(\beta^P\right)^2-\left(\beta^\phi\right)^2\right]^{-1} < U_S^2 , \tag{8.22c}$$

or

$$\left(\beta^P\right)^2 < U_S^2\left[1-\left(\beta^\phi\right)^2\right] . \tag{8.22d}$$

Equation (8.22d) yields the following inequality:

$$\gamma^{-2} > \left[1-U_S^2\right]\left[1-\left(\beta^\phi\right)^2\right] . \tag{8.22e}$$

Since $\gamma^2 > \left[1-\left(\beta^\phi\right)^2\right]^{-1}$ and $U_S^2 \leq \frac{1}{2}$, (8.22e) implies the constraint:

$$\left[1-\left(\beta^\phi\right)^2\right]^{-1} < \gamma^2 < 2\left[1-\left(\beta^\phi\right)^2\right]^{-1} . \tag{8.22f}$$

By (8.11) and (8.18), $\left[1-(\beta^{\phi})^2\right]^{-1} = \left[1-(\beta_F^{\phi})^2\right]^{-1} \gg 1$ just downstream of the shock. Thus (8.22f) implies that setting

$$\gamma_d^2 \approx \left[1-(\beta_F^{\phi})^2\right]^{-1} \gg 1\,, \tag{8.23}$$

is accurate to within a factor of 2, which is a good estimate by astrophysical standards.

From the definition of the Alfvén speed in (5.36) and (8.23)

$$U_I \approx \gamma_d^{-1} U_A c\,. \tag{8.24}$$

Using (8.24) to rewrite the mass flux parameter at the Alfvén point in (8.21c), we can estimate the thermal inertia to within a factor of 2 in terms of upstream parameters,

$$\left(\mu c^2\right)_d \approx \frac{F^{21}B^P}{4\pi n_u U_A} \approx m_e c^2 \left(\frac{F^{21}}{B^P}\right) U_A\,. \tag{8.25}$$

8.3.3 The Density of the Post Shock Gas

The conservation of poloidal momentum across the shock reduces to the relativistic version of pressure balance. The Rankine–Hugoniot relation for poloidal momentum conservation is found using (5.35), (3.21) and (3.23).

$$\begin{aligned} &\left[\left(u^P\right)^2 \mu n\right]_d + P_d + \frac{(F^{12})_d^2}{8\pi} - \frac{\left[1-(\beta_F^{\phi})^2\right](B^P)^2}{8\pi} \\ &\approx \left[\left(u^P\right)^2 \mu n\right]_u + P_u + \frac{(F^{12})_u^2}{8\pi} - \frac{\left[1-(\beta_F^{\phi})^2\right](B^P)^2}{8\pi}\,, \end{aligned} \tag{8.26}$$

where P is the pressure (combined gas and radiation) evaluated in the rest frame of the plasma. Since the incoming wind is magnetically dominated, the upstream gas pressure and ram pressure can be neglected. For a switch-off shock equation (8.26) then approximates to the following:

$$P_d \approx \frac{\left(F^{12}\right)^2}{8\pi} - \frac{k_- B^P \left(u^P \mu\right)_d}{\alpha}\,. \tag{8.27a}$$

Using (8.21c) to express $k_-\left(\mu\right)_d$, (8.27a) reduces to

$$P_d \approx \frac{\left(F^{12}\right)^2}{8\pi} - \frac{B^P \beta_d^P F^{12}}{4\pi}\,. \tag{8.27b}$$

Since the flow downstream is subslow and by (8.23), $\gamma_d^2 \gg 1$, we can ignore the second term in (8.27b) and to an excellent approximation the law of pressure balance across the shock is

$$\frac{\left(F^{12}\right)^2}{8\pi} \approx P_d = n_d k_B T \; , \tag{8.27c}$$

where k_B is the Boltzman constant.
The enthalpy of a gas with adiabatic constant Γ satisfies

$$n_d \left(\mu c^2\right)_d = n_d m_e c^2 + \frac{\Gamma}{\Gamma - 1}\left(n_d k_B T\right) \, . \tag{8.28a}$$

Since most of the disk resides at lapse functions $\alpha_S < 1/2$, (8.25) combined with the value of B^T of the minimum torque solution (5.50) and the defining relation (5.17) implies that the downstream plasma is relativistically hot,

$$\left(\mu c^2\right)_d \gg m_e c^2 \, . \tag{8.28b}$$

Combining this fact with the pressure balance (8.27c) and (8.25) yields an approximate expression (to within a factor of 2) for the downstream number density,

$$n_d \approx n_u u_A \frac{\Gamma}{2(\Gamma - 1)} \left(\frac{F^{21}}{B^P}\right) \, . \tag{8.28c}$$

8.3.4 The Downstream Poloidal Velocity

Using the value of n_d from (8.28c) and the definition of k_- in (5.12) yields

$$\left(u^P\right)_d = \frac{2(\Gamma - 1)}{\Gamma} \left(\frac{k_-}{\alpha}\right) \frac{\left(B^P\right)^2}{n_u u_A F^{21}} \, . \tag{8.29}$$

Eliminating γ_d and $\left(\mu c^2\right)_d$ using (8.21c) and (8.25) we can get an approximate expression for the poloidal three velocity downstream of the shock from (8.29),

$$\left(\beta^P\right)_d \approx 8\pi \left(\frac{\Gamma - 1}{\Gamma}\right) \left(\frac{k_-}{\alpha c}\right)^2 \left(\frac{m_e c^2 B^P}{n_u F^{12}}\right) , \tag{8.30a}$$

$$\approx 8\pi \left(\frac{\Gamma - 1}{\Gamma}\right) \left[\frac{\left(n_u m_e c^2\right)\left(\frac{U_I^2}{c^2}\right)}{F^{12} B^P}\right] . \tag{8.30b}$$

Thus $\mid \beta_d^P \mid \ll 1$ as shown in (8.22).

8.4 A Parametric Realization of Shock Parameters

Consider a black hole with a rest mass $\gtrsim 10^9\,\mathrm{M}_\odot$ and a magnetic field strength in the magnetosphere of the ergospheric disk, $B^P_M \sim 10^4\mathrm{G}$ as shown in Fig. 8.1. Then by (8.19) and (3.44) we can take an average value of $\Omega_F \sim \Omega_H/2$. Using (5.20) and (5.50), the Poynting flux driven by the ergospheric disk when $a/M = 0.9$ is

$$\int \alpha S^P \mathrm{d}A \sim 10^{46}\mathrm{ergs/sec}\,. \tag{8.31}$$

It is instructive to analyze the structure of the disk not only theoretically, but also numerically. Thus, as we derive theoretical expressions, we will also plug in parameters from this test model to quantify disk properties.

Consider a copious supply of disk and coronal γ-rays above 1 MeV, $L_c \gtrsim 10^{43}$ ergs/sec. Thus,

$$\frac{L_c}{L_{Edd}} \sim 10^{-4}\,, \tag{8.32a}$$

and from (6.1), we expect

$$n_u \sim 10^7\mathrm{cm}^{-3}\,. \tag{8.32b}$$

From (5.23b), a value of $B^P \sim 10^4\mathrm{G}$ combined with (8.32b) yields

$$U_A^2 \sim 10^6\,, \tag{8.32c}$$

and using (8.28c) this implies

$$n_d \sim 10^{10}\mathrm{cm}^{-3}\,. \tag{8.32d}$$

8.5 The Dynamics and Structure of the Disk

In this section we examine the structure and dynamics of the downstream pair plasma as it settles toward the equatorial plane. Since the pairs are prevented from falling into the hole by magnetic curvature stresses (see Fig. 8.6), every particle that enters the disk can be considered to be eventually annihilated to first approximation. According to (8.25), the plasma is relativistically hot so that synchrotron emission is an important radiative mechanism. Hence, the vertical structure of the disk is determined by the relative rates of synchrotron cooling and pair annihilation.

The particle collision time, τ_c, in the rest frame of the plasma is

$$\tau_c = \frac{1}{n\sigma\bar{u}_{\mathrm{rel}}}\,, \tag{8.33}$$

where σ is the collision cross section and $\bar{u}_{\mathrm{rel}}$ is the average relative four-velocity between the pairs, which by (8.25), is going to be relativistic. First note that

$$\bar{u}_{\mathrm{rel}} = \left[\overline{(u_0^2 - 1)}\right]^{1/2}_{\mathrm{rel}}\,. \tag{8.34a}$$

The relative Lorentz γ-factor between the two species is, $\gamma_{\rm rel} \equiv u^0_{\rm rel}$,

$$-\gamma_{\rm rel} = \left(u_{\rm pos}\right)_\mu \left(u_{\rm elec}\right)^\mu \ . \tag{8.34b}$$

We can express the four velocities of the two species, $u^\mu_{\rm pos}$ and $u^\mu_{\rm elec}$, in terms of the thermal Lorentz factor, $\gamma_{\rm th}$, as

$$u^\mu_{\rm pos} = \gamma_{\rm th}\left(1, \boldsymbol{\beta_p}\right) \ , \tag{8.34c}$$

$$u^\mu_{\rm elec} = \gamma_{\rm th}\left(1, \boldsymbol{\beta_e}\right) \ . \tag{8.34d}$$

The three velocity β in the frame of the plasma is primarily due to random thermal velocity although there is a non-random second order effect due to the magnetic field (gyro-orbits).

Since we are interested in order of magnitude estimates, we take the average of the inner product of the random velocities to vanish,

$$\overline{\boldsymbol{\beta_e}\cdot\boldsymbol{\beta_p}} \approx 0\,, \quad -\overline{u_{\rm pos}\cdot u_{\rm elec}} = \overline{\gamma}^2_{\rm th}\left(1+\overline{\boldsymbol{\beta_e}\cdot\boldsymbol{\beta_p}}\right) \approx \overline{\gamma}^2_{\rm th} \ . \tag{8.34e}$$

Because the motion is not entirely random due to the magnetic field as noted above, a Maxwellian distribution may not be representative of plasma in the disk. To find $\overline{\gamma}^2_{\rm th}$, instead of using the Maxwellian distribution, one can use (8.28a) to get

$$\overline{\gamma}_{\rm th} \approx \left(\frac{\mu c^2}{m_e c^2}\right) \ . \tag{8.34f}$$

Thus, by (8.34a), (8.34b) and (8.34f) for $\gamma_{\rm rel} \gg 1$

$$\overline{u}_{rel} \approx \left(\frac{\mu c^2}{m_e c^2}\right)^2 \ . \tag{8.34g}$$

Using (8.34g) in the expression for τ_c in (8.33),

$$\tau_c \approx \frac{1}{n\sigma}\left(\frac{m_e c^2}{\mu c^2}\right)^2 \ . \tag{8.35}$$

The cross section for high energy annihilation is given by [135] as

$$\sigma \approx \frac{\pi r_0^2}{\gamma_{\rm rel}}\left[\ln\left(2\gamma_{\rm rel}\right) - 1\right] \ , \tag{8.36}$$

where r_0 is the classical electron radius. Combining (8.34) - (8.36) the pair annihilation lifetime is

$$\tau_a \approx \frac{1}{n\pi r_0^2 \chi c} \ , \tag{8.37a}$$

$$\chi \equiv \ln\left[2\left(\frac{\mu c^2}{m_e c^2}\right)^2\right] - 1 \ . \tag{8.37b}$$

The synchrotron lifetime on the other hand in a frame where the electric field vanishes, such as the plasma rest frame is given by [33] as

$$\tau_S = \frac{3\times 10^8}{\overline{\gamma}_{\rm th}\beta_{\rm th}^2 \sin^2\alpha\, B_c^2}\ {\rm sec} = 9\,\overline{\gamma}_{\rm th}^{-1}\left(\frac{10^8{\rm G}^2}{F^{\mu\nu}F_{\mu\nu}}\right){\rm sec}\,. \tag{8.38}$$

During a synchrotron lifetime $B_c^2 = 1/2F^{\mu\nu}F_{\mu\nu}$ can be approximated by using $F^{12}=0$ in (5.35). One can simplify the magnetic field strength in (8.38) using (8.20) and (5.36) computed from (5.35) with $F^{12}=0$,

$$B_c^2 \approx \left(\frac{k_F}{k_S}\right)^2\left(\frac{c^2}{\Omega_F^2 g_{\phi\phi}}\right)U_A^2\,(B^P)^2 = 4\pi\left(\frac{k_F}{k_S}\right)^2 n_u m_e c^2\left(\frac{c^2}{\Omega_F^2 g_{\phi\phi}}\right). \tag{8.39a}$$

We can express the thermal Lorentz factor in (8.38) using upstream evaluated quantities for the specific enthalpy in (8.34f). Expression (8.25) for the specific enthalpy simplifies if we describe the toroidal magnetic field ahead of the shock using (5.50), (8.17) and (5.17),

$$F^{21} \approx \alpha^{-1}\left[\frac{\Omega_F\sqrt{g_{\phi\phi}}}{c}\right]\left[\frac{k_S}{k_F}\right]B^P\,. \tag{8.39b}$$

Inserting (8.25), (8.39b) into (8.34f) to find the thermal Lorentz factor and using this result with (8.39a) for the synchrotron lifetime in (8.38) we have

$$\tau_S \approx \frac{9}{2}\alpha\left(\frac{k_S}{k_F}\right)\left[\frac{\Omega_F\sqrt{g_{\phi\phi}}}{c}\right]U_A\left[\frac{10^4{\rm G}}{(B^P)}\right]^2. \tag{8.40}$$

If one sets $\Gamma = 4/3$ in the relativistic plasma, (8.40) and (8.37) yield the ratio of synchrotron to annihilation time scales in the relativistic plasma just downstream of the slow shock,

$$\frac{\tau_S}{\tau_a} \approx \frac{9n_d\pi r_0^2\chi}{2}\,\alpha\left(\frac{k_S}{k_F}\right)\left[\frac{\Omega_F\sqrt{g_{\phi\phi}}}{c}\right]U_A\left[\frac{10^4{\rm G}}{(B^P)}\right]^2. \tag{8.41}$$

Eliminate the downstream parameters in (8.41) by writing n_d in (8.28) with the aid of (8.39b) for the toroidal field as

$$n_d \approx n_u U_A\left[\frac{\Gamma}{2(\Gamma-1)}\right]\alpha^{-1}\left[\frac{\Omega_F\sqrt{g_{\phi\phi}}}{c}\right]\left[\frac{k_S}{k_F}\right]. \tag{8.42}$$

Inserting (8.42) back into (8.41) yields the ratio of synchrotron to annihilation lifetimes just downstream of the switch-off shock in terms of upstream parameters,

$$\frac{\tau_S}{\tau_a} \approx 0.69\left(\frac{k_S}{k_F}\right)^2\left[\frac{\Omega_F^2 g_{\phi\phi}}{c^2}\right]\chi\,. \tag{8.43}$$

Thus, $\tau_s \sim \tau_a$ and in the test model of Sect. 8.4, $\tau_s \sim 10^3$ s. It is interesting that (8.43) depends on the parameters of the incoming flow only mildly through χ which varies logarithmically.

As the plasma settles toward the center of the disk, it synchrotron cools. In the plasma rest frame, most of the synchrotron photons are emitted near the peak frequency, $\overline{\nu}$, [33]

$$\begin{aligned}\overline{\nu} &= \left(3\times 10^6\right) B_c\, \gamma_{\rm th}^2 \text{ Hz} \\ &\approx \left(3\times 10^6\right)\left[1-(\beta_F^\phi)^2\right]^{1/2} B^P \left(\frac{\mu c^2}{m_e c^2}\right)^2 \text{Hz}\,,\end{aligned} \tag{8.44}$$

where we used (8.34f) and $F^{12}=0$ in (5.35) to simplify B_c as before. In the ZAMO frame, the frequencies of the synchrotron photons appear to be clustered about $\overline{\nu}_Z$, where using (8.23) this can be expressed from (8.44) as

$$\overline{\nu}_Z = \gamma\overline{\nu} \approx \left(3\times 10^6\right)\left(\frac{\mu c^2}{m_e c^2}\right)^2 B^P\,. \tag{8.45a}$$

From (8.25) and (8.39b) we find

$$\overline{\nu}_Z \approx \left(3\times 10^6\right) U_A^2 \cdot B^P \left(\frac{k_S}{k_F}\right)^2 \alpha^{-2}\left[\frac{\Omega_F^2 g_{\phi\phi}}{c^2}\right]. \tag{8.45b}$$

For the test model of Sect. 8.4,(8.45b) corresponds to synchrotron radiation in the soft X-ray band as viewed by ZAMOs (of course this frequency band designation is observer dependent). From (8.43) we expect roughly equal energy fluxes in soft X-rays (synchrotron) and γ-rays (annihilation) coming from the disk.

The height of the disk can be estimated by balancing the particle flux through the shock front with the total annihilation rate in the disk. Since the disk has spatial extent, this balance must be evaluated in a coordinate system, not just a frame. The disk is very thin compared to the radius of curvature of spacetime, so we can choose a coordinate system with basis vectors coincident with those of the ZAMOs at the shock front and these will be approximately orthonormal over a set that includes the height of the disk. Define the annihilation time scale in the ZAMO frame (and therefore this coordinate system) and the ZAMO number density, as in (8.45b)

$$\tau_Z = \gamma\tau_a\,, \tag{8.46a}$$

$$n_Z = \gamma n\,. \tag{8.46b}$$

Balancing the inflow rate through the surface area of the disk, $A_\perp$, with the annihilation rate in a disk of height, h, we obtain from (3.36b),

$$n_d \mid \left(u^P\right)_d \mid A_\perp \approx A_\perp h\left(\tau_Z^{-1}\right)(n_Z)_d\,. \tag{8.47a}$$

Solving (8.47a) for h using (8.46) and (8.37),

$$h \approx \frac{|\, k_- \,|\, B^P}{\alpha n_{_d}} \tau_a = \frac{|\, k_- \,|\, B^P}{n_{_d}^2 \pi r_{_0}^2} \chi^{-1} \,. \tag{8.47b}$$

Applying (8.14), (8.17) and (8.42) we can get a rough estimate of the disk height in terms of upstream parameters

$$h \approx \frac{\left(3 \times 10^{24}\right) \alpha^2}{n_u U_{_A}^2} \left(\frac{k_{_F}}{k_{_S}}\right)^3 \left[\frac{c}{\Omega_{_F} \sqrt{g_{\phi\phi}}}\right]^3 \text{cm} \,. \tag{8.48}$$

For the test model of Sect. 8.4, from (8.48), the disk ranges in thickness from $\approx 10^{11}$ cm in the main body of the disk to $\lesssim 10^8$ cm near the inner edge. This is a very thin disk and is almost an electrodynamic current sheet, $h/M \sim 10^{-6} - 10^{-3}$.

8.6 The Global Energetics of the Disk

The global energetics of the flow are manifested through the radiative coupling between the hole and the ergospheric disk. From (5.50) and (5.20), the Poynting flux per unit magnetic flux that is generated by the disk is approximately

$$ke \approx \frac{\Omega_{_D}^2 \Phi}{4\pi c k_{_F}} \,, \tag{8.49}$$

where $\Omega_{_D}$ is the angular velocity of the disk as viewed from asymptotic infinity. In Sect. 8.3.1 and (8.19) we found that

$$\Omega_{_F} \approx \Omega_{_D} \gtrsim \Omega_{\min} \,. \tag{8.50}$$

At small lapse function, near the inner edge of the disk $\Omega_{\min} \lesssim \Omega_{_H}$ and near the stationary limit $\Omega_{\min} \approx 0$. Thus, by (8.49) and (8.50) the innermost flux tubes carry the largest Poynting fluxes.

Although the gravitational field forces the disk plasma to rotate, which in turn drives the wind, the conversion of black hole angular momentum to wind Poynting flux needs to be elucidated. The connection is made by the fate of the disk photons. The switch-off cross-field current in the slow shock torques the incoming plasma by $J^{\perp}\hat{\boldsymbol{e}}_2 \times \boldsymbol{B}^P$ forces onto trajectories that appear to rotate backwards with respect to the black hole as viewed by ZAMOs according to (8.11) and (8.18). Since this is a strong current driven in the low conductivity direction across the magnetic field, it is highly dissipative and heats the plasma according to (8.25) to relativistic temperatures (see Fig. 8.6). From (3.50) the shock has prepared the disk plasma to be on negative energy trajectories as viewed form asymptotic infinity. It also has negative angular momentum about the z-axis and is extremely hot. This is the most extreme of plasma states.

Since the plasma is moving relativistically in the $-\hat{\boldsymbol{e}}_\phi$ direction, $\beta^\phi \approx -1$, the annihilation and synchrotron radiation will experience a strong relativistic "headlight effect" that distorts the angular distribution of photons [87]. For this very pronounced "headlight effect" virtually all of the radiated photons will be beamed in the $-\hat{\boldsymbol{e}}_\phi$ direction. In terms of the four momentum of the photon field P^μ_{ph}, we have in the ZAMO frame

$$P^\phi_{ph} \approx -P^0_{ph} \,. \tag{8.51}$$

The relation (8.51) is true for virtually all photons radiated by the ergospheric disk. We can define the analog of the mechanical quantities m and ω, ℓ_{ph} and ω_{ph} from (1.36),

$$\ell_{ph} = \sqrt{g_{\phi\phi}}\, P^\phi \,, \tag{8.52a}$$

$$\omega_{ph} = -P^0 \left[\frac{\Omega_{\min}}{c} \sqrt{g_{\phi\phi}} - \left(1 + \frac{P^\phi}{P^0} \right) \frac{\Omega}{c} \sqrt{g_{\phi\phi}} \right] \,. \tag{8.52b}$$

The grouping of terms in (8.52b) is an intentional effort to make the expression look like (3.44). Inserting (8.51) into (8.52) and noting the positivity of photon energy in the physical, timelike ZAMO frame, $P^0 > 0$, we have

$$\ell_{ph} < 0 \,, \tag{8.53a}$$

$$\omega_{ph} \approx -P^0 \frac{\Omega_F}{c} \sqrt{g_{\phi\phi}} < 0 \,, \tag{8.53b}$$

$$\omega_{ph} \approx \frac{\Omega_F}{c} \ell_{ph} \,. \tag{8.53c}$$

One concludes that the photons are radiated from the pair plasma condensate on trajectories that are of negative energy as viewed from asymptotic infinity.

Negative energy trajectories never leave the ergosphere and will eventually be captured by the hole. This can be seen from the effective gravity term in (3.41), for $\beta^\phi \approx -1$, the Coriolis forces always counteract the centrifugal force, leaving the poloidal gravity extremely dominant. The negative energy photon orbits are precisely the superradiant modes of the electromagnetic wave equation [68]. The condition for superradiance is $0 < \omega_{ph}/\ell_{ph} < \Omega_H$ which is satisfied by (8.53c) since $\Omega_F = \Omega_{\min} < \Omega_H$. In summary, the disk radiates away the bulk of its inertia in an intense two-component (γ-ray annihilation and soft X-ray synchrotron) superradiant field toward the hole. The photon field spins down and extracts energy from the black hole. Ultimately, it is the rotational energy of the black hole that powers the magnetically dominated outgoing wind through this radiative coupling of the ergospheric dynamo to the hole. The global energetics are indicated in Fig. 8.6.

8.7 Near the Stationary Limit

So far we have concentrated on the main body of the ergospheric disk defined by $\alpha_S \sim 0.1$. As "r" increases, $\Omega_{\rm min}$ decreases and α_S becomes comparable to unity. Near the stationary limit $\Omega_F \ll \Omega_H$. Although these flux tubes must also be torqued, since they thread the ergosphere, by (8.49) their Poynting flux is negligible.

In the outer regions of the ergosphere, the disk plasma requires a smaller Lorentz factor in order to have enough inertia to anchor the field, this follows from (8.23) and (8.15) evaluated with the small values of $z \equiv \Omega_F/\Omega_H$. Very close to the stationary limit, $\Omega_{\rm min} \ll \Omega_H$ and Ω_F ceases to equal $\Omega_{\rm min}$, since from (8.15) $\beta_F^\phi \not\approx -1$ near the stationary limit, but is given by

$$\beta_F^\phi \approx -\left[\frac{\Omega_F\sqrt{g_{\phi\phi}}}{c}\right] U_A \left(\frac{k_S}{k_F}\right)\left(\frac{e-\dfrac{\Omega_F \ell}{c}}{m_e c^2}\right), \quad \left[\frac{\Omega_F\sqrt{g_{\phi\phi}}}{c}\right] \lll 1\,. \quad (8.54)$$

In this regime, the Poynting flux is so weak that the rest-mass inertia of the light pair plasma (with no Lorentz γ-factors) and its angular momentum help to power the flow. By (3.49), when $\beta^\phi \not\approx -1$ (as is the case near the stationary limit) the disk particles are at most just barely on negative energy trajectories. The annihilation photons have a very small spin-down effect (if any) on the hole.

For the ingoing wind near the stationary limit, the particle poloidal energy flux can be larger than the Poynting flux. A switch-off shock is still needed to slow the flow before it settles through the surface of the disk. Since the flow comes in almost parallel to the poloidal field ($F^{21} \ll B^P$), the shock is similar to a hydrodynamic shock. The disk plasma is heated by the kinetic energy of the incoming wind, although not much energy is extractable from the shock dissipation and the disk is not relativistically hot. Therefore, synchrotron radiation from this region is small compared with the annihilation radiation.

8.8 The Inner Edge of the Disk

The strong radial component of gravity at small lapse functions limits how close to the event horizon that a disk can exist. Consider the plasma that flows down a flux tube that crosses the equatorial plane at an extremely small lapse function, $\alpha_e \lll 1$. Since the flow in this flux tube experiences a radial acceleration due to gravity over a long length of flux tube that is accentuated just above the equatorial plane for $\alpha \ll 1$, the flow will pass successively through the Alfvén and fast critical surfaces. When this occurs, the plasma can no longer send MHD waves upstream to slow the flow.

If the plasma goes supermagnetosonic before reaching the equator, no disk will form. When the flow goes super fast, the plasma inertia dominates the magnetic field energy density. Hence, the magnetic pressure is insufficient to balance the ram pressure imparted to the plasma by the radial attraction of gravity. Magnetic stresses cannot force the flow to approach the equatorial plane vertically as in the ergospheric disk. By contrast, these flux tubes resemble those that thread the horizon in an electrodynamic sense.

As the plasma advances radially inward toward the hole, the field lines are stretched and pulled toward the hole by plasma inertia. However, as discussed in Sect. 8.1, the no hair theorem does not allow a charge neutral plasma flow to drag poloidal magnetic flux into the hole. The field lines must reconnect, allowing the plasma to enter the hole, without magnetic flux as discussed in Sect. 8.1.

8.9 Summary

The ergospheric disk described in this chapter is a useful tool for understanding ergospheric dynamo behavior. The biggest simplification is that the slow shock isolates the GHM interaction into a small region. Furthermore, we never needed to understand the plasma physics of the shock in order to find the global interaction. This simplicity is lost when one looks at the flux tubes that thread the event horizon.

The essence of the simplification is that the ergospheric disk stays in causal contact with the plasma source and the outgoing wind. The nonzero values of α mean that β_F^ϕ stays finite and the flow never reaches the inner light cylinder or crosses the Alfvén critical surface. This allows Alfvén waves to be radiated upstream and, as has been stressed throughout the book, and Chap. 7 in particular, this is a necessary condition for the causative agent driving the outgoing wind. Consequently, the ergospheric disk is a piston for both Alfvén and fast waves, hence it is a causal boundary for MHD paired winds. This circumstance does not occur on flux tubes that thread the event horizon as all winds go supermagnetosonic since $u^P \sim \alpha^{-1}$ is the horizon boundary condition.

It was noted in Sect. 8.6 that it is the rotational energy of the black hole that powers the wind. This is in contrast to magnetized accretion disks in which it is the energy of the accreting plasma that powers the wind. The extreme dynamics of the GHM interaction in the ergospheric disk was noted in Sect. 8.6. The exotic state of the plasma necessary for the black hole to power the wind is a consequence of a torsional tug of war between two very strong forces. The poloidal magnetic field is virtually rigid relative to the tenuous pair plasma energy density, $U_A^2 \gg 1$. However, because of the dragging of the inertial frames associated with the large rotational inertia of the black hole, the gravitational field has the capacity to make plasma rotate relative to infinity no matter what external forces are imposed. The only constraint on this condition is that the energy density of the imposed field makes negligible corrections to the Kerr metric, i.e.,

$$\frac{(B^P)^2}{8\pi}\left(\frac{GM}{c^2}\right)^3 \ll Mc^2 \,. \tag{8.55}$$

This is not much of a constraint as an astrophysical field of 10^{11}G around a $10^9\,M_\odot$ is not very realistic. Therefore, in practice the black hole has essentially a near infinite potential energy that can be imparted to the plasma as relativistic inertia. This is manifested both by the frame dragging condition $\Omega_{\min} > 0$ in the ergosphere and the horizon boundary condition (3.94). The result of the battle between the two forces is a compromise. The field yields by the creation of toroidal flux, but only enough to allow the plasma to rotate forward at its minimum angular velocity, $\Omega_{\min}$. This pushes the plasma onto $\beta^\phi \approx -1$, ultrarelativistic, negative energy trajectories. This cannot happen easily and occurs only at the expense of strong dissipation as a poloidal current is driven in the low conductivity cross-field direction. The existence of dissipation (entropy generation) is consistent with energy generation (Poynting flux) by the disk from the second law of thermodynamics. Since the two opposing forces are large, the plasma is relativistically hot. Consequently the pair plasma in the ergospheric disk is likely to be in the most extreme state of any matter in the Universe since the Big Bang.

We will show that these dynamics are universal in ergospheric dynamos for toroidal magnetic fields after we study flux tubes that thread the event horizon in Chap. 9. It is worth commenting as to why the ergospheric disk is potentially of great relevance astrophysically. From (1.38), rapidly rotating black holes with $a \approx M$ have the most extractable energy. Thus we are interested in these as power sources for radio loud AGN. However, by (4.90c) theses holes exclude magnetic flux from the horizon (unless the hole is charged which is probably a second order effect in large accretion systems such as a quasar, see Sect. 4.8). By contrast, the $a \lesssim M$ condition yields the largest ergospheres that can contain the largest amounts of poloidal magnetic flux. We will discuss in Chap. 10 the possibility that strong FR II radio sources are associated with ergospheric dynamos on field lines that thread the equatorial plane.

More details on the current structure of winds driven from the ergospheric disk can be found in [132].

Chapter 9
Winds From Event Horizon Magnetospheres

In this chapter we consider paired wind solutions driven by ergospheric dynamos on flux tubes that thread the event horizon. This is a more complicated flow problem than the ergospheric disk magnetosphere, since the flux tubes are doubly open ended (see Fig. 9.1). We showed in Chap. 6, in explicit detail, how there can be no meaningful MHD boundary condition near the horizon. The event horizon is completely describable in terms of an MHD asymptotic infinity. Thus, we no longer have the ergospheric disk or any boundary conditions that fix the field line angular velocity. There is no "solid" surface to anchor the magnetic flux tubes that thread the empty vacuum spacetime of the horizon. These are free floating magnetic flux tubes and the determination of the field line angular velocity is far from trivial. Thus, Ω_F can be estimated only by utilizing a deep understanding of the GHM causality of ergospheric dynamos as described in Chap. 7 and elucidated by the example of Chap. 8. The GHM solution is relevant in the limit that the plasma radiation time scale is much shorter than the MHD wave crossing time over a black hole scale length, i.e., perfect MHD can be grossly violated for certain initial conditions. Alternatively, if perfect MHD is an absolute restriction to the plasma state everywhere, Ω_F, and $k\ell$ can be determined by the boundary conditions (current sources in the dense bounding plasma) imposed by the accretion flow, combined with electrodynamic effects as indicated by the 3-D perfect MHD simulations described in Sect. 11.4.

9.1 Time Dependent Dissipative Winds

The study of winds from event horizon magnetospheres requires more tools than the perfect MHD, time stationary wind formalism of Chap. 5. We know from Chaps. 7 and 8 that the ergospheric dynamo involves strong cross-field currents driven by frame dragging. Since plasma is not free to flow across magnetic field lines in magnetically dominated perfect MHD winds, such large currents driven through such a low conductivity path require significant dissipation (see the detailed discussion in Sect. 2.10). In anticipation of this eventuality, we drop the perfect MHD assumption

B. Punsly, *Black Hole Gravitohydromagnetics, 2nd. ed.*,
Astrophysics and Space Science Library 355, doi: 10/1007/978-3-540-76957-6_9,

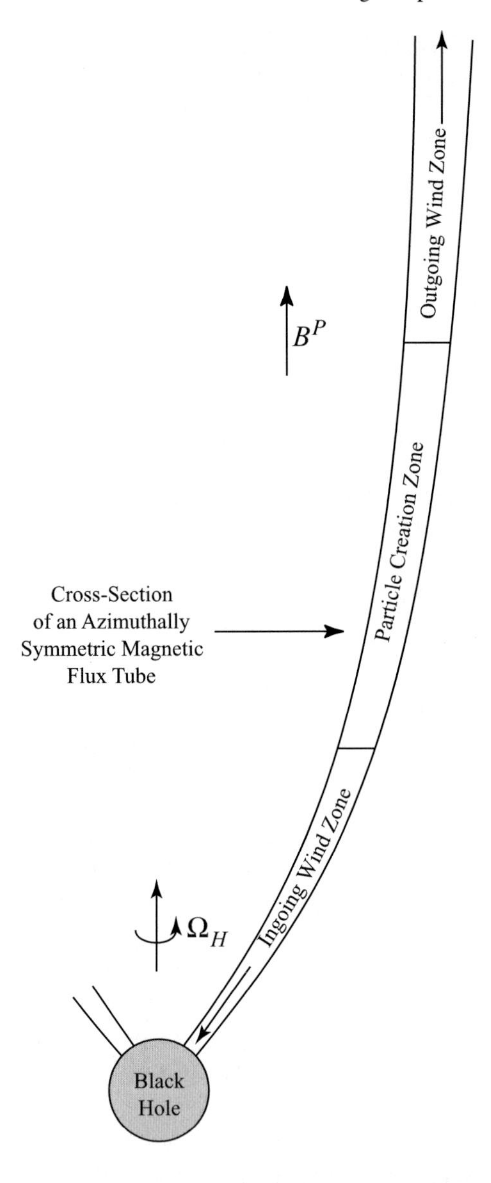

Fig. 9.1 The cross-section of an azimuthally symmetric flux tube that threads the event horizon of a rotating black hole. Most of the plasma injection occurs in a finite length of flux tube known as the particle creation zone. The flow divides in this region into an ingoing accretion flow (the ingoing wind) and an outgoing wind that is initiated by centrifugal force as well as being magnetically slung by $\boldsymbol{J}^P \times B^\phi \hat{\boldsymbol{e}}_\phi$ forces

in our wind formalism. Furthermore, we can no longer hide these dissipative effects in a thin shock layer (as in Chap. 8) since there is no piston for compressive magneto-acoustic waves associated with the vacuum spacetime infinity of the event horizon. In order to understand the physical determination of Ω_F, we need to describe some time dependent gedanken experiments that capture the essential GHM causality. Thus, we also drop the time stationarity condition. We keep only the axisymmetry assumption.

We begin the modification to perfect MHD wind theory by defining a field line angular velocity in the general case. To this end, we introduce a frame that "corotates" with the magnetic field in analogy to (5.13c):

$$F^{20} \equiv \beta_F^{\phi} B^P \,, \tag{9.1}$$

where β_F^{ϕ} is the "rotational velocity" of the field as viewed in the ZAMO frames. The field line angular velocity is then defined implicitly through (5.14). The voltage drop across the magnetic field lines as viewed in the stationary frames at asymptotic infinity is still the same as (5.15a),

$$\Delta V = -\int \frac{\Omega_F \sqrt{g_{\phi\phi}}}{c} B^P \mathrm{d}X^2 \,. \tag{9.2}$$

Relation (9.2) is the reason that the definition of Ω_F through (9.1) and (5.14) is still useful in the general case. However, Ω_F is no longer a constant in a flux tube as it is in the time stationary theory of perfect MHD winds, i.e., there can be voltage drops **along** the field lines.

The frozen-in condition (5.16b) is replaced by an expression for the poloidal component of the electric field in the rest frame of the plasma that is perpendicular to the ZAMO poloidal magnetic field, $E^{2'}$,

$$E^{2'} = F^{20} u_0 + F^{2\phi} u_\phi + F^{21} u_1 \,. \tag{9.3}$$

The proper electric field, $E^{2'}$, is the force that drives the cross-field current in the ergospheric dynamo. Thus, it is the fundamental quantity of interest. It represents the unbalanced rotationally induced EMF in the ergosphere that is the microscopic origin of dynamo behavior.

Similarly, the frozen-in condition can be rewritten as an expression for the toroidal magnetic field in the ZAMO frames,

$$F^{12} = \frac{\left[\beta^{\phi} - \beta_F^{\phi}\right] B^P}{\beta^P} - \frac{cE^{2'}}{u^1} \,. \tag{9.4}$$

We must also suitably modify the definition of the MHD wave critical surfaces discussed in Sect. 5.4. For a mildly dissipative plasma, $*F^{\mu\nu}F_{\mu\nu} \ll F^{\mu\nu}F_{\mu\nu}$, we can use the MHD wave speeds in a resistive plasma found in Sect. 2.7. Using (2.68) for the fast speed in a resistive plasma, we can rewrite it using (5.37) and (5.38) then substitute (5.35) and (5.38) as we did in Sect. 5.4. Then dividing by $\left(u^P\right)^2 = U_F^2$, we have

$$u_p^2 = U_F^2 = \frac{F^{\mu\nu}F_{\mu\nu}}{8\pi n\mu \left[1 + \dfrac{ik^2c^2\omega}{4\pi\sigma}\right]} + U_S^2 \left\{ 1 + \frac{n\alpha^2 \left[1 - \left(\beta_F^{\phi}\right)^2\right]}{k^2\mu \left[1 + \mathrm{i}\dfrac{k^2c^2 - \omega^2}{4\pi\sigma\omega}\right]} \right\} \,, \tag{9.5}$$

where σ is the scalar electrical conductivity (compare to 6.9).

Similarly, using (2.71) for the intermediate speed in a resistive plasma and regrouping as in (5.34) and expanding using (5.35), the Alfvén critical surface is defined by

$$u_p^2 = U_I^2 = \frac{\left[1-(\beta_F^\phi)^2\right](B^P)^2}{4\pi n\mu + \dfrac{i\omega n\mu}{\sigma}} - \frac{i\left[k^2c^2-\omega^2\right]\left[1-\beta^\phi\beta_F^\phi\right](B^P)^2}{4\pi n\mu\left[4\pi\sigma\omega + \dfrac{\omega^2-k^2c^2}{4\pi\sigma} + ik^2c^2\right]} \ . \tag{9.6}$$

As in Sect. 2.7, the lesson of this calculation is that as dissipation sets in, the wave speeds decrease. The resistivity impedes the current flow that is needed to support the electromagnetic fields in the waves. In terms of wind theory, it affects a displacement of the critical surfaces away from asymptotic infinity. Thus the Alfvén and fast surfaces are farther away from the event horizon than in perfect MHD.

In Sect. 2.7 we noted that for large σ (which is equivalent to the $*F^{\mu\nu}F_{\mu\nu} \ll F^{\mu\nu}F_{\mu\nu}$ condition used in this discussion) that the dispersive effects on the waves were of second order. Thus, we equated the phase velocities to the group velocities of the waves and that inaccuracy is carried over to the critical surface conditions (9.5) and (9.6) above. Even so, these relations still show dispersive effects and, unlike the nondispersive case, the location of the critical surface now depends on ω and k.

9.2 The Causal Determination of Ω_F

Finding the field line angular velocity is equivalent to determining the cross-field electrostatic potential by (9.2) and by Gauss's law also determines the Goldreich–Julian charge density in the magnetosphere. Since the oblique Alfvén wave is the mode associated with charge propagation in MHD wave theory (in perfect MHD, Sect. 2.5; resistive plasmas, Sect. 2.7, as well as high frequency waves in a pair plasma, Sect. 2.8), we expect that Ω_F is determined primarily by torsional Alfvén waves. In relativistic MHD wind theory, this is normally established by torsional Alfvén waves emitted by a unipolar inductor (e.g., the neutron star in an MHD pulsar, or the ergospheric disk of Chap. 6). The unipolar inductor is a tremendous simplification because it is a causal MHD boundary that sets Ω_F equal to angular velocity of the unipolar inductor at the point that the field line enters the conductor, i.e., it anchors the rotating the magnetic field. However, we showed in Chaps. 4 and 6 that there is no unipolar inductor on the free floating field lines that thread the event horizon.

We will show that Ω_F is determined to a large extent by the magnetic stresses associated with the plasma injection mechanism on magnetic flux tubes that thread the event horizon. In order to see this effect, consider the pair plasma injection mechanism, $\gamma + \gamma \rightarrow e^+ + e^-$, discussed in Sect. 6.1. Imagine that plasma is injected on perfectly straight field lines (i.e., $B^P \gg F^{21}$). The center of mass frame of the pairs moves with an azimuthal velocity, β_{cm}^ϕ, that is in general different than the azimuthal velocity of the field, β_F^ϕ, defined in (9.1). By (9.3), the injected pairs experience a Lorentz force,

$$qE^{2'} = \frac{q}{c}\left(u^0\right)_{cm}\left[\beta_{cm}^\phi - \beta_F^\phi\right]B^P\ . \tag{9.7}$$

The straight field line approximation, $F^{21} \ll B^P$, allows us to ignore $qF^{21}\left(u_1\right)_{cm}$ forces in (9.7). The Lorentz force in (9.7) accelerates electrons and positrons in opposite directions. This charge separation forms a macroscopic cross-field current, J^2. The particle acceleration in the $\hat{\boldsymbol{e}}_2$ direction creates an equal and opposite cross-field velocity, v^2, for the species that has an important second order effect. The $(qv^2/c)\hat{\boldsymbol{e}}_2 \times \boldsymbol{B^P}$ force is the same for both species of charge and the resulting torque exists until the pairs rotate with the $\boldsymbol{E}\times\boldsymbol{B}$ drift of the plasma (i.e., this sets $\beta_{cm}^\phi = \beta_F^\phi$ in (9.7), which makes the Lorentz force vanish). Physically, this shows how torsional stresses of the injected plasma can act to torque a flux tube.

The pair plasma is created with a distribution of azimuthal velocities. Clearly, this distribution and its mean azimuthal velocity is a strong function of the distance from the horizon to the location of the pair plasma injection. Thus, all of the created plasma cannot corotate with the magnetic field. Each fluid element of pair plasma injected into the flux tube creates a local $J^2\hat{\boldsymbol{e}}_2 \times \boldsymbol{B^P}$ force associated with the microscopic Lorentz force in (9.7). The $J^2\hat{\boldsymbol{e}}_2 \times \boldsymbol{B^P}$ forces are communicated up and down the magnetic flux tube primarily by torsional Alfvén waves. The MHD waves radiated from each element of freshly injected pair plasma are essentially a back reaction on the field as the plasma tries to minimize the stress it experiences, i.e., charges flow to cancel the proper electric fields, such as the one in (9.7), in a conductive medium. The torsional Alfvén wave communication up and down the flux tube equates to a global version of this phenomenon. The injected plasma spins up the free floating flux tubes in an effort to minimize the magnetic stresses in the system within the constraints of the system (MHD causality). Thus, the plasma spins up the magnetic flux tubes so it is as close as possible to corotation with the created pairs in this "average sense."

There also can be strong magnetic stresses experienced by plasma in the dynamo region as we expect this plasma to have a relativistic inertia imparted to it by the dragging of inertial frames as we found in Chap. 8. However, unlike the ergospheric disk there is very little plasma in this region that is in Alfvén wave contact with the upstream flow. There is no compressive MHD piston associated with the event horizon, however black hole gravity can act as a rarefaction piston for the MHD flow. The dynamo of the ergospheric disk is a slow compression wave. By contrast, in the event horizon magnetosphere, the dynamo has the MHD structure of a fast rarefaction wave (this is shown formally in Sect. 9.6). Thus, the flow passes out of Alfvén wave communication with the plasma source just outside the light cylinder as shown in (9.6). Large relativistic inertial effects initiate in the region of spacetime near the light cylinder as well. Thus, there is very little ultrarelativistic plasma near the light cylinder that is also in Alfvén wave contact with the plasma source. The total magnetic stresses associated with this thin layer ahead of the light cylinder is only a second order correction to the total magnetic stresses associated with the long lengths of flux tube in the plasma injection region. This is a fundamental distinction from the ergospheric disk in which all of the plasma downstream of the dynamo (the slow shock) is still within Alfvén wave contact with the plasma source and the upstream paired wind system.

For the sake of illustration we very crudely estimate Ω_F on flux tubes that thread the event horizon of a supermassive black hole in a radio loud extragalactic radio source. A more rigorous derivation of Ω_F in a charge starved Kerr–Newman magnetosphere is given in Chap. 11. Consider the γ-ray field to be sourced by the corona of an accretion disk about a rapidly rotating black hole, $a \lesssim M$. Most of the plasma that is created on a flux tube is made at distances $r > 5M$ from the hole if the main source of coronal γ-ray activity is at $r \sim 10M$ from the hole (see Fig. 9.2). Plasma created in the magnetic flux tube communicates torsional stresses to the field. If the created plasma is outside the inner critical surfaces defined by (9.5) and (9.6) this stress can be communicated to plasma upstream. The plasma exchanges torsional Alfvén waves up and down the flux tube and this process determines how rapidly the free floating flux tubes rotate.

In order to understand the causal structure of this process, consider that the coronal γ-rays at $r \sim 10M$ are suddenly turned on and pair creation begins on the vacuum Wald field of Sect. 4.6.2. The existence of the critical surfaces in the time dependent scenario through (9.5) and (9.6) plays a large role in the causal structure. The freshly created plasma sets up a local value of Ω_F in the flux tubes through (9.2). This initial value can then be subsequently affected by torsional Alfvén waves coming from upstream and downstream. Since the initial value of Ω_F is approximately $\mathrm{d}\phi/\mathrm{d}t$ of the pair producing corpuscular γ-rays, by frame dragging effects and angular momentum conservation, Alfvén waves arriving from downstream tend to spin up the local initial value of Ω_F and Alfvén waves arriving from upstream tend to spin down the local field rotation rate. Most of the plasma at $r > 5M$ creates initial values of $\Omega_F \ll \Omega_H$ in the magnetic flux tube and the plasma near the horizon has a local initial value of $\Omega_F \approx \Omega_H$.

Plasma near the event horizon has little effect in this time dependent tug of war. Firstly, we expect all outgoing plasma waves near the horizon to be highly redshifted as was found in Sect. 6.4. Globally, they would propagate outward slowly and carry extremely small torsional magnetic stresses. Thus, in the initial stages of the gedanken experiment, the plasma at $r > 5M$ has internally exchanged many Alfvén wave signals as the first Alfvén wave trains approach from the near horizon plasma. The plasma at $r > 5M$ starts to "agree" on a local value of Ω_F and sends this information inward in the form of Alfvén waves. Associated with this value of Ω_F there is a light cylinder and by (9.6), the Alfvén critical surface is nearby. For $\Omega_F \ll \Omega_H$, we expect $F^{21} \ll B^P$ in the strong field limit and the fast critical surface defined by (9.5) is close to the light cylinder as well. The bulk of the plasma upstream establishes MHD critical surfaces just inside the stationary limit, causally decoupling the downstream plasma. Furthermore, note from the geometry in Fig. 9.2, that we expect a very small percentage of the freshly created pair plasma to be injected into the flux tubes within the ergosphere. At this stage, plasma waves

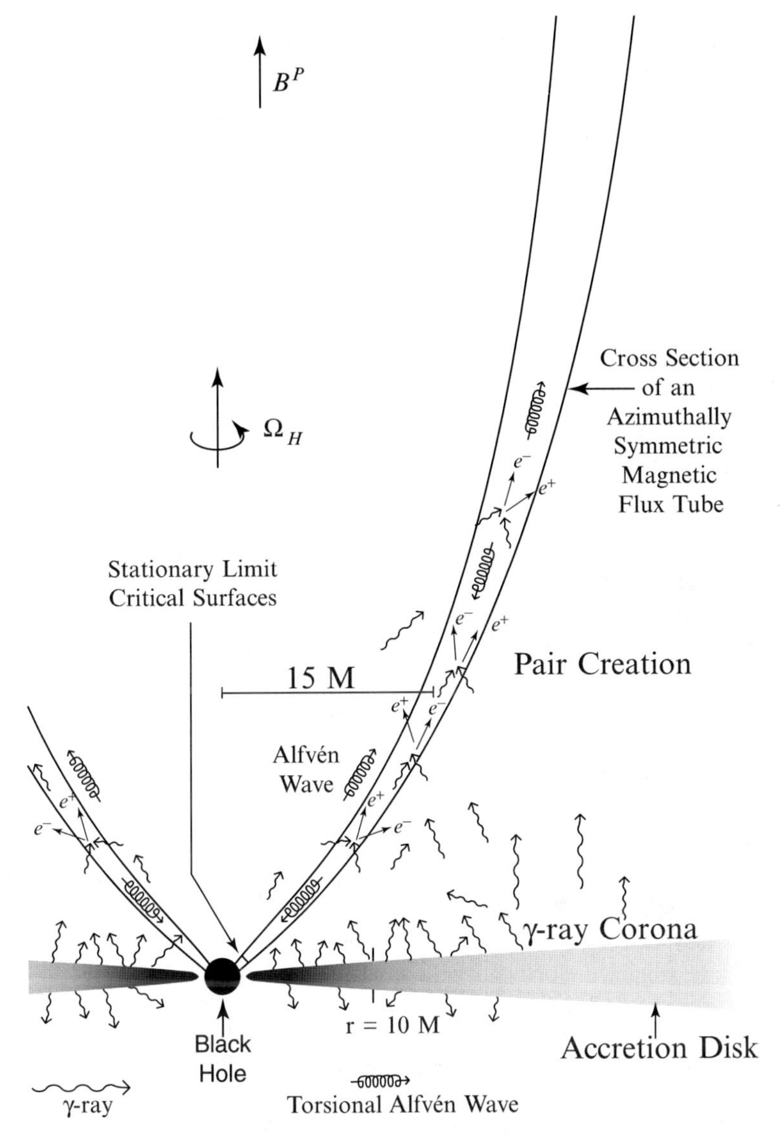

Fig. 9.2 A possible plasma injection geometry on an azimuthally symmetric magnetic flux tube as seen in cross section. Gamma rays are emitted from the hot corona of an accretion disk. The γ-ray field pair creates on the magnetic field background via the scattering process $\gamma + \gamma \rightarrow e^+ + e^-$. Each fluid element of positronic plasma tries to communicate its local rotational velocity (that is transferred from the local γ-ray field) to other plasma upstream and downstream in the flux tube by means of torsional Alfvén waves (represented by the helices with arrows). The resulting field line angular velocity, Ω_F is an "average" value of these locally determined angular velocities in the sense that final global value of Ω_F adjusts to minimize the total magnetic stresses in the flux tube between the inner and outer Alfvén points. For the geometry as drawn and $a \lesssim M$, $\Omega_F \approx 1/30\Omega_H$

emitted from near the horizon can no longer reach the plasma at $r > 5M$ because it is downstream of the MHD critical surfaces. In summary, we expect the bulk of the plasma that resides upstream to establish critical surfaces before the near horizon plasma has much to say in the matter.

Next we make a crude estimate of Ω_F by assuming that the γ-rays are created in the equatorial plane at $r \sim 10M$ and are emitted from a Keplerian disk with (see Fig. 9.2)

$$\left(\frac{d\phi}{dt}\right)_0 \approx \Omega_{Kep} \approx \frac{c}{M}\left[\frac{M}{r_0}\right]^{3/2} , \quad r_0 \approx 10M . \tag{9.8}$$

The angular momentum of the γ-rays, ℓ_{ph}, is conserved as they propagate from the corona. For $r > 10M$, we can approximate this conservation law as

$$\ell_{ph} \approx \frac{d\phi}{dt}\left(g_{\phi\phi}\right)\ell_0 = \text{constant} , \quad \ell_0 = \text{constant} . \tag{9.9}$$

Combining (9.8) and (9.9), the angular velocity of a corpuscular γ-ray is

$$\frac{d\phi}{dt} \approx \Omega_{Kep}\frac{\left(g_{\phi\phi}\right)_0}{g_{\phi\phi}} \approx \frac{c}{M}\left[\frac{M}{r_0}\right]^{3/2}\frac{r_0^2}{g_{\phi\phi}} = \frac{c\sqrt{Mr_0}}{g_{\phi\phi}} . \tag{9.10}$$

If the average location of a created pair on the flux tube is at $\sqrt{g_{\phi\phi}} = 15M$, we could expect from (9.10) when $a \lesssim M$,

$$\Omega_F \approx \frac{d\phi}{dt} \approx \frac{1}{30}\Omega_H . \tag{9.11}$$

Clearly, the result (9.11) is extremely model dependent. The only reason for generating this crude result is to suggest that $\Omega_F \ll \Omega_H$ seems likely for free floating flux tubes that thread the event horizon. From the geometry in Fig. 9.2, we expect the field line angular velocity to vary from flux tube to flux tube. It seems that $0.01\Omega_H < \Omega_F < 0.1\Omega_H$ is a reasonable range of plausible field line angular velocities for flux tubes threading the horizon of a rapidly rotating black hole at the center of an extragalactic radio source. We will use $\Omega_F = (1/30)\Omega_H$ as a fiducial value in the following.

9.3 The Ergospheric Dynamo in Free Floating Flux Tubes

In this section we describe the microphysics of the dynamo for toroidal magnetic field on flux tubes that thread the event horizon. We have already seen a stark contrast to a unipolar inductor in that Ω_F is not determined to first order by the dynamo region, but by plasma source. The plasma wave exchanges along the flux tube upstream of the inner critical surfaces that determine Ω_F (as discussed in the last section and illustrated in Fig. 9.2), also simultaneously determines the dynamo physics. As the MHD wave front from the bulk of the particle injection region at $r > 5M$ propagates inward, it transports the global potential (equivalently Ω_F) and the charge density (Goldreich–Julian charge density) necessary to support it. From Chap. 2, we know that this is primarily an Alfvén wave. The dynamo results from the reflection of the Alfvén wave off the rotating magnetosphere. The reflected Alfvén wave must be radiated at or beyond the inner Alfvén surface. In this section we describe the reflection process, the amplification of the reflected wave and compare and contrast the resulting dynamo with the ergospheric disk.

We explore the fundamental dynamo physics through a time dependent gedanken experiment similar to that of the last section. The basic components are verified through explicit model calculations in Sects. 9.4–9.7.

Recall the main results of Sect. 9.2 that motivate this discussion. The bulk of the plasma inertia is created within a particle creation zone (see Fig. 9.1). If one suddenly turns on the γ-ray source in the presence of Wald-like magnetic field lines, the pair creation process establishes a value of Ω_F on the field lines. The value of Ω_F is largely determined by plasma wave communications in the region where the bulk of the injected plasma resides, the particle creation zone. Before the portion of the flux tube above the light cylinder decouples from the near horizon plasma, MHD waves from downstream slowly increase Ω_F and change the position of the light cylinder. However, this is a second order effect and at later times the inner ergospheric plasma can no longer affect the determination of the global potential. Thus, we consider the approximate dynamics that ignores the slow evolution of the position of the light cylinder, before causal decoupling occurs, and consider a column of tenuous plasma extending far from the hole that has been created on Wald-like field lines at a time $t = t_0$ in Fig. 9.3. A steady source of plasma is created with $\mathrm{d}\phi/\mathrm{d}t \ll \Omega_H$. We assume that all plasma is created with the same angular velocity and the plasma injection occurs only above some inner boundary surface that is located just below the flow division point associated with this value of Ω_F (i.e., gravity is larger than centrifugal forces). Then, the plasma is allowed to accrete for times $t > t_0$.

We begin by determining the initial state of the plasma at $t = t_0$ just before accretion begins. Since the inertia of the plasma is small, $n\mu c^2 \ll \left(B^P\right)^2$, one expects initially that $B^T \approx 0$. Clearly, there will be some angular momentum transfer from the plasma to the field when the created plasma becomes threaded by the magnetic field. However, it is still an excellent approximation to (5.20) and (5.21) if we express the initial conditions at $t = t_0$ in the particle creation zone as

$$k\ell \approx km\,, \tag{9.12a}$$

$$ke \approx k\omega\,, \tag{9.12b}$$

$$ke \approx \frac{\Omega_F}{c} k\ell\,, \tag{9.12c}$$

$$\frac{\Omega_H g_{\phi\phi} B^P}{c} \gg |k\ell|\,. \tag{9.12d}$$

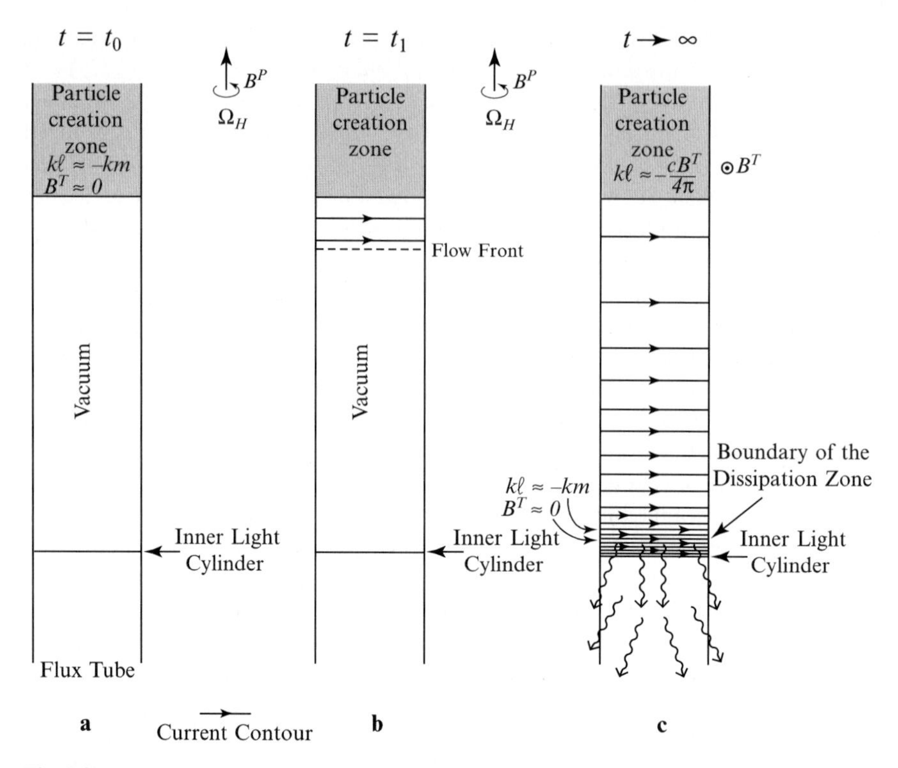

Fig. 9.3 The time evolution of the accretion of a tenuous plasma on strong magnetic field lines that thread the horizon. The accretion process can be thought of as an incident Alfvén wave propagated from the particle creation zone. The enhanced angular momentum flux, $k\ell$, in frame (**c**) compared to frame (**a**) represents a reflected Alfvén wave. The reflected wave is enhanced and as such represents superradiant scattering. The dynamo that supports the Poynting flux in the wave is a distributed flow of cross-field poloidal current that is strongly concentrated near the inner light cylinder [138]

At times $t > t_0$, an Alfvén waves propagates inward transporting the global cross-field associated with Ω_F of the particle creation zone through (9.2). There are two components of the electrodynamic currents that flow parallel to the poloidal field. Firstly, there is the advection of the Goldreich–Julian charge associated with the cross-field potential that we have already encountered in the plasma-filled waveguide (Sect. 2.9.4) and ingoing MHD winds (Sect. 6.3). Secondly, this is not a pure Alfvén wave, as a surface charge density, σ, exists at the flow front in order to shield the plasma from the Wald vacuum electric field (see Sect. 4.6.2). Since $\mathrm{d}\sigma/\mathrm{d}\tau \neq 0$, a field aligned current system must flow in the plasma to supply the surface charge. However, this flow front will eventually propagate inside the critical surfaces so that at late times it can have no effect on the global current system driven by a dynamo. These electrodynamic field aligned currents are inherently different than the inertial current that flows cross-field in the dynamo. In this analysis we ignore the contributions of these electrodynamic components of the current in Ampere's law for B^T, so as to separate out the dynamo behavior. This approximation is a good one as long as the Goldreich–Julian charge flows inward subrelativistically.

Once the flow begins to propagate inward (frame (**b**) of Fig. 9.3), we no longer have time stationarity, and Ω_F is a constant in the poloidal coordinate along the plasma-filled regions of a flux tube in this time dependent scenario, as long as perfect MHD can be maintained. Since $0 < \Omega_F < \Omega_H$, by (5.14) and (6.4), there will be an inner light cylinder where $\beta_F^\phi = -1$. This is where our perfect MHD assumption will no longer be valid. However, by (9.2) the voltage drop across a thin azimuthally symmetric flux tube is the same as (5.15b). This combined with perfect MHD allows us to time evolve the flow with a constant value of $\Omega_F = -2\pi\Delta Vc/\delta\Phi$ until just before it reaches the light cylinder.

We see the dynamo physics immediately in a subtle effect at a time $t = t_1$ just after t_0, $t_1 \gtrsim t_0$, when the plasma starts to accrete from the particle creation zone by the force of gravity (see Fig. 9.3b). The plasma is inertially so light that it will flow parallel to the field with, $\mathrm{d}\phi/\mathrm{d}t \equiv \Omega_p \approx \Omega_F$. By contrast, in free fall an ingoing geodesic ($m =$ constant, $\omega =$ constant) has a value of Ω_p that keeps increasing (see Figs. 3.1 and 3.2). Consequently, in order to keep $\Omega_p \approx \Omega_F$ during the inflow, there must be an electromagnetic force $\left(J^2\hat{\mathbf{e}}_2 \times \mathbf{B^P}\right)$ that decreases the angular momentum, m, of the plasma. This cross-field current is indicated in Fig. 9.3b. Ampere's law (3.61b) indicates that this cross-field current, J^2, creates a B^T upstream. Consequently, the current that flows to keep plasma corotating with the field is torsional Alfvén wave radiation (the reflected Alfvén wave we are looking for). It is an Alfvén wave as opposed to a fast wave since field aligned currents must flow upstream in order to support B^T by Ampere's law (3.62b). Note that the $J^2\hat{\mathbf{e}}_2 \times \mathbf{B^P}$ torque and the associated radiated Alfvén wave are manifestations of the torsional tug of war discussed in Chap. 7. Furthermore, it is frame dragging effects that drive the cross-field current.

Clearly, at a time $t = t_1$, $|B^T| \ll B^P\sqrt{g_{\phi\phi}}$, since the plasma has very little inertia, and is therefore ineffective at bending the strong magnetic field. The radiated Alfvén wave transports positive angular momentum $\left(B^T < 0\right)$ and energy up the flux tube into the particle creation zone. Thus, the quantum numbers ke and $k\ell$ of the plasma in the particle creation zone at time $t > t_1$ are different than they were at $t = t_0$ because of the new value of B^T from current sources downstream. As the flow approaches the inner light cylinder, more and more cross-field current, J^2 is generated. Thus, $-B^T$ keeps increasing in the particle creation zone as well.

Since the plasma source is steady, one expects that the plasma flow approaches a steady state above the inner light cylinder. We can get an idea of what the steady flow from the particle creation zone to the inner light cylinder is like by extending the previous discussion. First note that just above the inner light cylinder, $\Omega_p \gtrsim \Omega_F \gtrsim \Omega_{\rm min}$. The quantity Ω_p must be larger than Ω_F since gravity (the dragging of inertial frames) is always trying to increase Ω_p even though electromagnetic forces keep it close to Ω_F (the torsional "tug of war"). At this point of the flow, the angular momentum about the symmetry axis of the hole is much less than zero, and we have the following plasma state:

$$\beta^{\phi} \gtrsim -1\,, \tag{9.13a}$$
$$u^{0} \gg 1\,, \tag{9.13b}$$
$$m \ll 0\,, \tag{9.13c}$$
$$mk \gg 0\,. \tag{9.13d}$$

Similarly, since $\Omega_p \gtrsim \Omega_F$ and $\beta^P < 0$, $B^T < 0$ from the frozen-in condition (5.44). Combining this with (9.13) in the expression for the total angular momentum, $k\ell \gg 0$ near the inner light cylinder and in the steady state perfect MHD wind above the inner light cylinder, $k\ell$ is a constant in a flux tube,

$$\lim_{t\to\infty}(k\ell)_{P.C.} = \lim_{t\to\infty}(k\ell)_{L.C.} \gg 0\,. \tag{9.14}$$

where the subscript "P.C." means to evaluate in the particle creation zone and "L.C." means to evaluate just ahead of the light cylinder.

Consider what (9.14) means physically in the context of the global flow between the particle creation zone and the inner light cylinder. There is a cross-field current flow, J^2, associated with keeping the plasma in approximate corotation with the magnetic field ahead of the inner light cylinder. The current density has a strongly peaked maximum in the region near the light cylinder where the plasma starts to gain inertia (see 9.13) through rotation induced Lorentz γ-factors (see Fig. 9.3c). All of this cross-field poloidal current makes a B^T upstream such that the steady state quantum numbers have changed from (9.12),

$$k\ell \approx -\frac{c}{4\pi}\left(B^T\right)_{P.C.} \gg |km|_{P.C.}\,, \tag{9.15a}$$

$$ke \approx -\frac{\Omega_F}{4\pi}\left(B^T\right)_{P.C.} \gg |k\omega|_{P.C.}\,. \tag{9.15b}$$

The direction of J^2 indicates through Ampere's law (3.61b) that $|B^T|$ decreases as one moves down the flux tube. Dynamically, this can be understood through the frozen-in expression (5.44) for B^T. As one moves down the flux tube $\Omega_p - \Omega_F$ increases only slightly (since the plasma has very little inertia) and is very small, i.e., $\Omega_p - \Omega_F \ll \Omega_F$. On the other hand, $|\beta^P|$ increases drastically since the plasma is accelerated toward the hole by the J^2F^{21} force in (3.40) and the effective gravity "g" of (3.41) is inward directed and strong (especially for $\beta^{\phi} \approx -1$). This raises the interesting question of how small $|B^T|$ becomes in the flux tube. This relates back to causality; there is some causal physical process that creates B^T in the paired wind system. From our knowledge of plasma waves in Chap. 2, we know that the Alfvén wave is primarily responsible for transporting torsional stresses throughout the paired wind. Furthermore, causality dictates that all of the cross-field current that supports B^T in the paired wind system must be created before the wind passes through the wind critical surfaces. As we discussed in Sect. 6.2 (and this also follows from (9.5) and (9.6) in time dependent dissipative winds) for $\Omega_F \ll \Omega_H$, the fast and Alfvén critical surfaces of an ingoing magnetically dominated wind are located near the inner light cylinder. Thus, causality considerations demand that $B^T = 0$ at some point in the flow at or above the fast critical surface. We call this the anchor point in the flux tube (the topic of Sect. 9.6.2) defined by

$$\left(B^T\right)_{anc} \equiv 0\,. \tag{9.16}$$

We can learn even more from this simple gedanken experiment. Note that the magnetically dominated inflow attains significant relativistic inertia imparted by approximate corotation with the field only when it is extremely close to the light cylinder. The change in u^ϕ corresponds to a proper acceleration a^ϕ of the plasma near the light cylinder. Note that pure corotation implies

$$\lim_{\beta_F^\phi \to -1} a^\phi = -\infty\,, \quad \lim_{\beta_F^\phi \to -1} u^\phi = -\infty\,. \tag{9.17}$$

Thus, the acceleration can grow very abruptly near the light cylinder and large radiation losses are expected even though pure corotation will never occur due to plasma inertial effects. Large radiation losses equate to a dissipative plasma and a breakdown of perfect MHD. This is expected from the second law of thermodynamics. The strong cross-field currents (in the low conductivity direction of the plasma) requires strong dissipation, as we found in the study of the slow shock bounding the ergospheric disk in Chap. 8. Dissipation is a fundamental property of ergospheric dynamos.

We explore the inertial effects that prevent corotation and the dissipative plasma physics at the light cylinder and inward in the remainder of the chapter.

9.4 Perfect MHD Paired Outgoing Minimum Torque Winds: $\Omega_F \ll \Omega_H$

In this section, we model a paired wind system on an azimuthally symmetric flux tube that threads the event horizon as depicted in Fig. 9.1. The plasma injection mechanism sets $\Omega_F = (1/30)\Omega_H$ as in (9.11). We consider the outgoing wind to be the minimum torque solution described in Sect. 5.6. The perfect MHD assumption is imposed everywhere. The calculation shows that there is no paired ingoing perfect MHD wind that can connect the particle creation zone to the asymptotic horizon infinity for this value of Ω_F. This ingoing perfect MHD solution is the analog of the subcritical outgoing wind solution described in Sect. 5.5 and Fig. 5.2. In the magnetically dominated winds of interest from (5.20) and (5.21) the energy and angular momentum fluxes are virtually purely electromagnetic,

$$ke \approx -\frac{\Omega_F B^T}{4\pi}\,, \tag{9.18a}$$

$$k\ell \approx -\frac{cB^T}{4\pi}\,. \tag{9.18b}$$

The ingoing wind is subcritical by (9.18) (i.e., $k\ell$ and ke are less than a minimum value required to connect the plasma source and outgoing wind to the horizon infinity) since $|B^T|$ is "too small" in the wind.

In this section, we demonstrate and discuss whether the ingoing wind is extendible beyond the environs of the light cylinder. We make strong connections to the gedanken experiment of Fig. 9.3. It is the dissipative physics near the light cylinder as perfect MHD breaks down that allows large cross-field poloidal currents to flow in the ergospheric dynamo. The fact that $|B^T|$ is "too small" to allow smooth passage through the environs of the inner light cylinder is of fundamental physical significance. This circumstance was central to the discussion of the plasma accretion on the initially purely poloidal field in Sect. 9.3. Thus, the gedanken experiment was a preview of the relevant physics near the light cylinder.

9.4.1 Mathematical Formulation of Paired Wind as a Boundary Value Problem

Consider the axisymmetric, time stationary, paired winds emanating from the particle creation zone as indicated in Fig. 6.1. Each wind can be represented mathematically as being integrated from the initial surfaces I_+ and I_-. Formally, as discussed in Sect. 5.1, there are seven constants of motion for each wind. In this section, we choose an arbitrary flux tube and ignore the Grad–Shafranov equation discussed in Sect. 5.7. Thus, Φ and $\delta\Phi$ are arbitrary and equal in both the outgoing and ingoing winds. This reduces the total number of constants from 14 to 10 that need to be specified on the disjoint union, $I_- \cup I_+$.

Added simplification occurs due to the magnetically dominated condition. By (9.7) there is dissipation, $J^{2'} \cdot E^{2'}$, associated with the plasma injection mechanism. Since J^2 scales with the number of created particles and $U_A^2 \gg 1$, $J^{2'} \cdot E^{2'} \ll \left(B^P\right)^2/r$. Consequently, we can ignore the dissipation in the particle creation zone and we implement the perfect MHD condition

$$(\Omega_F)_+ \approx (\Omega_F)_- \equiv \Omega_F = \frac{\Omega_H}{30} , \tag{9.19}$$

where the subscripts "+" and "−" refer to the outgoing and ingoing winds, respectively.

The winds to be considered are warm winds with a specific enthalpy $\mu \sim m_e$. In the magnetically dominated perfect MHD winds to be considered, the entropy generation has a negligible effect of the wind dynamics. Thus, we simply take

$$\mathbb{S}_+ \approx \mathbb{S}_- . \tag{9.20}$$

Recall from Sect. 6.1 that the winds that propagate away from I_+ and I_- (the boundary of the particle creation zone) will initiate as perfect MHD flows since the number density will exceed the Goldreich–Julian charge density by many orders of magnitude.

The laws of conservation of angular momentum and energy, (5.21) and (5.20), in the particle creation zone are (see Fig. 6.1)

$$(k\ell)_- = (k\ell)_+ + (k\ell)_{inj} , \tag{9.21a}$$

$$(ke)_- = (ke)_+ + (ke)_{inj} , \tag{9.21b}$$

where $(k\ell)_{inj}$ and $(ke)_{inj}$ are the mechanical angular momentum and energy fluxes, respectively, of the created pairs that are injected into the particle creation zone. Note that the magnetically dominated condition requires (as shown in the time evolved gedanken experiment of Sect. 9.3)

$$(k\ell)_- \approx (k\ell)_+ , \quad |k\ell|_- \gg |k\ell|_{inj} , \tag{9.22a}$$

$$(ke)_- \approx (ke)_+ , \quad |ke| \gg |ke|_{inj} . \tag{9.22b}$$

Consequently, the paired wind system is determined by seven constants: Ω_F, $(ke)_{inj}$, $(ke)_+$, $(k\ell)_{inj}$, $(k\ell)_+$, k_-, and k_+. Five of the constants Ω_F, k_+, k_-, $(ke)_{inj}$ and $(k\ell)_{inj}$ are determined by the plasma injection mechanism.

9.4.2 The Outgoing Minimum Torque Wind

Choosing the outgoing minimum torque wind is consistent with the principal implemented in Sect. 9.2 to determine Ω_F, that the system will adjust to minimize magnetic torsional stresses. The minimum stress is given by the minimum toroidal magnetic field which typifies the minimum torque solution as discussed in Sect. 5.6. From (5.50) and (9.18), the angular momentum and energy fluxes in a magnetic flux tube that supports a minimum torque outgoing wind satisfy

$$(ke)_+ \approx \frac{\Omega_F^2 \Phi}{4\pi c k_F} , \tag{9.23a}$$

$$(k\ell)_+ \approx \frac{\Omega_F \Phi}{4\pi k_F} . \tag{9.23b}$$

From the discussion at the end of the last section, the plasma injection mechanism determines five and the outgoing minimum torque wind determines two (through 9.23) of the seven constants of motion necessary to determine a paired, warm magnetically dominated wind system. Thus we can proceed to designate a complete set of initial conditions on $I_- \cup I_+$, as the boundary conditions on the injection surface of the ingoing wind, I_-, are constrained by the wind constants of the outgoing minimum torque solution through (9.21).

At this point we are ready to parameterize an actual paired wind model on a flux tube that threads the event horizon. We assume a $2 \times 10^9\,\mathrm{M}_\odot$ central black hole and a magnetic flux,

$$\Phi = 10^{34}\ \mathrm{G} - \mathrm{cm}^2\,, \tag{9.24}$$

which corresponds to a horizon magnetic field $B^P_H \approx 10^4$ G. We choose $L_c \approx 10^{44}$ ergs/sec in (6.1), thus

$$n \sim 10^7 - 10^8\mathrm{cm}^{-3}\,. \tag{9.25}$$

We assume an initial velocity on I_+ of $0.1c$. Then by (9.25) and (5.12),

$$k_+ = 3 \times 10^{13}\mathrm{cm}^{-2}\,\mathrm{sec}^{-1}\,\mathrm{G}^{-1}\,. \tag{9.26}$$

In relativistic wind theory it is customary to define the magnetization parameter, σ, the ratio of electromagnetic energy flux to inertial flux [113]. From (9.24)–(9.26), we can express σ in parametric form,

$$\sigma \approx 10^4 \frac{\left(B^P_H\right)_4}{\left(\frac{k_+}{c}\right)_3}\,, \tag{9.27}$$

where the poloidal magnetic field strength near the horizon, $\left(B^P_H\right)_4$, is measured in units of 10^4 G and the mass flux parameter divided by the speed of light, $(k_+/c)_3$, is measured in units of 10^3 cm^{-1} G^{-1}.

It has been shown [113] that the magnetization parameter is useful for describing the asymptotic wind

$$u^P_\infty = \sigma^{1/3} c \approx 10^{4/3} c\,, \tag{9.28a}$$

$$m_\infty = \frac{\sigma^{1/3}}{\Omega_F} m_e c^3 \approx 10^{4/3} \frac{m_e c^3}{\Omega_F}\,, \tag{9.28b}$$

where the subscript "∞" means to evaluate at asymptotic infinity, $r \to \infty$. More details on the outgoing wind can be found in [136].

9.4.3 Initial Data for the Ingoing Wind

From (9.19)–(9.23) the initial data on I_- for the ingoing wind on the azimuthally symmetric flux tube is

$$(k\ell)_{-} = (k\ell)_{inj} + k_{+} m_{\infty} + \frac{\Omega_F \Phi}{4\pi k_F}, \tag{9.29a}$$

$$(ke)_{-} = (ke)_{inj} + k_{+} \omega_{\infty} + \frac{\Omega_F^2 \Phi}{4\pi k_F}, \tag{9.29b}$$

$$\Omega_F = \frac{\Omega_H}{30}, \tag{9.29c}$$

$$\Phi = 10^{34} \mathrm{G-cm}^2, \tag{9.29d}$$

$$\delta\Phi = \text{arbitrary and small}, \quad \delta\Phi \ll \Phi, \tag{9.29e}$$

$$\mathbb{S}_{-} \approx 0. \tag{9.29f}$$

The remaining parameter is the mass flux parameter, k_{-}. Note that k_{-} is independent from k_{+}. Considering the geometry in Fig. 9.2, the pair producing γ-rays from the disk have a net outward momentum, thus $|k_{+}| > |k_{-}|$. The final physical results are fairly insensitive to the value of k_{-} as long as the magnetically dominated condition holds. So we arbitrarily impose $|k_{+}| = 2|k_{-}|$,

$$k_{-} = -1.5 \times 10^{13} \mathrm{cm}^{-2} \sec^{-1} \mathrm{G}^{-1}. \tag{9.29g}$$

Relations (9.29) provide a complete set of initial data for the inward integration of the wind. Note that the Grad–Shafranov equation can change Φ in a self consistent solution since it describes how inertial stresses create J^{ϕ} and hence a B^P. The initial data, (9.29), assumes that the Grad–Shafranov equation has already been solved and the resulting Φ value is given by (9.29d).

9.4.4 The Force Free Limit of the Ingoing Wind

In this section we illustrate the fundamental physics that does not allow the perfect MHD assumption to be satisfied with the conservation laws of energy and angular momentum in the ingoing wind deep in the ergosphere. Consider the force free limit that plasma inertia is negligible in the MHD wind equations. We still assume that there are enough charges to short out the proper electric fields in the ingoing wind zone, so perfect MHD relations still apply. These assumptions reduce to the exact equality (as opposed to "approximate equality" that can incorporate plasma inertia and the cross-field inertial currents that can slightly modify B^T for nonzero plasma inertia),

$$B^T = B^T_{\infty} = \text{constant}. \tag{9.30}$$

In analogy to the definition k_F in (5.50b) we define a geometrical factor that is a function along the flux tube, k_i,

$$k_i \equiv k_i(r, \theta) \equiv \frac{\Phi}{B^P g_{\phi\phi}}. \tag{9.31}$$

For the ingoing wind, we expect $k_i/k_F \sim 1$ for well behaved field topologies. At this point, we introduce the quantity Z that describes the global cross-field electrostatic potential. This parameter is surprisingly fundamental for characterizing the ingoing wind,

$$Z \equiv \frac{\Omega_F \sqrt{g_{\phi\phi}}}{\alpha c} \left[\frac{k_i}{k_F} \right] . \tag{9.32}$$

Using the definition of Z in the force free condition (9.30), with B^T_∞ from (5.50b) and the frozen-in condition (5.16b), we have in the force free limit,

$$\frac{\beta^\phi - \beta^\phi_F}{\beta^P} = -Z . \tag{9.33}$$

One can also rewrite this expression in terms of Boyer–Lindquist angular velocities as in the frozen-in condition (5.44),

$$\Omega_p - \Omega_F = -\beta^P \Omega_F \left(\frac{k_i}{k_F} \right) . \tag{9.34}$$

The ergospheric physics applies to (9.34) through the dragging of inertial frames in (3.43),

$$\Omega_{\min} < \Omega_p = \Omega_F \left[1 + \left| \beta^P \right| \left[\frac{k_i}{k_F} \right] \right] . \tag{9.35}$$

However, combining (3.44) and (9.29c) with Figs. 9.4 and 9.5 we have

$$\lim_{r \to r_+} \Omega_{\min} \gg \Omega_F . \tag{9.36}$$

Thus, there is a difficulty in satisfying (9.35) for physical trajectories. In order to make this more precise we try to keep $\Omega_p > \Omega_{\min}$ as close to the hole as possible, so that both the frame dragging condition (3.43) and the force free condition (9.34) are satisfied. The tendency is for the frozen-in value of Ω_p to be less than $\Omega_{\min}$ as $r \to r_+$. Thus, as the violation of (9.35) is approached, $\Omega_p \gtrsim \Omega_{\min}$ or $\beta^\phi \gtrsim -1$, and therefore one must have $|\beta^P| \ll 1$, since $\beta^2 < 1$. Thus,

$$\Omega_F \left[1 + \left| \beta^P \right| \left(\frac{k_i}{k_F} \right) \right] \ll \Omega_H . \tag{9.37}$$

Clearly, the inequalities (9.37) and (9.35) are incompatible with (3.44) deep in the ergosphere. There is no value of plasma four velocity (equivalently the pair Ω_p and β^P) that can satisfy the B^T = constant condition as $r \to r_+$.

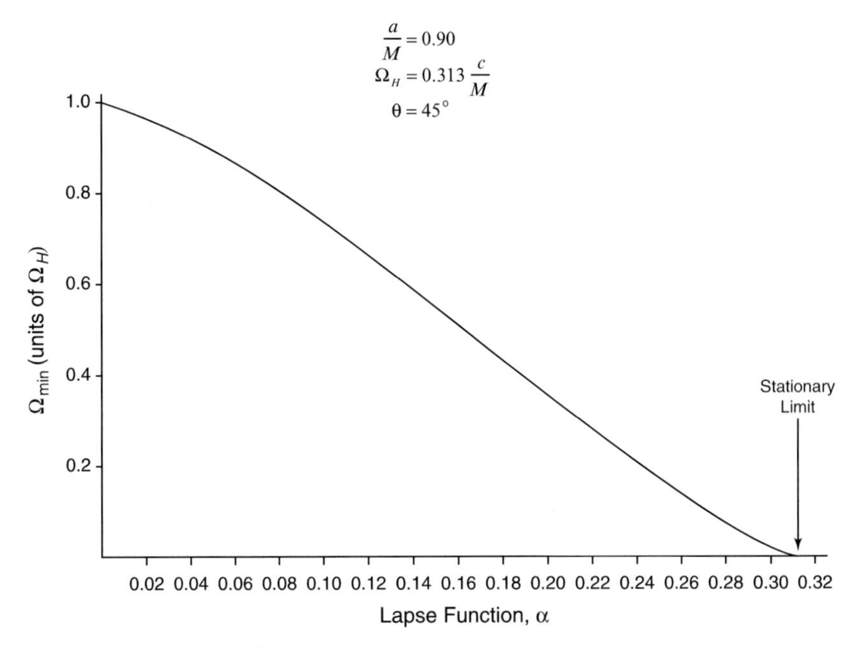

Fig. 9.4 The minimum allowed angular velocity of a physical particle as viewed from asymptotic infinity, $\Omega_{\min}$, as a function of lapse function, α. The value of $\Omega_{\min}$ represents the effects of the dragging of inertial frames associated with a rotating black hole. In the ergosphere, $\Omega_{\min} > 0$ and this is the region plotted above for a black hole rotating with $a/M = 0.90$ and along a radial trajectory defined by $\theta = 45°$. This plot can be used to determine the lapse function at the inner light cylinder for a given value of Ω_F, since $\Omega_F = \Omega_{\min}$ at the light cylinder

We elaborate on this ubiquitous condition for ingoing magnetically dominated winds along flux tubes that thread the horizon. Equation (9.37), the equality on the right hand side of (9.35), Figs. 9.4, and 9.5 imply that the value of Ω_p needed to maintain perfect MHD, $(\Omega_p)_{\text{MHD}}$, at small values of "$r - r_+$" satisfies $(\Omega_p)_{\text{MHD}} \ll \Omega_{\min}$, $r - r_+ \ll r_+$. This constraint on Ω_p violates the definition of $\Omega_{\min}$ imposed by the dragging of inertial frames through (3.43). The angular velocity of the plasma will eventually become too large as the flow proceeds inward (i.e., $\Omega_p > (\Omega_p)_{\text{MHD}}$) to satisfy both the frozen-in condition (5.44) and the angular momentum conservation condition (9.30). If Ω_F were larger, or equivalently by (5.50b) if $|B^T|$ were larger, there would be no such contradiction above because $\Omega_F \sim \Omega_{\min}$ and (9.37) would no longer be true. In terms of the wind formalism of Sect. 5.5, the wind does not have a large enough angular momentum, $|B^T|$, to reach the horizon as a perfect MHD flow.

Surprisingly including finite plasma inertia makes the contradiction more extreme as we will see in the remainder of Sect. 9.4. The cross-field inertial currents described in the gedanken experiment of Sect. 9.3 and illustrated in Fig. 9.3 actually switch off B^T ahead of the flow, exacerbating the situation.

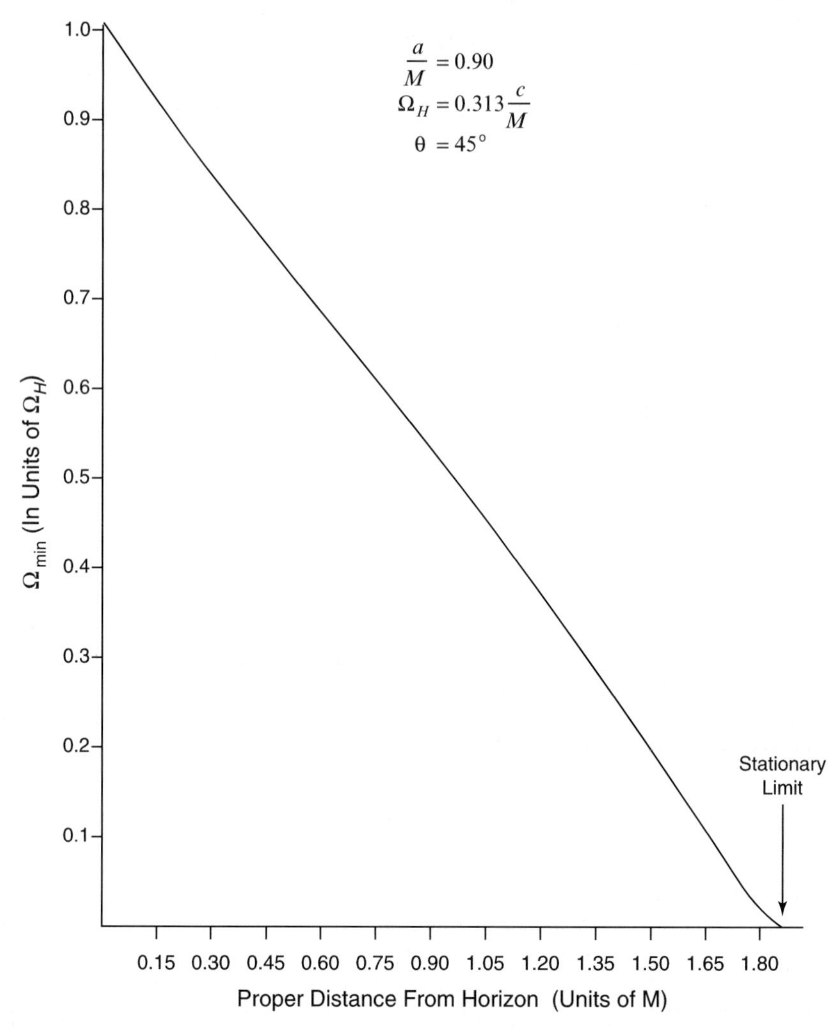

Fig. 9.5 The minimum allowed angular velocity of a physical particle as viewed from asymptotic infinity, $\Omega_{\rm min}$, as a function of proper distance $L = \int \sqrt{g_{rr}}\,\mathrm{d}r$, from the event horizon. $\Omega_{\rm min}$ is plotted in the ergosphere for a black hole rotating with $a/M = 0.90$ along a radial trajectory at $\theta = 45°$. This plot can be used to determine the proper distance from the horizon to the inner light cylinder

9.4.5 The Poloidal Equation of Motion of the Ingoing Wind

In order to explore the effects of plasma inertia on the conservation equations, we compute a useful equation of motion for β^{P}. This relation can be used to integrate the perfect MHD solution in the physically interesting environs of the light cylinder.

From the energy and angular momentum conservation laws for the ingoing wind, (5.20) and (5.21), the initial data on $I_{\_}$, (9.29) and the frozen-in relation (5.16b), we have an expression for the ZAMO four velocity defined by the plasma source and outgoing wind,

$$k_{\_}u^0 = \frac{\beta_F^\phi \Phi}{\mu\sqrt{g_{\phi\phi}}}\left[\frac{\Omega_F}{4\pi k_F c} + \frac{(\beta^\phi - \beta_F^\phi)\,\alpha}{4\pi\sqrt{g_{\phi\phi}}\,k_i\beta^P}\right] + \frac{k_+\left[\omega_\infty - \frac{\Omega}{c}m_\infty\right]}{\mu\alpha c} + \frac{(ke)_{\text{inj}} - \frac{\Omega}{c}(k\ell)_{\text{inj}}}{\mu\alpha c}, \tag{9.38a}$$

$$k_{\_}u^\phi = \frac{\Phi}{\mu\sqrt{g_{\phi\phi}}}\left[\frac{\Omega_F}{4\pi k_F c} + \frac{(\beta^\phi - \beta_F^\phi)\,\alpha}{4\pi\sqrt{g_{\phi\phi}}\,k_i\beta^P}\right] + \frac{k_+ m_\infty}{\mu c\sqrt{g_{\phi\phi}}} + \frac{(k\ell)_{\text{inj}}}{\mu c\sqrt{g_{\phi\phi}}}. \tag{9.38b}$$

Note that the force free condition, B^T = constant, of (9.30) is equivalent to the vanishing of the quantity inside the bracket in the first term on the right hand side of (9.38ab).

Combining (9.38a) and (9.38b) we get the desired expression in terms of the parameter defined in (9.32),

$$\beta^P = (\beta^\phi - \beta_F^\phi) \times \left\{ -Z + \frac{4\pi}{\alpha^2\beta_F^\phi B^P c}\left[k_+\left(\omega_\infty - \frac{\Omega m_\infty}{c}\right) + \left[(ke)_{\text{inj}} - \frac{\Omega}{c}(k\ell)_{\text{inj}}\right] - \frac{k_{\_}\left[e - \frac{\Omega_F \ell}{c}\right]}{\left(1 - \beta^\phi\beta_F^\phi\right)}\right]\right\}^{-1}. \tag{9.39}$$

For the values of ω_∞ and m_∞ from the minimum torque solution found in (9.28), we have from the definition, (9.32), from the inner light cylinder outward,

$$|Z| \gg \left|\frac{4\pi}{\alpha^2\beta_F^\phi B^P c}\left\{-k_+\left(\omega_\infty - \left(\frac{\Omega}{c}\right)m_\infty\right)\right\}\right|. \tag{9.40a}$$

From the magnetically dominated condition we have,

$$|Z| \gg \left|\frac{4\pi}{\alpha^2\beta_F^\phi B^P c}\left[(ke)_{\text{inj}} - \frac{\Omega}{c}(k\ell)_{\text{inj}}\right]\right|, \tag{9.40b}$$

and

$$|Z| \gg \left|\frac{4\pi k_{\_}}{\alpha^2\beta_F^\phi B^P c}\left[e - \frac{\Omega_F \ell}{c}\right]\right|. \tag{9.40c}$$

From (9.40) and (9.39) as long as $1-\beta^{\phi}\beta_{F}^{\phi}\not\approx 0$, we recover the force free condition (9.33),

$$\beta^{P}\approx\frac{(\beta^{\phi}-\beta_{F}^{\phi})}{Z}\,,\quad 1-\beta^{\phi}\beta_{F}^{\phi}\not\approx 0\,. \tag{9.41}$$

We can use (9.41) to integrate the poloidal trajectory as long as the condition, $1-\beta^{\phi}\beta_{F}^{\phi}\not\approx 0$, is satisfied.

9.4.6 Numerically Quantifying the Wind Near the Inner Light Cylinder

The most important quantity for integrating the trajectory given by (9.39) near the inner light cylinder is "Z." In order to determine Z we need to accurately compute the metric coefficients. We take $a/M=0.9$ for the $2\times 10^{9}\,\mathrm{M}_{\odot}$ black hole and its metric is given by (1.24). Take a characteristic magnetic flux tube to be located at a Boyer–Lindquist coordinate $\theta=45°$ at its inner light cylinder. Then by (9.32) and (9.29),

$$Z=4.63\times 10^{-2}\left[\frac{k_i}{k_F}\right]\,,\quad \theta=45°\,. \tag{9.42a}$$

Consider a second flux tube with $\theta=30°$ at its light cylinder, then

$$Z=4.15\times 10^{-2}\left[\frac{k_i}{k_F}\right]\,,\quad \theta=30°\,. \tag{9.42b}$$

The quantities (9.42a) and (9.42b) are very similar. This suggests that typical Z values at the light cylinder for the initial data (9.29) are given by a flux tube at $\theta=45°$.

We do not know Z because the Grad–Shafranov equation has not been solved to determine k_i and k_F. We can get an idea of what k_i is from the two following examples. From the axisymmetric divergence equation (3.55b) we know that $B^P\rightarrow B^r$ near the horizon. We can compute the flux through the horizon for two interesting configurations and then compute k_i with (9.31). The total flux of course vanishes, but we can compute the flux through a hemisphere or a portion of a hemisphere. Since $B^P\rightarrow B^r$, it is suggestive to look at a split monopole magnetic field (i.e., B^r is independent of θ and ϕ) in the ZAMO frame,

$$\Phi_{B}=\int B^{P}\mathrm{d}X^{\theta}\wedge\mathrm{d}X^{\phi}=2\pi\int B^{P}\sqrt{g_{\phi\phi}}\sqrt{g_{\theta\theta}}\,\mathrm{d}\theta\,. \tag{9.43a}$$

One might argue that you cannot compute flux at the horizon in the ZAMO frames since they are pathological at the horizon. However, $\tilde{F}_{\theta\phi}=\sqrt{g_{\phi\phi}}\sqrt{g_{\theta\theta}}\,F^{\theta\phi}\equiv$

$\sqrt{g_{\phi\phi}}\sqrt{g_{\theta\theta}}\,B^P$ by (3.4) and (3.7), so (9.43a) can be written equivalently as in (4.73) in the well behaved Boyer–Lindquist coordinates,

$$\Phi_B = \int \tilde{F}_{\theta\phi}\, \mathrm{d}\theta\, \mathrm{d}\phi \,. \tag{9.43b}$$

Thus, (9.43a) is well-defined and we insert the expression into (9.31) with B^P a constant to get

$$k_i\left(r_+,\, \theta_0\right) = \frac{2\pi \int_0^{\theta_0} \sqrt{g_{\phi\phi}}\sqrt{g_{\theta\theta}}\, \mathrm{d}\theta}{g_{\phi\phi}\left(r_+,\, \theta_0\right)}$$
$$= 2\pi \left[\frac{r_+^2 + a^2\cos^2\theta_0}{\left(r_+^2 + a^2\right)\left(1+\cos\theta\right)}\right] . \tag{9.43c}$$

For a rapidly rotating hole $r_+^2 \gtrsim a^2$ and (9.43c) implies for a split monopole just outside the horizon,

$$k_i\left(r,\, \theta\right) \approx \frac{1+\cos^2\theta}{\left(1+\cos\theta\right)}\pi \,. \tag{9.43d}$$

The function $k_i\left(r,\, \theta\right)$ has a minimum at $\theta = 65^\circ$ and maxima at $\theta = 0$ and $\theta = \pi/2$.

$$0.84\pi < k_i\left(r,\, \theta\right) < \pi \,, \quad \frac{\partial}{\partial\theta} B^P = 0 \,. \tag{9.43e}$$

Next consider the Wald magnetic field of Sect. 9.6.2. From (4.84a) at the horizon

$$B^r\left(r = r_+\right) = \frac{B_0\left(r_+^4 - a^4\right)}{\left[r_+^2 + a^2\cos^2\theta\right]^2}\cos\theta \,. \tag{9.44a}$$

The magnetic flux in a hemispherical cap above latitude θ near the event horizon is

$$\Phi\left(\theta\right) = \frac{\pi B_0\left(r_+^4 - a^4\right)}{r_+^2 + a^2\cos^2\theta}\sin^2\theta \,. \tag{9.44b}$$

For $r_+^2 \gtrsim a^2$, from (9.44b) and (9.31)

$$k_i\left(r_+,\, \theta\right) \approx \frac{\pi}{2}\frac{\left(1+\cos^2\theta\right)^2}{\cos\theta} \,. \tag{9.44c}$$

By (9.44b), 90% of the flux is concentrated between $0 < \theta < 70^\circ$ in each hemisphere. In this range of latitude

$$1.5\pi < k_i\left(r_+,\, \theta\right) < 2\pi \,, \quad 0 < \theta < 70^\circ \,. \tag{9.44d}$$

Note that $k_i = 1.59\pi$ at the horizon in the Wald field at $\theta = 45°$ and $k_i = 0.88\pi$ in the split monopole under similar conditions.

Asymptotically, we expect a cylindrical wind with $k_{_F} \approx \pi$ based on the desired high collimation that is observed. Thus (9.44d) and (9.43e) combined with (9.42a) and (9.42b) suggest

$$0.01 < Z^2 < 0.001\ , \quad r = r_{_{L.C.}} \tag{9.45}$$

We use these values of Z to numerically order the parameters in (9.39). From the outgoing minimum torque wind described in Sect. 9.4.2 defined by $\sigma = 10^4$, we have at the light cylinder,

$$\left|\frac{4\pi}{\alpha^2\beta_{_F}^{\phi}B^{P}c}\left[-k_{_+}\left(\omega_\infty - \frac{\Omega m_\infty}{c}\right)\right]\right| = 4.98\times 10^{-3}\ , \tag{9.46a}$$

$$\left|\frac{4\pi}{\alpha^2\beta_{_F}^{\phi}B^{P}c}k_{_-}\left[e - \frac{\Omega_{_F}\ell}{c}\right]\right| = 5.48\times 10^{-6}\ . \tag{9.46b}$$

Thus, (9.45) and (9.46ab) yield the following ordering of parameters near the light cylinder at $\theta = 45°$,

$$1 >> |Z| >> \left|\frac{4\pi}{\alpha^2\beta_{_F}^{\phi}B^{P}c}\left[k_{_+}\left(\omega_\infty - \frac{\Omega m_\infty}{c}\right)\right]\right| >> \left|\frac{4\pi k_{_-}}{\alpha^2\beta_{_F}^{\phi}B^{P}c}\left(e - \frac{\Omega_{_F}\ell}{c}\right)\right|\ . \tag{9.47}$$

Equations (9.46a) and (9.46b) demonstrate that the poloidal equation of motion (9.39) reduces to the force free relation (9.41) as long as

$$1 - \beta^{\phi}\beta_{_F}^{\phi} >> \frac{5.84\times 10^{-6}}{Z}\ . \tag{9.48}$$

9.4.7 Accessibility of the Inner Alfvén Point

As was found in Sect. 9.3, the flow obeys the force free poloidal velocity law (9.41) until it nears the light cylinder. The main interest of this chapter is to analyze the flow in this region where dynamically interesting effects appear. By (5.36), just before the light cylinder, the flow will cross the Alfvén critical surface as a perfect MHD wind and if it does what are the effects of plasma inertia?

From the Alfvén speed (5.36) and the Alfvén point condition (5.42b), one has

$$\left(U_{_I}\right)_{_A} = -\frac{c}{\alpha\left(\beta_{_F}^{\phi}\right)_{_A}}Z_{_A}\left(\frac{mc}{\mu}\right)\ . \tag{9.49}$$

where the subscript "A" means to evaluate at the Alfvén critical surface in the flux tube.

Note that at the light cylinder the lapse function $\alpha\left(r_{L.C.}\right)$ is (see Fig. 9.4)

$$\alpha\left(r_{L.C.}\right) = 0.296\,. \tag{9.50}$$

By (5.42b), we have $\left(\beta_F^\phi\right)_A \approx -1$ and therefore

$$\left(U_I\right)_A \approx \begin{cases} 33.78c\,, & Z^2 = 10^{-2} \\ 106.7c\,, & Z^2 = 10^{-3} \end{cases}. \tag{9.51a}$$

As in the analysis of [113] and Fig. 5.2, it is convenient to describe the location of the Alfvén critical surface by treating β_F^ϕ as a variable that indicates poloidal displacement along the flux tube. From (9.49), (5.36) and the definition of k in (5.12),

$$\left[1-(\beta_F^\phi)^2\right]_A \approx \frac{4\pi k_{-n} m_e c}{\alpha^2 B^P \beta_F^\phi Z}\,. \tag{9.52}$$

In the numerical model being considered, (9.52) implies

$$\left[1-(\beta_F^\phi)^2\right]_A \approx \begin{cases} 5.61\times 10^{-5}\,, & Z^2 = 10^{-2} \\ 1.78\times 10^{-4}\,, & Z^2 = 10^{-3} \end{cases}, \tag{9.53a}$$

$$\left(\beta_F^\phi\right)_A = \begin{cases} -0.999972\,, & Z^2 = 10^{-2} \\ -0.999911\,, & Z^2 = 10^{-3} \end{cases}. \tag{9.53b}$$

Combining (9.53a), (9.51) and the definition of the Alfvén speed at the critical surface, (5.36), we find the pure Alfvén speed,

$$U_A = \begin{cases} 4.52\times 10^{3}\,, & Z^2 = 10^{-2} \\ 8.09\times 10^{3}\,, & Z^2 = 10^{-3} \end{cases}, \tag{9.53c}$$

at the Alfvén point.

We can get an expression for β^P analogous to (9.39) that is valid only at the Alfvén critical surface by multiplying (9.38a) by β^P and setting $u^P = U_I$ in (5.36),

$$\left(\beta^P\right)_A = \left[1-\beta^\phi\beta_F^\phi\right]_A \times \Bigg\{ \left(\beta_F^\phi\right)_A Z_A + \frac{4\pi}{B_A^P \alpha_A^2 c}\Big\{\left[k_+\omega_\infty + (ke)_{\rm inj}\right] \\ -\frac{\Omega_A}{c}\left[k_+ m_\infty + (k\ell)_{\rm inj}\right]\Big\}\Bigg\}^{-1}. \tag{9.54}$$

Using the scalings in (9.47), we have the approximate simplified version of (9.59),

$$\left(\beta^P\right)_A \approx \frac{\left[1-\beta^\phi\beta_F^\phi\right]_A}{\left(\beta_F^\phi\right)_A Z_A}\,. \tag{9.55}$$

The normalization condition, $u \cdot u = -1$, can be used to eliminate β^ϕ in (9.55) at the Alfvén point. We find that

$$\left(\beta^P\right)_A = \frac{U_I^2}{c^2 \beta_F^\phi} \times \frac{\left[Z \pm \sqrt{Z^2 - \left(\frac{U_I}{c}\right)^{-2} - (1+Z^2)\left[1 - \left(\beta_F^\phi\right)^2\right]}\right]}{1 + (1+Z^2)\left(\frac{U_I}{c}\right)^2}, \tag{9.56}$$

where all quantities are to be evaluated at the Alfvén point. The negative root in (9.56) is incompatible with the frozen-in condition (5.16b) and $F^{12} < 0$ that requires $\beta^\phi > \beta_F^\phi$. Taking the positive root in (9.56) and using (9.51) and (9.53a), we have an approximate accurate expression for $\left(\beta^P\right)_A$,

$$\left(\beta^P\right)_A \approx \frac{2Z_A}{\left(\beta_F^\phi\right)_A \left[1 + Z_A^2\right]}, \tag{9.57}$$

and in the numerical model,

$$\left(\beta^P\right)_A \approx \begin{cases} -0.2\,, & Z^2 = 10^{-2} \\ -0.063\,, & Z^2 = 10^{-3} \end{cases}. \tag{9.58}$$

Inserting the value $\left(\beta^P\right)_A$ of (9.50) into the value of the poloidal four velocity in (9.51), we have the Lorentz factor at the Alfvén point,

$$u_A^0 = \begin{cases} 169c\,, & Z^2 = 10^{-2} \\ 1{,}690c\,, & Z^2 = 10^{-3} \end{cases}. \tag{9.59}$$

The parameter "$1 - \beta^\phi \beta_F^\phi$" is very useful for describing the flow near the Alfvén point and beyond, downstream of the light cylinder. From the Alfvén point condition (5.42b) and (5.36) along with the positivity of ZAMO energy, we know that

$$e - \frac{\Omega_F \ell}{c} > 0\,. \tag{9.60}$$

Expanding the wind constant in (9.60) and implementing the conservation laws (5.20) and (5.21), we have

$$1 - \beta^\phi \beta_F^\phi = \frac{e - \frac{\Omega_F \ell}{c}}{\mu u^0 \alpha c} \approx \frac{c}{\alpha u^0}, \tag{9.61a}$$

Equation (9.61a) establishes the positive definite condition,

$$1 - \beta^\phi \beta_F^\phi > 0\,, \tag{9.61b}$$

in perfect MHD and is very accurate for warm wind initial conditions created within the particle creation zone (i.e., rotational inertia is negligible to rest mass when the injected plasma is threaded onto the field lines in the particle creation zone).

Using (9.59) in (9.61a), in the numerical model we have

$$\left[1-\beta^{\phi}\beta_F^{\phi}\right]_A \approx \begin{cases} 2.00\times 10^{-2}\,, & Z^2 = 10^{-2} \\ 2.00\times 10^{-3}\,, & Z^2 = 10^{-3} \end{cases}. \tag{9.62}$$

From (9.48), the plasma inertia is negligible and the force free equation of motion (9.41) is applicable. Using the value of $\left(\beta_F^{\phi}\right)_A$ in (9.53b) we can determine the final unknown parameter in the wind at the Alfvén point,

$$\left(\beta^{\phi}\right)_A \approx \begin{cases} -0.98\,, & Z^2 = 10^{-2} \\ -0.998\,, & Z^2 = 10^{-3} \end{cases}. \tag{9.63}$$

The value of u^0 in (9.59) shows that special relativistic effects are important near the Alfvén point. The large Lorentz factor is attributed to the plasma being in approximate corotation with the field as evidenced by (9.63). Relation (9.48) and the values of $1-\beta^{\phi}\beta_F^{\phi}$ in (9.62) shows that the plasma at the Alfvén point, even with this relativistic Lorentz factor, does not have enough inertia to bend the field lines azimuthally a significant amount. Thus, there is not enough toroidal magnetic field to allow the plasma to slide forward relative to the (corotating frame of) magnetic field near the light cylinder and thus avoid relativistic rotation velocities. The flow downstream of the Alfvén point has similar properties.

9.4.8 Accessibility of the Inner Fast Point

It was shown in Sect. 6.2 that when $0 < \Omega_F \ll \Omega_H$, the inner fast point is near the light cylinder as well (see Fig. 9.6). We can get an expression for the poloidal three velocity at the fast point (designated by a subscript "f" in this section to distinguish it from the corotating frame of the magnetic field designated by the subscript "F"), $\left(\beta^P\right)_f$, by multiplying (9.38a) by β^P and using (6.9) for the poloidal velocity at the fast point. First note that, in the magnetically dominated limit,

$$U_F^2 \approx \left\{\left[1-\left(\beta_F^{\phi}\right)^2\right]+\bar{Z}^2\right\}U_A^2\,, \tag{9.64}$$

$$\bar{Z} \equiv \frac{F^{12}}{B^P} = \frac{\beta_F^{\phi}-\beta^{\phi}}{\beta^P}\,, \tag{9.65}$$

where we made use of (5.35) and $U_A^2 \gg U_S^2$ in (9.64). The identification of $\bar{Z}$ is made to look like Z defined in (9.32), since $Z=\bar{Z}$ in a force free wind and $\bar{Z}\approx Z$ throughout a magnetically dominated perfect MHD wind. Proceeding as described

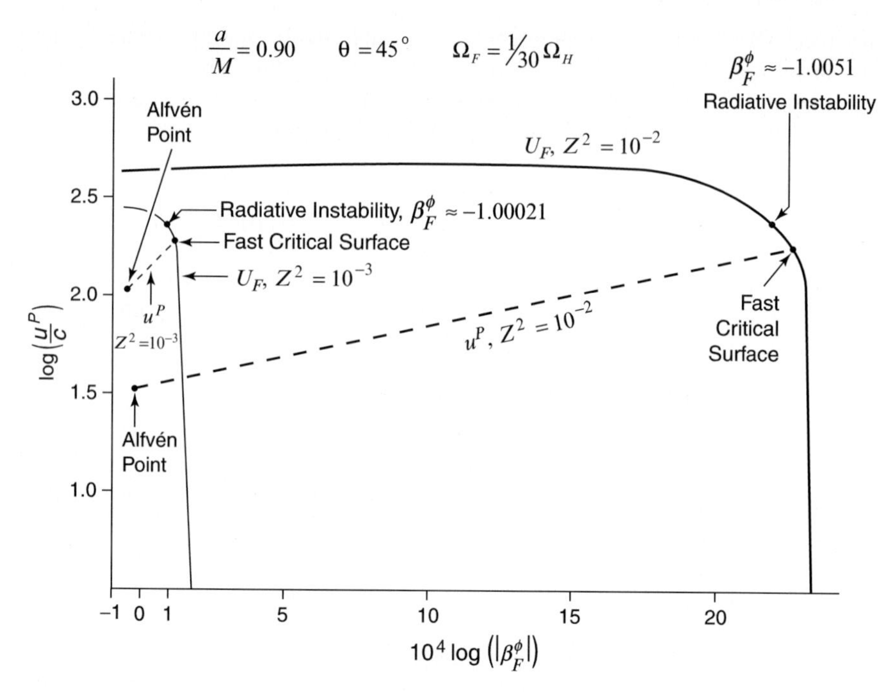

Fig. 9.6 The plots of u^P/c (the dashed curves) and U_F/c (the solid curves) as a function of the field line azimuthal velocity, β^ϕ_F, are given in logarithmic units for the model perfect MHD flow discussed in Sect. 9.4. The poloidal four velocity, u^P, is approximated by a power law fit between the Alfvén and fast critical surface. This expedience should be fairly accurate since the two critical surfaces almost coalesce for $\Omega_F \ll \Omega_H$. The fast speed U_F/c is plotted, based on this value of u^P (since by mass conservation, u^P determines the proper number density in the expression (9.64) for U_F). The fast critical surface is defined by the intersection of u^P and U_F. Both u^P and U_F are plotted for the two values of Z^2 ($Z^2 = 10^{-2}$ is in bold and $Z^2 = 10^{-3}$ is in normal font) used in the text. Note that $U_F \to U_S$ just beyond the inner light cylinder when $\Omega_F \ll \Omega_H$. The plot is also used to determine the location of the onset of the radiative instability described in Sect. 9.5. By (9.109) we expect $\delta(k\ell)_{ph} \gtrsim |k\delta m|$ as perfect MHD begins to break down. Thus, by (9.121) and (9.111b) we expect the linear leading edge of the radiative instability to initiate when $u_1^2 \approx \frac{1}{2}U_F^2$

above, using the approximate form of (6.9) as expressed in (9.64) we find

$$\left(\beta^P\right)_f \approx \left[1-\beta^\phi\beta^\phi_F+\bar{Z}^2\right]\times\left\{\left(\beta^\phi_F\right)_f Z_f+\frac{4\pi}{B^P_f\alpha_f^2 c}\left\{\left[k_+\omega_\infty+(ke)_{\text{inj}}\right]\right.\right.$$
$$\left.\left.-\frac{\Omega_f}{c}\left[k_+ m_\infty+(k\ell)_{\text{inj}}\right]\right\}\right\}^{-1}. \tag{9.66}$$

We can numerically estimate the inertial terms in the denominator of (9.66) near the fast point by noting that at the light cylinder,

$$\Omega \approx \frac{0.212c}{M}\,, \quad r = r_{L.C.}\,, \tag{9.67a}$$

$$\Omega_F = \frac{1}{30}\Omega_H = 1.09 \times 10^{-n2}\frac{c}{M}\,. \tag{9.67b}$$

Furthermore, by writing the pure Alfvén speed as

$$U_A^2 = \frac{\alpha\left(\frac{u^P}{c}\right)B^P}{4\pi\frac{k_-}{c}\mu}\,, \tag{9.67c}$$

then, using (9.53c) and (9.51) at the Alfvén point,

$$B^P \approx 1.13 \times 10^4\mathrm{G}\,, \quad r = r_{L.C.}\,. \tag{9.67d}$$

Noting the values of m_∞ and ω_∞ in (9.28) and k_+ and k_- in (9.26) and (9.29g) for a warm injected plasma, $(ke)_{\mathrm{inj}} \approx \left[|k_+| + |k_-|\right] m_e c^2$, and using the values from (9.67), we have the good approximation to (9.66),

$$\beta_F^\phi \beta^P \approx \frac{\bar{Z}^2 + (1 - \beta^\phi \beta_F^\phi)}{Z + 4.8 \times 10^{-3}}\,. \tag{9.68}$$

Combining (9.64) and (9.67c) we have

$$\left(\frac{u^P}{c}\right)_f = \frac{\left[1 - (\beta_F^\phi) + \bar{Z}^2\right] B^P \alpha}{4\pi\left(\frac{k_-}{c}\right) m_e c^2}\,. \tag{9.69a}$$

From (9.29g) and (9.67d) we can rewrite (9.69) as

$$\left(\frac{u^P}{c}\right)_f \approx 6.44 \times 10^5 \left[1 - (\beta_F^\phi)^2 + \bar{Z}^2\right]. \tag{9.69b}$$

Combining (9.66) for β^P and (9.61a) for u^0 and inserting these expressions into (9.69a) for $\left(u^P\right)_f$, we get

$$\frac{\bar{Z}_f^2}{1 - \beta^\phi \beta_F^\phi} \approx \alpha_f \left[Z_f + \frac{4\pi}{B_f^P \alpha_f^2 c}\left\{\left[k_+\omega_\infty + (ke)_{\mathrm{inj}}\right]\right.\right.$$
$$\left.\left. - \frac{\Omega_f}{c}\left[k_+ m_\infty + (k\ell)_{\mathrm{inj}}\right]\right\}\right] \times \left[1 - (\beta_F^\phi)_f^2 + \bar{Z}_f^2\right] - 1\,. \tag{9.70a}$$

or in approximate form with $\alpha_f = 0.293$ (as solved self consistently with $\alpha = 0.296$ at the Alfvén point)

$$\frac{\bar{Z}_f^2}{1-\beta^\phi\beta_F^\phi} = \begin{cases} 6.9\times 10^3\left[1-(\beta_F^\phi)_f^2+\bar{Z}_f^2\right]-1\,, & Z^2=10^{-2} \\ 2.04\times 10^4\left[1-(\beta_F^\phi)_F^2+\bar{Z}_f^2\right]-1\,, & Z^2=10^{-3} \end{cases}\,, \tag{9.70b}$$

where Z is evaluated at the light cylinder (Z evaluated at the fast point is slightly larger).

Using $u\cdot u=-1$, (9.68) for β^P and (9.61a) for u^0, we get a second equation:

$$\alpha^2\left(1-\beta^\phi\beta_F^\phi\right)^2 = 1-\left(\beta^\phi\right)_f^2 - \frac{\left[\bar{Z}_f^2+\left(1-\beta^\phi\beta_F^\phi\right)\right]^2}{\left[Z+4.8\times 10^{-3}\right]^2} \tag{9.71}$$

We have three equations (9.65), (9.70b) and (9.71) that must be solved simultaneously for three quantities, β^ϕ, β_F^ϕ and β^P at the fast point.

First consider the case $Z=0.1$ at the light cylinder. We find a solution for the fast critical surface just inside the light cylinder. Expanding geometrical quantities about the light cylinder we solve self consistently, (9.65), (9.70b) and (9.71):

$$Z_f = 0.103\,, \tag{9.72a}$$

$$\bar{Z}_f = 0.104\,, \tag{9.72b}$$

$$\left(1-\beta^\phi\beta_F^\phi\right)_f = 2.2\times 10^{-3}\,, \tag{9.72c}$$

$$\left[1-\left(\beta_F^\phi\right)^2\right]_f = -1.05\times 10^{-2}\,, \tag{9.72d}$$

$$\left(\beta_F^\phi\right)_f = -1.00524\,, \tag{9.72e}$$

$$\left(\frac{U_F}{c}\right)_f = 187\,, \tag{9.72f}$$

$$\left(\frac{u^0}{c}\right)_f = 1{,}551\,, \tag{9.72g}$$

$$\left(\beta^P\right)_f = -0.1205\,, \tag{9.72h}$$

$$\left(\beta^\phi\right)_f = -0.9927\,. \tag{9.72i}$$

Next consider our other example with $Z^2=10^{-3}$, we self consistently solve (9.65), (9.70b) and (9.71) as

$$\left(\frac{\bar{Z}}{Z}\right)_f = 0.94\,, \tag{9.73a}$$

$$\left(1-\beta^\phi\beta_F^\phi\right)_f = 8.3\times 10^{-4}\,, \tag{9.73b}$$

$$\left[1-\left(\beta_F^\phi\right)^2\right] = 5.79\times 10^{-4}\,, \tag{9.73c}$$

$$\left(\beta_F^{\phi}\right)_f = -1.00029\,, \tag{9.73d}$$

$$\frac{U_F}{c} = 192\,, \tag{9.73e}$$

$$\frac{u^0}{c} = 4,112\,, \tag{9.73f}$$

$$\beta^P = -0.0466\,, \tag{9.73g}$$

$$\beta^{\phi} = -0.9989\,. \tag{9.73h}$$

One can compare (9.72) and (9.73) to the same quantities evaluated at the Alfvén point. The extreme conditions in the plasma are accentuated as it flows toward the horizon, β^{ϕ} keeps getting closer to -1, the ultrarelativistic Lorentz factors, u^0, keep increasing and the magnitude of β^P actually decreases (see Fig. 9.7). By (3.50) the plasma at the Alfvén and fast points is on highly negative energy trajectories.

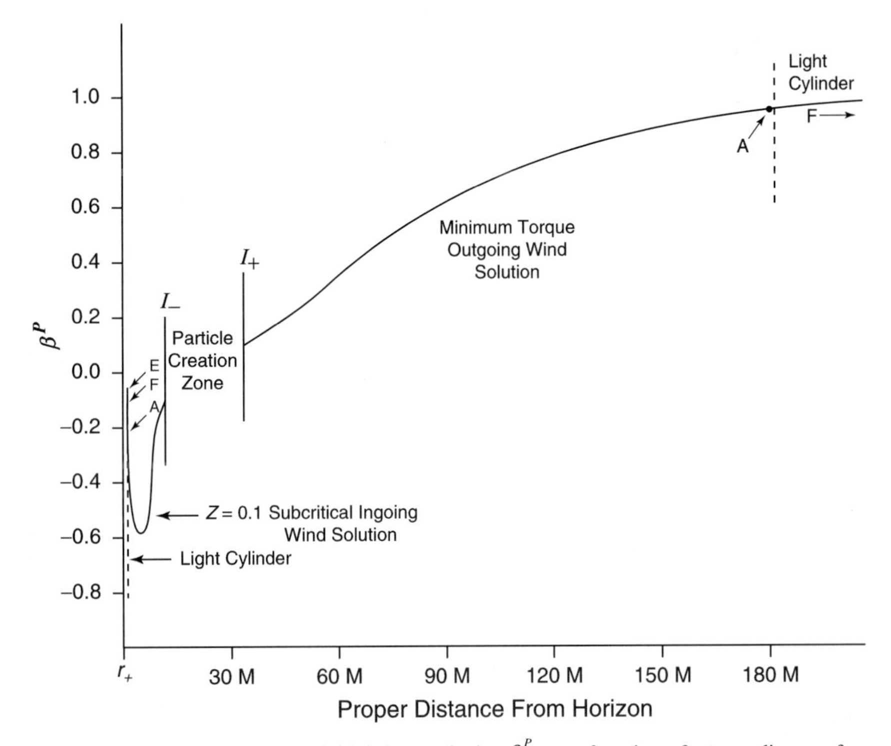

Fig. 9.7 The ZAMO evaluated poloidal three velocity, β^P, as a function of proper distance from the event horizon of the paired wind system that is modeled in Sect. 9.4. The Alfvén point is labeled "A," the fast point is labeled "F" and the terminus of the perfect MHD ingoing wind is labeled "E." Compare the ingoing wind to the subcritical outgoing wind in Fig. 5.2. The plot is linear in proper distance, as opposed to logarithmic as in Fig. 5.2, in order not to further compress the interesting structure of the ingoing wind. As a result, the fast point of the outgoing wind is far off to the right of the figure. The three velocity, β^P, is small near the inner light cylinder since the plasma is in approximate corotation with the magnetic field, $\beta^{\phi} \gtrsim -1$, and the four velocity is normalized, $u \cdot u = -1$

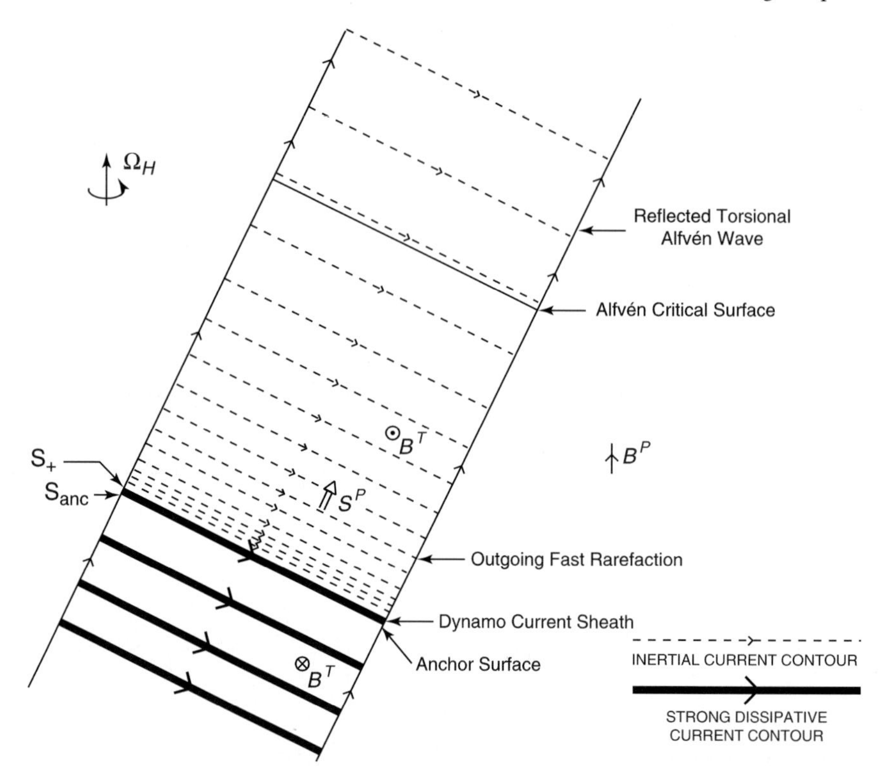

Fig. 9.8 The inertial current system in the dynamo region of the ingoing wind. An incoming Alfvén wave reflects from the rotating magnetosphere in the ergosphere. The incoming wave is absorbed by the magnetosphere in an MHD skin depth between the Alfvén critical surface and the anchor at S_{anc}. The MHD structure of the skin depth is a fast rarefaction wave. The strong cross-field current flow in this region is indicative of a fast wave as discussed in Chaps. 2 and 6. A radiative instability initiates at S_+. The flow causally decouples from the plasma source and outgoing wind at S_{anc}. However, the dissipative dynamo currents continue to flow in the deflagration wind below the anchor. The reflected Alfvén wave transports a poloidal Poynting flux S^P to infinity that is supported primarily by inertial currents flowing between the Alfvén critical surface and S_{anc}

Since the plasma was injected with positive angular momentum from the γ-ray production process (the γ-rays came from a disk corona that is in a near Keplerian orbit), the fact that $\beta^\phi \approx -1$ in the wind implies that significant cross-field poloidal currents have been driven across the magnetic field ahead of the Alfvén point. These currents are greatly enhanced between the Alfvén and fast points (see Fig. 9.8). This is a manifestation of the "torsional tug of war" described in Chap. 7 and was foreshadowed in the gedanken experiment of Fig. 9.3.

Consider the large azimuthal accelerations between the Alfvén and fast point. To quantify this value, we first evaluate the distance between the fast and Alfvén points, ΔX^1, by

$$\frac{\partial \beta^\phi_F}{\partial X^1} \approx \frac{\Delta \beta^\phi_F}{\Delta X^1} \,. \tag{9.74}$$

From the definition of β_F^ϕ in (5.14) and the connection coefficients in (3.32), we have

$$\frac{\partial \beta_F^\phi}{\partial X^1} = -\beta_F^\phi \Gamma^1{}_{00} - 2\Gamma^1{}_{0\phi} - \beta_F^\phi \Gamma^1{}_{\phi\phi} , \tag{9.75a}$$

$$\frac{\partial \beta_F^\phi}{\partial X^1} \approx g_{eff} \left(\beta^\phi = -1\right) , \tag{9.75b}$$

where g_{eff} is the poloidal gravity as defined in (3.41). Thus, if we approximate X^1 by X^r, we have from (9.75a),

$$\frac{\partial \beta_F^\phi}{\partial X^1} \approx 0.521\,(M)^{-1} + 2\,(0.184)\,M^{-1} - 0.131M^{-1} = 0.76M^{-1} . \tag{9.76}$$

Thus, between the fast and Alfvén points we have a proper distance of

$$\Delta X^1 \approx 2.0 \times 10^{12}\text{cm} , \quad Z^2 = 10^{-2} , \tag{9.77a}$$

$$\Delta X^1 \approx 1.5 \times 10^{11}\text{cm} , \quad Z^2 = 10^{-3} . \tag{9.77b}$$

From (9.72), (9.73) and (9.59) we can find the proper acceleration near the light cylinder, $a^\phi \approx u^1 \partial/\partial X^1 \left(u^\phi\right)$,

$$a^\phi \approx 1.17 \times 10^{14}\text{cm/sec}^2 , \quad Z^2 = 10^{-2} , \tag{9.78a}$$

$$a^\phi \approx 2.79 \times 10^{15}\text{cm/sec}^2 , \quad Z^3 = 10^{-3} . \tag{9.78b}$$

These accelerations are enormous and we expect tremendous amounts of radiation from the plasma. These radiation losses will cause a breakdown of the perfect MHD assumption and this is the topic of the next section. Notice that the flow is just starting to break the force free condition (9.48) as evidenced particularly by (9.73b).

9.4.9 The Terminus of the Perfect MHD Wind

Even ignoring the effects of radiative dissipation, the perfect MHD assumption cannot persist very far inward of the fast point. Perfect MHD requires a vanishing proper electric field, or $F^{\mu\nu}F_{\mu\nu} \geq 0$. Using the expansion of the proper field in (5.35) and the definition $\bar{Z}$ in (9.65), this condition becomes

$$1 - \left(\beta_F^\phi\right)^2 + \bar{Z}^2 \geq 0 . \tag{9.79}$$

As more cross-field current is driven across the magnetic field lines (equatorward) inside the fast point, by Ampere's law (3.61b), $\bar{Z}$ will decrease. We can approximate

$\bar{Z}$ by $\bar{Z}_f$, noting that in actuality $\bar{Z}_f$ might be slightly larger. The end of the perfect MHD flow can be found from (9.79),

$$\left(\beta_F^\phi\right)_e^2 = 1 + \bar{Z}_e^2\,, \tag{9.80a}$$

$$\left(\beta_F^\phi\right)_e \approx -1.00539\,, \quad Z^2 = 10^{-2}\,, \tag{9.80b}$$

$$\left(\beta_F^\phi\right)_e \approx -1.00044\,, \quad Z^2 = 10^{-3}\,. \tag{9.80c}$$

Thus, the fast point is located within a short distance, ΔX_{ef}, of where perfect MHD breaks down by (9.80), (9.72e), (9.73d) and (9.76),

$$\Delta X_{ef} \lesssim 5.9 \times 10^{10}\text{cm}\,, \quad Z^2 = 10^{-2}\,, \tag{9.81a}$$

$$\Delta X_{ef} \lesssim 5.9 \times 10^{10}\text{cm}\,, \quad Z^2 = 10^{-3}\,. \tag{9.81b}$$

We indicate the behavior of the perfect MHD wind near the inner light cylinder in Fig. 9.7. The figure is only qualitative in that it captures the nature of the wind in a manner that is comparable to the outgoing wind solution space that is depicted in Fig. 5.2. The arbitrariness arises since the Grad–Shafranov equation was not solved to find B^P, thus we picked a functional dependence to reproduce the outgoing minimum torque solution of Fig. 5.2 in a qualitative sense. Note that in this linear scale, the domain of the ingoing wind is small and the structure near the light cylinder is below the resolution of the figure (in a logarithmic scale as is used in Fig. 5.2, the resolution would be even worse). Even so, the fast point of the outgoing wind is way off the right side of the page (as indicated in Fig. 5.2). The important aspect of Fig. 9.7 is that due to approximate corotation of the plasma with the magnetic field, the dynamics change drastically near the inner light cylinder. The poloidal three velocity, β^P, decreases abruptly as the plasma spirals around at relativistic velocities (i.e., $|\beta|^2 < 1$). Even though β^P decreases, u^P experiences a marked increase as a consequence of the large rotational inertia acquired by the plasma.

This solution is the analog of the subcritical outgoing MHD wind solution described in Sect. 5.5. In the magnetically dominated limit, the energy and angular momentum flux are given essentially by the electromagnetic terms in (5.20) and (5.21). For a given value of Ω_F, the critical solution (minimum torque solution) described in Sect. 5.5 can be labeled by a critical value of $|B^T|$, $|B^T|_c$. The value of $|B^T|_c$ is the minimum toroidal magnetic field that links a plasma source to asymptotic infinity via a perfect MHD wind. The degeneracy of (5.20) and (5.21) in the magnetically dominated limit yields three equivalent designations for subcritical winds:

$$|B^T| < |B^T|_c\,, \tag{9.82a}$$

$$|ke| < |ke|_c\,, \tag{9.82b}$$

$$|k\ell| < |k\ell|_c\,. \tag{9.82c}$$

The subcritical solutions in Fig. 5.2 proceed beyond the light cylinder until the perfect MHD solution becomes over constrained. This is manifested by the fact that the extension of the subcritical solution extends beyond this region as an "unphysical branch" where $u^r > 0$, but $dr/dt < 0$ [113].

The problem encountered here is that Ω_F is not established by the dynamics of a unipolar inductor as in a star, but by plasma injection on free-floating magnetic flux tubes. Consequently, one is not guaranteed a value of Ω_F that yields a sufficient toroidal magnetic field by (5.50), so that the plasma can pass uneventfully (by sliding forward in the azimuthal direction, relative to black hole rotation, along the field lines) through the region near the light cylinder. Mathematically, this can be traced back to our force free analysis of (9.35) which can be satisfied all the way to the horizon only if $\Omega_F \sim \Omega_H$. The value of $|B^T|$ in the wind equates to energy and angular momentum fluxes that are subcritical for the ingoing wind by (9.82). By contrast, a unipolar inductor radiates an Alfvén wave that transports the appropriate B^T into a wind and nothing eventful happens at the light cylinder. Without a unipolar inductor, only plasma inertia can bend the magnetic field lines azimuthally (through inertial cross-field currents) to create the proper B^T for smooth passage through the light cylinder. However, in a magnetically dominated wind, the plasma inertia is small and it is therefore incapable of bending the magnetic field to a degree that can substantially alter B^T.

9.4.10 The Ingoing Extension of the Subcritical Solution

Consider the poloidal momentum equation (3.40) at the point of the maximal extension of the perfect MHD ingoing wind discussed in the last section (labeled "E" in Fig. 9.7). The effective gravity "g" in (3.41) was found in (9.75) and (9.76) to equal $0.76M^{-1}$. This value of "g" represents a strong inward force of gravity on a surface fixed at the terminus of the wind, $r = r_E$.

Next consider the electromagnetic force in the poloidal momentum equation, $F^{12}J^2$. First of all, the ZAMO energy and azimuthal momentum equations are virtually identical near the inner light cylinder for these winds. The large gradient in the specific mechanical ZAMO azimuthal momentum, u^ϕ, and the ZAMO specific energy, u^0, illustrated in (9.78), near the terminus of the perfect MHD regime are nearly identical $\left(\beta^\phi \approx -1\right)$, as a consequence of the $J^2\hat{e}_2 \times \boldsymbol{B^P}$ and $\boldsymbol{J} \times \boldsymbol{E} = J^2E^2$ forces, being nearly equal in this region. Thus, the rapid increase in the u^0 contribution to u^P is essentially purely electrodynamic in nature. Near the inner light cylinder, J^2, is relatively large (see Fig. 9.8) and is directed equatorward. By Ampere's law (3.61b), this cross-field current density slowly decreases F^{12}. The current density, J^2, also torques the plasma to large values of negative angular momentum, $u^\phi \ll 0$. Thus, the $F^{12}J^2$ force in the poloidal momentum equation is strong and inwards directed at the terminus of the perfect MHD wind, $F^{12}J^2/c \ll 0$.

All of the forces in the poloidal momentum equation (3.40) are large and inward directed. The is no slight modification of this flow that will come to rest near

"E." The force of gravity is going to pull the plasma inward all of the way to the horizon. The inward force of gravity is a huge distinction from an outgoing subcritical MHD wind. Outgoing subcritical winds cannot make it to asymptotic infinity, they are free expanding flows with insufficient energy. By contrast, the ingoing subcritical wind does not decelerate to a stop because the powerful force of the black hole gravity drags it inexorably toward the horizon. This dynamic is unavoidable and as we showed in this section the flow must proceed without the perfect MHD constraint. This demonstrates the inequivalence of the outgoing and ingoing wind solution spaces. The wind can proceed as a deflagration wind that we analyze in the remainder of the chapter.

9.5 The Radiative Instability Near the Light Cylinder

One might wonder why the plasma cannot attain an arbitrarily large relativistic inertia by corotating with the field near the light cylinder and therefore be able to bend the field lines as needed (i.e., increase $|B^T|$) to allow passage to the black hole with mild dissipative effects from the plasma. Relativistic velocities are achieved only very close to the light cylinder (only $\sim 10^{-3}M$ away by 9.77). Thus, there are huge proper accelerations associated with approximate corotation as the light cylinder is approached (see 9.78), and large radiation losses. The radiative losses drain the plasma of the relativistic azimuthal momentum necessary to bend the field. In fact, this circumstance initiates an instability that accentuates the behavior near the light cylinder seen in the perfect MHD analysis of the last section (see Fig. 9.9 at the end of this section).

This instability is explored through a linear stationary point analysis that describes the linear leading edge. We perturb a time stationary axisymmetric wind away from a perfect MHD initial state with radiation resistance. If the resulting plasma state contains forces that tend to restore the perfect MHD state then the flow is stable to radiation losses. If the perturbation creates a plasma state with forces that tend to accentuate the deviation from perfect MHD, then the flow is unstable. We can explore variations from perfect MHD, through the proper electric field, $\mathbf{E}'$ (by definition). In particular, we are interested in $E^{2'}$. Since $E^{2'} = 0$ in the initial state, the proper electric field exists only as a perturbation,

$$\delta E^{2'} \equiv E^{2'} = \delta\left(F^{2\nu}u_\nu\right) = \delta F^{21}u_1 + \delta u_1 F^{21} + \delta u^\phi F^{2\phi} + u_\phi \delta F^{2\phi} - u^0 \delta F^{20} - \delta u^0 F^{20}\,. \tag{9.83}$$

We can find the various terms in (9.83) by perturbing the energy and angular momentum equations of the perfect MHD fluid. The energy and angular momentum equations are particularly simple in Boyer–Lindquist coordinates because the connection terms add to zero. From (3.39) we have,

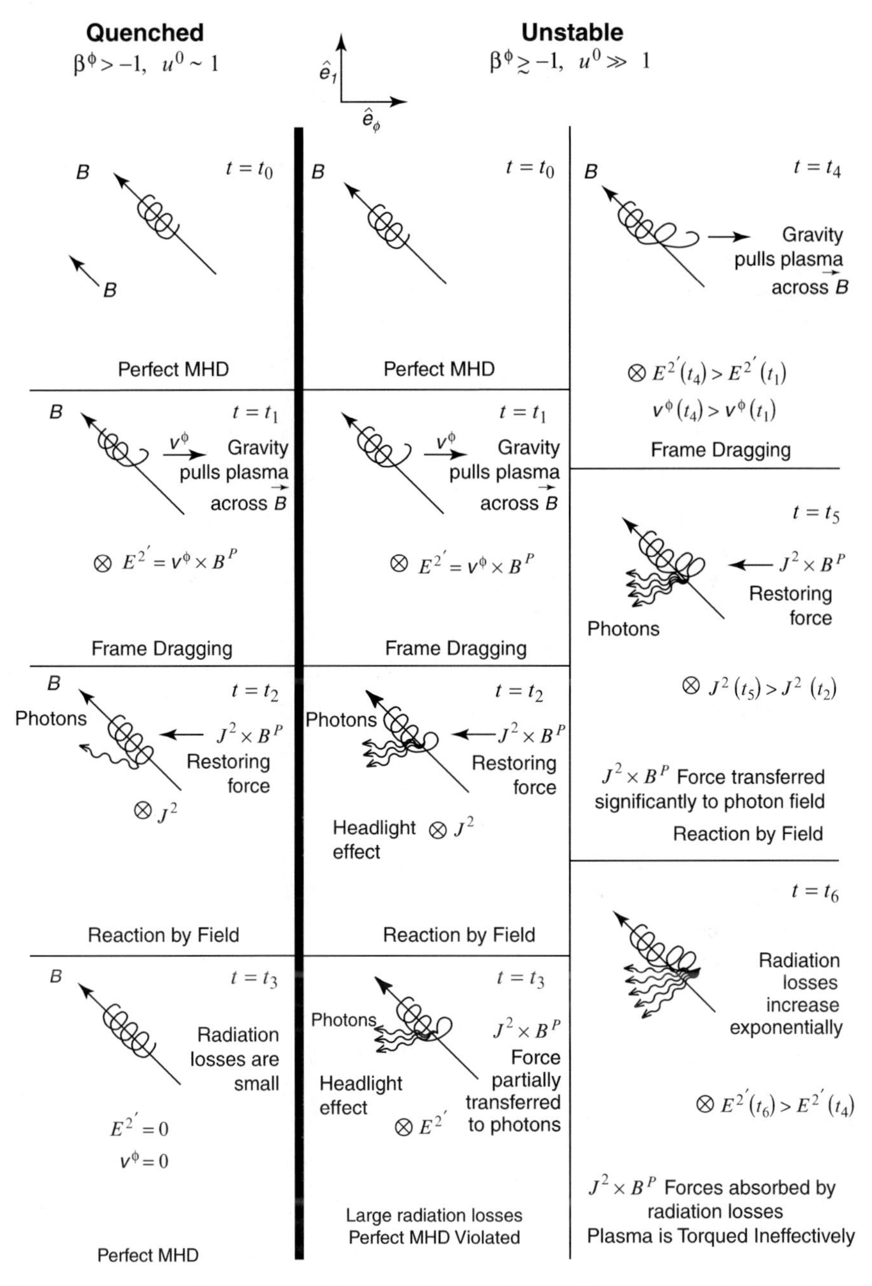

Fig. 9.9 This figure illustrates the differences between the dynamics of a quenched perturbation and an unstable perturbation

$$\frac{\mathrm{d}m}{\mathrm{d}\tau} = \frac{\tilde{F}_{\phi\alpha}J^{\alpha}}{nc} + \left(\tilde{T}_{\phi}{}^{\alpha}{}_{;\alpha}\right)_{r}, \tag{9.84a}$$

$$\frac{\mathrm{d}\omega}{\mathrm{d}\tau} = -\frac{\tilde{F}_{t\alpha}J^{\alpha}}{nc} + \left(\tilde{T}_{t}{}^{\alpha}{}_{;\alpha}\right)_{r}, \tag{9.84b}$$

$$\frac{\mathrm{d}}{\mathrm{d}\tau} = u^{\alpha}\frac{\partial}{\partial X^{\alpha}}. \tag{9.84c}$$

In the following calculations, the radiation reaction force, $\left(\tilde{T}_{\nu}{}^{\alpha}{}_{;\alpha}\right)_{r}$, exists only as a perturbation to the initial perfect MHD state.

9.5.1 The Initial Unperturbed State

The initial unperturbed state is obtained from (9.84), (3.3) and (3.8) as, $\omega = \omega_{0} + \delta\omega$ and $m = m_{0} + \delta m$,

$$\frac{\mathrm{d}m}{\mathrm{d}\tau} = -\frac{J^{2}B^{P}\sqrt{g_{\phi\phi}}}{nc}, \tag{9.85a}$$

$$\frac{\mathrm{d}\omega}{\mathrm{d}\tau} = -\frac{\Omega_{F}J^{2}B^{P}\sqrt{g_{\phi\phi}}}{nc^{2}}. \tag{9.85b}$$

The initial state, near the light cylinder, was found in the last section to be in relativistic azimuthal motion, $\beta^{\phi} \approx -1$. Thus, by (3.14) and (3.50),

$$\omega_{0} \approx \frac{\Omega_{\min}}{c}m_{0}. \tag{9.86a}$$

Since we are evaluating the instability near the inner light cylinder, $\Omega_{F} \approx \Omega_{\min}$, and (9.86a) becomes

$$\omega_{0} \approx \frac{\Omega_{F}}{c}m_{0}. \tag{9.86b}$$

Finally, the perfect MHD condition requires as in (5.2),

$$u_{0}^{2} = 0. \tag{9.87}$$

9.5.2 The Radiation Resistance Perturbation

The radiation resistance term is obtained from special relativity [87] with the covariant modifications to general relativity,

$$(T^{\mu\nu}{}_{;\nu})_{r} = \frac{2e^{2}}{3m_{e}c^{3}}\left[(P^{\mu}{}_{;\nu}u^{\nu})_{;\alpha}u^{\alpha} - \frac{P^{\mu}}{m_{e}^{2}c^{2}}\left(P^{\alpha}{}_{;\nu}u^{\nu}P_{\alpha;\delta}u^{\delta}\right)\right]. \tag{9.88}$$

Note the change in sign in the second term compared to [87]. This arises because the metric in that book is of signature (1, 3); ours is (3, 1). Normally, one would think that for small perturbations that the first term on the right hand side of (9.88) is dominant, since it is linear in the force and the second term is quadratic. But for ultrarelativistic momenta, the second term is comparable (in the range of Lorentz γ-factors that are relevant in this analysis) and for large γ-factors the second term actually becomes dominant. From the initial state (9.85) and noting that the connection terms cancel out in Boyer–Lindquist coordinates for the "ϕ" and "t" components,

$$\left(\tilde{P}_{\phi;\nu}u^{\nu}\right)_{;\alpha}u^{\alpha} = \frac{e}{c}\frac{\mathrm{d}}{\mathrm{d}\tau}\left[\sqrt{g_{\phi\phi}}\,B^{P}u_{D}^{2}\right] , \tag{9.89a}$$

$$\left(\tilde{P}_{t;\nu}u^{\nu}\right)_{;\alpha}u^{\alpha} = -\left(\frac{e}{c}\right)\left(\frac{\Omega_{F}}{c}\right)\frac{\mathrm{d}}{\mathrm{d}\tau}\left[\sqrt{g_{\phi\phi}}\,B^{P}u_{D}^{2}\right] , \tag{9.89b}$$

where we have introduced the drift velocity between the two species,

$$u_{D}^{\mu} = \frac{n_{+}u_{+}^{\mu} - n_{-}u_{-}^{\mu}}{n_{+} + n_{-}} . \tag{9.90}$$

The inertial effects of the plasma in perfect MHD are manifested by the cross-field drift between the species in the $\hat{\boldsymbol{e}}_{2}$ direction.

Combining (9.89), (9.88) and (9.86b), we have the simplifying relation,

$$\left(\tilde{T}_{t}{}^{\nu}{}_{;\nu}\right)_{r} = \frac{\Omega_{F}}{c}\left(\tilde{T}_{\phi}{}^{\nu}{}_{;\nu}\right)_{r} . \tag{9.91}$$

9.5.3 The Perturbed Four Velocity

Perturbing the energy and angular momentum equations (9.84), using (9.86b) and (9.91), we have

$$\delta\omega = \frac{\Omega_{F}}{c}\delta m . \tag{9.92}$$

This result simplifies the perturbed quantities to follow in this section.

We take the adiabatic approximation in this linear leading edge analysis (the heat flow only shows up as a perturbed quantity). Note that $n\mu = \rho + P$ and adiabatically one has the relation,

$$\frac{\mathrm{d}\rho}{\mathrm{d}\tau} = \mu\frac{\mathrm{d}n}{\mathrm{d}\tau} . \tag{9.93}$$

The sound three speed is defined from

$$\frac{\delta\mu}{\mu} = c_{S}^{2}\frac{\delta n}{n} . \tag{9.94}$$

By the mass conservation law (3.36b),

$$\frac{\delta n}{n} = -\frac{\delta u^1}{u^1} \,. \tag{9.95}$$

Combining (9.95a) and (9.94) one finds that

$$\delta u^1 = -\frac{u^1}{c_S^2}\frac{\delta\mu}{\mu} \,. \tag{9.96}$$

Noting that $u \cdot u = -1$, (9.87) implies

$$-u^0\delta u^0 + u^1\delta u^1 + u^\phi \delta u^\phi = 0 \,. \tag{9.97}$$

Perturbing (3.14) and using (9.92),

$$\delta u^0 = \frac{\beta_F^\phi \delta m}{\sqrt{g_{\phi\phi}}\,\mu} - u^0\frac{\delta\mu}{\mu} \,, \tag{9.98a}$$

$$\delta u^\phi = \frac{\delta m}{\sqrt{g_{\phi\phi}}\,\mu} - u^\phi\frac{\delta\mu}{\mu} \,. \tag{9.98b}$$

Combining (9.95)–(9.98), we find the perturbed specific enthalpy,

$$\delta\mu = \left[\frac{\left(u^\phi - \beta_F^\phi u^0\right) c_S^2 \delta m}{\left(u^1\right)^2 + c_S^2\left(u_\phi^2 - u_0^2\right)}\right] \,. \tag{9.99a}$$

Inserting (9.99a) into (9.96) and (9.98), one can compute the perturbed four-velocity,

$$\delta u^1 = \frac{u^1\left(u^0\beta_F^\phi - u^\phi\right)\delta m}{\mu\sqrt{g_{\phi\phi}}\left[\left(u^1\right)^2\left(1-c_S^2\right) - c_S^2\right]} \,, \tag{9.99b}$$

$$\delta u^0 = \left[\frac{\delta m}{\mu\sqrt{g_{\phi\phi}}}\right]\left\{\beta_F^\phi + \frac{u^0 c_S^2\left(u^0\beta_F^\phi - u^\phi\right)}{\left[\left(u^1\right)^2\left(1-c_S^2\right) - c_S^2\right]}\right\} \,, \tag{9.99c}$$

$$\delta u^\phi = \left[\frac{\delta m}{\mu\sqrt{g_{\phi\phi}}}\right]\left\{1 + \frac{u^\phi c_S^2\left(u^0\beta_F^\phi - u^\phi\right)}{\left[\left(u^1\right)^2\left(1-c_S^2\right) - c_S^2\right]}\right\} \,. \tag{9.99d}$$

9.5.4 The Perturbed Field Strengths

In order to find the perturbed field strengths in (9.83), we define the following components of the vector potential, A^μ,

$$\mathcal{E} = -A \cdot \frac{\partial}{\partial t}\,, \tag{9.100a}$$

$$\mathcal{L} = A \cdot \frac{\partial}{\partial \phi}\,. \tag{9.100b}$$

The components of the electromagnetic vector potential in the ZAMO frame can be expressed in terms of these quantities as

$$A^{\phi} = \sqrt{g_{\phi\phi}}\,\mathcal{L}\,, \tag{9.101a}$$

$$A^{0} = \alpha^{-1}\left(\mathcal{E} - \frac{\Omega}{c}\mathcal{L}\right). \tag{9.101b}$$

ZAMO evaluated fields can be derived from $\mathcal{E}$ and $\mathcal{L}$ as follows:

$$\frac{\partial \mathcal{L}}{\partial X^1} = \frac{\partial}{\partial X^1}\left[\sqrt{g_{\phi\phi}}A^{\phi}\right] = \sqrt{g_{\phi\phi}}\left[\frac{\partial A^{\phi}}{\partial X^1} - \Gamma^{\phi}{}_{1\phi}A^{\phi}\right], \tag{9.102a}$$

where we used (3.32c) to define the connection. We can write (9.102a) as

$$\frac{\partial \mathcal{L}}{\partial X^1} = \sqrt{g_{\phi\phi}}\left[A_{\phi;1} - A_{1;\phi}\right] = \sqrt{g_{\phi\phi}}\,F^{1\phi}\,. \tag{9.102b}$$

Other expressions can be derived in a similar manner,

$$F^{2\phi} = \sqrt{g_{\phi\phi}}\,\frac{\partial \mathcal{L}}{\partial X^2}\,, \tag{9.102c}$$

$$F^{20} = \alpha^{-1}\left[\frac{\partial \mathcal{E}}{\partial X^2} - \frac{\Omega}{c}\frac{\partial \mathcal{L}}{\partial X^2}\right], \tag{9.102d}$$

$$F^{10} = \alpha^{-1}\left[\frac{\partial \mathcal{E}}{\partial X^1} - \frac{\Omega}{c}\frac{\partial \mathcal{L}}{\partial X^1}\right] \tag{9.102e}$$

The other electric field component vanishes:

$$F_{\phi 0} = \frac{\partial A_{\phi}}{\partial X^0} - \frac{\partial A_0}{\partial X^{\phi}} + \Gamma^{\alpha}{}_{\phi 0}A_{\alpha} - \Gamma^{\alpha}{}_{0\phi}A_{\alpha} = 0\,, \tag{9.103a}$$

by time stationary and axisymmetry combined with the connection symmetries,

$$\Gamma^{2}{}_{\phi 0} = \Gamma^{2}{}_{0\phi}\,, \tag{9.103b}$$

$$\Gamma^{1}{}_{\phi 0} = \Gamma^{1}{}_{0\phi}\,, \tag{9.103c}$$

and for all other values of μ, $\Gamma^{\mu}{}_{0\phi} = 0$.

The frozen-in equation (5.13c) applied to (9.102d) yields

$$\frac{\partial \mathcal{E}}{\partial X^2} = \frac{\Omega_F}{c}\frac{\partial \mathcal{L}}{\partial X^2}\,. \tag{9.104}$$

The perfect MHD condition expressed as the vanishing of the proper electric field in the "ϕ" and "0" direction of the ZAMO basis become

$$\frac{\mathrm{d}\mathcal{L}}{\mathrm{d}\tau} = 0\,, \tag{9.105a}$$

$$\frac{\mathrm{d}\mathcal{E}}{\mathrm{d}\tau} = 0\,. \tag{9.105b}$$

Consider the variation of $\mathcal{E}$ as a frozen-in piece $\mathcal{E}_0$ and a perturbation $\delta\mathcal{E}$ and similarly for $\mathcal{L}$:

$$\mathcal{E} = \mathcal{E}_0 + \delta\mathcal{E} \quad , \quad \mathcal{L} = \mathcal{L}_0 + \delta\mathcal{L}\,. \tag{9.106}$$

From the definition of Ω_F in a dissipative plasma, (9.1), it follows from (9.104) that

$$\frac{\partial \delta\mathcal{E}}{\partial X^2} = \frac{\Omega_F}{c}\delta\mathcal{L} \tag{9.107}$$

Combining (9.107) and (9.102d), we have

$$-\delta F^{20}u^0 + \delta F^{2\phi}u^\phi = u^\phi\left(\beta^\phi - \beta^\phi_F\right)\delta B^P \approx 0\,. \tag{9.108}$$

The approximate equality in (9.108) results from the fact that $\beta^\phi \approx \beta^\phi_F \approx -1$ in the initial state as was found in the last section. Furthermore, we expect δB^P to be small for an inertially light plasma, $U_A^2 \gg 1$.

We can find the perturbed toroidal magnetic field by varying the angular momentum conservation law (5.21),

$$\delta F^{12} = \frac{k\delta m}{c\alpha\sqrt{g_{\phi\phi}}} + \frac{\delta(k\ell)_{ph}}{c\alpha\sqrt{g_{\phi\phi}}}\,, \tag{9.109}$$

where $\delta(k\ell)_{ph}$ is the flux of angular momentum that is radiated into the photon field. The plasma has $\beta^\phi \approx -1$ and $u^0 \gg 1$, so the radiated photons are beamed onto negative angular momentum trajectories. These trajectories are of negative energy globally, and always approach the horizon, thus

$$\delta(k\ell)_{ph} > 0\,. \tag{9.110}$$

9.5.5 The Perturbed Proper Electric Field

Inserting (9.99), (9.108) and (9.109) into the expression for $E^{2'}$ in (9.83), we find with the aid of (5.35),

$$\delta E^{2'} = -\frac{n\delta m}{B^P\sqrt{g_{\phi\phi}}} \frac{u_1^4 - u_1^2\left(U_S^2 + \frac{F^{\mu\nu}F_{\mu\nu}}{8\pi n\mu}\right) + \frac{\left[1-\left(\beta_F^\phi\right)^2\right]\left(B^P\right)^2}{4\pi n\mu} U_S^2}{c^2\left[u_1^2 - U_S^2\right]} - \frac{u_1\,\delta(k\ell)_{ph}}{c^2\alpha\sqrt{g_{\phi\phi}}} \,. \tag{9.111a}$$

This expression reduces with the aid of the wind magneto-acoustic critical speeds defined in (5.38) to

$$\delta E^{2'} = -\frac{n\delta m}{B^P\sqrt{g_{\phi\phi}}} \frac{\left[u_1^2 - U_F^2\right]\left[u_1^2 - U_{SL}^2\right]}{c^2\left[u_1^2 - U_S^2\right]} - \frac{u_1\,\delta(k\ell)_{ph}}{c^2\alpha\sqrt{g_{\phi\phi}}} \,. \tag{9.111b}$$

Physically, we can understand the direction of $\delta E^{2'}$ through the following discussion. As the plasma approaching the light cylinder is torqued by cross-field inertial currents, it radiates an outward directed angular momentum flux (see 9.110). For the inward directed mass flux this loss of angular momentum means through (5.21) that km is too small for the plasma to stay frozen-in, thus (remember $k < 0$)

$$\delta m > 0 \,. \tag{9.112}$$

Near the light cylinder, $u^0/c \gg 1$, thus we have a supersonic flow ahead of the fast critical surface,

$$U_F > u_1 > U_S > U_{SL} \,. \tag{9.113}$$

Combining this with (9.112), the first term on the right hand side of (9.111b) is positive ahead of the inner fast critical surface. The second term on the right hand side of (9.111b) is positive as well, but it is on the order of $(u_1/U_F)^2$ of the first term, so is negligible until the fast point is approached. In any event, (9.110), (9.112) and (9.113) yield the direction of $E^{2'}$ (see Fig. 9.9 at the end of this section),

$$\delta E^{2'} > 0 \,. \tag{9.114}$$

This proper electric field drives a current across the magnetic field lines, increasing J^2 in the ingoing wind. Physically, the sign of $\delta E^{2'}$ is clear. As plasma radiates away negative angular momentum (toward the asymptotic infinity of the event horizon), it has too much angular momentum to stay frozen-in and keep $E^{2'} = 0$ (in terms of the torsional "tug of war," the dragging of inertial frames pulls the plasma in the forward azimuthal direction off of the magnetic field lines). The electric field, $\delta E^{2'}$, develops in order to drive the charges back into their frozen-in state by inducing accelerations to make a cross-field particle drift u_D^2 as in (9.90), which in turn creates a $q\left(v^2\hat{\mathbf{e}}_2 \times \mathbf{B^P}\right)$ force (the torque) that is negative for both species of charge (in terms of the torsional "tug of war" this effect is the field trying to counteract the tug induced by the dragging of inertial frames).

9.5.6 Stationary Point Analysis

We can use (9.111b) to analyze the stability of the wind to radiative losses as it approaches the inner light cylinder. The relevant effect is seen in second order. The the radiation reaction induced perturbation from perfect MHD requires $\delta E^{2'} > 0$ by (9.114). The question is whether $\delta E^{2'} > 0$ alters the plasma state so that the second order proper electric field, $\left(\delta E^{2'}\right)_{(2)}$, is larger or smaller than $\delta E^{2'}$ (we will denote second order perturbed quantities by the subscript "(2)"). If $\delta E^{2'}$ alters the plasma so that the proper electric field decreases, the perturbation is self quenching,

$$\delta E^{2'} > \left(\delta E^{2'}\right)_{(2)} \quad : \quad \text{quenched} \,. \tag{9.115a}$$

If $\delta E^{2'}$ increases in second order, the plasma flow is unstable to radiation losses,

$$\delta E^{2'} < \left(\delta E^{2'}\right)_{(2)} \quad : \quad \text{unstable} \,. \tag{9.115b}$$

9.5.6.1 Quenched Perturbations

Consider the case ahead of the fast point,

$$U_{S}^2 < u_{1}^2 \ll U_{F}^2 \,. \tag{9.116}$$

By (9.114) the perturbed proper electric field makes a perturbed current,

$$\delta J^2 > 0 \,. \tag{9.117}$$

We are interested in the effect of δJ^2 on the perturbed plasma state in order to find the second order perturbation. By the expression (9.88) for the radiation resistance, the perturbed current produces stronger electromagnetic forces than radiation reaction forces,

$$\left| \frac{\tilde{F}_{\phi\alpha} J^{\alpha}}{nc} \right| \gg \left| \left(T^{\alpha}{}_{\phi;\alpha} \right)_{,r} \right| \,. \tag{9.118}$$

Inserting (9.118) into the azimuthal momentum equation (9.84a), implies that the primary effect of the perturbed cross-field current is to torque the plasma, thus

$$(\delta m)_{(2)} < \delta m \,. \tag{9.119}$$

Since $(u_{1}/U_{F})^2 \ll 1$ by assumption in (9.116), we have from (9.119) and (9.111b),

$$\left(\delta E^{2'}\right)_{(2)} < \delta E^{2'} \,. \tag{9.120}$$

Thus, condition (9.116) yields a quenched perturbation.

9.5.6.2 Unstable Perturbations

Consider the case where the flow is approaching the fast point in the azimuthally symmetric flux tube,

$$\left|u_1^2 - U_F^2\right| < u_1^2 \left|\frac{\delta(k\ell)_{ph}}{k\delta m}\right| . \tag{9.121}$$

Note in (9.109) that

$$\delta F^{12} \geq 0 . \tag{9.122a}$$

This follows from two facts. First, we know from Sect. 9.4 that $\left|B^T\right|$ is too small to allow smooth passage through the environs of the light cylinder and $B^T < 0$. Thus, F^{12} is too positive for perfect MHD to hold. Secondly, $\delta E^{2'}$ drives a current that switches off $\left|B^T\right|$, so

$$\left(\delta F^{12}\right)_{(2)} > \delta F^{12} \geq 0 . \tag{9.122b}$$

From (9.109) and (9.122), we have

$$\delta(k\ell)_{ph} \geq -k\delta m . \tag{9.123}$$

Combining (9.123) and our assumed four velocity (9.121), the dominant term in (9.111b) is now the second term and this is the distinction from the quenched perturbation discussed above.

As before, the cross-field current produces second order forces dominated by electrodynamics as in (9.118), and the primary effect in the azimuthal momentum equation, (9.84a), is to torque the plasma. Thus, again we have (9.119), which we transcribed below,

$$(\delta m)_{(2)} < \delta m . \tag{9.124}$$

However, since the plasma is torqued by a larger J^2 in second order, the acceleration and Lorentz factor, $u^0 \approx -u^\phi$, are larger as discussed in Sect. 9.4,

$$\left|a^\phi + \delta a^\phi\right| > \left|a^\phi\right| , \tag{9.125a}$$

$$\left|u^0 + \delta u^0\right| > \left|u^0\right| . \tag{9.125b}$$

In second order, (9.125) applied to (9.88) implies that

$$\left[\delta(k\ell)_{ph}\right]_{(2)} > \delta(k\ell)_{ph} \tag{9.126}$$

By (9.121), the second term is dominant in (9.111b), thus (9.126) yields

$$\left(\delta E^{2'}\right)_{(2)} > \delta E^{2'} . \tag{9.127}$$

Equations (9.127) and (9.120) show that the ingoing perfect MHD wind is stable to radiation losses until it approaches the fast point. Unstable growth of $E^{2'}$ occurs just before the fast point. Since $\delta F^{12} \approx 0$, $\delta(k\ell)_{ph} \approx -k\delta m$ by (9.109), until the instability initiates. Using (9.64), (9.72f), (9.73c) and (9.51), we can approximate the instability condition in (9.121) to find the location of the linear leading edge of the plasma instability (see Fig. 9.6),

$$\beta_F^\phi \approx -1.0051 \; , \quad Z^2 = 10^{-2} \; , \tag{9.128a}$$

$$\beta_F^\phi \approx -1.00021 \; , \quad Z^2 = 10^{-3} \; . \tag{9.128b}$$

Physically, the instability can be described in terms of the torsional "tug of war." Near the fast point, the dragging of inertial frames drags plasma across the magnetic field, azimuthally. An $E^{2'}$ is generated and the field tries to pull the plasma back onto the field lines with a $J^2\hat{\boldsymbol{e}}_2 \times \boldsymbol{B^P}$ force (see Fig. 9.9). However, unlike the situation upstream, this "tug" by the magnetic field is translated largely into radiation losses, making the $J^2\hat{\boldsymbol{e}}_2 \times \boldsymbol{B^P}$ force only marginally effective at torquing the plasma. Thus, perfect MHD cannot be restored and $E^{2'}$ grows as the flow propagates downstream, creating more $J^2\hat{\boldsymbol{e}}_2 \times \boldsymbol{B^P}$ force and more radiation losses and so on.

The microphysics of the radiative instability that distinguishes the quenched perturbations from the unstable perturbations is delineated in Fig. 9.9. The four frames on the left hand column indicate the microphysics upstream of S_+ in Fig. 9.8 during the radiation process. The flow is stable in this domain to radiation losses. This is contrasted to the seven frames on the right hand side of Fig. 9.9 that depict the microphysics of the radiative instability below S_+. This figure summarizes the details of the discussion of this section.

The time evolution of the plasma is defined by infinitesimal time changes, $t_0 < \cdots < t_i < t_{i+1} < \cdots$. At $t = t_0$, a perfect MHD state exists with plasma on its gyro-orbits. As the plasma flows closer to the hole, frame dragging instantaneously perturbs the MHD state by dragging plasma across the magnetic field in the $+\phi$ direction at $t = t_1$, thus creating a proper electric field, $E^{2'}$. This $E^{2'}$ drives a cross-field current, J^2, and the field reacts to the frame dragging force in the torsional tug of war by $J^2\hat{\boldsymbol{e}}_2 \times \boldsymbol{B^P}$ forces at time $t = t_2$. Simultaneously, this electrodynamically torqued plasma (which is accelerated in the $-\phi$ direction) radiates. Above, S_+, the radiative losses are small and the $J^2\hat{\boldsymbol{e}}_2 \times \boldsymbol{B^P}$ force is very effective at torquing the plasma back onto the gyro-orbits that thread the magnetic field lines at $t = t_3$, as shown at the bottom left of the figure.

Below S_+, the plasma state is defined by $\beta^\phi \gtrsim -1$ and $u^0 \gg 1$, thus the radiation losses at $t = t_2$ are extremely large and are beamed by the headlight effect into the $-\phi$ direction. The loss of angular momentum to the photon field reduces the efficiency of the $J^2\hat{\boldsymbol{e}}_2 \times \boldsymbol{B^P}$ forces to torque plasma back onto the magnetic field lines (i.e., some of the $J^2\hat{\boldsymbol{e}}_2 \times \boldsymbol{B^P}$ force is dissipated in the photon field) at $t = t_3$. Consequently, perfect MHD is not restored and $E^{2'} > 0$. Furthermore, at $t = t_4$, as the plasma flows down the field lines, frame dragging forces continue to pull plasma across the magnetic field lines. Thus, both $E^{2'}$ and the azimuthal velocity relative to the frozen-in frame, v^ϕ, are larger than in the initial perturbation at $t = t_1$. The larger

$E^{2'}$ means a larger J^2 at $t = t_5$ and therefore a larger $J^2\hat{\boldsymbol{e}}_2 \times \boldsymbol{B^P}$ force. However, as u^0 keeps increasing, the radiation losses increase and more of the electrodynamic torque is transferred to the photon field than was the case at $t = t_2$. Consequently, the $J^2\hat{\boldsymbol{e}}_2 \times \boldsymbol{B^P}$ force at $t = t_6$ is less efficient at torquing the plasma back onto the field lines than it was at $t = t_3$. The plasma continues to be pulled farther and farther across the field lines in the $+\phi$ direction due to frame dragging effects as the wind propagates toward the hole (gravity is winning the torsional tug of war).

9.6 The Dynamo Region

Once the radiative instability is initiated, the plasma can flow more freely across the magnetic field lines and the proper electric field, $E^{2'} > 0$, can drive much stronger cross-field currents than occur in the perfect MHD region of the wind. This region includes the dynamo for the toroidal magnetic field in the wind. The dynamo is bounded from above by the location of the initiation of the instability given (9.128) and it is bounded from below by the surface on which $B^T = 0$. By causality, B^T must be created somewhere within the wind. We call the $B^T = 0$ surface the anchor for the magnetic wind. Furthermore, the anchor must occur outside the fast magnetosonic surface for a causal global structure to be attained. By (9.5), the flow will immediately go superfast once the radiative instability (a dissipative instability) initiates. The dynamo region is therefore a thin current sheet to first approximation.

9.6.1 Resistivity and the Saturation of the Instability

The cross-field current instability will grow rapidly until a source of resistivity develops that can counteract the driving forces that create the cross-field particle drift, u^2_D. One can hypothesize radiation reaction [137], Compton drag or plasma wave scattering as the source of Ohmic resistance. In the thin dynamo layer Compton drag is not viable based on geometrical considerations. The plasma at this point of the flow and downstream (as we shall see below) radiates photons on negative energy trajectories that necessarily approach the horizon. Even though there is an intense radiation field from the dissipative plasma, it is directed inward away from the dynamo current layer. Thus, Compton drag is a negligible force in the momentum equation for the individual species in the dynamo.

Radiation drag is only comparable to electrodynamic forces in the wind if [138]

$$u^2_D \sim 10^7 \left(\beta^P\right)^{-9/8} u^1 . \tag{9.129}$$

In a charge separated plasma, radiation drag could damp the instability; however in a two fluid plasma, a two stream instability will be incurred long before the cross-field plasma drift attains velocities anywhere close to those required in (9.129).

Expressing the cross-field current in terms of u_D^2 $(J^2 = neu_D^2)$, in the angular momentum equation, (9.85a),

$$u_D^2 = \frac{u^P}{\Omega_L}\frac{\partial u^\phi}{\partial X^1} = \frac{a^\phi}{\Omega_L}\,, \tag{9.130}$$

in a perfect MHD plasma, where Ω_L is the Larmour frequency,

$$\Omega_L = \frac{eB^P}{m_e c}\,. \tag{9.131}$$

Equation (9.130) can be used to show that strong particle drifts occur in the cross-field direction between the Alfvén and fast points. From (9.78) inserted into (9.130), we can find the cross-field drift ahead of the fast point in the ZAMO frames,

$$u_D^2 \approx 7.1 \times 10^2 \mathrm{cm/sec}\,, \quad Z^2 = 10^{-2}\,, \tag{9.132a}$$

$$u_D^2 \approx 1.7 \times 10^4 \mathrm{cm/sec}\,, \quad Z^2 = 10^{-3}\,. \tag{9.132b}$$

We are interested in the cross-field drift in the proper frame of the plasma, $(\beta_D^2)_{proper}$, in order to understand the two stream instability. From time dilation effects, we can relate the drifts in the proper and ZAMO frames:

$$(\beta_D^2)_{proper} = \frac{u^0}{c}\beta_D^2 \approx \left[\frac{u_D^2}{c}\right]\,. \tag{9.133a}$$

Thus, (9.132) represents the proper frame cross-field drifts ahead of the dynamo region. In the dynamo region, J^2 is enhanced so we expect u_D^2 to be much larger than the values in (9.132). Hence, we expect a two stream instability to occur in the $\hat{\boldsymbol{e}}_2$ direction [77].

The two stream instability grows until there is enough charge separation in the electrostatic waves to produce an electric potential which is large enough to impede the kinetic energy of the streams which flow across the magnetic field lines. When this particle trapping occurs, bulk kinetic energy of the streams is converted into thermal energy in the potential well. This turbulently heated plasma will synchrotron radiate. In steady state, the momentum which is lost to the photon field through synchrotron radiation will approximately balance the electromagnetic dynamo forces in the $\hat{\boldsymbol{e}}_2$ direction for each species, separately (i.e., $\partial J^2/\partial t = 0$). Equivalently, the momentum equation in the $\hat{\boldsymbol{e}}_2$ direction for the two species can be subtracted to form an Ohm's law in which this balance will be true as well. This can be modeled by a contribution to $\nabla \cdot T_r$ which represents the distribution of four momentum in the synchrotron radiation field. However, one can only get this term by modeling the streaming instability to obtain the plasma temperature. This source of dissipation provides the resistivity in a complicated Ohm's law.

Another contribution to $\nabla \cdot T_r$ is inverse Compton scattering. This will be important if the photon energy density is on the order of $\left(B^P\right)^2$. Until the photon emission from turbulent heating is calculated numerically, we do not know the energy density.

Normally, resistivity is understood in terms of collisions. In this excited state of the plasma, plasma wave scattering provides the collisions. In a normal two stream instability analysis these collisions occur between the electrostatic or Langmuir modes. In this analysis, one would want to model the collision using electrostatic waves that propagate across the magnetic field, the Bernstein modes [77].

9.6.2 The Anchor Point

We consider the downstream side of the dynamo for the paired wind system, the anchor point defined by the condition $B^T = 0$. We can crudely estimate the exit parameters as the flow passes through the dynamo region of the flux tube. Since we do not model the two stream instability and radiative instability of Sect. 9.5 due to the very complicated plasma physics involved, we use the geometrically thin aspect of the dynamo current layer to make approximations. Firstly, since the layer is thin, we assume that the energy of the plasma obtained from the torsional tug of war is contained primarily in the specific enthalpy, i.e., the transit time through the layer is so short that the plasma does not have time to synchrotron radiate away its inertia by the time it reaches the anchor (as in the slow shock atop the ergospheric disk in Chap. 8). Thus, even though we must drop all of our perfect MHD relations, we still have our conservation equations (5.20), (5.21) and (5.50b).

At the anchor, all of the energy flux is in mechanical form by the $B^T = 0$ condition,

$$\left(\mu u^0\right)_{anc} = \frac{\left[e - \dfrac{\Omega \ell}{c}\right]_{anc}}{c\alpha_{anc}}, \tag{9.134}$$

where the subscript “anc” means to evaluate at the anchor.

From (5.20) and (5.21), if all of the energy is in mechanical form as in (9.134),

$$1 - \left(\beta^\phi\right)_{anc}\left(\beta^\phi_F\right)_{anc} = \frac{e - \dfrac{\Omega_F \ell}{c}}{e - \dfrac{\Omega \ell}{c}}, \tag{9.135}$$

where Ω_F is the perfect MHD value in the wind and β^ϕ_F is computed as in (5.14) from this perfect MHD value at the location of the anchor. We note this clarification because the anchor is not a perfect MHD region and we still have an actual value of Ω_F at the anchor as defined in (9.1) and (9.2). The nonMHD (actual) value of Ω_F at the anchor differs from Ω_F in the wind (slightly) and we designate its value by $\bar{\Omega}_F$. Similarly, we compute the field line azimuthal velocity, $\bar{\beta}^\phi_F$, from $\bar{\Omega}_F$ using (5.14).

We have at the anchor,

$$\beta_F^\phi < \bar{\beta}_F^\phi < \beta^\phi \, . \qquad (9.136)$$

From the initial conditions (9.23), (9.24) and (9.26) on $I\_$ inserted in (9.135),

$$1 - \left(\beta^\phi\right)_{anc} \left(\beta_F^\phi\right)_{anc} = 5.61 \times 10^{-5} \qquad (9.137)$$

For a thin dynamo layer, we use (9.128) for β_F^ϕ at the anchor to find

$$\left(\beta^\phi\right)_{anc} = 0.9949 \, , \quad Z^2 = 10^{-2} \, , \qquad (9.138a)$$

$$\left(\beta^\phi\right)_{anc} = 0.9997 \, , \quad Z^2 = 10^{-3} \, . \qquad (9.138b)$$

The angular momentum conservation law in a magnetic flux tube, (9.29a), implies that at the anchor, one has

$$k\_m_{anc} \approx \frac{\Omega_F \Phi}{4\pi k_F} \, . \qquad (9.139)$$

For $\beta^\phi \approx -1$, this reduces to an expression for the Lorentz factor in the ZAMO frames,

$$\left(\frac{u^0}{c}\right)_{anc} \approx -\frac{\alpha \bar{Z} B^P}{4\pi k\_\mu c} \, , \qquad (9.140a)$$

$$\left(\frac{u^0}{c}\right)_{anc} \approx 5.93 \times 10^4 \left[\frac{m_e c^2}{\mu c^2}\right] \, , \quad Z^2 = 10^{-2} \, , \qquad (9.140b)$$

$$\left(\frac{u^0}{c}\right)_{anc} \approx 1.76 \times 10^4 \left[\frac{m_e c^2}{\mu c^2}\right] \, , \quad Z^2 = 10^{-3} \, . \qquad (9.140c)$$

Using the condition $u \cdot u = -1$, (9.140) and (9.138) yield

$$\left(\beta^P\right)_{anc} \approx -0.101 \, , \text{ if } \left[\frac{(\mu c^2)_{anc}}{m_e c^2}\right]^2 \ll 3.6 \times 10^7 \, , Z^2 = 10^{-2} \, , \qquad (9.141a)$$

$$\left(\beta^P\right)_{anc} \approx -0.023 \, , \text{ if } \left[\frac{(\mu c^2)_{anc}}{m_e c^2}\right]^2 \ll 1.7 \times 10^4 \, , Z^2 = 10^{-3} \, . \qquad (9.141b)$$

The constraints on the specific enthalpy at the anchor in (9.141) equate to the condition $1 - \left(\beta^\phi\right)^2 \gg \left(u^0\right)^{-2}$ or $\left(\beta^P\right)^2 \approx 1 - \left(\beta^\phi\right)^2$. When $Z^2 = 10^{-2}$ this is clearly a good approximation, for $Z^2 = 10^{-3}$ it is less certain. Combining (9.140) and (9.141),

$$\left(\frac{u^P}{c}\right)_{anc} = 5.99 \times 10^3 \left[\frac{m_e c^2}{\mu c^2}\right] , \quad \left[\frac{\mu}{m_e}\right]^2_{anc} \ll 3.6 \times 10^7 \, , \; Z^2 = 10^{-2} \, , \tag{9.142a}$$

$$\left(\frac{u^P}{c}\right)_{anc} = 4.11 \times 10^2 \left[\frac{m_e c^2}{\mu c^2}\right] , \quad \left[\frac{\mu}{m_e}\right]^2_{anc} \ll 1.7 \times 10^4 \, , \; Z^2 = 10^{-3} \, . \tag{9.142b}$$

Expressing U_A as in (9.67) and using (9.142),

$$(U_A)_{anc} \approx 6.08 \times 10^4 \left[\frac{m_e c^2}{\mu c^2}\right] , \quad \left[\frac{\mu}{m_e}\right]^2_{anc} \ll 3.6 \times 10^7 \, , \; Z^2 = 10^{-2} \, , \tag{9.143a}$$

$$(U_A)_{anc} \approx 1.59 \times 10^4 \left[\frac{m_e c^2}{\mu c^2}\right] , \quad \left[\frac{\mu}{m_e}\right]^2_{anc} \ll 1.7 \times 10^4 \, , \; Z^2 = 10^{-3} \, . \tag{9.143b}$$

Thus, comparing (9.143) and (9.140),

$$\left(\frac{u^0}{c}\right)_{anc} \approx (U_A)_{anc} \, . \tag{9.144}$$

Equation (9.144) shows that the ingoing wind experiences a transition from being magnetically dominated (see 9.53c and 9.59) above the dynamo region to being inertially dominated below the dynamo. The torsional tug of war is fought primarily in this dynamo layer and by the time that the flow reaches the anchor point, gravity has won the battle.

We would also like to quantify some parameters in the anchor that typify the dissipative process such as $\bar{\beta}^\phi_F$. From (9.3) we find an expression for $\bar{\beta}^\phi_F$:

$$\left(E^{2'}\right)_{anc} = \left(\frac{u^0}{c}\right)_{anc} \left[(\beta^\phi)_{anc} - (\bar{\beta}^\phi_F)_{anc}\right] B^P \, . \tag{9.145}$$

Consider a small azimuthally symmetric volume, V, ahead of the anchoring point, where the switch-off current J^2 flows. The volume V is the dynamo. The dynamo is bounded from below by a cross-sectional surface in the flux tube, S_{anc}, and from above by a surface, S_+ (see Fig. 9.8). In this region, energy conservation in the ZAMO frames yields

$$\int J^2 F^{02} \mathrm{d}V \approx \int \frac{\beta^\phi_F F^{12} B^P}{4\pi} \mathrm{d}S_+ \, . \tag{9.146}$$

Approximate equality in (9.146) is achieved as a consequence of the fact that above the surface, S_+, the wind in the flux tube carries relatively small cross-field current densities (see 9.72ab and 9.73a).

In the volume through which the switch-off current flows, the plasma is resistively heated as discussed in Sect. 9.6.1. This is primarily a consequence of cross-field dissipative currents. In the ZAMO frame, this circumstance is represented by

$$\int J^2 E^{2'} \mathrm{d}V \approx \int \frac{\mathrm{d}}{\mathrm{d}X^0} \left[n u^0 \mu c\right] \mathrm{d}V \, . \tag{9.147}$$

Considering that there is far more thermal inertia at S_{anc} than at S_+, for a time stationary flow, the integral form of the energy conservation law applied to (9.147) yields

$$\int J^2 E^{2'} \mathrm{d}V \approx -\int \left[n u^P \mu c \right]_{anc} \mathrm{d}S_{anc} \,. \tag{9.148}$$

For a narrow flux tube, we can crudely approximate the flow as homogeneous in the dynamo. Then using (9.1) and $\bar{\beta}_F^\phi \approx -1$, we have

$$\frac{\int J^2 E^2 \mathrm{d}V}{\int J^2 E^{2'} \mathrm{d}V} \approx \frac{B^P}{E^{2'}} \,. \tag{9.149}$$

Combining (9.145), (9.146), (9.148), (9.149) and (9.140a) one obtains the relation

$$\left(\beta^\phi\right)_{anc} - \left(\bar{\beta}_F^\phi\right)_{anc} \approx \left[\frac{4\pi k_- \left(\mu c\right)_{anc}}{\alpha Z_{anc} \left(B^P\right)_{anc}} \right] \approx \left(\frac{c}{u^0} \right)^2_{anc} \,. \tag{9.150}$$

The point of this digression is that Ohmic heating requires in our analysis that the inequality in (9.136) holds and more specifically,

$$\left(\beta^\phi\right)_{anc} \approx \left(\bar{\beta}_F^\phi\right)_{anc} \quad \text{if} \quad \left(\frac{\mu c^2}{m_e c^2} \right) < 10^3 \,. \tag{9.151}$$

From (9.141) - (9.143) and (9.151) we have an interesting approximate anchor condition for $\bar{\beta}_F^\phi$,

$$\left(u^P\right)_{anc} \approx \left[1 - \left(\bar{\beta}_F^\phi\right)^2 \right]^{1/2}_{anc} \left(U_A\right)_{anc} \,. \tag{9.152}$$

9.6.3 Causal Structure of the Dynamo

We are interested in understanding the microphysics that drives the dynamo current. By (9.3), $E^{2'} > 0$ results because there is not enough of a contribution from F^{21} to make $E^{2'}$ vanish (i.e., $\left|F^{12}\right|$ is too small in the subcritical wind). The $F^{2\phi} u_\phi + F^{21} u_1$ term in (9.3) is the unbalanced EMF that we were looking for in the ergosphere that drives the dynamo in analogy to the Faraday wheel. However, unlike the Faraday wheel, the electrostatic force, $F^{20} u_0$, exceeds the EMF. In a unipolar inductor, the EMF is the causative agent and if an electrostatic equilibrium cannot be achieved (i.e., due to the existence of an external conduction path) a current flows. By contrast, an incoming Alfvén wave reflects off the ergosphere, imposing a global potential and therefore $F^{20} u_0$. If the plasma is incapable of creating a large enough EMF to cancel $F^{20} u_0$, a current will flow to cancel off some of the electrostatic field.

When this current flows in an effort to cancel off the surplus electrostatic force, charges are necessarily separated to establish a modified Goldreich–Julian charge density. Alternatively stated the value of Ω_F in the magnetic flux tube changes in the dynamo layer, as defined by (9.2). In order for $F^{20}u_0$ to be decreased, (9.1) shows that $|\bar{\beta}_F^\phi|$ must decrease. In terms of the global potential, (5.14) indicates that Ω_F must increase to be closer to Ω in the dynamo layer in this eventuality. From (9.151), (9.138) and (9.128), the field line angular velocity changes through the dynamo layer,

$$\frac{\Delta\Omega_F}{\Omega_F} =\approx \times 10^{-1}\,, \quad Z^2 = 10^{-2}\,, \tag{9.153a}$$

$$\frac{\Delta\Omega_F}{\Omega_F} \approx 1.10 \times 10^{-2}\,, \quad Z^2 = 10^{-3}\,. \tag{9.153b}$$

One can interpret the dynamo region in terms of MHD wave structure. First note that u^P increases through the dynamo layer and therefore n decreases (mass conservation). In Sect. 2.6, we showed that a fast rarefaction wave decreases the magnetic field and the density. Consequently, black hole gravity acts like a fast rarefaction piston in the ergospheric dynamo (see Fig. 9.8). The incoming Alfvén wave from the particle creation zone excites a fast rarefaction wave as it reflects off the ergosphere. The interaction is not localized, but extends from above the Alfvén point to the anchor. As expected, the fast rarefaction carries strong cross-field currents as pointed out in Sect. 2.6. Above the Alfvén point, a field aligned current system is attained which represents a reflected Alfvén wave as discussed in Sect. 2.5 (see Fig. 9.10). One might think of the region between the Alfvén point and the anchor ($\sim 10^{11} - 10^{12}$ cm thick) as a skin depth in which the incoming MHD wave is absorbed by the ergosphere.

One should note that frame dragging is responsible for the driving force for the dynamo, $E^{2'}$ in (9.3). If u^ϕ could be more negative (even though it is already less than "-10^4c") then $E^{2'}$ could be set to zero. However, frame dragging forces attempt to make u^ϕ more positive.

9.6.4 The Global Energetics of the Dynamo

The gedanken experiment of Sect. 9.3 showed that the reflected Alfvén wave from the particle creation zone has more energy flux than the incident Alfvén wave. This is a phenomenon expected when an Alfvén wave reflects off a rotating conductor. For a magnetically dominated Alfvén wave, the ingoing wave is primarily a transverse electromagnetic disturbance. It has been shown [139] that electromagnetic waves that reflect off a rotating conductor are amplified. In the black hole case, the conductor is the magnetosphere inside the Alfvén point that rotates with an angular velocity, Ω_F. However, the electrical conductivity is very poor in the cross-field direction and current flow only proceeds with large dissipation.

The global energetics are exactly the same as the ergospheric disk dynamo and are delineated in Fig. 9.11. The dynamo is located near the light cylinder associated

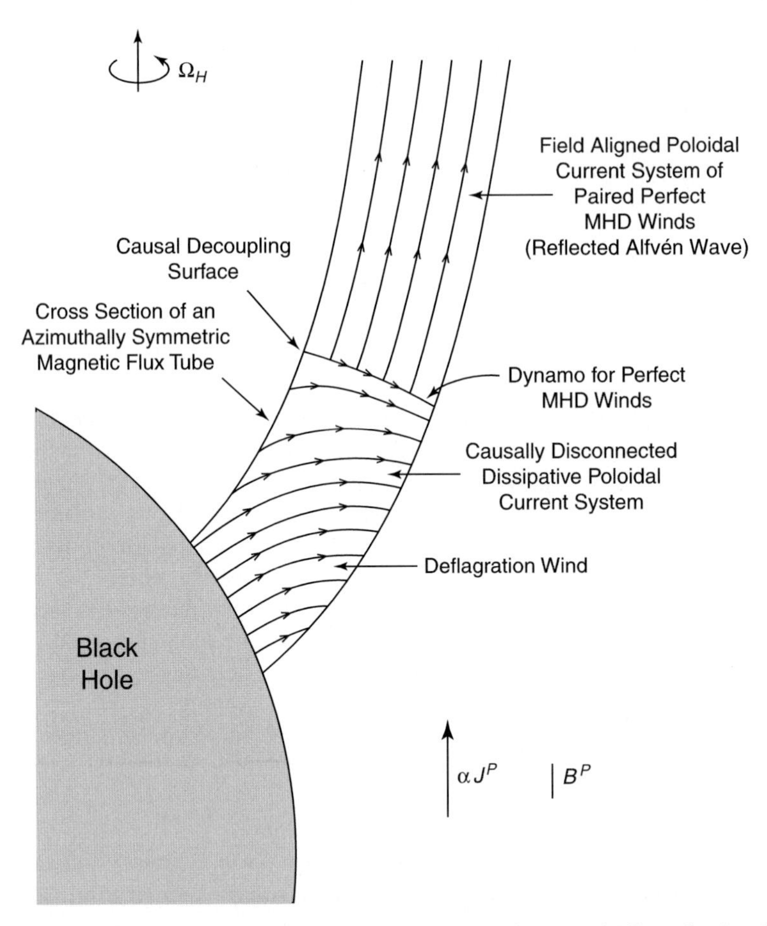

Fig. 9.10 The poloidal current distribution in an axisymmetric magnetic flux tube that threads the event horizon. In the perfect MHD wind zone, the poloidal current system is virtually field aligned as is characteristic of a magnetically dominated wind. Strong cross-field poloidal currents exist only in the dissipative regions of the wind as a frozen-in plasma is not free to cross strong magnetic field lines and this greatly inhibits the cross-field electrical conductivity. The dissipative region includes the dynamo for the paired wind system and the causally decoupled deflagration wind below the dynamo. Note how the dynamo current appears to flow in a thin skin depth in this global view, consistent with the reflected Alfvén wave interpretation of the dynamics. Strong dynamo currents exist throughout the deflagration wind. Note that B^T increases from zero at the beginning of the deflagration wind to a maximum at the horizon, because of the dynamo currents, J^2, and Ampere's law. As more B^T is generated, the gradients $\partial B^T/\partial X^2$ become larger as well near the horizon. Thus, J^1 increases near the horizon to support the cross-field gradient in B^T in accord with Ampere's law. This explains why the poloidal current density is purely cross-field in nature at the top of the deflagration wind $\left(B^T = 0\right)$, and why it transitions to being nearly field aligned as the wind approaches the horizon

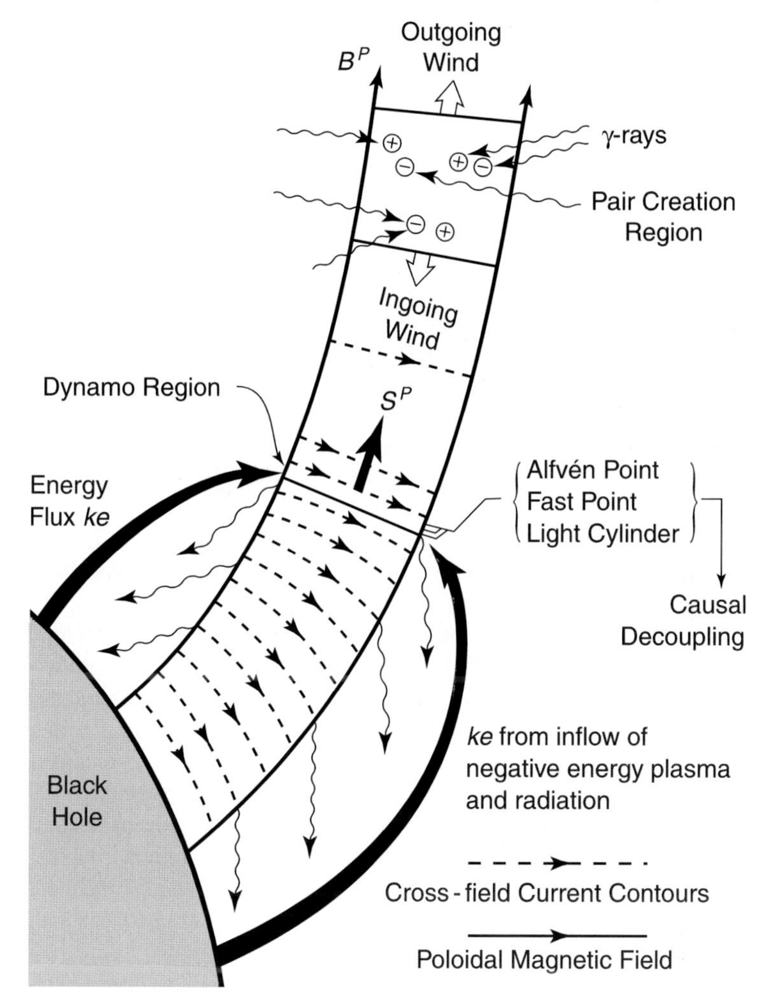

Fig. 9.11 The global energetics of a dynamo that powers an outgoing magnetically dominated wind in a magnetic flux tube that threads the event horizon of a rotating black hole. Plasma is prepared hot and on negative energy trajectories in the dynamo. The hot plasma eventually radiates away its energy in a highly beamed distribution of photons (by the headlight effect) onto negative energy trajectories (see Sect. 8.6 to review). This negative energy photon field spins down the hole and extracts its reducible mass. Similarly, the plasma is of negative energy as well and all of the inertia will eventually be absorbed by the hole, spinning it down in the process. Ultimately, it is the rotational energy of the hole that powers the outgoing MHD wind [130]

with Ω_F of the ingoing perfect MHD wind. The magnetic field rotates backward, with respect to the black hole rotation direction, at the speed of light at the inner light cylinder as viewed by all local physical observers. The dragging of inertial frames pulls plasma forward across the large scale magnetic field in the azimuthal direction near the light cylinder. This ensures subluminal rotational velocities of the plasma in this region at the expense of perfect MHD. The resulting proper electric field drives strong cross-field dynamo currents which support the poloidal Poynting flux, S^P, in the magnetically dominated paired perfect MHD wind system. The cross-field currents in the dynamo region affect the conversion of energy (angular momentum) flux, $ke\,(k\ell)$, from mechanical to electromagnetic form by $\boldsymbol{J} \cdot \boldsymbol{E^P}$ force ($\boldsymbol{J} \times \boldsymbol{B^P}$ torque). The plasma in the dynamo region rotates backward at nearly the speed of light as a consequence of approximate corotation with the magnetic field as viewed by a ZAMO. Thus, $\beta^\phi \gtrsim -1$ and $u^0 \gg 1$. The plasma has a huge relativistically induced inertia and a highly negative angular momentum about the symmetry axis of the black hole. Such states were shown in Chap. 3 to have a large negative specific global energy, $\omega \ll 0$, and are well defined in the ergosphere (where the dynamo resides). The plasma is relativistically hot, so it radiates violently. However, since $\beta^\phi \gtrsim -1$ for the emitting plasma, the relativistic headlight effect beams the radiation along negative energy ($\omega < 0$), negative angular momentum trajectories. The influx of negative energy (angular momentum), plasma and radiation is effectively an outflow of energy (angular momentum) $ke\,(k\ell)$, from the hole. Therefore, in a global sense, it is the rotational energy of the black hole (the reducible mass) that powers the dynamo and outgoing wind. Notice the incredible similarity with Fig. 8.6 describing the global energetics of the ergospheric disk. As discussed in Chap. 7 this is not a coincidence, but is mandated by GHM.

From (9.23) and the initial data on I_+, the bipolar winds from the horizon magnetosphere can deliver $\sim 10^{43} - 10^{44}$ ergs/sec in Poynting flux to asymptotic infinity. This is two to three orders of magnitude weaker than a strong ergospheric disk wind. The difference in energy can be thought of as arising because ingoing horizon magnetospheric winds go superfast before the gravitational field has a chance to get a "good grip" on the wind and establish a large field line angular velocity $\sim \Omega_H$.

The resulting ergospheric dynamo establishes a global current system in the paired wind system. The topology of the poloidal current system is indicated qualitatively in Fig. 9.12.

9.7 The Deflagration Wind

As the flow exits the anchor region of the flux tube it becomes causally decoupled from the paired perfect MHD wind system since it passes through the dissipative fast critical surface defined by (9.5). Even though any dynamo behavior in this dissipative deflagration wind cannot be communicated to the outgoing wind, we note some of its general properties below.

9.7.1 The Near Zone

As the wind exists the flow decoupling surface (the anchor), it becomes an ingoing wind that is determined by initial data on this boundary surface, $r(\theta)_{anc}$,

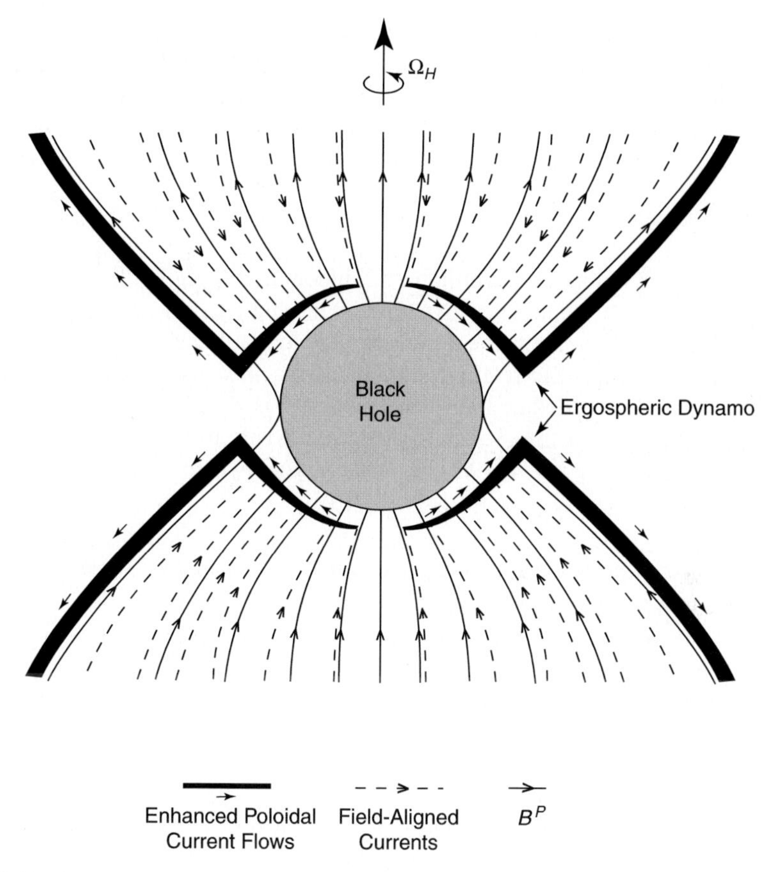

Fig. 9.12 The topology of the poloidal current system of a paired wind emanating from the horizon magnetosphere of a rotating black hole. Note that the return current flows outside of the perfect MHD wind zone [138]

$$B^T = 0\,, \quad r = r(\theta)_{anc}\,, \tag{9.154a}$$

$$k\ell = km\,, \quad m \ll 0\,, \quad r = r(\theta)_{anc}\,, \tag{9.154b}$$

$$ke = k\omega\,, \quad \omega \ll 0\,, \quad r = r(\theta)_{anc}\,, \tag{9.154c}$$

$$J_n = 0\,, \quad r = r(\theta)_{anc}\,, \tag{9.154d}$$

$$F^{20} < 0\,, \quad r = r(\theta)_{anc}\,. \tag{9.154e}$$

The ingoing deflagration wind has its own current system that is causally independent of that in the outgoing wind (see Figs. 9.8 and 9.10). The dynamo that exists above the anchor also exists in the near zone of the deflagration wind just below the anchor. From Ampere's law applied to (9.154a), one knows the poloidal current density normal to the wind decoupling surface, J_n, must vanish as indicated in

(9.154d), note: J_n is not J^1 in general. Since the frame dragging induced dynamo effects continue to operate at $r \lesssim r(\theta)_{anc}$, more cross-field current flows in an attempt to cancel the electrostatic force $F^{20}u_0$ in (9.3). Then by Ampere's law (3.61b),

$$F^{12} \gtrsim 0 \,, \quad r \lesssim r(\theta)_{anc} \,. \tag{9.155}$$

The toroidal magnetic field actually switches sign at the onset of the ingoing deflagration wind.

We can see the radiative instability in full bloom in this region. Recall that even at $r \gtrsim r(\theta)_{anc}$, the substantial negative toroidal magnetic field was not bent enough (not negative enough) to allow the plasma to slide forward relative to the corotating frame of the field and avoid relativistic rotation velocities and the associated large radiation losses. In the deflagration wind, not only is B^T not negative enough, it is of the wrong sign for the plasma to avoid relativistic rotation for $r \lesssim r(\theta)_{anc}$! Furthermore, the term $F^{21}u_1$ in (9.3) becomes another dynamo term for the driving force $E^{2'}$ in the deflagration wind (see Fig. 9.13).

In order for the plasma to have enough inertia to cross the field lines it needs $\beta^\phi \gtrsim -1$ in the near zone. Since we assume that the plasma has some conductive properties in the near zone, we expect

$$\left|E^{2'}\right| \ll B^P \,, \quad r \lesssim r(\theta)_{anc} \,, \tag{9.156a}$$

and from (9.4) since F^{12} is still small we have

$$\beta^\phi \gtrsim -1 \,, \quad r \lesssim r(\theta)_{anc} \,, \tag{9.156b}$$

$$\Omega_F \gtrsim \Omega_{\min} \,, \quad r \lesssim r(\theta)_{anc} \,. \tag{9.156c}$$

9.7.2 The Breakdown of Near Zone Physics

The approximations used to describe the near zone physics in the last section break down because of the additional dynamo term $F^{21}u_1$ in (9.3). As the deflagration wind propagates inward, its dynamo keeps pumping cross-field inertial currents, J^2. By Ampere's law (3.61b), F^{21} keeps growing. As long as $F^{21} \ll B^P$ the approximations in the last section are justified. As $F^{21} \to B^P$ for $r < r(\theta)_{anc}$, (9.4) and (9.156a) imply that

$$\beta^\phi - \bar{\beta}^\phi_F < 0 \,, \quad F^{12} \gg 0 \,, \quad r < r(\theta)_{anc} \,. \tag{9.157}$$

These relations are reversed from the ingoing perfect MHD wind and are opposite to that found in Chap. 7 during the analysis of the torsional tug of war. However, that analysis depended on β^ϕ_F representing the rotation rate of rigid magnetic field lines. In the dissipative flow, there is no such interpretation of rotating magnetic field lines. By contrast, $\bar{\beta}^\phi_F$ is merely a quantity defined implicitly by (9.1) and is related to the electrostatic potential drop across the magnetic field lines by (5.14) and (9.2).

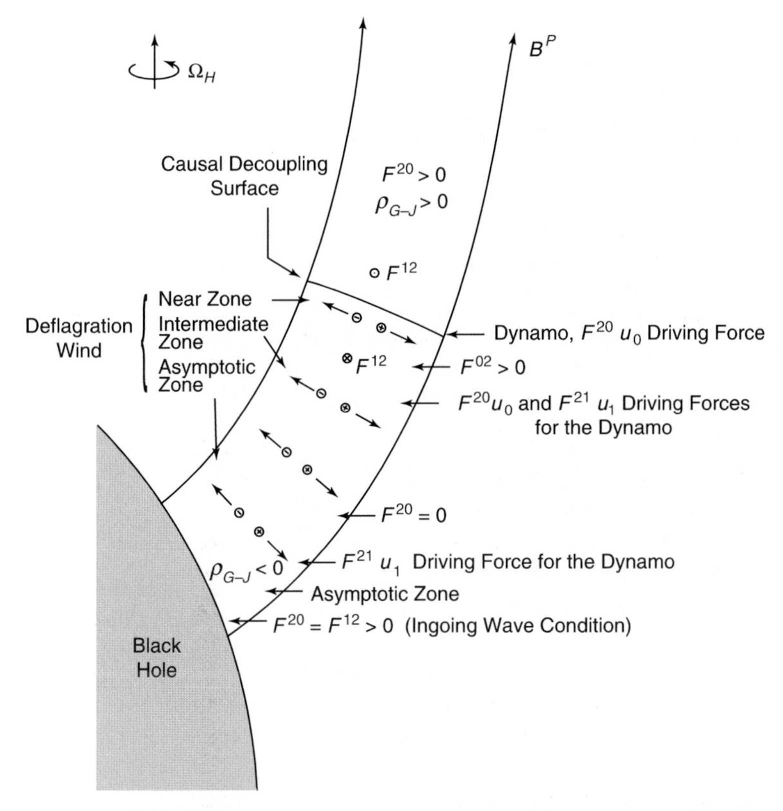

Fig. 9.13 The dynamics of charge separation in the ingoing deflagration wind. The dynamo force in the near zone of the deflagration wind is the same as in the adjacent dynamo for the paired MHD wind system. The electrostatic field F^{02} is larger than the $F^{2\phi}\beta^{\phi}$, opposing EMF, in magnitude since β^{ϕ} is limited in magnitude to less than one (equivalently frame dragging imposes $\mathrm{d}\phi/\mathrm{d}t > \Omega_{\rm min}$). Charges separate to cancel the excess electrostatic field F^{02}, as indicated in the figure. There is a small additional dynamo driving term in this same direction, $F^{21}\beta^{P}$. The charge separation and the consequent cross-field current continues to flow downstream due to this dynamo action. By Ampere's law, F^{12} increases downstream and in the intermediate zone of the deflagration wind, $F^{21}\beta^{P} \sim F^{20}$. The existence of two dynamo forces requires that charge continue to separate across the magnetic field lines. Eventually, the electrostatic field F^{02} is canceled off by this charge separation. However, a strong $F^{21}\beta^{P}$ force persists in the asymptotic zone as charges are dragged radially across the toroidal magnetic field near the horizon by the powerful force of black hole gravity. The charge separation continues near the horizon and the electrostatic field direction is reversed from its initial condition in the near zone, $F^{02} < 0$. As $\alpha \to 0$, the ingoing wave condition $F^{12} \approx F^{20}$ is achieved and the sign of the Goldreich–Julian charge density, ρ_{G-J}, is opposite its initial value as a consequence of the ubiquitous cross-field dynamo which separates charge across the magnetic field the entire length of the deflagration wind

The reversal in sign of (9.157) relative to the upstream ingoing wind reflects the fact that the plasma wind which was initially magnetically dominated has been transformed into an inertially dominated wind by black hole gravity as anticipated in Chap. 7.

9.7.3 The Asymptotic Wind Zone

As is typical of complicated flows, it is much easier to understand the asymptotic zone and near zone than the intermediate zone in the region between them. From (9.4) and the horizon boundary conditions (3.94),

$$\lim_{r \to r_+} F^{12} \approx \bar{\beta}^{\phi}_{F} B^{P} = F^{20} . \tag{9.158}$$

We noted in the last section that for large positive values of F^{12} there are two significant dynamo terms, $F^{21}u_1 > 0$ and $F^{20}u_0 > 0$. The instability generates more and more J^2, thus there is nothing to limit the growth of the $F^{21}u_1$ term (since Ampere's law keeps increasing F^{12} downstream in the deflagration wind). By contrast, the charge separation proceeds under the action of $E^{2'}$ in a sense that tends to negate the electrostatic force $F^{20}u_0$. Thus, we expect that eventually (see Fig. 9.13),

$$F^{21}u_1 > F^{20}u_0 , \quad r \gtrsim r_+ . \tag{9.159}$$

The dynamo effect near the horizon is dominated by the radial force of gravity as opposed to the dragging of inertial frames. Plasma is pulled radially across the toroidal magnetic field creating a cross-field dynamo as indicated in (9.159). The effect of (9.159) is a dynamo that keeps increasing B^T all the way to the hole with

$$\lim_{\alpha \to 0} J^2 \sim +\alpha^0 . \tag{9.160}$$

From the law of current conservation in time stationary form in the ZAMO frames, as given by the divergence law in (3.36b), we can find J^1 as well,

$$\nabla^{(3)} (\alpha \boldsymbol{J}) = 0 . \tag{9.161}$$

The derivatives in (3.36) are expressed in (3.25) and (3.33). Using the asymptotic form of the differential operators in (3.25) and (3.33) as $\alpha \to 0$, we find from (9.161) and (9.160) that

$$\lim_{\alpha \to 0} J^1 \approx J^r \sim +\alpha^{-1} . \tag{9.162}$$

Inserting (9.160) into Ampere's law (3.62b) we obtain the asymptotic scaling,

$$\lim_{\alpha \to 0} F^{12} \sim +\alpha^{-1} . \tag{9.163}$$

From (9.163) and (5.17), the toroidal magnetic field density is positive at the horizon,

$$\lim_{\alpha \to 0} B^T > 0 , \tag{9.164a}$$

and by (9.158) and (5.14),

$$\lim_{\alpha\to 0} \bar{\beta}_F^{\phi} = +\infty \,, \tag{9.164b}$$

$$\lim_{\alpha\to 0} \Omega_F > \Omega_H \,. \tag{9.164c}$$

Furthermore, (9.158) and (9.163) imply that

$$\lim_{\alpha\to 0} F^{20} = F^{12} > 0 \,. \tag{9.165}$$

In the near zone of the deflagration wind and the ingoing perfect MHD, F^{20} was negative. In order to understand the sign change in (9.165), note that by (9.159), we expect the dynamo term $F^{21}u_1$ to dominate the dynamo forces deep in the ergosphere. The net result of this force is to separate charges across the field lines that will increase F^{20} (i.e., cancel off the initial value of F^{20} in (9.154e)). At some point, F^{20} is canceled off, but at this stage the dynamo does not operate due to electrostatic forces, but to the motion induced EMF, $F^{21}u_1$. The dynamo force keeps separating charges (irrespective of the electrostatic force) in the same sense eventually reversing the direction of F^{20} as indicated in (9.165) and Fig. 9.13.

9.8 The Unique Physical Solution

It was shown in [136] that another paired wind solution exists for the initial data on I_+ and I_- defined in Sect. 9.4.1. For that paired wind system, both the ingoing and outgoing winds obey perfect MHD everywhere. This is the analog of the Blandford–Znajek solution [66] for $\Omega_F = (1/30)\Omega_H$. The outgoing wind is subrelativistic, $\beta^P \sim 0.1$, and it is a supercritical wind as discussed in Sect. 5.5 and Fig. 5.2.

The paired outgoing minimum torque/ingoing deflagration wind described in this chapter is a viable physical solution, if the accretion flow is passive and the event horizon magnetosphere is seeded by gamma ray pair production, for three reasons:

1. It extracts the minimum energy and angular momentum from the black hole of any paired wind system for $\Omega_F = (1/30)\Omega_H$ (i.e., it is the minimum torque outgoing wind).
2. There is a causal structure consistent with the properties of MHD waves that drives the current system in the ergospheric dynamo.
3. The second law of black hole thermodynamics equates the horizon surface area, A, in (1.34) to entropy (see [80] for a proof). Thus, for the paired wind system we can write using (5.20) and (5.21),

$$\mathrm{d}\mathbb{S}_H = \frac{k_B}{4\hbar}\,\mathrm{d}A_H = \frac{2\pi k_B}{\kappa\hbar}\,[\Omega_F - \Omega_H]\,\mathrm{d}(Ma) \,, \tag{9.166}$$

where "$\mathrm{d}(Ma)$" is the angular momentum deposited into the hole, $\mathrm{d}(Ma) < 0$, and "$\mathbb{S}_H$" is identified with horizon entropy in [80]. Thus, by (9.166) the minimum torque solution has an order of magnitude less entropy generation in the horizon than the paired perfect MHD wind system (see 5.52 with $\beta^P \sim 0.1$). Consequently, in a global context, the full general relativistic theory of the paired wind system identifies the paired outgoing minimum torque/ingoing deflagration wind as the minimum entropy generating solution (surprisingly, it is not the "dissipation-less" paired perfect MHD winds).

Point "3" above is significant based on the "principle of minimum entropy production." In Chap. 35 of [140] is a textbook statement of the principle for irreversible process such as the one described by (9.166) for $\Omega_F < \Omega_H$ (see [15]). The steady state of the system achieves a minimum rate of entropy production consistent with the external constraints that prevent equilibrium (if the constraints were removed, equilibrium is achieved and $d\mathbb{S} = 0$).

Based on points 1–3 above, in certain circumstances, the paired outgoing minimum torque/ingoing deflagration wind solution described in this chapter is a physical solution. More importantly, it shows that if the perfect MHD assumption is violated in a tenuous plasma with pair production, large departures from the family of perfect MHD solutions are indicated.

Chapter 10
Applications to the Theory of Extragalactic Radio Sources

This chapter is the most speculative part of the book. We attempt to apply the idealized theory of black hole GHM to actual astrophysical situations. Remember, from the Introduction, that the motivation for studying black hole GHM was to describe radio loud AGNs. That is the intent of this chapter. One can always adjust parameters such as Ω_H, M, a and Φ so that a dimensional analysis yields the observed luminosity of extragalactic radio sources. However, this would not be a convincing argument that black hole GHM has any physical relevance in radio loud AGN. We begin the chapter with a look at evidence for a structure to the central engine in relation to the unified scheme for radio loud AGN that was discussed in the Introduction. The remainder of the chapter shows how these spectral diagnostics of the central engines occur naturally in the black hole GHM theory of extragalactic radio sources.

10.1 Spectral Diagnostics of Blazar Central Engines

In principle, there must be some fine structure associated with the central engines of extragalactic radio sources since there are physically three distinct components: the accretion disk, the ergospheric disk and the horizon magnetosphere. Each has a unique structure replete with its unique poloidal magnetic flux and distribution of magnetic field line angular velocities. Thus, it is hard to envision how a single jet (single pair of bipolar jets) is driven by three distinct engines. Unfortunately, most of the work in this field has concentrated on single jet theories of radio loud AGN. Notable exceptions are found in [141], [142] and more recently [143]. Strong arguments against the possibility of a single jet in certain sources are presented in [141] and [142]. In particular, certain quasars detected with the Compton Gamma Ray Observatory have strict bounds on the annihilation radiation emitted from the source. This was used to bound the number of positron-electron pairs in the jet in [141]. For quasars with strong FR II radio emission, [141] and [142] noted a problem with the single jet theory. Specifically, the high energy gamma ray emission measured by

B. Punsly, *Black Hole Gravitohydromagnetics, 2nd. ed.*,

Astrophysics and Space Science Library 355, doi: 10/1007/978-3-540-76957-6_10,

the EGRET telescope in the space based observatory as well as the energy in the extended lobe emission (as discussed in Table 1.2) provide two additional independent constraints. Combining the three constraints above, it was shown in [142] that if the high energy γ-rays are from an external Compton scattering (ECS) process [144] or a self synchrotron Compton (SSC) process [145] then there is not enough internal energy in the magneto-fluid in the γ-ray emitting region to support the energy flow to the FR II radio lobes of many sources (3C 279, for example). Thus, it appears based on this analysis alone that the FR II radio structures are supported by a jet that is largely distinct from the EGRET detected radiating plasma (which is commonly believed to be from a relativistic jet as well).

The broadband synchrotron spectrum of magnetized jets can tell us something about the structure of the central engine which created them, if we can quantify how the emissivity of such jets has evolved cosmologically (i.e., how high redshift jets differ from jets in low redshift AGNs) or if the properties of the spectrum are correlated with other observable characteristics of AGNs. A possible breakthrough in this regard was made in [146]. These observers were interested in making quasi-simultaneous broadband spectral measurements of blazars (in this chapter we define a blazar as an AGN with a strong compact radio core, i.e., strong radio emission concentrated in a region less than 0.25 at 5 GHz, which is the resolution of the VLA in A-array). Since blazar emission is highly variable in the radio and submillimeter bands, one needs simultaneous observations in order to get the detailed shape of blazar broadband spectra. Logistically, this is very difficult and quasi-simultaneous spectra are often sampled on time frames spanning as much as two weeks. Brown et al [146] noted that many blazar spectra are clearly "double humped" and decompose naturally into two distinct synchrotron components (see Fig. 10.1), with one peak at cm wavelengths and the other at mm wavelengths.

This finding was pursued in [142] with a larger sample of 118 quasi-simultaneous blazar spectra. To properly interpret the spectra one needs to note that a "double humped" structure is the ultimate simplification of the physical phenomenon. Each peak probably involves multiple components that are synchrotron self absorbed at low frequencies (see [34] for a discussion of synchrotron self absorption). The two peak simplification requires that in each of the two clusters of synchrotron components, one component is clearly dominant. Quasi-simultaneous, VLBI measurements (measurements that can resolve the 0.25 VLA radio core) at more than two frequencies, [147–151], show that strong VLBI knots and the unresolved core (see the jet structure in Fig. 1.10 for example) have spectral flux peaks staggered across the microwave band (i.e., 5–40 GHz). Thus, it is likely in principle (and true in practice) that a source could have two or more peaks at centimeter wavelengths without violating the basic concepts of the "double humped" model of a cm peak and a mm peak.

The nearby blazar 3C 273 provides the best possibility of sampling the radio to submillimeter band with a quasi-simultaneous spectra and a good signal to noise ratio above 300 GHz. Figure 10.2 provides an excellent example of the double humped spectra of 3C 273 [152]. The centimeter peak is identified at the frequency $\nu_1 = 45$ GHz (in the quasar rest frame) and the mm peak is identified at $\nu_2 = 320$ GHz

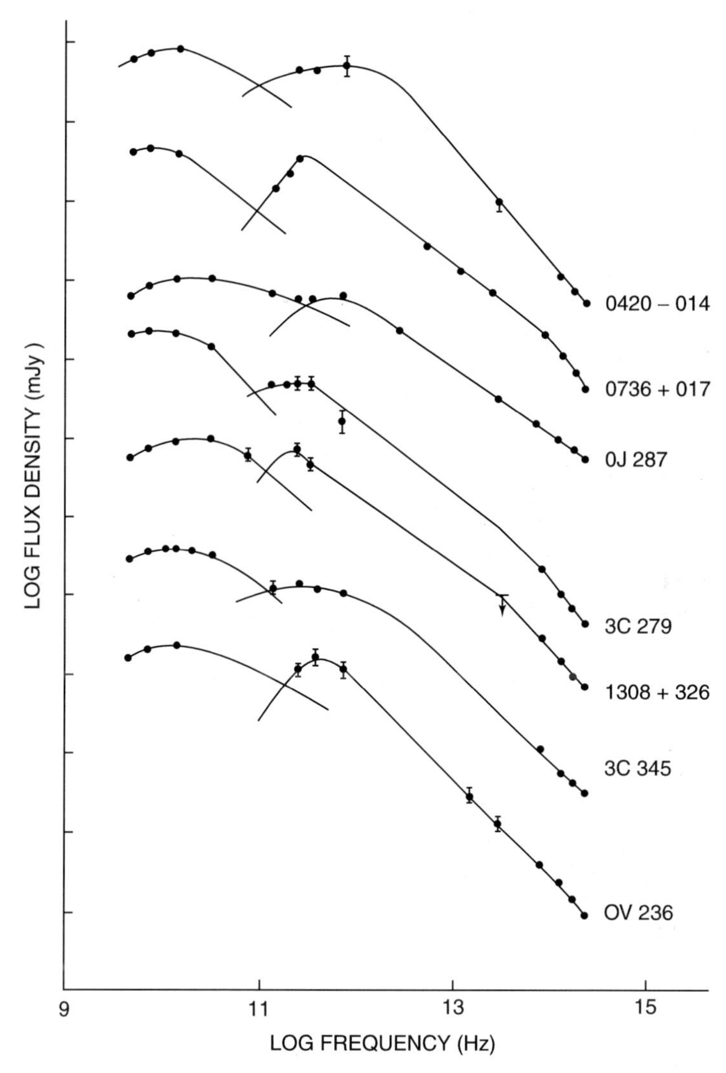

Fig. 10.1 A sample of quasi-simultaneous broad band spectra of blazars from [146]. Note the ubiquitous "double hump" structure. There is one spectral peak at centimeter wavelengths and another at millimeter wavelengths. The two peaks were fit with synchrotron self absorbed spectra for all of the sources [146]. Note that the optical data seems to be depicted far more accurately as the high frequency tail of the mm peaked synchrotron component rather than the high frequency tail of the cm peaked synchrotron component for each source. The sample includes HPQs (0420-014, 3C 279, 3C 345 and OV 236), an FR II BL Lac (1308+326), an FR I BL Lac (OJ 287) and an unusual LPQ (0736+017). The plots are published with the permission of Ian Robson

(in the quasar rest frame). Note that the mm peak is not necessarily a true local flux maximum, but can appear as an inflection point in the spectral flux density (i.e., for a strong cm peaked synchrotron component, the flux density from the high frequency

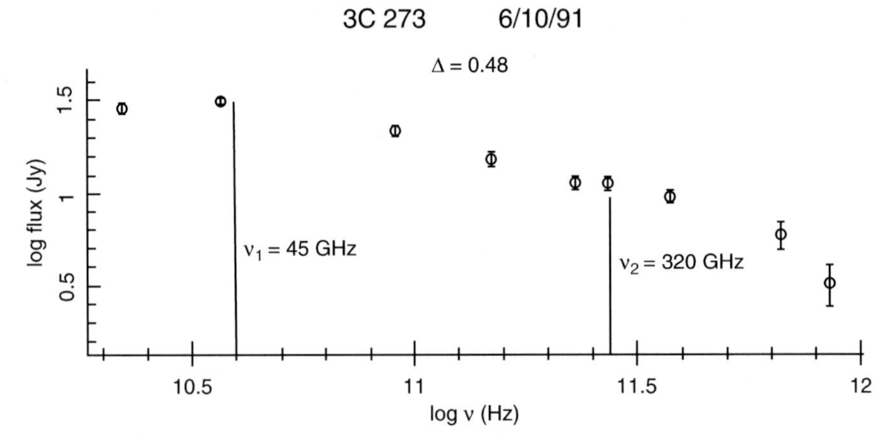

Fig. 10.2 The best quasar compact core for obtaining a high quality quasi-simultaneous spectrum is the nearby blazar 3C 273. Being "so close" to earth allows for a measurable flux of high frequency submillimeter radiation as in this spectrum from [152]. This figure is useful for depicting the location of the cm and mm peaks of the spectra of blazars, ν_1 and ν_2, respectively due to its excellent frequency coverage. Note that as a consequence of cosmological redshifting that the AGN rest frame frequencies, ν_1 and ν_2, differ from the earth based frequencies that are used to label the horizontal axis in the figure. The spectrum is characteristic of an LPQ with a large Δ value of 0.48. Note that the mm peak is not necessarily a local flux maximum, but can be manifested as an inflection point in the high frequency tail of the cm peak. This figure is printed with the permission of Ian Robson

tail at mm wavelengths can exceed that of the mm peaked synchrotron component, thus the superposition of the mm peak will only cause the high frequency tail of the cm peaked synchrotron component to flatten out in the mm band). The shapes of blazar spectra are continually changing due to flaring effects. Thus, a snapshot of a blazar spectra at just one instance of time might not be representative. Each blazar does seem to have a quiescent background state on which the flares are superimposed [153]. The quiescent background state and flaring behavior characterize the synchrotron emissivity of the jet.

In order to quantify the flaring and quiescent behavior of the jet from a blazar we need many quasi-simultaneous spectra and a numerical description. To this end, [142] introduced a parameter, Δ, that is the ratio of the flux density at the cm peak, $(F_\nu)_{cm} \equiv F_\nu(\nu_1)$, to the flux density at the mm peak, $(F_\nu)_{mm} \equiv F_\nu(\nu_2)$,

$$\Delta \equiv \log\left[\frac{(F_\nu)_{cm}}{(F_\nu)_{mm}}\right] . \tag{10.1}$$

We note some of the astoundingly strong correlations of this parameter with other blazar properties in the remainder of the section.

10.1.1 BL Lacs and Quasars

One of the strengths of this analysis is the clear distinction that it makes between BL Lac objects and core dominated radio loud quasars. These two classes of objects have very similar radio core morphologies, angular sizes and radio fluxes. However, BL Lacs have weak or no broad emission lines and quasars have strong broad emission lines. Furthermore, quasars have a strong UV thermal component (the big blue bump) that, by contrast, is generally much weaker in BL Lacs. Thus, it is commonly assumed that the accretion states of BL Lacs (weak accretion systems) are vastly different than quasars (strong accretion systems). Hence, the Δ values of these two classes of objects relate physically to the accretion state.

It was found in [142] that quasars have Δ values larger than those of BL Lacs at the 99.9997% significance level according to a Wilcoxon rank sum test (see the histogram of Fig. 10.3). Even more significantly, one can distinguish the high polarizations quasars, HPQs, (HPQs are defined by optical continuum polarizations >3% while low polarizations quasars, LPQs, have optical continuum polarizations <3%) from the BL Lacs (which are also a class of high optical polarization objects) with this same parameter. These two classes of objects are considered the closest in characteristics of any two in the unified scheme. Yet, Δ is larger in HPQ spectra compared to BL Lac spectra at the 99.997% significance level according to a Wilcoxon rank sum test [142]. Thus, Δ is an excellent diagnostic of the fine structure of the central engine.

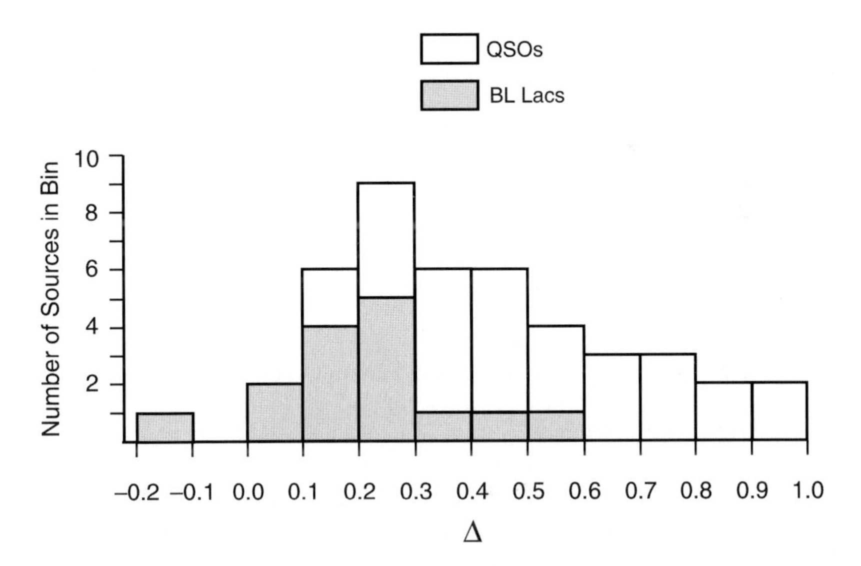

Fig. 10.3 A histogram of the distribution of BL Lacs and QSOs as a function of the parameter, Δ, the logarithm of the ratio of the flux density at the cm peak to the flux density of the mm peak. The BL Lacs are indicated in the histogram by the shaded regions [141]

Table 10.1 Average spectral properties of core-dominated radio sources

Class	$\overline{\log \nu_1}$	$\overline{\log \nu_2}$	%F	$<\Delta>$
QSOs	10.46 ± 0.28	11.68 ± 0.20	23.3	0.490 ± 0.213
HPQs	10.53 ± 0.30	11.70 ± 0.20	24.9	0.424 ± 0.162
LPQs	10.38 ± 0.26	11.66 ± 0.24	21.2	0.537 ± 0.241
BL Lacs	10.28 ± 0.27	11.48 ± 0.16	70.2	0.212 ± 0.156

Table 10.1 (from [142]) delineates these distinctions further. Column (1) is the class of object and columns (2) and (3) are the averages of the logarithm of the cm peak and mm peak frequencies, respectively. Column (4) is the percent of the time that the mm peak is actually a flux density local maximum as opposed to an inflection point in the spectrum. Column (5) is the average Δ factor for each class of blazar. The result of the Δ factor analysis and Table 10.1 indicate that BL Lacs have prominent mm peaks in their spectra and core dominated radio loud quasars have prominent cm peaks in their spectra. In an absolute sense, [142] shows that BL Lacs are blazars with weak cm peaks in their spectra. Similarly, this infers a correlation between accretion luminosity and the strength of the cm emission in blazar radio cores.

10.1.2 Other Correlations

We note some other correlations of physical parameters with Δ in blazar spectra that were found in [142].

1. Large values of Δ are correlated with strong extended radio emission, P_E, at the 99.97% significance level (see the histogram in Fig. 10.4).
2. Large values of Δ are correlated with large redshifts at the 99.997% significance level, which is likely a redundancy with the correlation with radio power in “1” (see the histogram in Fig. 10.5).
3. Small values of Δ are correlated with high continuum optical polarization at the 99.99% significance level (see the histogram in Fig. 10.6).
4. QSOs that have been detected as γ-ray sources by EGRET have smaller Δ values than other core dominated QSOs at the 98.5% significance level (see the histogram in Fig. 10.7). However, if one just looks at the HPQ subpopulation, the correlation is weaker with a significance level of only 92.3% (see Fig. 10.8).

We note one other important finding of [142] that was deduced from the analysis of multi-frequency, quasi-simultaneous VLBI maps. The plasma responsible for the cm peak is near the based of the VLBI jet or in the unresolved core (distances <10 pc from the central engine), and the mm peak-emitting plasma is buried deep inside of the VLBI core.

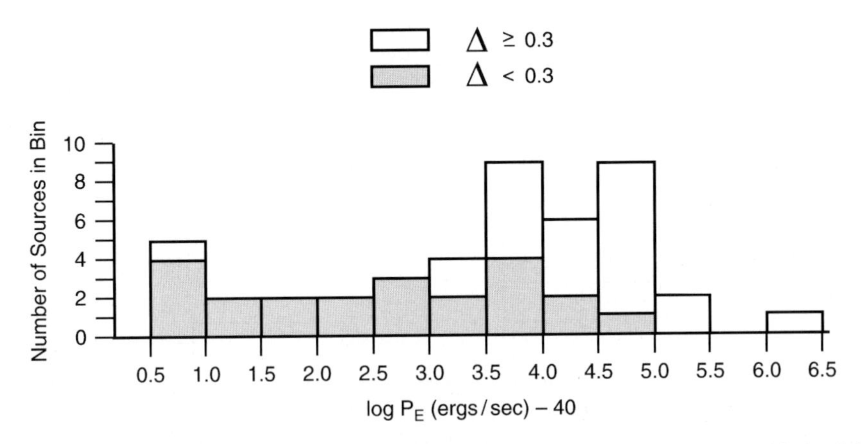

Fig. 10.4 A histogram of the logarithm of the extended radio emission, P_E, for blazars with $\Delta < 0.3$ (the shaded regions) and $\Delta \geq 0.3$ (the unshaded regions) [141]

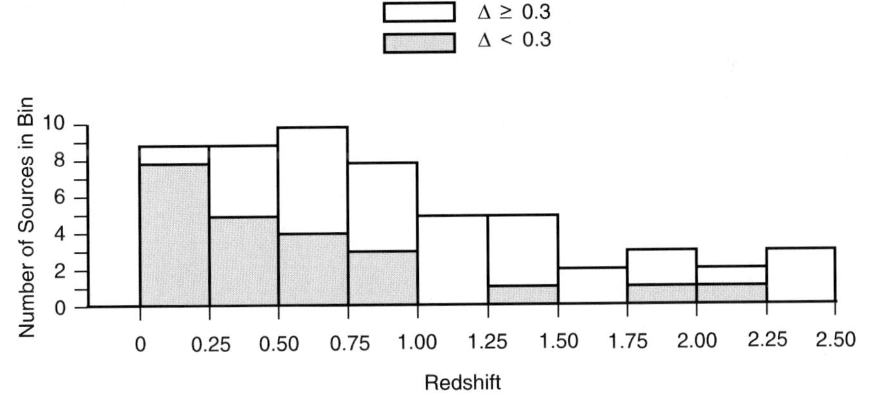

Fig. 10.5 A histogram of the redshift distribution for the sample of blazar cores in [142] with $\Delta < 0.3$ (the shaded regions) and $\Delta \geq 0.3$ (the unshaded) [141]

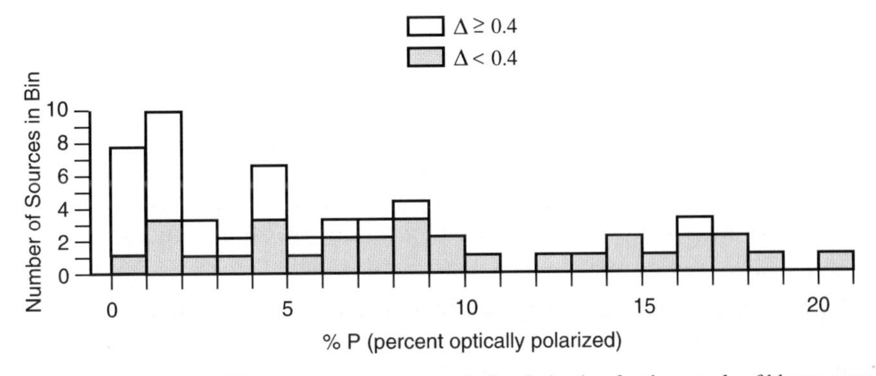

Fig. 10.6 A histogram of the percent continuum optical polarization for the sample of blazar cores in [142] with $\Delta < 0.4$ (the shaded regions) and $\Delta \geq 0.4$ (the unshaded) [141]

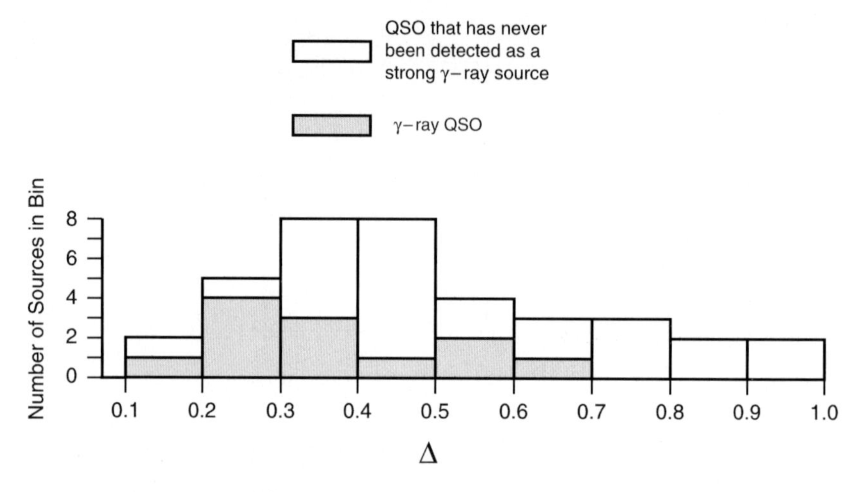

Fig. 10.7 A histogram of the distribution of γ-ray loud quasars detected by EGRET (shaded) and quasars that have never been detected as strong γ-ray emitters (unshaded) from the sample of blazars in [142] as a function of Δ [141]

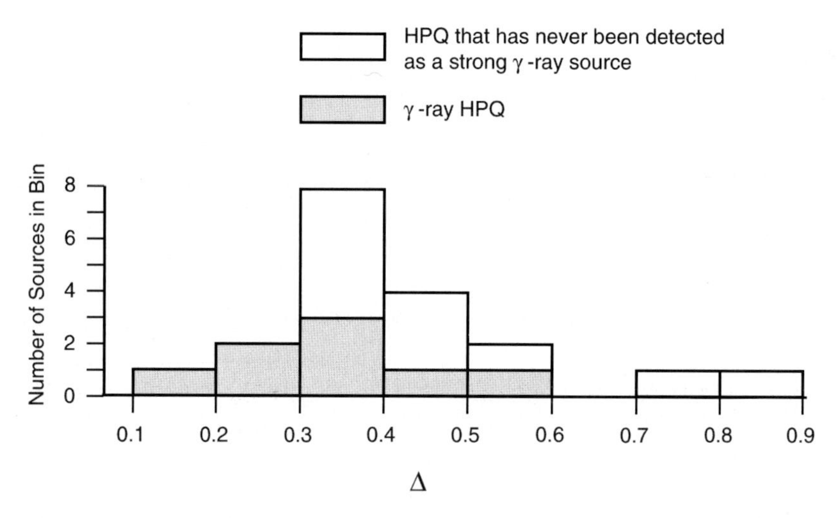

Fig. 10.8 A histogram of the distribution of γ-ray loud HPQs detected by EGRET (shaded) and HPQs that have never been detected as strong γ-ray emitters (unshaded) from the sample of blazars in [142] as a function of Δ [141]

10.2 The Black Hole GHM Theory of the Central Engine

We describe the correlations found in Sect. 10.1 in terms of the black hole GHM dynamos described in Chaps. 8 and 9. This section describes a model in which the horizon magnetospheric jet is nested within an ergospheric jet as it must be if they exist (see Fig. 10.9). The horizon jet could have been chosen to be largely electrodynamic

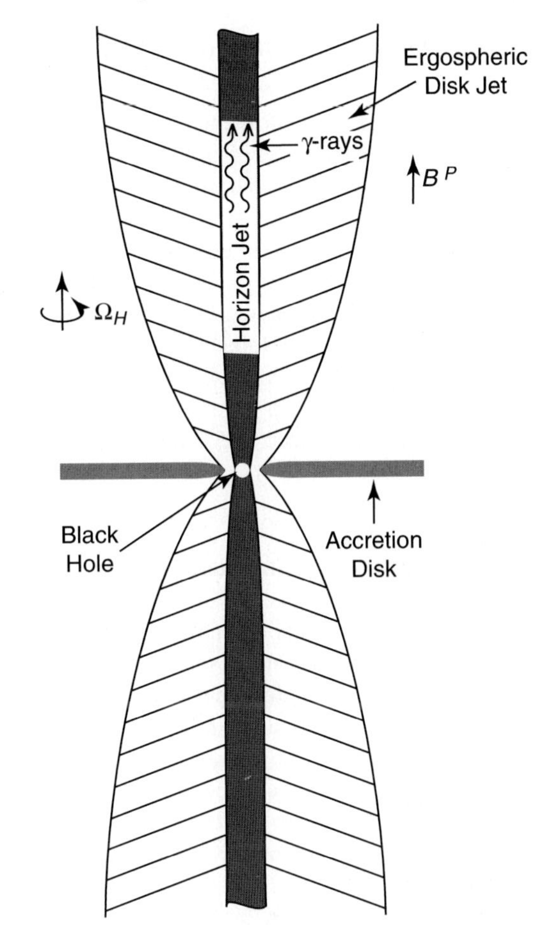

Fig. 10.9 The dual jet GHM model of the blazar central engine. The relativistic jet driven by the event horizon magnetosphere is nested within the jet driven by the ergospheric equatorial plasma. The diversity of observed blazar properties is associated with the relative strengths of the two black hole driven components. Quasar central engines are predominantly the ergospheric jet and BL Lacs have a prominent horizon magnetospheric jet. The ergospheric jet produces a cm peaked synchrotron self absorbed spectrum and the horizon jet radiates a mm peaked synchrotron self absorbed spectrum. This dichotomy of jet properties allows for an interpretation of the strong correlations with Δ displayed in Figs. 10.3–10.7

in nature as suggested in the perfect MHD solutions of [111], however this exercise is designed to show how to parameterize all the GHM dynamo regions. There is no justification that the GHM horizon jet is more reasonable than a perfect MHD electrodynamic jet. Furthermore, the two jets are physically distinct in this dual jet scenario. The horizon magnetosphere driven jet will be called the "blazar jet" since we will empirically associate it with properties commonly used to distinguish blazars (i.e., high optical continuum polarization, γ-ray emission and rapid variability). The horizon magnetosphere driven jet (see Chap. 9) is identified as the main contributor to the mm peak of the blazar spectra. A BL Lac object can be an almost naked horizon jet. Quasars on the other hand are identified with rapidly rotating central black holes which therefore have large ergospheres. The cm peaked synchrotron component is associated with an ergospheric disk driven wind (see Chap. 8) in quasars and this dominates the emission on VLBI (parsec) scales.

10.2.1 The Distribution of Poloidal Magnetic Flux

The structure of the central engine depends strongly on the distribution of poloidal magnetic flux since the energy flux in the minimum torque wind $\sim \Phi^2$ (see 5.51). The essential physics is that the accretion disk is a pathway for magnetic flux toward the black hole, not a sink for magnetic flux. This is true for any flux created within the disk by dynamo effects or that which is accreted from large distances. Secondly, as discussed in Sect. 8.1, the horizon is not a likely to be an efficient sink for magnetic flux either. However, we will discuss the possibility of magnetic flux being temporarily trapped between the inner edge of the accretion disk and the event horizon (see Fig. 8.1). A large uncertainty in these models is the fate of the accreted flux in the inner ergosphere in the presence of a realistic resistive, reconnecting plasma (see Sect. 11.5.7 for 3-D numerical results that support this posited evolution of accreted flux).

10.2.1.1 The Horizon Boundary Condition

Figure 10.10 depicts a poloidal field distribution for an extreme case, $a/M = 0.996$, with strong fields built up by substantial accretion. The figure is particularly interesting because it illustrates the details of the fate of accreted flux (see Sect. 8.1) near the horizon in a global context. The horizon boundary condition on charge neutral flows of "attracting" the plasma and "repelling" the magnetic flux is responsible for the magnetic field reconnection near the horizon as discussed in Sect. 8.1. This is a consequence of the "no hair" theorem derived in Chap. 4. That theorem was derived with the vacuum Maxwell's equations and was extended to the case of a plasma-filled magnetosphere in Chap. 8 as is the circumstance here. In Chap. 8, it was stressed that this horizon boundary condition is merely an assumption and it is not true if large charge separation occurs in the inner accretion flow. One can cast these results in framework of the global geometry of Fig. 10.10 as well. Recall from the discussion of Sect. 8.1 that the azimuthal current in the stationary frames, $\tilde{I}_\phi \sim \alpha$, in the intermediate zone (roughly defined by $0.55M < L < 1.0M$ in Fig. 10.10) and $\tilde{I}_\phi \sim \alpha^2$ in the asymptotic zone ($L < 0.55M$ in Fig. 10.10). Thus, the total azimuthal current in the equatorial accretion flow is finite. The total $\tilde{I}_\phi$ is the source of the $l = 1$ moment of the magnetic field. By (4.89) we know that all other moments of the magnetic field must vanish at the horizon for a charge neutral accretion flow, otherwise the poloidal magnetic field would diverge at infinity. The magnetic flux through the event horizon is therefore given by (4.90c) and does not grow in time for time stationary accretion flows. The dynamics that yield the frame dragging effects in the no hair theorem, $\tilde{I}_\phi \sim \alpha^2$, are a result of the horizon boundary conditions (3.94) and (3.95). Thus, they exist in any magnetosphere whether it be vacuum or plasma-filled. This is the essential physics behind the horizon boundary condition for accreting magnetic flux described in Sect. 8.1.

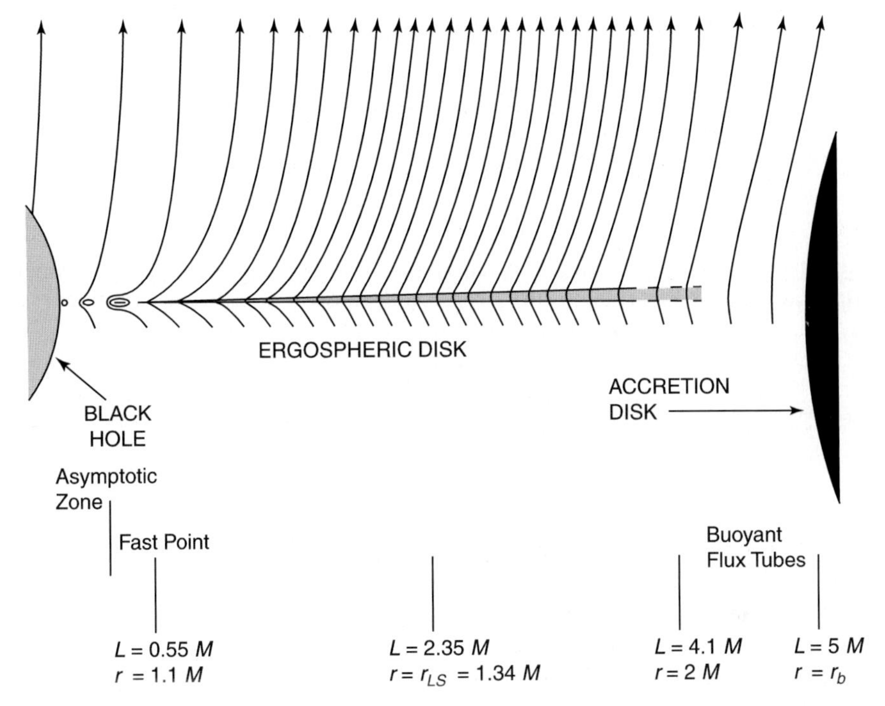

Fig. 10.10 The magnetic field structure in the environment of an extremely rapidly rotating black hole, $a/M = 0.996$. For the sake of clarity, only the magnetic field lines in the northern hemisphere are extended away from the equatorial plane. The magnetic field lines are indicated by the family of curves with arrowheads. Notice that the magnetic field strength of the flux permeating the event horizon is much weaker than in the ergospheric disk, B^P_E. Various points of interest that are discussed in the text are indicated by both their Boyer–Lindquist radial coordinate, r, and their proper distance from the event horizon, L [153]

10.2.1.2 Buoyant Magnetic Flux Tubes

The reconnection process (described in Sect. 8.1) strips accreting magnetic flux tubes of their plasma as they approach the event horizon. The flux tubes stripped of their plasma become buoyant and are pushed outward by magnetic pressure at the Alfvén speed (see Fig. 8.5). The buoyant flux tubes get stuck in the annular gap (between the horizon and accretion disk) because surface currents along the highly conductive inner edge of the accretion disk or flow prevents re-entry into the disk for as long as diffusion timescales. As this process continues, the magnetic field strength in the ergosphere, B^P_E, can grow substantially. Eventually, the magnetic pressure, $\left(B^P_E\right)^2/8\pi$, can push the inner edge of the accretion disk out beyond the last stable orbit at r_{LS}.

10.2.1.3 The Magnetosphere-Accretion Flow Interface

The radial momentum equation integrated across the boundary between the magnetosphere of trapped flux tubes (designated by the subscript M) and the accretion flow (designated by the subscript D) at $r \equiv r_b$ yields the pressure balance condition,

$$(P_r + P_g + P_{ram} + P_B)_M = (P_r + P_g + P_{ram} + P_B)_D \,, \tag{10.2}$$

where P_r, P_g, P_{ram} and P_B are the radiation, gas, ram and magnetic pressures in the radial direction, respectively. At the boundary, we expect a fairly isotropic radiation pressure,

$$(P_r)_M \approx (P_r)_D \,. \tag{10.3}$$

Furthermore, the gas density is much smaller at $r < r_b$ than in the disk, so

$$(P_g)_M \ll (P_g)_D \,. \tag{10.4}$$

When accretion advects predominantly the same direction of B^P_D for any length of time, the flux trapped inside r_b will continue to grow so that

$$(P_B)_M \gg (P_B)_D \,. \tag{10.5}$$

Equations (10.2)–(10.5) imply that a magnetized accretion flow in the equatorial plane satisfies the approximate pressure balance at r_b,

$$(P_B + P_{ram})_M \approx (P_G + P_{ram})_D \,. \tag{10.6}$$

The critical magnetic field strength, B^P_c, is defined through (10.6) as the magnetic pressure necessary to choke the accretion flow, i.e., $(P_{ram})_M = 0$:

$$B^P_c \equiv \sqrt{8\pi}\,(P_g + P_{ram})_D^{\frac{1}{2}} \,. \tag{10.7}$$

When $B^P_M \ll B^P_c$, one can have an accretion disk with a well defined inner boundary at r_b. The interface resembles a classic Rayleigh–Taylor instability (gas pressure balancing the effective gravity including centrifugal and Coriolis forces); a gravitational instability in which the magnetic field plays a negligible role. The accretion proceeds nonuniformly inward of the boundary as a result of instabilities, resulting in the sporadic inflow of large plasma clouds that cross the inner boundary of the accretion disk and fall toward the horizon.

When $B^P_M \approx B^P_c$ (as defined in 10.7), the Kruskal–Schwarzschild hydromagnetic analog of the Rayleigh–Taylor instability is attained. Magnetic flux accretes into the ergospheric gap in the form of flutes as in flat spacetime [77]. A full general relativistic calculation shows that the flutes spin about their symmetry axis due to Coriolis forces in their rest frame induced by the rotating geometry (the dragging of inertial frames).

10.2.1.4 The Critical Magnetic Field Strength

According to (10.7) the magnetic field strength in the ergosphere can not exceed B^P_c, hence the term "critical magnetic field." The most interesting point concerning B^P_c is that if the accretion of magnetic flux persists long enough for B^P_M to attain B^P_c then B^P_M and B^P_E in the inner regions of the ergosphere will remain approximately constant as more flux is accreted. In this scenario, the Kruskal–Schwarzschild interface is dynamic as magnetic pressure pushes it farther outward and one can have $r_b > r_{LS}$. Figure 10.10 is an extreme example of this circumstance. The scenario sketched above would result from the intense accretion of a mildly magnetized plasma for long periods of time (i.e., $> 10^6$ years), as is likely in quasar central engines. In such a scenario there is probably no Keplerian disk with a well defined inner edge, but a continuous accretion flow toward the horizon. Even though poloidal flux of both signs is likely to be accreted, over long periods of time, the value of the magnetic field experiences stochastic variations and a substantial nonzero field in the ergosphere can be attained. The magnetic flux in the ergosphere, Φ_E, is much less than the total magnetic flux accreted, Φ_{acc}, during the accretion history of the AGN,

$$\Phi_E^2 \ll \int \frac{\mathrm{d}}{\mathrm{d}t} [\Phi_{acc}]^2 \, \mathrm{d}t \, . \tag{10.8}$$

If Φ_E becomes large in a luminous quasar, it can create enough magnetic pressure to push the interface with the accretion flow out beyond the last stable orbit, r_{LS}, as shown in Fig. 10.10. Such a scenario requires large accretion of angular momentum onto the central black hole and we expect a/M to approach its maximum value. This is reflected in the choice of $a/M = 0.996$ in Fig. 10.10. The geometry in the figure represents the most powerful central engine attainable in black hole GHM. This extreme accretion state should be a rare occurrence in AGN.

The value of B^P_c depends on the details of the accretion process. Consider an accretion flow with a bolometric luminosity L_0. The efficiency ε is defined in terms of the mass accretion rate by

$$L_0 \equiv \varepsilon \dot{M} c^2 \, . \tag{10.9}$$

For an "α disk model" [6] (to avoid confusion with the lapse function, α of general relativity in the discussion of accretion disks we denote this measure of viscous dissipation as α_v), the gas pressure is small compared to the ram pressure at r_b and [154] estimates for $L_0 > 10^{45}$ ergs s^{-1},

$$B^P_c \approx 7.0 \times 10^3 \left[L_{46} \left(\frac{0.1}{\varepsilon} \right) \right] \sqrt{\alpha_v} \, (M_9)^{-3/2} \left(\frac{2M}{r_b} \right)^{3/4} \mathrm{G} \, , \tag{10.10}$$

where L_{46} is L_0 in units of 10^{46} ergs s^{-1} and M_9 is the black hole mass in units of $10^9 M_\odot$. For $L_0 < 10^{45}$ ergs s^{-1}, (10.10) is probably an underestimate because $(P_g)_D$ is likely to be significant in the pressure balance at the interface given by (10.7).

For very large accretion rates, a Keplerian disk at r_b is probably unrealistic and the accretion flow across r_{LS} is most likely continuous. For $L_0 \gtrsim 10^{46}$ ergs s^{-1}, the flow inside of r_{LS} probably does not have centrifugal forces that can approximately balance radial gravity, implying almost a free-fall accretion flow. Pressure balance at the interface to the magnetosphere becomes

$$\frac{\left(B_c^P\right)^2}{8\pi} \equiv \frac{\left(B_M^P\right)^2}{8\pi} \approx \left(P_{ram}\right)_A , \tag{10.11}$$

where $\left(P_{ram}\right)_A$ is the ram pressure of the accretion flow. The accretion flow can compress the magnetosphere of the ergospheric disk until a balance is achieved allowing the radial momentum equation (10.2) to be satisfied. Such flows are considered in [154] where it is found that

$$B_c^P \approx \sqrt{L_{46}\left(\frac{0.2}{\varepsilon}\right)\left(\frac{\beta^P M}{h}\right)}\, M_9^{-1}\left[1.5\times 10^4\,\mathrm{G}\right] , \tag{10.12}$$

where h is the “height” of the accretion flow at r_b. For example, if the flow were to resemble an accretion disk then h is the thickness of the disk. The value of β^P is the poloidal velocity just upstream of the interface.

Both expressions (10.10) and (10.12) are extremely model dependent. Consequently, it is not worth polluting this book with the adhoc details. What is interesting is that the two results agree for the interesting case $L_0 \sim 10^{47}$ ergs s^{-1} (i.e., a luminous quasar). For $\varepsilon \approx 0.2$, $\beta^P M/h \sim 1$ and $M \sim 10^9 M_\odot$, we find in either scenario (either 10.10 or 10.12) that

$$B_c^P \sim 3.5 - 5.0\times 10^4\,\mathrm{G} . \tag{10.13}$$

10.2.2 The Structure of the Ergospheric Disk

There are two sources of plasma in the ergospheric disk. One is a pair plasma condensate that is present regardless of the accretion dynamics (see Chap. 8). The other component is related to accretion. Note, as a consequence of geometrical thickness differences most of the accretion flow passes above and below the thin ergospheric disk (see Figs. 10.10 and 8.48). The thin equatorial flow that impacts the outer edge of the ergospheric disk is probably braked by a shock. However, as accretion plasma flows by the magnetic field lines above and below the ergospheric disk, a Kelvin–Helmholtz instability develops (see [155] for the neutron star analogy). The plasma becomes shredded into droplets. After further cycles of Kelvin–Helmholtz instabilities generates smaller and smaller “cat’s eyes,” the plasma is able to diffuse onto magnetic field lines on timescales less than the infall time to the hole. Nearly one-tenth of the mass inflow can, in principle, become threaded on the flux tubes before reaching the horizon. This plasma will condense into the ergospheric disk by the same physical mechanism that draws pair plasma into the disk that was described in Chap. 8.

The extreme case of Fig. 10.10 is useful for elucidating the fine structure of the ergospheric disk. The key elements are the following:

1. An interface with the accretion disk or flow at which magnetic flux accretes via the Rayleigh–Taylor instability at r_b, a proper distance $L = 5.0M$ from the event horizon.
2. A gap filled with buoyant flux tubes, in which there is very little power generated, located in the region $4.1M < L < 5.0M$.
3. Plasma-threaded flux tubes anchored in an ergospheric disk, extending from $L \lesssim 1.0M$ to $L = 4.1M$ in Fig. 10.10. Most of the energy flux is generated in this region.
4. An intermediate zone (as discussed in Sect. 8.1) in which gravity is approximately balanced by $\mathbf{J} \times \mathbf{B}$ forces in the disk and B^P starts dying off like α as the flow approaches the horizon. Significant energy flux is emitted from this region of the disk located a proper distance $0.5M$ to $1.0M$ from the horizon.
5. The vicinity of the fast critical point of the accreting ergospheric disk plasma. As in the wind theories of Chap. 5, the plasma causally decouples from the paired wind system inside of the fast critical point. The inner edge of the ergospheric disk is effectively the fast point. Only plasma waves emitted from $L > 0.55M$ (outside of the fast point) can affect the energy flux driven by the ergospheric disk.
6. A reconnection zone in which the large scale B^P from the accreting plasma dies off faster than α^2, denoted by $0.2M < L < 0.5M$.
7. An asymptotic accretion zone near the horizon in which all closed magnetic loops from the reconnection zone are annihilated (see Fig. 8.4).

10.3 The Electromagnetic Power From the Three Component Central Engine

The fine structure of the central engine will be described in terms of power emitted from the three components (the horizon magnetosphere, the ergospheric disk and the accretion disk) as a function of black hole rotation, a/M and poloidal magnetic flux. Again, we could have also included an electrodynamic perfect MHD jet from the horizon magnetosphere instead of the GHM jet. However, our interest here is to give examples of GHM at work and assess the relative strengths of the various putative dynamos. We assume axisymmetry and quasi-stationary configurations.

The two relevant physical facts are the following:

1. In Sect. 10.2 we discussed that the accretion disk is not a sink for magnetic flux but a pathway toward the black hole. The only place where flux can be temporarily captured is in the gap between the horizon and inner edge of the accretion flow. Thus, in general, we expect the magnetic field in the disk to be much weaker than in the ergosphere $B^P_D \ll B^P_E$ (see Figs. 8.1 and 10.10).

2. The horizon magnetosphere driven wind gets weaker with fast rotation rates due to flux exclusion (see 4.90c and Sect. 10.2.1). By contrast, the ergosphere grows in volume as $a/M \to 1$, thus the ergospheric disk wind becomes much more powerful than the horizon wind as $a/M \to 1$. It is important to note that the perfect MHD simulations that are described in Chap. 11 show no evidence of the flux exclusion effect as $a/M \to 1$.

In order to quantify the power driven by the ergospheric disk we need to model the radial flow as was done in [154]. First, it was found that B^P_E was roughly a factor of two larger in the main body of the disk than at the Rayleigh–Taylor interface (see Fig. 10.10) from the relativistic Euler equation. The most important parameter is $\Omega_F \approx \Omega_{min}$ in the disk. This quantity is plotted for the example of $a/M = 0.996$ in Fig. 10.10, as a function of lapse function, α, in Fig. 10.11 and as a function of proper distance from the horizon, L, in Fig. 10.12.

The most significant result of this section is the comparison of the electromagnetic luminosities of the three components of the central engine as a function of a/M that are plotted in Fig. 10.13 (based on the modelings of [154]). The one difference in this book from [154], is that the horizon magnetospheric outgoing wind is assumed to be the minimum torque solution.

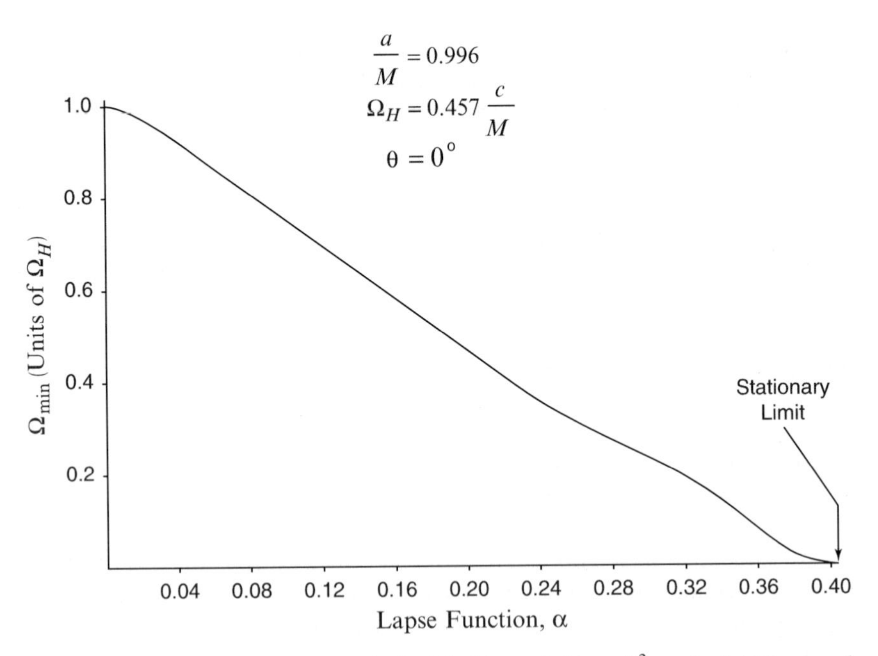

Fig. 10.11 The energy flux in the ergospheric disk driven wind is $\propto \Omega_F^2$, so the field line angular velocity is an extremely important parameter. The theory of the ergospheric disk that was developed in Chap. 8 sets $\Omega_F \gtrsim \Omega_{min}$. This figure is a plot of Ω_{min} as a function of lapse function, α, in the equatorial plane of the geometrical configuration that is depicted in Fig. 10.10

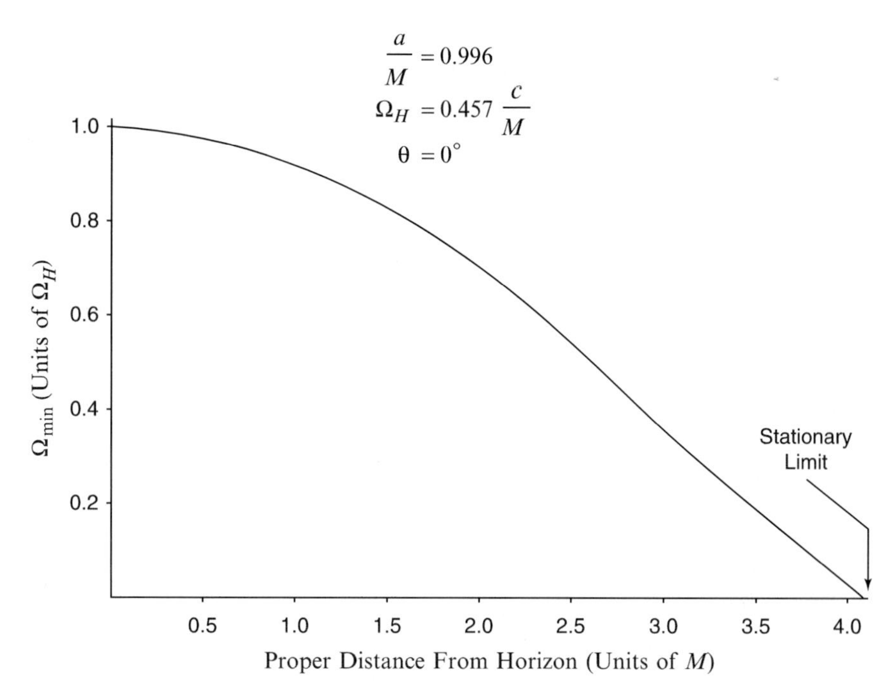

Fig. 10.12 The minimum angular velocity as viewed from asymptotic infinity that is allowed by the dragging of inertial frames, Ω_{min}, as a function of proper distance from the event horizon, L, in the equatorial plane of the geometrical configuration that is depicted in Fig. 10.10

The dashed curves for S^P_D represent the **maximum** Poynting flux from an accretion disk driven wind consistent with pressure balance at the interface to the annular gap at r_b. The poloidal field in the disk, B^P_D, can not be made arbitrarily large, relative to B^P_E in the ergosphere. This is the important aspect of modeling all three central engines simultaneously. S^P_D is plotted for two cases that were modeled in [154] where $r_b = 2M$ and $r_b = 10M$. However, it should be emphasized that physically one should expect S^P_D to be much less than these upper bounds. As mentioned earlier, we could have also included an electrodynamic jet as envisioned in [66, 111] emanating from the event horizon magnetosphere, instead of a GHM horizon jet. The strength of the Poynting flux, S^P_{B-Z}, of a putative electrodynamic jet can be compared to the other power sources in Fig. 10.13 by inspection of Fig. 11.25. The upper bound for S^P_{B-Z} is the same as the upper bound for S^P_D, $r_b = 10M$ to within a factor of 1.5 for $a/M > 0.5$. To understand how an ergospheric disk can coexist with the [66, 111] electrodynamic jet, the reader should consult the numerical simulations in Chap. 11, in particular Sect. 11.4. The bottom line is that at high spin rates the ergospheric disk jet is $\approx$ 10 - 100 times as powerful as the electrodynamic jet.

The Poynting flux of the horizon driven wind, S^P_H, was calculated using $\Omega_F = (1/10)\Omega_H$ which is at the high end of plausible field line rotation rates (see Chap. 9). Note the effect of flux exclusion as $a/M \to 1$. If the perfect MHD simulations of

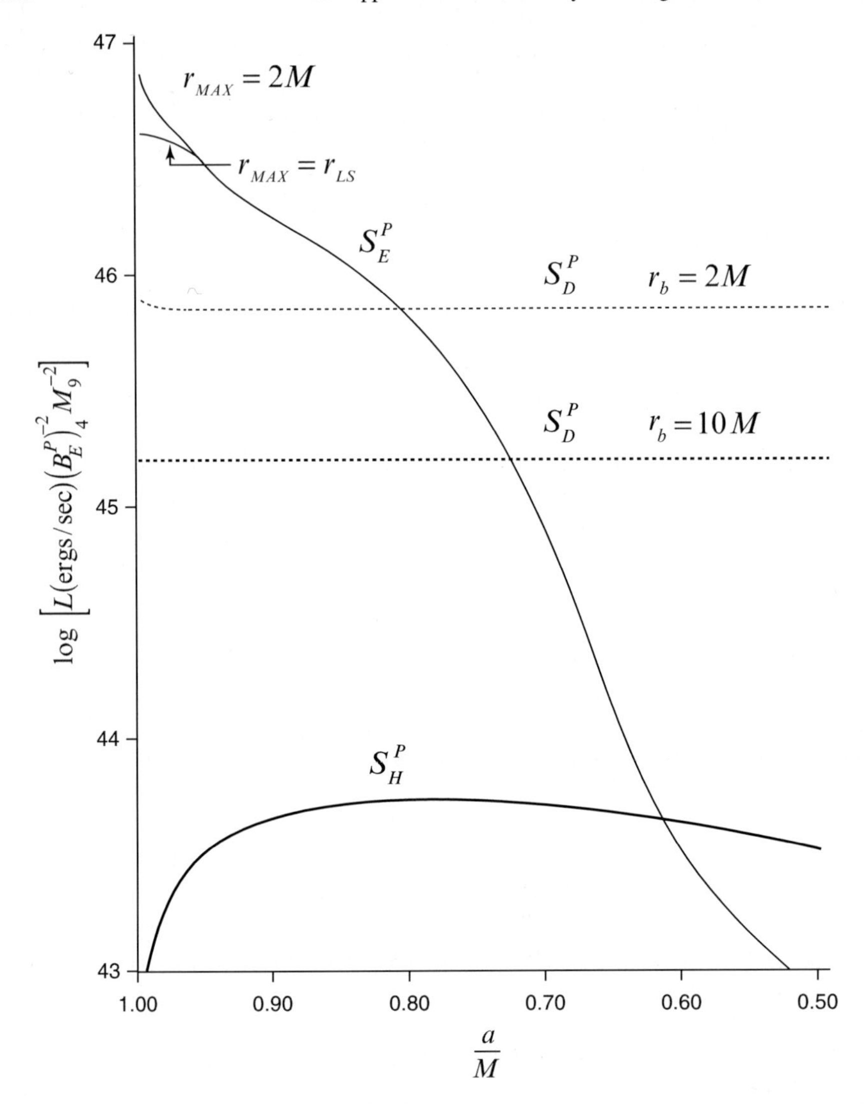

Fig. 10.13 The logarithm of the luminosity, L, of the three components of the black hole central engine as a function of the rotation rate of the black hole, a/M. The luminosity, L, has been scaled to an ergospheric poloidal magnetic field strength of 10^4 G and a black hole mass of $10^9\,M_\odot$. This is accomplished by dividing L by $\left(B_E^P\right)_4^2$ (the square of the ergospheric poloidal field strength in units of 10^4 G), as well as M_9^2 (the square of the black hole mass in units of $10^9\,M_\odot$), as indicated by the label on the vertical axis. The Poynting flux that is transported by the ergospheric disk wind is designated by the curve labeled, S_E^P. Similarly, the wind driven by the event horizon magnetosphere has a Poynting flux, S_H^P. The Poynting flux from the magnetized accretion disk, S_D^P, is in general far below the upper bounds (the dashed lines) indicated in the figure (see the text for details). Fig. 10.13 is the "Rosetta Stone" for interpreting the correlations with the parameter, Δ, noted in Sect. 10.1. It is interesting to compare this figure to the numerically generated data in Fig. 11.25

Chap. 11 are any indication then the vacuum flux exclusion as $a/M \to 1$ is not a valid assumption when a plasma is present. It is straightforward to see the effect of dropping this assumption in Fig. 10.13, the high spin region of S_H^P does not tail downward, but stays approximately level as $a/M \to 1$.

As the rotation rate increases, the size of the ergosphere increases rapidly (compare Figs. 10.12 and 9.5). Consequently, since Ω_H increases as well, we see a rapid increase in the Poynting flux from the ergospheric disk, S_E^P, as $a/M \to 1$. Note that there are two branches to the curve describing S_E^P as $a/M \to 1$. For $a/M > 0.94$, the last stable orbit, r_{LS}, moves inside of the ergosphere at $r = 2M$. Only flux rubes at $r < 2M$ participate in the ergospheric disk dynamics. There is no ambiguity when $a/M < 0.94$. However, when $a/M > 0.94$, there are two possibilities. For low or moderate magnetized accretion, the outer boundary of the ergospheric disk is at $r = r_{LS}$. However, for strong magnetized accretion as in Fig. 10.10, magnetic pressures can grow in the ergosphere that push the Rayleigh–Taylor interface out beyond r_{LS} and $r = 2M$. In this situation the outer edge of the ergospheric disk is at $r = 2M$. Hence, the two branches for S_E^P at the upper left corner of Fig. 10.13 corresponding to the outer boundary of the ergospheric disk being either at $r = r_{LS}$ or at $r \geq 2M$.

Since there is a critical magnetic field strength (or maximum B^P) in the ergosphere by the discussion of Sect. 10.2, there is a critical luminosity $L_c \equiv (S^P)_{max}$ associated with the three central engines in Fig. 10.13. For a high accretion system with a corresponding viscous bolometric luminosity of $L_0 > 10^{46}$ ergs s^{-1}, we expect the hole to be spun up to $a/M \lesssim 1$. In the process, B_c^P can be accreted and the inner edge of the accretion flow is likely to be pushed out beyond r_{LS} in such a circumstance. This essentially makes $\varepsilon \approx 0.2$ in the relation for the critical field (10.12) (see [6]). Using (10.12) in conjunction with Fig. 10.13, the critical luminosity of the ergospheric disk is

$$(L_E)_c \approx 20L_0\,, \quad L_0 > 10^{46}\,\text{ergs s}^{-1}\,. \tag{10.14}$$

This relation should hold in the most powerful radio loud quasars such as 3C 9 or 1318+113 which have total mechanical powers $Q \sim 10^{48}$ ergs s^{-1} supporting the radio lobes (see the long discussion in Chap. 1).

The critical luminosity of the horizon magnetospheric wind requires some attention. There is both the high luminosity case, $L_0 > 10^{45}$, ergs s^{-1} and the low accretion luminosity case, $L_0 < 10^{45}$ ergs s^{-1}. For prolonged strong accretion, $a/M \lesssim 1$ and $(L_H)_c$ is very small due to flux suppression as indicated in Fig. 10.13. The maximum of S_H^P is at $a/M \approx 0.8$ in Fig. 10.13. Strong accretion for a short period of time, moderate accretion, or electromagnetic spin down of the hole from $a/M \lesssim 1$ with slow accretion are three distinct ways of obtaining $a/M \approx 0.8$. In the moderate accretion scenario we pick $\varepsilon \approx 0.1$ in (10.12) for B_c^P, then Fig. 10.13 implies that

$$(L_H)_c \approx 10^{-2}L_0\,, \quad L_0 > 10^{46}\,\text{ergs s}^{-1}\,. \tag{10.15}$$

In the low disk luminosity case as in a BL Lac object, the lower accretion rates imply that the plasma state is in the regime known as an advected-dominated accretion flow [156]. Such two temperature plasmas are inefficient radiators and much of the viscous dissipation stays trapped in the plasma as it accretes toward the horizon. As such, low luminosity systems have strong ram pressures at the inner edge, therefore (10.12) is still a reasonable approximation for B^P_c. For BL Lac objects, estimating L_0 is a guess because it is swamped by the high frequency tail of the synchrotron component [157]. The quiescent states seen in the long term optical variability studies of [158] and [159] are consistent with BL Lacs having accretion luminosities similar to Seyfert 1 galaxies. Thus, we pick a moderate bolometric luminosity typical of a Seyfert 1 galaxy to compute $(L_H)_c$, $L_0 \approx 10^{44}$ ergs s^{-1} or $L_0 \approx 10^{-3} L_{Edd}$. The typical advection dominated accretion rate is given in [156] to be $\dot{M}c^2 \approx 10^{-1} L_{Edd}$. Thus, from (10.9) we use $\varepsilon \approx 10^{-2}$ in (10.12) for the critical field strength,

$$B^P_c \approx 6.7 \times 10^3 M_9^{-1}\,\mathrm{G}\,, \quad L_0 \approx 10^{44}\,\mathrm{ergs\,s^{-1}}\,. \tag{10.16}$$

Applying (10.16) to Fig. 10.13 for advection dominated accretion, we obtain,

$$(L_H)_c \approx 2.5 \times 10^{43}\,\mathrm{ergs\,s^{-1}}\,, \quad L_0 < 10^{45}\,\mathrm{ergs\,s^{-1}}\,. \tag{10.17}$$

The results of this section are extremely model dependent. The basic fact that as $a/M \to 1$ (if there is a substantial magnetic flux in the ergosphere), the ergospheric disk (which is powered by black hole rotation) should be the dominant energy source in the central engine. This fact is independent of modeling. It results from the large ergospheric disk surface area as $a/M \to 1$ over which large values of $\Omega_F \approx \Omega_{min} \sim \Omega_H$ exist (see Fig. 10.12).

10.4 Applications of the Theory

In this section, we use the models that were constructed in Sects. 10.2 and 10.3 to describe the unified scheme for radio loud AGNs (see Chap. 1), the correlations of physical observables with the shape of blazar broadband spectra (through the parameter Δ that was introduced in Sect. 10.1) and the cosmological evolution of extragalactic radio sources.

10.4.1 Interpreting the Unified Scheme

This subsection, develops a four parameter model of AGN central engines, $\dot{M}$ (the mass accretion rate), a/M (the black hole rotation rate), B^P_E (the magnetic field strength near the black hole) and M (the mass of the black hole). When B^P_E is small (i.e., less than ~ 10G), one has a radio quiet AGN. From the discussion of black hole masses in Chap. 1 and the association of radio loud AGNs with large elliptical host galaxies, we choose a fiducial black hole mass, $M \approx 10^9\,\mathrm{M}_\odot$. Also, a/M and $\dot{M}$ are not independent. A large accretion rate implies $a/M \sim 1$. The parameter space analysis is illustrated through Figs. 10.14 and 10.15 with $M \approx 10^9\,\mathrm{M}_\odot$ assumed throughout for simplicity.

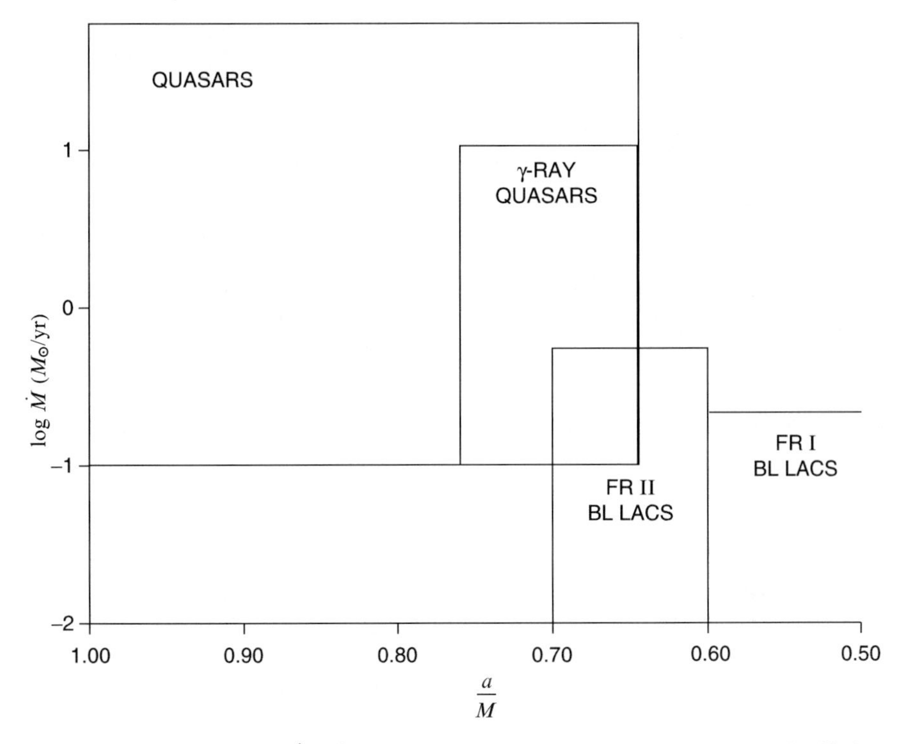

Fig. 10.14 A two parameter, ($\dot{M}$, a/M), plot of the various classes of objects that are identified as powerful compact extragalactic radio sources when they are observed along the jet axis. Cosmological evolution proceeds from the upper left to the lower right in the diagram [153]

10.4.1.1 FR I BL Lac Objects

The vast majority of BL Lac objects are associated with FR I radio structures. There are a few with FR II morphologies and luminosities and we will discuss these interesting objects later in this subsection. FR I jets tend to be tightly collimated within 100 pc - 1 kpc of the source and are mildly relativistic. On kiloparsec scales the jets are subrelativistic and are loosely collimated (see Figs. 1.2 and 1.5). By contrast, FR II jets remain tightly collimated (see Figs. 1.6–1.8) far from the central engine, on scales $\sim$ 10–100 kpc, and are likely to be mildly relativistic as well. We use these distinctions to probe the fine structure of the central engines of FR I BL Lac objects.

In order for the inn er jet in the nested jet system in Figure 10.9 to be loosely collimated requires that the ergospheric jet be weak compared to the horizon jet. This condition can be described in terms of the radiated powers of the two jets (see Fig. 10.15d),

$$L_H > L_E \, . \tag{10.18}$$

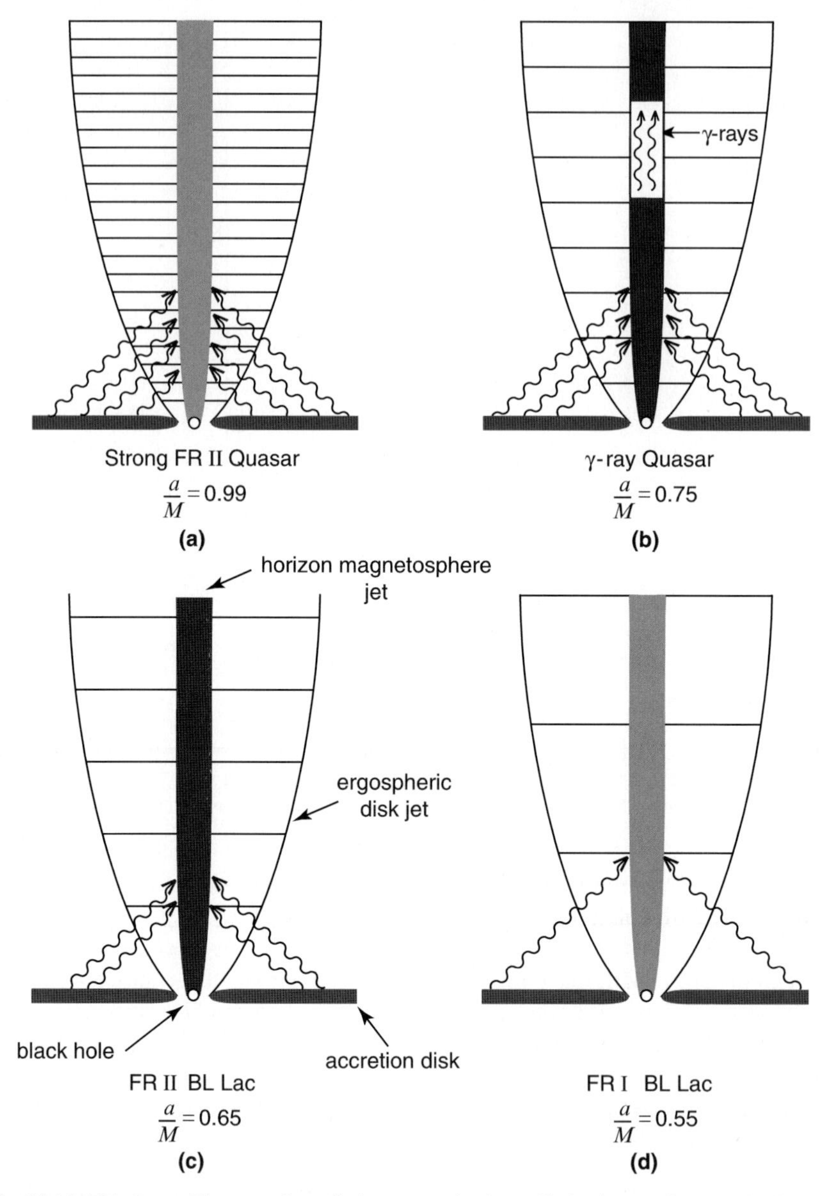

Fig. 10.15 This figure illustrates four distinct states in the unified scheme of extragalactic radio sources in the context of the dual jet model of black hole GHM. The strength of the ergospheric disk jet, L_E, is indicated by the density of horizontal hatches and the strength of the horizon magnetosphere driven jet, L_H, is indicated by the intensity of the shading. The luminosity of the accretion disk due to viscous dissipation, L_0, is represented by the number of photons emanating from the disk. The accretion rate, $\dot{M}$, is directly related to L_0. Cosmological evolution proceeds from **(a)** to **(d)**

The plots in Fig. 10.13 indicate that the condition, (10.18), is satisfied for $a/M < 0.60$. Note that the collimation condition is empirical and is not derived from first principles. However, consider the following suggestive argument. If the ergospheric disk drives tightly collimated relativistic jets then the collimation is most likely a consequence of a toroidal magnetic field applying hoop stresses [52, 53]. Furthermore, the toroidal magnetic field is a direct measure of the Poynting flux (see 5.21 and 5.51) and therefore, L_E. If $L_E > L_H$, one would expect that hoop stresses in the enveloping ergospheric disk wind would collimate the inner horizon magnetosphere driven jet in Fig. 10.9.

Equation (10.17) for the maximum horizon jet luminosity, $(L_H)_c$, and (10.18) are consistent with the maximum mechanical power, Q, transported through the jet from the central engine to the extended radio lobes in FR I radio galaxies. It was found in [42], in analogy to the results in Table 1.2, that $Q < 5 \times 10^{43}\,\mathrm{ergs\,s^{-1}}$ for FR I radio galaxies. By setting the total power from the ergosphere equal to the energy transported to the lobes, we have

$$L_H + L_E = Q\,. \tag{10.19}$$

Equation (10.19) combined with (10.18) and (10.17) imply a maximum central engine luminosity in the black hole GHM model of a BL Lac (for a black hole rotation rate of $a/M \approx 0.6$) of $L_E + L_H \approx 2L_H \approx 4 \times 10^{43}\,\mathrm{ergs\,s^{-1}}$. Significantly, this is roughly the maximum energy flux transported to the lobes in an FR I radio galaxy. Thus, it is proposed that the central engine in a BL Lac object (or FR I radio galaxy, depending on viewing angle) is primarily an event horizon magnetosphere driven jet.

It is possible that the accretion disk driven magnetized wind can be competitive in luminosity with the event horizon driven wind in FR I radio galaxies. However, this would require poloidal magnetic field strengths $\sim 500\,\mathrm{G}$ in the accretion disk of a strong FR I radio galaxy (see Fig. 10.13). Physically, it is hard to understand why poloidal large scale fields of such magnitude would exist in the accretion disk.

Note that the condition (10.18) combined with Fig. 10.13 describes what is known as the "FR I/FR II break." The morphology of radio loud AGNs is roughly bimodal and the segregation point is approximately given by the luminosity criterion that if $Q < 5 \times 10^{43}\,\mathrm{ergs\,s^{-1}}$ (or equivalently, by the results of Table 1.2 and [42], an extended radio power of $P_E < 5 \times 10^{42}\,\mathrm{ergs\,s^{-1}}$) then the source has an FR I morphology and is an FR II radio source otherwise.

10.4.1.2 FR II Radio Sources (Quasars)

When a galactic nucleus experiences quasar activity, one expects large amounts of angular momentum to be accreted toward the black hole from the luminous accretion flow. This is consistent with large values of a/M and, by Fig. 10.13, a strong ergospheric disk wind that dominates the other winds, energetically. There is a wide range of quasar radio luminosities. If one selects radio quasars based on the core luminosity as in [46], one finds that the extended luminosity of these sources, if they were viewed in the sky plane, P_{sky}, is much smaller than what is normally selected in low frequency radio flux-limited samples (these are the catalogs where lobe dominated quasars are generally found). The point of defining P_{sky} in [46] is to subtract out the Doppler enhancement of the one-sided jet on VLA scales that dominates the resolved radio power in these sources. The residual radio structure, after the subtraction of the jet contribution, represents the luminosity of the lobes. Thus, P_{sky}, can be used to find the energy in the lobes, Q, as in Table 1.2. Most radio loud quasars have central engine luminosities of 5×10^{43} ergs s^{-1} $< Q < 5 \times 10^{44}$ ergs s^{-1} with $Q \approx 10P_{sky}$. The strongest quasar central engines have $Q \sim 10^{48}$ ergs s^{-1}.

We decompose this huge range of central engine strengths into different categories for illustrative purposes. First, consider the common, weak FR II radio galaxies and quasars that were described above, 5×10^{43} ergs s^{-1} $< L_E < 5 \times 10^{44}$ ergs s^{-1}. For a "slowly rotating" black hole, Fig. 10.13 implies that,

$$2.85 \times 10^3\,\mathrm{G} < B^P_E < 8.99 \times 10^3\,\mathrm{G}\,, \quad \frac{a}{M} = 0.70\,, \tag{10.20a}$$

$$4.3 \times 10^{42}\,\mathrm{ergs\,s^{-1}} < L_H < 4.3 \times 10^{43}\,\mathrm{ergs\,s^{-1}}\,, \quad \frac{a}{M} = 0.70\,. \tag{10.20b}$$

The magnetic field strength is near its critical value in this case.

For a rapid rotator, a weak FR II radio source requires a weaker field strength than was found for a slow rotator as a consequence of Fig. 10.13,

$$4.28 \times 10^2\,\mathrm{G} < B^P_E < 1.35 \times 10^3\,\mathrm{G}\,, \quad \frac{a}{M} = 0.90\,, \tag{10.21a}$$

$$8.2 \times 10^{40}\,\mathrm{ergs\,s^{-1}} < L_H < 8.2 \times 10^{41}\,\mathrm{ergs\,s^{-1}}\,, \quad \frac{a}{M} = 0.90\,. \tag{10.21b}$$

The most likely configuration is the one that requires the weakest magnetic field strengths in the ergosphere. The hardest thing to justify physically in the model are the existence of large poloidal magnetic field strengths in the ergosphere. A near extreme rotator is parameterized by $a/M = 0.996$ as in Fig. 10.10. However, now $B^P_E \ll B^P_c$, so the outer edge of the ergospheric disk is at r_{LS}. We find the following parameters from Fig. 10.13,

$$2.87 \times 10^2\,\mathrm{G} < B^P_E < 9.07 \times 10^2\,\mathrm{G}\,, \quad \frac{a}{M} = 0.996\,, \tag{10.22a}$$

$$4.1 \times 10^{39}\,\mathrm{ergs\,s^{-1}} < L_H < 4.1 \times 10^{40}\,\mathrm{ergs\,s^{-1}}\,, \quad \frac{a}{M} = 0.996\,. \tag{10.22b}$$

The typical FR II radio sources found in deep low frequency surveys have $Q \sim 10^{45}$ ergs s^{-1}. We tabulate the parameter values of these sources and the rest of the FR II population in Table 10.2. Column (1) is the energy flux of the ergospheric disk wind which is roughly equal to the mechanical power in the radio lobes, $L_E = Q$. The second column is the parameter, a/M, representing black hole rotation. Column (3) is the power in the horizon magnetosphere driven jet, L_H, and column (4) is the poloidal magnetic field strength in the ergosphere, B^P_E.

Table 10.2 Parameter space of FR II radio sources

L_E (ergs s^{-1})	aM^{-1}	L_H (ergs s^{-1})	B_E^P (G)	Notes
5×10^{43}	0.70	4.3×10^{42}	2.85×10^{3}	
5×10^{43}	0.90	8.2×10^{40}	4.28×10^{2}	
5×10^{43}	0.996	4.1×10^{39}	2.87×10^{2}	a
5×10^{44}	0.70	4.3×10^{43}	8.99×10^{3}	
5×10^{44}	0.90	8.2×10^{41}	1.35×10^{3}	
5×10^{44}	0.996	4.1×10^{40}	9.07×10^{2}	
10^{45}	0.75	1.9×10^{43}	5.90×10^{2}	
10^{45}	0.996	8.3×10^{40}	1.29×10^{3}	
10^{46}	0.85	3.6×10^{43}	8.04×10^{3}	b
10^{46}	0.996	8.3×10^{41}	4.07×10^{3}	
10^{48}	0.95	6.2×10^{44}	4.33×10^{4}	c
10^{48}	0.996	6.3×10^{43}	3.54×10^{4}	c

a) A highly probable configuration
b) A Cygnus A type of radio source
c) The most powerful known radio sources such as 3C 9 and 1318 + 133

The final entry in Table 10.2 is compelling evidence for the model. By (10.13), the maximum magnetic field strength allowed in the ergosphere of a maximally rotating black hole yields the maximum observed intrinsic power in radio sources.

10.4.1.3 Intermediate Objects

Thus far, we have described the extreme ranges of parameter space, the slow rotators (FR I radio galaxies and BL Lacs, $a/M < 0.60$) and rapid rotators (FR II radio galaxies and quasars, $a/M > 0.70$). However, there are interesting objects within the transitional region $0.8 > a/M > 0.6$. These include BL Lac objects with FR II radio luminosities and morphologies and the γ-ray loud quasars detected by EGRET.

Recall the contradiction with a single jet model for EGRET sources with strong FR II emission noted in the first section of this chapter. The stronger ergospheric disk jet is associated with powerful radio emission. Consequently, according to the discussion of Sect. 10.1, hard γ-rays must come from the horizon jet in these EGRET sources. An example of such a source is 3C 279 with a mechanical power of $Q \approx 2\times10^{45}$ ergs s^{-1}, in the radio lobes.

The intrinsic γ-ray luminosity, L_γ, produced by inverse Compton scattering has the following scaling property,

$$L_\gamma \propto U_r N \gamma_{th}^2 \,, \tag{10.23}$$

where U_r is the radiation density, N is the pair density and γ_{th} is the Lorentz factor from thermal motion evaluated in the rest frame of the γ-ray emitting plasma. There are many possible physical processes that can convert the Poynting flux of the horizon jet into thermal inertia (e.g., shocks and plasma instabilities). In order to get most of the jet (thermal) energy released as hard γ-rays requires a strong soft radiation source U_r in (10.23) that is inverse Compton upscattered.

There are two plausible sources of soft radiation. First, there are the synchrotron photons from the jet itself, the SSC scenario. Second, there are disk photons (possibly reprocessed in the dusty torus, accretion disk corona or broad emission line gas), the ECS scenario discussed in Sect. 10.1. A strong U_r in the SSC scenario requires a large B^P_E as does a strong horizon wind (i.e., the $N\gamma^2_{th}$ factor in (10.23) represents internal energy in the jet that was created from the electromagnetic energy reserves). The large value of B^P_E required for a strong horizon jet implies $B^P_E \sim B^P_c$ which is much more likely for high accretion rates. However, intense accretion radiates large soft photon fluxes and this circumstance clearly favors the ECS process as being the dominant cooling mechanism of the horizon jet in γ-ray loud quasars. Thus, it seems to be an unavoidable conclusion that ECS processes are the γ-ray sources in EGRET detected quasars.

Finding the bolometric luminosity of the accretion flow in a blazar is difficult because its IR/optical emission is masked by the high frequency synchrotron tail of the jet. In [142], the optical/UV disk luminosity was crudely extracted from the broadband spectrum of various blazars. To this result we add the IR and broad emission line luminosities to estimate $L_0 \approx 5 \times 10^{45}$ ergs s^{-1} for 3C 279. Using $\varepsilon = 0.1$ (in this moderate luminosity system) in (10.12) yields

$$B^P_c \approx 1.5 \times 10^4\,\text{G}\,. \tag{10.24a}$$

According to Fig. 10.13, if $B^P_E \approx B^P_c$ and $a/M \gtrsim 0.70$,

$$L_E = Q \approx 2 \times 10^{45}\,\text{ergs s}^{-1}\,, \tag{10.24b}$$

as observed for 3C 279, and

$$L_H = 1.3 \times 10^{44}\,\text{ergs s}^{-1}\,. \tag{10.24c}$$

The apparent gamma ray luminosity of 3C 279 observed at earth averaged over time is [142],

$$\left(L_\gamma\right)_{app} \lesssim 10^{48}\,\text{ergs s}^{-1}\,. \tag{10.25}$$

However, for γ-rays emitted from a Doppler enhanced knot or compact region of the jet, the apparent luminosity is related to the intrinsic γ-ray luminosity, L_γ, by [72],

$$\left(L_\gamma\right)_{app} = \delta^4 L_\gamma\,. \tag{10.26a}$$

The Doppler enhancement factor, δ, is given in terms of the velocity of the emitting plasma, β, propagating at an angle θ relative to the line of sight to earth by the equation,

$$\delta = \frac{\sqrt{1-\beta^2}}{1-\beta\cos\theta}\,. \tag{10.26b}$$

Comparing (10.26a) with (10.25) and (10.24c) implies that $\delta \lesssim 10$ in 3C 279 is required in the γ-ray emitting plasma by the black hole GHM model of the central engine. (This compares favorably with most independent estimates of δ in 3C 279 [31,160].)

Within the GHM model, γ-ray quasars are associated with the slowest rotating black holes that are consistent with rapid accretion. This allows for a large value of B^P_E to be established near the black hole over time (without making L_E enormous, i.e, $\sim 10^{47}$ ergs s^{-1}, for instance), which in turn creates a strong horizon jet according to Fig. 10.13. The large accretion rate provides both the magnetic flux and the seed photons that are necessary for the ECS process (see Fig. 10.15). From Table 10.2, the extreme accretion systems at the bottom of the table are viable γ-ray sources as well. However, these sources are very rare, especially at $z < 2$. Thus, there would not be many (if any) with a γ-ray flux above EGRET threshold of sensitivity. These sources are significant γ-ray emitters (if viewed end on by 10.26b) simply because they are strong in every observing band.

Physically, the γ-ray quasars could be AGNs in which accretion has just increased so that a/M is still low and a significant poloidal flux is advected in the process. Another possibility are AGNs in which there is so much magnetic flux that the electromagnetic torques on the black hole compete with the accretion of angular momentum, thereby keeping a/M moderate.

An interesting class of blazars are BL Lacs with FR II luminosities and morphologies [30]. They are generally at high redshift (for a BL Lac), $0.5 < z < 1.0$, and they include 0235+164, 0954+658, 1308+326, 1538+149, 1803+784 and 1823+568. They typically have $Q \approx 10^{44}$ ergs s^{-1}. The strongest central engine resides in 1308+326 with $Q \approx 5 \times 10^{44}$ ergs s^{-1}. These objects have the largest optical polarizations of any class of blazar including FR I BL Lacs and HPQs [154]. Thus, we expect very strong blazar or horizon jets within the GHM model. Since the jets are tightly collimated, $L_E \gtrsim L_H$ in order to provide enough magnetic hoop stresses on kiloparsec scales (see the related discussion in Sect. 10.4.1). Applying this observation to Fig. 10.13, we expect $a/M \gtrsim 0.60$ in FR II BL Lac objects.

Consider an FR II BL Lac with $a/M = 0.65$ and a value of $L_O \lesssim 10^{45}$ ergs s^{-1}. The Doppler boosted synchrotron emission from the horizon jet swamps this value of L_O making it difficult to detect in the optical band. For $\varepsilon \lesssim 0.05$, (10.12) and Fig. 10.13 imply,

$$B^P_c \approx 10^4\,\mathrm{G}\,, \quad \frac{a}{M} = 0.65\,, \tag{10.27a}$$

$$L_E \approx 10^{44}\,\mathrm{ergs\,s^{-1}}\,, \quad \frac{a}{M} = 0.65\,, \tag{10.27b}$$

$$L_H \approx 5 \times 10^{43}\,\mathrm{ergs\,s^{-1}}\,, \quad \frac{a}{M} = 0.65\,. \tag{10.27c}$$

Note that $L_H/L_O \gtrsim 5 \times 10^{-2}$ for typical FR II BL Lac and from (10.24), $L_H/L_O \approx 2.5 \times 10^{-2}$ for 3C 279. The FR II BL Lacs have the most prominent horizon jets of any FR II radio source, hence their high optical polarizations.

The taxonomy of the blazar family is indicated qualitatively in the two parameter $(\dot{M}, a/M)$ plot in Fig. 10.14. The values of $\dot{M}$ for quasars are taken from the estimates of [11]. The BL Lac object $\dot{M}$ values are difficult to estimate since the accretion disk luminosity and broad emission lines are swamped by the beamed synchrotron component from the jets. Figure 10.14 is a summary of the discussions of this section. HPQs would be concentrated in parameter space close to where the γ-ray quasars are found, since they require strong horizon jets to produce their optical polarization. This common property of HPQs and γ-ray quasars explains why the Δ factors of γ-ray quasars are more similar to those of HPQs (Fig. 10.8) than they are to the those of the core dominated quasar population as a whole (Fig. 10.7). Note the loose correlation of $\dot{M}$ with a/M in Fig. 10.14.

The panoply of blazar classes is described within the black hole GHM theory of extragalactic radio sources in Fig. 10.15. Frame (**a**) of Fig. 10.15 represents a strong FR II quasar with a large value of $\dot{M}$ and therefore a rapid black hole rotation rate, $a/M \lesssim 1$. The ergospheric disk jet is extremely powerful and L_H is moderate. Frame (**b**) represents a state that is typical of many γ-ray loud quasars in the black hole GHM model. The accretion luminosity, L_0, and $\dot{M}$ are at the low end of the range for a quasar, hence a/M is at the low end of the quasar range as well. The γ-ray luminosity from the horizon magnetosphere driven jet and L_H are near maximal in strength. An FR II BL Lac object is another intermediate radio source in the GHM theory and this blazar state is represented by frame (**c**). They are similar to γ-ray loud quasars in that L_E is moderate and both $\dot{M}$ and a/M are just slightly smaller than the typical values found for γ-ray loud quasars. The distinguishing characteristic of this class of blazar is that L_H is very large for an FR II radio source. The fourth frame represents an FR I radio galaxy (i.e., a BL Lac object if viewed along the jet axis) in the dual jet model. The accretion rate, $\dot{M}$, is small, so L_0, a/M and L_E are small. The horizon jet has a moderate luminosity, $L_H > L_E$. The role of the intermediate objects in the unified scheme is highlighted in the black hole GHM theory in Fig. 10.15. One of the strengths of the theory is the ability to describe the rare hybrid objects such FR II BL Lacs.

10.4.2 Correlations with Blazar Spectra

In Sect. 10.1, it was pointed out that the shape of the radio to submillimeter blazar spectrum was correlated with various observables. These results can be interpreted within the context of the black hole GHM model in a straightforward manner. The correlations were described in terms of the parameter, Δ, defined in (10.1) as the logarithm of the ratio of the flux density at the cm peak of the spectrum to the flux density at the mm peak.

The horizon magnetosphere driven jet tends to have a spectral peak at mm wavelengths. It is nested within the larger ergospheric disk driven jet that tends to have a spectral peak at cm wavelengths. Only the ergospheric disk jet is strong enough to power FR II radio emission. The horizon magnetospheric jet is associated with the attributes that are commonly ascribed to the blazar phenomenon. Thus it is also called the blazar jet. It is highly variable and can be highly polarized (>10%) in the high frequency optical tail of the spectrum. These properties can be used to understand the correlations found in Sect. 10.1. We list these in the order presented there.

1. QSOs and HPQs have larger values of Δ than BL Lac objects because the latter have smaller accretion rates (as evidenced by smaller disk luminosities, L_O) and therefore smaller a/M values of the central black hole. This translates, by Fig. 10.13, to a weak ergospheric disk wind and therefore a weak VLBI jet and cm peak in BL Lac objects.
2. Large values of Δ are correlated with strong extended radio emission, P_E. A large value of Δ implies a strong cm spectral peak from the VLBI jet and unresolved core. Since the cm peak is associated with the ergospheric disk jet, this implies $L_E \gg L_H$. Consequently, by Fig. 10.13, a/M is large. For these black hole rotation rates, the ergospheric disk can power strong radio luminosities.
3. Large values of Δ are correlated with large redshift. This is explained by cosmological evolution of galactic mergers and black hole accretion rates. It is believed that large accretion rates onto supermassive black holes were far more common in the past as evidenced by the quasar luminosity function. At large z, accretion was more likely to deposit large amounts of angular momentum into the black hole (which means large values of a/M) and large magnetic fluxes into the ergosphere. From Fig. 10.13, large values of B^P_E and a/M correspond to very strong ergospheric disk winds. These power strong VLBI jets and the associated unresolved cores. Knots (regions of local dissipation) in the wind produce a very strong cm peak and therefore large values of Δ. It is important to note that the cm peak in a quasar spectrum is not solely a function of the central engine power but depends on the nuclear environment as well (radiation results from dissipative interactions such as shocks, that occur from jet collisions with circumnuclear gas). Gas densities and therefore the dissipation (synchrotron radiation peaked at cm frequencies) of the ergospheric disk wind were probably much higher in the nuclear regions of QSOs in the distant past. The denser circumnuclear gas and larger $\dot{M}$ values explains the correlation of Δ with z. This correlation is essentially the same as "2" above for P_E, but is stronger because it depends on two factors (dissipation on parsec scales and L_E) that correlate with z not just L_E (99.997% statistical significance versus 99.97% statistical significance).
4. Small values of Δ are correlated with high optical polarization. Within the theory, the blazar region of the horizon magnetosphere driven wind produces a copious supply of optically polarized radiation in the high energy tail of the mm peak. Thus a strong mm peak, and therefore small Δ, implies a prominent horizon magnetospheric jet compared to the ergospheric disk wind (which is assumed to have a lower optical polarization) and high optical polarization.
5. Gamma ray quasars have smaller Δ values than other core dominated quasars. It was shown that it is energetically reasonable to associate the horizon jet with γ-ray emission. A strong horizon jet has $L_H \approx 10^{44}$ ergs s^{-1} and $\delta \approx 10$ yields $(L\gamma)_{app} \sim 10^{48}$ ergs s^{-1}. In order to obtain L_H large and moderate to small FR II

lobe energies (as observed) requires intermediate values of a/M in Fig. 10.13. These intermediate values of a/M are at the low end of the allowed values of a/M associated with large $\dot{M}$ and the quasar phenomenon. Thus, L_H/L_E is larger than it is for most quasars, hence Δ is smaller. Decoupling the γ-ray region from the strong VLBI jet driven by the ergospheric disk has several theoretical advantages. First, the Doppler factor inferred for the VLBI jet by observation need not be the same as it is in the "blazar component" : the two jet Doppler factors are independent. Second, as discussed in Sect. 10.1, if the ergospheric disk supports the FR II level extended emission in strong γ-ray quasars then there are no contradictions from energy constraints on the large scale wind set by the γ-ray emitting region, since it is decoupled.

10.4.3 Radio Source Evolution

The radio source population evolves with cosmological redshift. If one looks at the sources in the 3CR catalog [161], one can make a strong statement about the relative evolutionary rates of the optical continuum and the extended radio emission. Consider strong FR II sources with $P_{sky} > 10^{44}$ ergs s^{-1}. In the 3CR catalog, for $z < 1$, approximately 75% of these sources are radio galaxies. By contrast, for $z > 1$, approximately 60% are QSOs. The few remaining unidentified sources can not significantly alter the magnitude of this evolutionary effect. Since these sources are selected by their radio flux alone, this implies a strong evolution of the optical component relative to the radio component. Note that evolution is beyond the scope of the unified scheme of [20] which is restricted to 3CR sources with $0.5 < z < 1.0$.

Secondly, [162] showed that the "blazar component" or beamed optical component evolves more slowly than the isotropic optical component and is a necessary modification to unification schemes for radio loud AGNs based on the beaming hypothesis. This fact can not be explained by an increased boosting efficiency as $z \to 0$. This can be illustrated from Doppler factor (the quantity δ of 10.26b) estimates in BL Lac VLBI jets relative to QSO VLBI jets. BL Lac objects are more common than radio loud quasars at low z and the opposite appears to be true at high z (although this second point could be a result of BL Lacs being intrinsically weaker radio sources and harder to detect at large z). The study in [31] finds δ to be smaller in BL Lac jets than QSO jets. Similarly, using an independent determinant, [55, 163] find that the average jet Lorentz factors, $\bar{\Gamma}_Q$ and $\bar{\Gamma}_B$, are 11 and 7 for QSOs and radio-selected BL Lac objects, respectively. Thus, the beaming efficiency appears to be diminishing with cosmological time. Consequently, the slower evolution of the blazar component relative to the isotropic disk emission must be a consequence of an increase in the intrinsic jet power relative to the accretion power as $z \to 0$.

The following ordering of evolutionary rates exists in radio loud AGNs:

1. Isotropic optical emission;
2. Steep-spectrum extended radio emission;
3. Beamed optical/radio component.

The evolutionary sequence above is described by diagonal movement in Fig. 10.14 from the upper left corner of the diagram to the lower right corner. In the distant past, $z \sim 1$–3, large accretion rates onto black holes appears to have been more common in this dynamic state of cosmological evolution. This equates to more quasars (as is well known) and the tendency for AGNs to be located at the upper left corner of Fig. 10.14. This cosmological state corresponds to the observation that 60% of the strong FR II sources in the 3CR catalog are quasars at high redshift, $z > 1$. The optical components in AGN are often so powerful due to very large $\dot{M}$ values at $z > 1$ that they can be viewed as being bright enough to be classified as a quasar from a wide range of angles between the symmetry axis of the hole and the line of sight ($\theta < 65°$, as indicated in Fig. 10.16).

An important aspect of evolution in the GHM model is that the ergospheric disk continues to operate when accretion diminishes or stops, as long as B^P_E remains trapped in the annular gap between the black hole and accretion disk by surface currents on the inner boundary of the accretion disk. In fact, the dynamics of the ergospheric disk were worked out in this limit in Chap. 8. Thus, as the Universe evolved to $z \sim 1$ and the dynamics became less conducive for galactic cannibalism and the fueling of central black holes, $\dot{M}$ values diminished in many AGNs. However, the ergospheric disk will continue to support FR II radio emission if $B^P_E \sim 10^3$ G. This magnetic flux can probably be supported with nominal accretion rates considering the long re-entry times into the accretion disk for buoyant flux tubes. This explains two evolutionary circumstances. First, the existence of FR II BL Lacs at $0.5 < z < 1.0$ is explained because a/M is the relevant parameter associated with strong extended radio power, not $\dot{M}$. Second, at $z < 1$, only 25% of the strong FR II 3CR radio sources are quasars. Thus, in the unified scheme for lines of sight relative to the black hole rotation axis of $\theta < 40°$ one sees a quasar, and for $\theta > 40°$ one sees an FR II radio galaxy (see Fig. 10.16). Comparing this to the differentiating angle of $\theta = 65°$ at $z > 1$, one can deduce that the isotropic optical luminosity, L_O, is smaller as $z \to 0$ in the FR II population. Yet, the FR II radio luminosity can still be supported in low $\dot{M}$ sources that are common at $z < 1$, since the ergospheric disk still operates. This is likely the case for Cygnus A, since the central engine appears to be very weak for a quasar, $L_O \lesssim 10^{45}$ ergs s^{-1} [32]. The GHM theory provides a natural explanation of a/M being the dominant parameter for determining extended radio luminosity, not $\dot{M}$ (review the strong observational evidence for this circumstance in Sect. 1.3).

During the course of cosmological evolution, one would expect episodic accretion of gas and magnetic flux in a radio loud AGN. The central black holes would be spun up by accretion and then spun down by electromagnetic torques when large amounts of magnetic flux are present in the ergosphere. The accretion history of these objects are unknown, but it is reasonable to expect the following evolutionary track. A rapidly spinning central black hole begins to be torqued down by a large influx of accreted poloidal magnetic field. For lines of sight of less than 65° to the black hole symmetry axis the accretion luminosity is observed as bright enough to be called a quasar. At all angles of observation an FR II radio lobe structure is detected (the upper left corner of Fig.10.14). Strong electromagnetic torques persist as

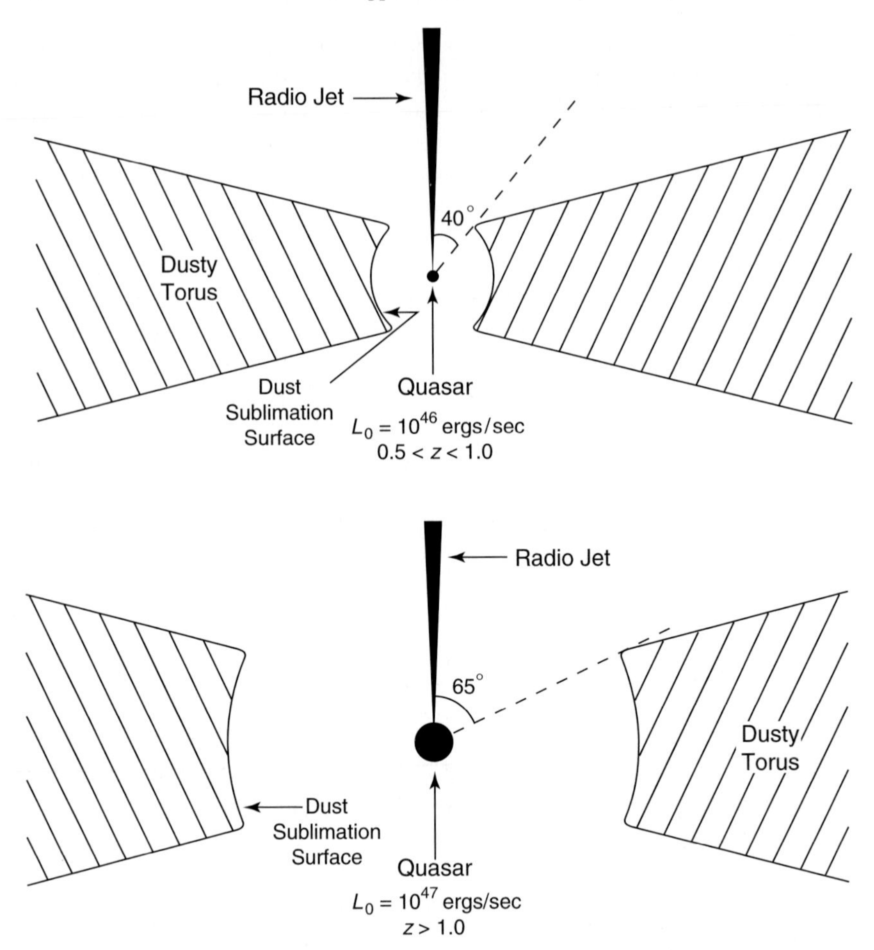

Fig. 10.16 As a quasar's luminosity increases, so does its ability to sublimate dust. If the spectral shape of the quasar emission remains relatively constant then the dust sublimation radius (the radius inside of which all dust grains are sublimated), r_d, will scale with luminosity in a simple way, $r_d \sim \sqrt{L_0}$. Thus, a more luminous accretion flow can photo-evaporate more dust and blow a bigger hole out of the center of the dusty torus (this is indicated in the bottom frame relative to the top frame, above). This provides a second order evolutionary correction to the unified scheme for extragalactic radio sources. The figure indicates a simple explanation of the evolution found in the 3CR catalog of radio sources. For $z > 1$, 60% of the FR II radio sources appear to be quasars (the bottom frame) and for $0.5 < z < 1.0$, 75% of the "strong" FR II radio sources are radio galaxies (the top frame). Even for $L_0 \sim 10^{46}\,\mathrm{ergs\,s^{-1}}$, a/M can still be large (i.e., $a/M \sim 0.9$), thus the ergospheric disk wind can still power strong FR II radio emission in the black hole GHM theory of extragalactic radio sources

$\dot{M}$ diminishes lowering a/M. The ergospheric disk continues to support strong FR II radio emission. The lower $\dot{M}$ means that L_0 is lower and less dust is sublimated from the inner edge of the molecular torus (see Figs. 1.9 and 10.16), thus a quasar will only be viewed for lines of sight less than 40° from the black hole symmetry axis (this is the center of Fig. 10.14, $a/M \sim 0.8$). In general, the ergospheric disk wind will start to diminish with a/M evolution, but it can lag the L_0 evolution. The source might progress through a γ-ray loud quasar or FR II BL Lac state briefly as $\dot{M}$ diminishes. Finally, for small $\dot{M}$, electromagnetic torques will decreases the black hole rotation so that $a/M < 0.60$ and an FR I BL Lac is obtained (lower right corner of Fig. 10.14). The evolutionary track through Fig. 10.14 increases the ratio of L_H/L_0 hence the beamed optical component increases relative to the isotropic component as observed.

Consider the time scale for such an evolutionary scheme from an FR II quasar at $a/M = 0.90$ to a BL Lac object at $a/M = 0.60$ in Fig. 10.14. From (1.38), the reducible mass at these two rotation rates are given for a $10^9 M_\odot$ black hole by,

$$M_{red}(a = 0.9M) = 0.152M \,, \tag{10.28a}$$

$$M_{red}(a = 0.6M) = 0.051M \,. \tag{10.28b}$$

The spin down time from $a/M = 0.90$ to $a/M = 0.60$ is found from (10.28) to be

$$t_{sd} \approx \frac{0.1Mc^2}{L_E + L_H} \,. \tag{10.29a}$$

and the corresponding redshift for this look back time is,

$$\Delta z \approx H_0 t_{sd} \,. \tag{10.29b}$$

For $L_E + L_H \approx 5 \times 10^{44}$ ergs s^{-1}, we find $\Delta z \gtrsim 0.5$ from (10.29). Thus, black hole GHM provides the observed evolution in radio loud AGNs on the appropriate time scales.

10.5 The GHM Theory of Extragalactic Radio Sources

In this chapter we applied the theory of black hole GHM to the study of extragalactic radio sources. It is a four parameter model: a/M, M, B^P_E and $\dot{M}$. The model was motivated in the Introduction by the observation that radio loud and radio quiet quasars have indistinguishable optical/UV thermal spectra and UV broad emission lines. If (as is commonly believed in the astrophysical community) these are powered by accretion generated viscous dissipation then it makes no sense for a quasar with strong radio power, Q, ($Q > L_0$) to have both the radio and UV energy coming from accretion. Clearly in this scenario, the $Q > L_0$ condition would modify accretion dynamics significantly from the $Q \ll L_0$ condition of a radio quiet quasar and this should show up in the viscous dissipation generated emission (and it does not). Secondly, if the radio power is a function of $\dot{M}$, it is impossible to describe the FR II BL Lac objects and the weak central engine in Cygnus A which is a strong FR II radio source. A resolution to these contradictions is black hole GHM. Black hole GHM describes radio power as primarily a function of a/M (not $\dot{M}$), B^P_E and M, and its evolution can lag the $\dot{M}$ evolution significantly (P_E can be decoupled from $\dot{M}$ variations on time scales $\sim 10^7 - 10^8$ years). Thirdly, B^P_E is independent of the disk magnetic field strength, B^P_D. In fact, it was argued that in general B^P_D is small (the accretion flow is a pathway for magnetic flux) when B^P_E is large. Thus, the accretion properties of the disk are largely unmodified even when a powerful black hole GHM central engine exists.

The black hole GHM theory has many more successes that were described in this chapter. Firstly, it provides a natural explanation of the "FR I/FR II break." For $a/M > 0.6$, an FR II radio structure is attained as the ergospheric disk jet becomes more powerful than the horizon jet (see Fig. 10.13). The maximum mechanical power (extended radio luminosity), Q (P_{sky}), at $a/M = 0.6$ is 5×10^{43} ergs s^{-1} (5×10^{42} ergs s^{-1}) and this represents the most extreme FR I state according to black hole GHM theory. Similarly, black hole GHM explains the maximum observed mechanical powers (extended radio luminosity), Q (P_{sky}), of FR II radio sources to be $\gtrsim 10^{48}$ ergs s^{-1} ($\gtrsim 10^{47}$ ergs s^{-1}) when the quasar disk luminosity, $L_0 \sim 10^{47}$ ergs s^{-1}.

The theory was also used to explain correlations with the radio to submillimeter spectra of compact radio cores (Δ parameter) with the physical state of the AGN. It was shown why the Δ parameter was strongly correlated with the BL Lac/quasar distinction (or equivalently $\dot{M}$), extended radio power, redshift, optical polarization and hard γ-ray luminosity, all as a consequence of the black hole GHM theory of the central engine.

In the process, we resolved the hard γ-ray/FR II radio luminosity conundrum of quasars. Observations limit the energy in the jet that produces γ-ray emission to be too small to support the FR II radio structures seen in some γ-ray loud quasars (such as 3C 279 and 1156+295). In black hole GHM, the γ-rays arise from inverse Compton scattering in the horizon magnetosphere driven jet, and FR II radio structures can only be supported by the ergospheric disk jet. Thus, black hole GHM decouples the γ-ray emitting region from VLBI jet and FR II radio emission in quasars.

Black hole GHM also explains the weak correlation between the time evolution of VLBI jets and γ-ray flares. A single jet, single central engine theory mandates a strong correlation of these two properties in either the ECS or SSC model. However, this expected result of the single jet theory was not found in the only large sample of time monitored parsec scale radio maps of γ-ray loud quasars [164]. Black hole GHM allows for individual flares in the horizon magnetosphere driven jet (γ-ray flares) to be associated with time delayed flares in the radio emission from the VLBI jet. However, the decoupling of the engines driving the two jets does not mandate a one to one association of γ-ray and radio events. There are many physical processes that would produce large dissipation at the base of the horizon jet (the γ-ray region), yet make only a small impact on the radio fluxes from a strong ergospheric jet on parsec scales. Hence, the correlation between γ-ray flares and time delayed VLBI detections of plasmoid ejections would be only weak to moderate. Similarly, both [164, 165] find no difference in VLBI measured apparent velocities, Doppler beaming or brightness temperature between γ-ray loud blazars and blazars that were not detected by EGRET. Furthermore, [166] find no evidence that

VLBI jet morphology such as jet bending is associated with γ-ray activity in blazars. These observations are all explained by the black hole GHM theory since the VLBI maps of quasars are detecting primarily emission from the ergospheric disk driven jet and the γ-rays come from the horizon jet, thus any correlations with the above mentioned quantities in a large sample of objects would be weak. The descriptive power of black hole GHM with regard to the enigmatic γ-ray quasars is a major accomplishment of the theory.

Finally, the black hole GHM theory of radio loud extragalactic radio sources explains the cosmological evolution of the observed properties of these objects. The theory explains why FR II radio sources are a fairly equal mix of quasars and radio galaxies at $z > 1$ and are predominantly radio galaxies at $z < 1$. Furthermore, it explains the lack of FR II radio sources at low redshift and the abundance of FR I radio galaxies and BL Lacs at $z < 0.5$, through an evolutionary track that goes from radio loud quasar to FR I radio galaxy as first $\dot{M}$ diminishes, then electromagnetic torques reduce a/M. The time scale for evolution from FR II quasar to FR I BL Lac corresponds to $\Delta z \gtrsim 0.5$. This is consistent with the cosmological evolution of the FR II population seen in deep radio surveys.

Much of this chapter is speculative because it depends on model building and therefore does not attain the rigor of the theoretical treatments in previous chapters that are the backbone of this book. However, black hole GHM has far more descriptive power than other theories of radio loud AGNs that currently exist. Most importantly, many of the qualitative properties and successes of the theory described in this section are independent of the exact model of the black hole ergosphere and its associated magnetosphere.

Chapter 11
Numerical Results

Numerical simulations provide a virtual laboratory for investigating the theories discussed in this book. Typically, numerical results are highly sensitive to the initial conditions and the assumptions of the simulation. Thus, one must be cognizant of the limitations of the assumptions and avoid the temptation of over interpreting the results. For example, the GHM solution discussed in Chap. 9 could never be found in a perfect MHD simulation. This condition does not allow for radiation losses of ultra-relativistic accelerating particles nor for small proper electric fields. For example, one could model magnetic field lines, loaded with a tenuous plasma and threading the event horizon (the event horizon magnetosphere). If the lateral boundary surfaces are passive and the particles reach large outward velocities then the unique perfect MHD solution is the Blandford–Znajek solution by definition. Even in such a simplified configuration, the perfect MHD assumption runs into conflict. A charge starved black hole magnetosphere quickly attains regions of low density in which any perfect MHD code will fail and these types of simulations always require the artificial injection of plasma (by hand). This expedience goes directly against one of the primary deductions of this work, the plasma injection mechanism is not independent of the physics that is ultimately responsible for driving the jet. Unfortunately, the only simulations at our disposal are perfect MHD and the even more suspect force-free simulations. Even so, all simulations, no matter how simplified, introduce numerical error. The biggest concern is numerical diffusion with a magnitude and ramifications that are difficult to assess. Numerical diffusion can over-ride realistic physics when reconnection is involved. The point of this introductory diatribe, is to caution the reader that numerical data needs to be considered judiciously. A simulation might look beautiful, but it can be irrelevant to any astrophysical environment.

It is also difficult to measure the degree of conformance of an idealized theoretical model in a simple geometry to the results of complicated accretion evolution in a simulation. For example, the ergospheric disk described in Chap. 8 was developed in complete isolation of the enveloping accretion flow of the quasar. There are no protons anywhere in the problem. Yet, in Chap. 10, the underlying physics was applied to simplified accretion flows of protonic matter that might be representative

B. Punsly, *Black Hole Gravitohydromagnetics, 2nd. ed.*,
Astrophysics and Space Science Library 355, doi: 10/1007/978-3-540-76957-6_11,

of a quasar. The underlying physics of the ergospheric disk was assumed to survive. In this chapter, 3-D simulation are presented that are strongly supportive of this assumption.

The dimensionality of the simulations seems to also be an important consideration. It was indicated in Chap. 8 that the details of the ergospheric disk structure depends strongly on the reconnection of poloidal magnetic flux. This is primarily governed by a realistic resistivity for the plasma and the perfect MHD results are going to be misleading. Also, it is clear from Chap. 8 that 3-D is necessary for a realistic description of the ergospheric disk. The buoyant flux tubes need a pathway to move back outward (i.e., interchange instabilities), they need displacement into the azimuthal dimension in order to "swim" around the flux tubes that are anchored in the ergospheric disk as they move outward.

The philosophy of this chapter is to start from the simplest manifestations of GHM (relativistic strings) and proceed to the complicated 3-D accretion simulations. The relativistic string simulations clearly show the microscopic details of GHM at work in 3-D around a spinning black hole. The 3-D simulations allow us to establish:

- The existence of a GHM driven ergospheric disk jet in the presence of a strong accretion flow (the primary result of Chap. 8)
- For high spin black holes the ergospheric disk output will swamp the energy output of the horizon magnetosphere (the primary result of Chap. 10)
- The spacetime near the event horizon is a passive acceptor of electromagnetic information imposed by the physical boundaries of the magnetosphere and thc plasma source (the primary result of Chap. 4–6)

11.1 The Current State of Numerical Simulations

There has been tremendous progress in the development of numerical simulations since thew first edition of this book. The seminal work was led by Shinji Koide and collaborators, who studied the time evolution of an initial perfect MHD state based on the Wald poloidal magnetic field (described in Chap. 4), in the presence of a rotating black hole. The initial magnetic field was threaded with plasma at rest with respect to the local ZAMO [167,168]. As the hole rotated, plasma accreted and was torqued back onto negative energy trajectories, just as in GHM. Plasma was depleted from localized regions as a consequence of accretion. Very low particle densities appeared in short order. Since the code was adapted form hydrodynamics, the numerical method was being used too far out of its realm of applicability. The code generated large numerical errors and the simulation was halted before an outgoing jet could form. Around the same time [169] ran simulations of the perfect MHD Wald field magnetosphere from similar initial conditions. Using diagnostics that are useful to understanding the physics, they actually kept track of the current flow. What they found was a strong cross-field poloidal current in the equatorialplane and across the field lines threading the event horizon, deep in the ergosphere. These seem to be the GHM currents for the toroidal magnetic field dynamo that were discussed in Chap. 7.

Working independently, Serguei Komissarov began creating numerical simulations based on the force-free assumption. He introduced ingoing Kerr–Schild coordinates to the field instead of Boyer–Lindquist coordinates. This is conceptually superior since there is no coordinate singularity at the horizon in these coordinates. He also introduced the notion of using fully conserved numerical steps (i.e., the code was evolving $T^{\mu\nu}$ and not the field components). These were great improvements, but don't lend themselves to numerical efficiency when adapted to perfect MHD and Komissarov has been working in 2-D as a consequence. The first simulation was of a split monopole magnetic field in the force-free limit. Since there is only one solution that is force-free by definition, the Blandford–Znajek solution, it is not surprising that the code found this solution [170]. This effort was criticized in [171] as it was shown to be based on large un-physical waves emanating from near the event horizon in the early stages of the simulation. In line with the detailed discussion of Chap. 6, it was demonstrated that such waves would cause un-physically large accelerations of the local plasma in order to support the electromagnetic content. No real plasma could achieve these accelerations and the associated radiation resistance would severely damp their amplitudes. Regardless, this ambitious effort introduced important calculational tools to the field.

Komissarov then turned his attention to a much more difficult problem, the time evolution of a force-free magnetosphere based on the Wald poloidal magnetic field [172]. The problem is difficult to pose properly in the initial state and again one must start from an un-physical initial condition. There are large transients in the early stages and the simulation ultimately breaks-down due to the seeds of a very non-force free GHM interaction in the equatorial plane in the ergosphere. The vertical flux through the equatorial plane impedes the accretion of plasma. The inability of an ergospheric plasma to fall freely into the black hole due to the impediment of an externally imposed field is the fundamental ingredient of a GHM dynamo. A large electric field develops near the equatorial plane of the ergosphere that will grow until the electromagnetic field actually is transformed in character from magnetic to electric, unless its growth is saturated by an ad hoc resistivity as was done in [172]. Komissarov ended up with a GHM like ergospheric disk on the field lines that thread the equatorial plane and a Blandford–Znajek solution on the field lines that thread the event horizon. This example is the ultimate simplification of the ergospheric disk, a resistive equatorial current sheet in which the negative energy is associated with Ohmic dissipation, $\mathbf{J} \cdot \mathbf{E}$. Presumably the created negative energy flux emerges as a super-radiant photon field that extracts the black hole rotational energy, thereby powering the Poynting flux.

In the same year, Komissarov adapted his conservative Kerr–Schild approach to perfect MHD in the simplified example of a monopolar magnetic field [173]. He was exploring the low density limit of the perfect MHD version of the Blandford–Znajek solution as described in [111]. Thus, the same problem that confronted [168] plagued these simulations. The density will tend to zero as plasma is accreted and ejected outward in certain locations, i.e., the flow division points, causing the code to crash. This was remedied by implementing a mass floor, when the density reaches a minimum value more mass with the same characteristics of the previously existing mass is injected into the grid cells. Essentially all of these perfect MHD simulations technically violate perfect MHD in localized regions where the mass injection occurs. The hope is that this adhoc injection method did not alter the final state. Presumably this simulation, like the force-free version, suffered form large un-physical transients right after the initial state, but few details were provided. The simulation reached the critical solution (the Blandford–Znajek solution) similar to the force-free result. The only other possible MHD solutions were the supercritical ones which actually require a redistribution of poloidal flux [136]. The details of how the poloidal field is maintained were not addressed in this monopolar solution. However, it is shown in Sect. 11.4 that if one does not assume an un-physical monopolar source inside the horizon or an equatorial current sheet that is a pure mathematical boundary condition then a real MHD source for the poloidal field will control the wind parameters causing significant departures from the Blandford–Znajek solution.

The most recent contribution of Komissarov involved a low density perfect MHD equivalent of the force-free Wald poloidal magnetic field simulation [174]. The simulation ran much longer than [168] and reached a steady state. The ergospheric disk appeared as an early transient, but as plasma accumulated in the equatorial plane it started dragging the field lines inward. Since, the simulation was 2-D the plasma could not accrete by moving around islands on strong buoyant field by interchange instabilities. The gravitational force increased with plasma accumulation until the radial field was stretched completely radial near the equatorial plane and dragged into the horizon (see Sect. 11.5.7 for a discussion of the contrast between simulations of 2-D and 3-D vertical flux accretion). The end result is a simple radial accretion flow along the field lines in the equatorial plane. As for the field lines that threaded the horizon there was no mass outflow anywhere inside the outer calculational boundary, a pure accretion flow. The stationary state is an accretion flow, yet the Blandford–Znajek field line angular velocity mysteriously appeared, $\Omega_F \approx (1/2)\Omega_H$ in contradiction to the work of [111] and the discussions in Chap. 9 that this parameter value derives in the perfect MHD version of the Blandford–Znajek solution as a direct consequence of energy conservation and the condition that at large distances the plasma is moving outward approximately at the speed of light (not slowly inward).

In [175], it was shown that a thin perfect MHD magnetic flux tube evolving in a background pressure distribution was mathematically equivalent to a relativistic string. The authors proceeded to develop 3-D simulations of relativistic strings around rotating black holes. The string approximation greatly improves the numerical efficiency. The string representation of a thin magnetic flux tube is characterized by the slow and Alfven modes, but ignores fast mode propagation orthogonal to the string. The advantage of this expedience is that thousands of points can be used to characterize a single flux tube, providing much higher numerical resolution than is achievable with fully self consistent perfect MHD numerical schemes. Thus, the method can capture large gradients in field parameters with much less distortion due to numerical diffusion. In [175], these authors working without any knowledge of the first edition of this book, independently discovered the GHM interaction that launches a relativistic jet. These simulations were developed further in [176]. The main virtue of this method is that the details of the jet launching are shown extremely clearly. This will be illustrated in Sect. 11.2 in which the accretion of a magnetic flux tube is shown to evolve into a relativistic Poynting flux dominated jet.

The most sophisticated simulations presently in existence are the 3-D fully self consistent perfect MHD numerical work of [177–183]. Sections 11.3 and 11.4 are dedicated to the relevance of the high spin, $a/M \geq 0.95$, 3-D simulations to this book. Section 11.3 shows the existence of powerful ergospheric disk driven jets in the high spin simulations even in the presence of an intense equatorial accretion flow (as discussed in Chaps. 8 and 10). Section 11.4 shows how the spacetime near the event horizon passively accepts information flowing into the system from plasma in the causally connected boundary plasma (the subject of Chaps. 4–6).

All of the 3-D simulations begin with a torus of gas surrounding the black hole which would be stable if not for the ad hoc introduction of loops of poloidal magnetic field along the equal pressure contours. All the loops are oriented in the same direction. The loops destabilize the torus as the shearing of the loops in the differentially rotating plasma creates magnetic torques on the plasma that initiates an accretion flow. The angular momentum removal by the field is sustained by the inward flow of gas that approaches a centrifugal barrier near the black hole. This barrier creates an "inner edge" of the accretion flow that forms a funnel roughly along the gravitational equipotential surface. The subsequent accretion of gas is restricted primarily to the equatorial plane. Magneto-rotational instabilities (MRI) permeate the inflow and regulate the accretion rate. As the magnetic flux loops accrete, the upper portion gets stretched vertically by gas pressure gradients and electromagnetic forces away from the hole and the inward part of the loop gets severely twisted azimuthally as it approaches the horizon. The field lines become inextricably tangled. Therefore, the rotating solution is not axisymmetric and this effect, in of itself, creates a non-time stationary magnetosphere. The net result is the formation of magnetosphere of poloidal flux that is highly twisted azimuthally, in the region near the black hole, restricted to the funnel. The strong transients that set up this initial state die off by $t = 2,000$ M. These strong transients generate a magnetic tower that evacuates the funnel interior [183]. As with Komissarov's work, the plasma density must be controlled by an adhoc mass floor. The early time magnetic tower at $t < 1,000M$ is a powerful transient Poynting jet that can be thought of as a strong GHM transient that is the long term extension of what [168] found. At $t > 2,000M$, a funnel filled with trapped magnetic flux is achieved. This is the black hole magnetosphere that supports a Poynting jet.

This final configuration resulting from simulations of magnetized tori, although derived from first principles is strongly dependent on the initial conditions. For example purely toroidal loops do not produce the black hole magnetosphere. Flux loops that do not symmetrically thread the circular centerline of the torus (i.e., a quadrupolar distribution of loops) produce a very weak poloidal field in the black hole magnetosphere [183]. Regardless of this arbitrariness, the simulations provide an excellent virtual laboratory for understanding which 3-D MHD structures can self-consistently exist around a rapidly rotating black hole.

The most promising technique for studying complicated black hole accretion systems is the method of [184, 185]. These authors begin their simulations from initial conditions that are virtually identical to DeVilliers and Hawley except in 2-D (magnetized torii with all loops oriented in the same direction). This method uses the ingoing Kerr–Schild coordinates which do not have the coordinate singularity at the event horizon which is a concern for the Boyer–Lindquist coordinates used by DeVilliers and Hawley. Furthermore, the [184] method is conservative (i.e., evolves the components of $T^{\mu\nu}$) which is aesthetically appealing, but numerically very inefficient compared to the method of DeVilliers and Hawley which evolves field components and the plasma momentum. The 2-D expedience allows for better numerical resolution in principle. This is true for quantities that diverge near the horizon in Boyer–Lindquist coordinates, such as g_{rr}. Even with the dense radial grid spacing employed by DeVilliers and Hawley the divergent metric derived quantities (evaluated outside of the inner calculational boundary) are only 1/4 as resolved, deep in the ergosphere, as the corresponding nonsingular Kerr–Schild derived quantities in the highest resolution simulation of [184]. However, not all aspects of the numerical resolution are superior. For example, the fundamental frame dragging quantity $g_{\phi t}$, or equivalently the ZAMO angular velocity, Ω, is more resolved in the DeVilliers and Hawley simulations by a factor of 2–3 in the ergosphere than in the high resolution simulation in [184]. Thus, even though the inner calculational boundary is placed inside of the horizon in [184], numerical diffusion has a propensity to allow information on the local rotation rate to evolve outward (acausally) from the horizon. For example, Ω_{min} (the primary driver of a strong GHM interaction), changes by a factor of 10, in only ≈ 15 zones. By contrast, a similar variation occurs over ≈ 40 zones in the Hawley and DeVilliers simulations. Another concern with the 2-D simulations is that there are strong MRI modes called "channel solutions" which are the primary driver of the accretion rate within the simulations. Conversely, these modes are highly damped in 3-D. Secondly, the MRI dies off quickly in 2-D (the anti-dynamo theorem). By $t = 2{,}000$ M, the MRI is negligible in contrast to the [183] 3-D simulations in which they consider $t < 2{,}000M$ to be filled with initial transients and their late time data dumps begin after $t = 2{,}000$ M. Clearly the accretion history of the 2-D solutions needs to be viewed cautiously. Also, sampling of data after $t = 2{,}000$ M in [184, 185] should be considered judiciously. The reader is reminded that these 2-D simulations also get depleted of mass in regions of the black hole magnetosphere within the funnel. Perfect MHD is formally broken by the ad hoc mass floor used to remedy this issue.

Based on the claims of [184, 185], it appears that they have found Blandford–Znajek solutions in the funnel. The claim is based on the fact that there is outward directed Poynting flux at the event horizon and $\Omega_F \approx 0.45\Omega_H$ in a time averaged sense for $a/M = 0.938$, which is close to the Blandford–Znajek value [184]. However, they then proceed to show that this value of Ω_F persists into the inertially dominated regime, $F^{\mu\nu}F_{\mu\nu} < 16\pi n\mu c^2$. This is a very curious result that does not follow from any of the analysis of the MHD version of the Blandford–Znajek solution that assumes $F^{\mu\nu}F_{\mu\nu} \gg 16\pi n\mu c^2$. For similar spin rates, the [183] simulations produce a time and azimuth averaged funnel value of $\Omega_F \approx 0.35\Omega_H$. There are also large variations about this average, $-0.5\Omega_H < \Omega_F < \Omega_H$ in the individual time and azimuth slices. The data in [184, 185] has not been explored with the same level of scrutiny (especially individual time slices of the high spin simulations) as the DeVilliers and Hawley data, so the precise details of the internal physics is unclear, especially considering some of the curious particulars. There is definitely electrodynamic energy extraction, but the role of inertial sources of Poynting flux on the funnel boundary has not been explored in the high spin simulations. The details of the simulation are important because it was the method of solution and the resultant field parameters in [66, 111] that was called into question in [67], not the notion of electrodynamic energy extraction.

There are two important physical constraints on the existing 2-D and 3-D simulations of magnetized tori that limit their applicability to realistic AGN models. First of all, there is no radiation in the accretion flow, thus much of the enthalpy generation will actually be lost to a radiation field and not all of it equates to the increase in gas pressure. This will manifest itself in two ways at the boundary of the funnel. The accretion disk and coronal gas will not be as pressurized so the poloidal magnetic field pressure in the funnel will be smaller in this pressure balance. Furthermore, radiation should fill the funnel and this will be an important component of the total electromagnetic pressure in balance with the gas pressure at the funnel boundary. Both effects should weaken the field strength in the funnel and therefore the total power output. It is important to note that all these simulations increase black hole energy. More energy is accreted than emitted in the jet, thus there really is no black hole energy extraction. This is likely to be inconsistent with some of the powerful jets coming from moderate and weakly accreting AGN [189, 190].

Another recent development in the field might rewrite all of the above. C. Fragile has found that just by tilting the magnetized torus by 15°, the inner accretion flow is primarily deposited near the poles of the black hole as opposed to the equator [188]. The simulations are 3-D, of course, and are performed in Kerr–Schild coordinates. These simulations also require high resolution near the polar axis. Consequently, the numerical efficiency is extremely low and it is unlikely that simulations with high resolution in the ergosphere will be performed in the near future.

11.2 Simulations of Relativistic Strings

The simulations of the relativistic string representation of thin poloidal flux tubes, at the time of printing, are still the only numerical models that produce a relativistic jet and actually extract black hole energy [176]. They use a method that exploits the simplification that the full set of perfect MHD equations in curved spacetime indicate that a magnetized plasma can be regarded as a fluid composed of non-linear strings in which the strings are mathematically equivalent to thin magnetic flux tubes [175]. In this treatment, a flux tube is thin by definition if the pressure variations across the flux tube are negligible compared to the total external pressure (gas plus magnetic), P, that represents the effects of the enveloping magnetized plasma (the magnetosphere). By concentrating the calculation on individual flux tubes in a magnetosphere, one can focus the computational effort on the physical mechanism of jet production (on all the field lines). Thus, this technique is able to elucidate the fundamental physics of black hole driven jets without burying the results in the effort to find the external pressure function, P. The goal of the approach is to understand the first order physics of jet production not all of the dissipative second order effects that modify the efficiency. For these purposes, the string depiction of perfect MHD is adequate.

Consider a vertical flux tube in which the initial velocity of the plasma is that of the local ZAMO observer. The flux tube will begin to accrete toward the black hole. The simulation is performed on the background spacetime of a rapidly spinning black hole with a/M = 0.995. The equation of state is chosen as $P \sim \rho^2$ and $P \sim (r - r_+)^{-2}$, where P is the total pressure, gas and magnetic. These parameterizations were chosen for computational simplicity. The fundamental physics of energy extraction and jet production was found in [176] to be quite independent of the pressure function and the initial conditions. When the flux tube first penetrates the ergosphere it appears twisted as in Fig. 11.1.

Initially, $d\phi/dt = \Omega_0 \equiv \Omega \ll \Omega_H$, since the flux tube is far from the event horizon. The flux tube accretes towards the black hole under the influence of the gravitational force. The "natural state" of plasma motion (geodesic motion) induced by frame dragging is to spiral inwards faster and faster as the plasma approaches corotation with the event horizon (frame dragging). By contrast, the "natural state" of plasma motion in a magnetic field is a helical Larmor orbit that is threaded onto the field lines. In general, these two "natural states" of motion are in conflict near a black hole. The torsional struggle between these two strong forces is the dynamical effect that drives the simulation depicted in Figs. 11.1–11.6. The plasma far from the hole is still rotating slowly near $d\phi/dt \approx \Omega_0$ in Fig. 11.1. However, inside the ergosphere, Ω_{min} is necessarily a significant fraction of Ω_H, by (3.43b) and Ω_p must exceed Ω_0 in short order. Thus, the ergospheric plasma gets dragged forward, azimuthally, relative to the distance portions of the flux tube, by the gravitational field. The back reaction of the field is an attempt to keep the plasma threaded on the field lines (Larmor helices) by torquing the plasma back onto the field lines with $\mathbf{J} \times \mathbf{B}$ forces (the cross-field current density, $\mathbf{J}$, driven by this torsional struggle is sunk within the enveloping magnetosphere). By Amperes law, the current driven by the global torsional struggle also makes a negative azimuthal magnetic field, B^{ϕ}, upstream of the current flow.

The $\mathbf{J} \times \mathbf{B}$ back reaction forces driven by the global torsional struggle provide a torque on the plasma in the flux tube in the ergosphere. Figures 11.1 and 11.2 show that the B^{ϕ} created in the ergosphere propagates upstream in the form of an MHD plasma wave at later times, as more and more negative energy is created in the ergospheric region of the flux tube. The negative energy (indicated by the red portion of the magnetic flux tube) is the total plasma energy including both the electromagnetic and the mechanical components of the plasma. In Figs. 11.1–11.6, a

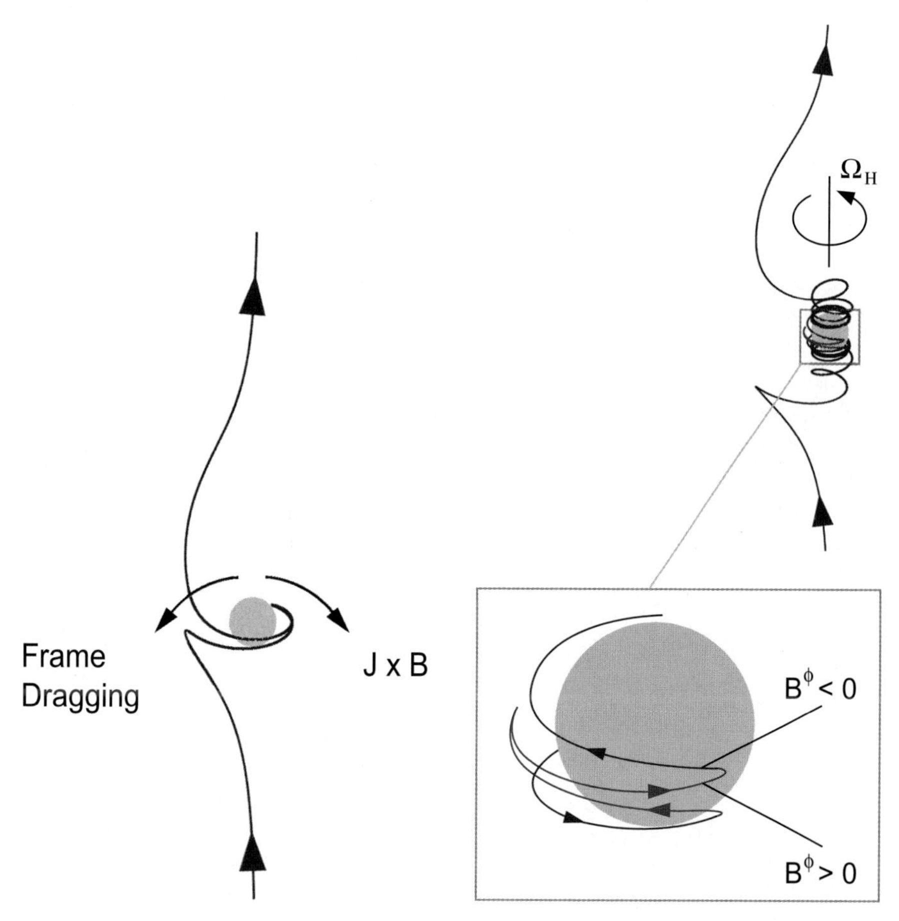

Fig. 11.1 The magnetic flux tube experiences a torsional struggle between inertial forces and electromagnetic forces as it enters the ergosphere

Fig. 11.2 The red portions of the field line indicate plasma with negative energy, as viewed globally. The back reaction of the field in the torsional battle torques plasma onto negative energy trajectories while simultaneously creating an outgoing Poynting flux in the jet. The details are described in the text. The time lapse between frames, as measured by a distant stationary observer from Fig. 11.1 to 11.2 is $t = 85.7GM/c^3$. Note that the flux tube rotates in the same sense as the black hole. The bottom frame is a close-up of the dynamo region for the toroidal flux in the jet

jet emerges from the ergosphere. The magnetic tower created by B^ϕ in combination with the poloidal field component, B^P, naturally provides stable hoop stresses that are the only known collimation mechanism for the jet morphology of quasars.

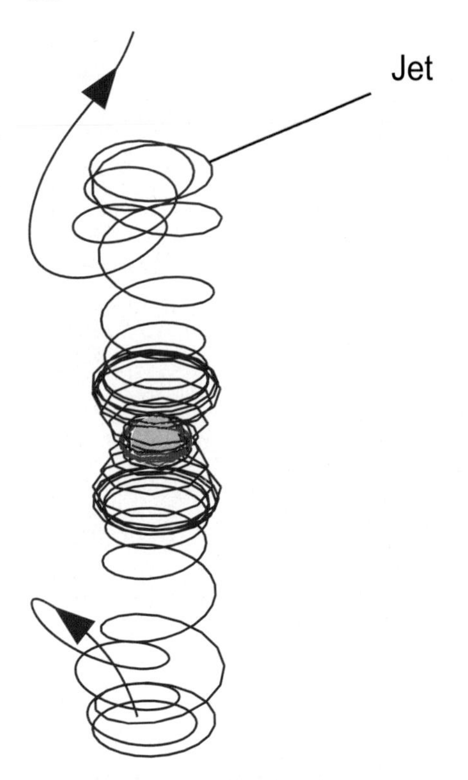

Fig. 11.3 A jet is produced on the magnetic flux tubes that experience the ergospheric torsional struggle between frame dragging forces and $\mathbf{J} \times \mathbf{B}$ forces. The Boyer–Lindquist time lapse from Fig. 11.1 to 11.3 is $t = 133.2GM/c^3$

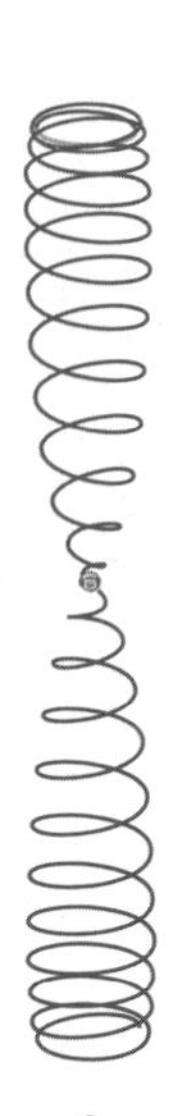

Fig. 11.4 A well-formed jet emerges from the horizon at late times. The Boyer–Lindquist time lapse from Fig. 11.1 to 11.4 is $t = 265GM/c^3$ with a pair of jets with lengths of over $60GM/c^2$. The plasma has attained an outflow Lorentz factor of a little less than 2 in the late stages

The dynamo region for B^ϕ in the ergosphere is expanded in the bottom frame of Fig. 11.2. Since $B^\phi < 0$ upstream of the dynamo, from the frozen-in condition expressed in (5.13c) and the transformation (5.14), there is an electromagnetic energy flux ($\sim -\Omega_F B^\phi B^P$) and an electromagnetic angular momentum flux ($\sim -B^\phi B^P$) along B^P, away from the hole in the jet. The red portion of the field line indicates the total plasma energy per particle, $E < 0$,

$$E \approx \omega + S^P/k\,, \tag{11.1}$$

downstream of the dynamo, where S^P is the poloidal component of the Poynting flux along the magnetic field, k is the poloidal particle flux downstream of the dynamo and the mechanical energy is ω. Since $B^\phi > 0$ downstream of the dynamo in Fig. 11.2, the field transports energy and angular momentum towards the hole with the inflowing plasma. Thus, $S^P/k > 0$ in the downstream state. Consequently,

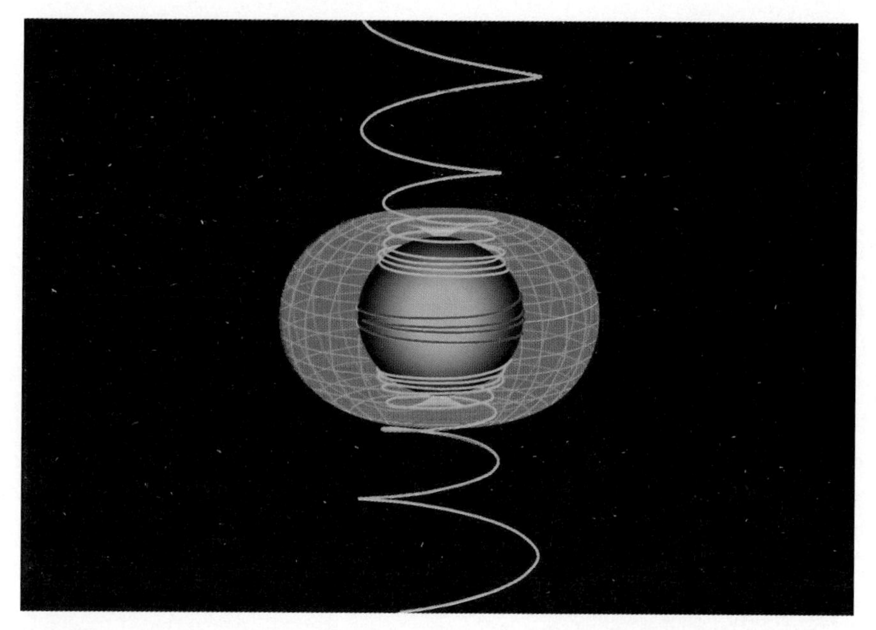

Fig. 11.5 A closeup view of the dynamo region at $t = 265GM/c^3$. The hatched region is the stationary limit surface

$\omega \ll 0$ in order for $E < 0$, downstream. The $\mathbf{J} \times \mathbf{B}$ back reaction forces in the torsional battle torque the plasma onto trajectories with $\Omega_p \approx \Omega_{min}$. The ingoing $\omega \ll 0$ plasma extracts the rotational energy of the hole since $\beta^\phi \approx -1$ as noted in (3.51) and global mechanical angular momentum of the plasma is simply expressed in the ZAMO frames as

$$m = u^0 \beta^\phi \sqrt{g_{\phi\phi}} , \tag{11.2}$$

which implies that $m \ll 0$ when $\beta^\phi \approx -1$. Thus, black hole rotational inertia is powering the jet in the simulation.

This is precisely the physics of the Penrose process [16]. Penrose envisioned that a particle could be split into two pieces in the ergosphere. A negative energy ingoing particle extracts the rotational energy of the hole and an outgoing particle goes off to infinity. Thus, energy is extracted from the black hole. In the GHM process the negative energy particles is the torqued plasma in the global torsional struggle. The outgoing particle is a nonlinear MHD wave that can be almost pure Poynting flux. A closeup of the dynamo region of the jet is shown in Fig. 11.5. The Figs. 11.4 and 11.5 illustrate the utility of this numerical method, the microphysics of relativistic jet production is clearly displayed. To see the shortcomings and limitations of this calculational technique, one should consult the Methods section of [176]. It is also instructive to see the entire GHM interaction in Figs. 11.1–11.3 captured in one image as in Fig. 11.6 from [176].

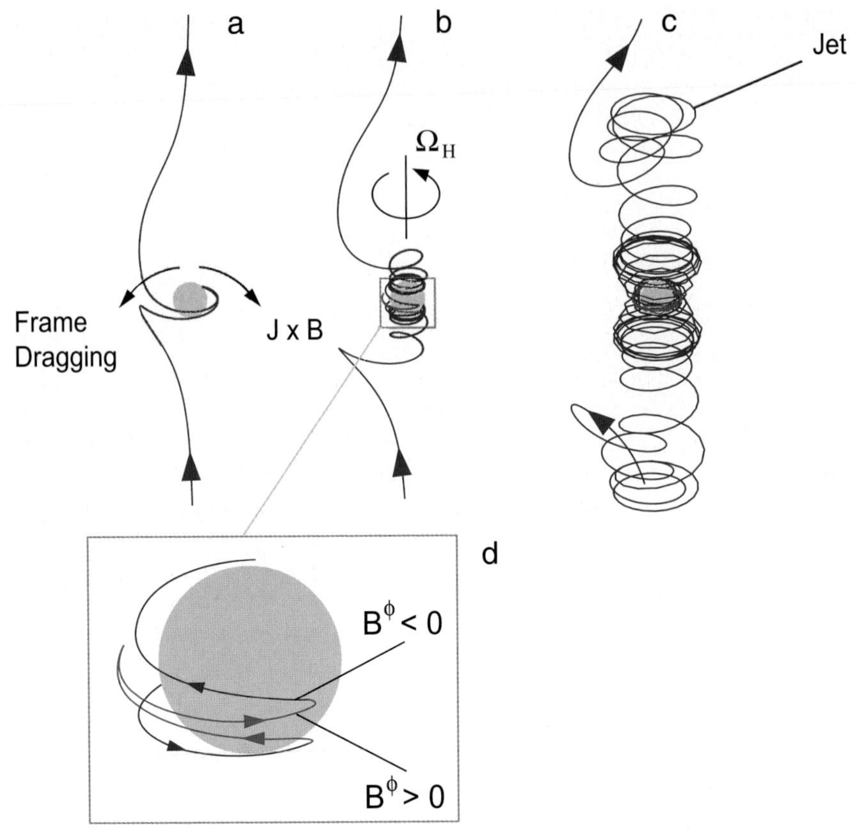

Fig. 11.6 A summary of the GHM interaction. Figures 11.1–11.3 are frames a–c above, respectively

The physics that was found is related to the fact that the spacetime around a black hole is essentially rotating and so are the particles. This frame dragging force is actually very strong, to overcome it one must overwhelm the rotational inertia of the black hole $\sim 0.1Mc^2$ (see equations (1.38) and (1.39)). Thus, the black hole can impart a huge relativistic inertia to any plasma in order to enforce the constraints of frame dragging, the field is overwhelmed regardless of its strength and is twisted and spun-up, as necessary, generating the Poynting flux that powers the jet.

It is expected that a typical black hole magnetosphere would drive a jet with the same GHM physics. The dynamics of the simulations are not a consequence of any the simplifying assumptions used in the string formalism. The main effect of the enveloping magnetosphere would be to compress or rarify the flux tubes and change their inclination. Semenov and Dyadechkin have created a collection of simulations indicating that the same basic physics of jet production exists largely independent of flux tube inclination, the external pressure function and the value of the pure Alfven speed, $U_A = B^P / \sqrt{4\pi n \mu}$. For example, in the simulation of Figs. 11.1–11.6, the Poynting flux emerges from a region which is magnetically dominated with $U_A = 12 - 13$. Web links to movies of many of these simulations can be found in [176]. Thus, the entire magnetosphere that is found by solving the transverse force equations should be many thin flux tubes that are pieced together at various inclination angles and with various Alfven speeds and pressures.

11.3 Ergospheric Disk Jets in 3-D MHD Accretion Flow Simulations

In Chap. 8, the ideal ergospheric disk was described. It represents GHM in its purest form. The ergospheric disk model was based on plasma that was purely positronic. In Chap. 10, it was assumed that the fundamental GHM physics of black hole energy extraction still persists in the presence of accreting equatorial protonic plasma, i.e., a quasar accretion flow. The 3-D perfect MHD simulation known as KDJ (KDJ is distinguished by a/M =0.99) in [183] shows that this is likely the case. The discovery of a powerful ergospheric disk within the enormous database of KDJ was made in [186]. The interesting new aspect of the ergospheric disk that emerged from the simulation was that the GHM jet did not have to be initiated on buoyant magnetic flux tubes that thread the equator and extend to large distances from the hole as presupposed in Chaps. 8 and 10. The simulation realized a highly turbulent and unsteady accretion flow in which strong loops of poloidal flux with internal magnetic pressures comparable to the ram and gas pressures of the accretion flow get generated by MRI instabilities. The vertical flux comprising the ergospheric disk magnetosphere is small in extent, $\sim$1 M–2 M and episodic. The magnetic pressure is so high that the flux patches of twisted loops are buoyant near the upper boundary of the equatorial accretion flow. As such, the loops tend to sporadically bubble out of the ergospheric accretion flow into the accretion vortex. All the magnetospheric features are very twisted by the frame dragging of spacetime and the rotating gas. An irregular feature, such as a strong twisted loop of magnetic flux extending vertical upward, will promptly be twisted up more and become inextricable tangled with the large scale magnetic flux in the funnel. Thus, there is effectively a causal connection between the GHM dynamo forming in the equatorial plane and large scale plasma-filled flux tubes flux far way, i.e., a pathway that can transfer magnetic stresses from the dynamo to the outgoing jet. This simulation indicates that realistic quasars are probably not well described by homogeneous simple structures, but are comprised of many complicated small features that are synthesized into an active dynamo and magnetosphere. This complexity was unexpected, but somewhat obvious in hindsight since the magnetosphere is generated by highly turbulent gas.

In this section, an ergospheric disk within the inner accretion flow in the high spin 3-D simulations is established by exploring the following points:

1. Just as in the ergospheric disk, the Poynting flux emerges from the ergospheric equatorial accretion flow.

2. The GHM dynamo is triggered by the ergospheric plasma accretion towards the black hole being impeded by a large scale poloidal magnetic flux barrier. Within KDJ there are strong patches of vertical flux coincident with the base of the Poynting jets.
3. For a putative Blandford–Znajek process within a magnetosphere shaped by the accretion vortex, the field line angular velocity is, $\Omega_F \approx \Omega_H/2$ (where Ω_H is the angular velocity of the horizon) near the pole and decreases with latitude to $\approx \Omega_H/5$ near the equatorial plane of the inner ergosphere [111]. In a GHM ergospheric disk, since the magnetic flux is anchored by the inertia of the accretion flow in the inner ergosphere, frame dragging enforces $d\phi/dt \approx \Omega_H$. One therefore has the condition, $\Omega_F \approx \Omega_H$ in the inner regions of the ergosphere. The GHM condition holds in KDJ in regions that are spatially and temporally coincident with the base of the Poynting jet.
4. The torsional tug of war between the vertical flux and the equatorial plasma creates ergospheric disk plasma with negative mechanical energy (the Penrose process). It is shown that this occurs as the plasma accretes through the dynamo at the base of the Poynting jet, even in the presence of an intense bath of accreting positive energy protonic plasma!

Perhaps the most important point arising from these simulations from an astrophysical standpoint is that the ergospheric disk jet dominates the power output from the black hole. An even more powerful ergospheric disk jet was found in the highest spin simulation, KDE with a/M $= 0.998$ in [181, 187]. The KDE results are described in Sect. 11.3.5.

Since these results are such an outstanding corroboration of the theory, it is prudent to critique the numerical technique. Numerically, the problem is formulated on a grid that is 192 x 192 x 64, spanning $r_{in} < r < 120M$, $8.1° < \theta < 171.9°$ and $0 < \phi < 90°$. The inner calculational boundary, r_{in}, is located close to, but just outside of the event horizon, r_+, where the coordinates are singular. The ϕ boundary condition is periodic and the θ boundary conditions are reflective. Zero-gradient boundary conditions are employed on the radial boundaries, where the contents of the active zones are copied into the neighboring ghost zones. As discussed in Chap. 9, MHD waves propagate slower than the speed of light, therefore the gravitational redshift creates a fast magneto-sonic critical surface outside of r_+ from which no MHD wave can traverse in the outward direction, even in a nonaxisymmetric, nonstationary magnetosphere. The philosophy was to choose r_{in} to lie inside the fast magneto-sonic critical surface, thereby isolating it from the calculational grid. There are also steep gradients in the metric derived quantities as r_+ is approached. This is handled by increasing the resolution of the grid near r_{in} with a cosh distribution of radial nodes. The validity of the numerics of this method was verified, near r_{in}, in [177] by comparing simulations to solutions with simple analytic forms. Even so, the simulations are closely monitored to look for unnatural boundary reflections. We also note the 3-D simulations in Kerr–Schild coordinates (which are nonsingular on the horizon) in [188]. To test the code, in preparation of the paper [188], they ran simulations of magnetized tori that were initiated from identical input parameters to those used by Hawley et al. In the words of C. Fragile (private communication), the results were "remarkably similar." Even though this was only verified for a/M=0.9, it is compelling. Consequently, for the purposes of this study it was concluded that the numerics were reliable inside the ergosphere.

11.3.1 The Equatorial Poynting Flux Source in KDJ

The 3-D perfect MHD simulation KDJ, $a/M = 0.99$, is characterized by strong flares of electromagnetic energy that originate near the equatorial plane. In order to understand the source of the strong flares of radial Poynting flux, one needs to merely consider the conservation of global, redshifted, or equivalently the B-L coordinate evaluated energy flux [80]. In general, the divergence of the time component of the stress-energy tensor in a coordinate system can be expanded as (note that the tilde notation is dropped from the Boyer–Lindquist evaluated $T_{\mu}^{\ \nu}$ in this chapter in order to keep the expressions that are implemented later in this chapter from becoming too cluttered),

$$T_{t\ ;\nu}^{\ \nu} = (1/\sqrt{-g})[\partial(\sqrt{-g}\,T_t^{\ \nu})/\partial(x^\nu)] + \Gamma^{\mu}_{t\,\beta} T_{\mu}^{\beta} \,. \tag{11.3}$$

However, the Kerr metric has a Killing vector (the metric is time stationary) dual to the Boyer–Lindquist time coordinate. Thus, there is a conservation law associated with the time component of the divergence of the stress-energy tensor. Consequently, if one expands out the inhomogeneous connection coefficient term in the expression above, it will equate to zero. The conservation of energy evaluated in Boyer–Lindquist coordinates reduces to,

$$\partial(\sqrt{-g}\,T_t^{\ \nu})/\partial(x^\nu) = 0 \,, \tag{11.4}$$

where the four-momentum $-T_t^{\ \nu}$ has two components: one from the fluid, $-(T_t^{\ \nu})_{\mathrm{fluid}}$, and one from the electromagnetic field, $-(T_t^{\ \nu})_{\mathrm{EM}}$. The reduction to a homogeneous equation with only partial derivatives is the reason why the global conservation of energy can be expressed in integral form in (3.70) of [80] (see also equation (11.8)). It follows that the poloidal components of the redshifted Poynting flux are

$$S^\theta = -\sqrt{-g}\,(T_t^{\ \theta})_{\mathrm{EM}} \,, \tag{11.5a}$$

$$S^r = -\sqrt{-g}\,(T_t^{\ r})_{\mathrm{EM}} \,. \tag{11.5b}$$

We can use these simple expressions to understand the primary source of the Poynting jet in KDJ. J. Krolik and J. Hawley have generously shared the data for the last three time slices of KDJ, at $t = 9{,}840$ M, $t = 9{,}920$ M and $t = 10{,}000$ M. Figure 11.7 show plots of S^θ (left) and S^r (right) in KDJ at $t = 9{,}840$ M, $t = 9{,}920$ M and $t = 10{,}000$ M, respectively. The inside of the inner calculational boundary ($r = 1.203$ M) is black. The calculational boundary near the poles is at 8.1° and

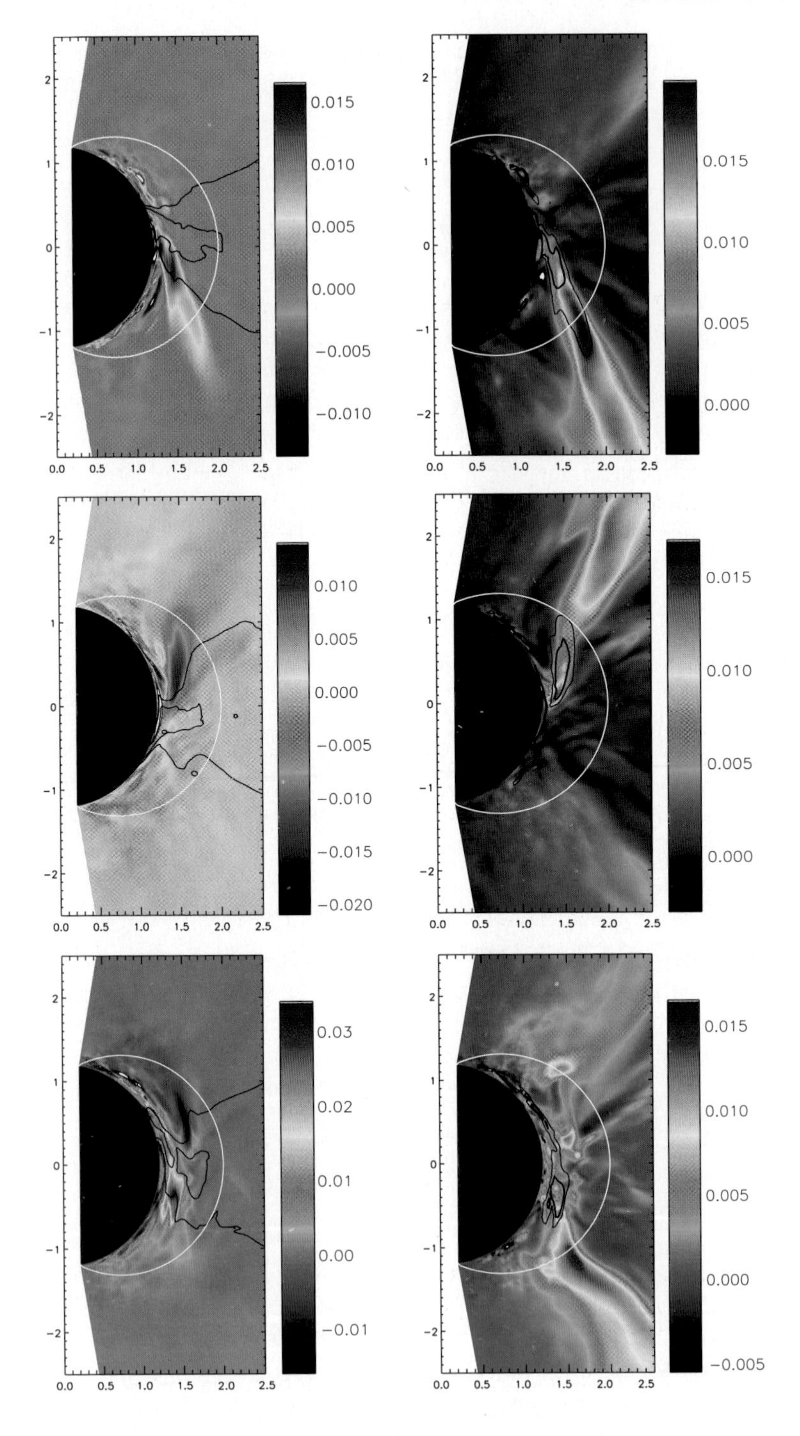
0.015
0.010
0.005
0.000
−0.005
−0.010
0.015
0.010
0.005
0.000
0.010
0.005
0.000
−0.005
−0.010
−0.015
−0.020
0.015
0.010
0.005
0.000
0.03
0.02
0.01
0.00
−0.01
0.015
0.010
0.005
0.000
−0.005
0.0 0.5 1.0 1.5 2.0 2.5
2
1
0
−1
−2

171.9°. Each frame is the average over azimuth of each time step. This greatly reduces the fluctuations as the accretion vortex is a cauldron of strong MHD waves. The individual $\phi =$ constant slices show the same dominant behavior, however it is embedded in large MHD fluctuations. On the left hand column of Fig. 11.7 left, density contours have been superimposed on the images to indicate the location of the equatorial accretion flow. The density is evaluated in B-L coordinates with contours at 0.5 and 0.1 of the peak value within $r < 2.5M$. Notice that in Fig. 11.7 within the left frames, S^{θ} is created primarily in regions of very high accretion flow density. In all three of the right frames of Fig. 11.7, there is an enhanced S^{r} that emanates from the ergosphere. There are 40 grid points between $r_{in} = 1.203M$ and r_s at $\theta = \pi/2$. These features are well resolved, they are clearly coherent physical structures and are not numerical artifacts and it is meaningful to discuss their origin. Notice that the radial energy beam diminishes precipitously just outside the horizon, near the equatorial plane in all three time steps. The region in which S^{r} diminishes is adjacent to a region of strong S^{θ} that originates in the inertially dominated accretion flow in the inner ergosphere, $1.2M < r < 1.6M$ (this region is resolved by 28 radial grid zones). In fact, if one looks at the conservation of energy equation, the term $\partial(S^{\theta})/\partial\theta$ is sufficiently large to be the source of $\partial(S^{r})/\partial r$ at the base of the radial beam in all three frames. This does not preclude the transfer of energy to and from the plasma. It merely states that the magnitude is sufficient to source S^{r} (in general, the hydrodynamic energy flux is negligible in the funnel and these terms can be ignored in a discussion of equation (11.4) to first order). To illustrate energy conservation in the electromagnetic field, contours of S^{θ} are superimposed on the color plots of S^{r} in Fig. 11.7. The contour levels are chosen to be 2/3 and 1/3 of the maximum value of S^{θ} emerging from the dense equatorial accretion flow. One clearly sees S^{θ} switching off where S^{r} switches on. We conclude that a vertical Poynting flux created in the equatorial accretion flow is the source of the strong beams of S^{r}. This establishes condition 1 of the GHM interaction that drives an ergospheric disk that was noted in the introductory remarks of this section.

11.3.2 The Vertical Flux in the Equatorial Dynamo

Figure 11.8 shows plots of S^{θ} (left) and the magnetic field component, $B^{\theta} \equiv \tilde{F}_{r\phi}$ (right) in KDJ at $t = 9,840$ M, $t = 9,920$ M and $t = 10,000$ M, respectively. At

Fig. 11.7 The source of Poynting flux. The left frame is S^{θ} and the right frame is S^{r} in KDJ, both averaged over azimuth, at *top*: $t = 9840$ M, *middle*: $t = 9920$ M, *bottom*: $t = 10000$ M. The relative units (based on code variables) are in a color bar to right of each plot for comparison of magnitudes between the two plots. The contours on the S^{θ} plots are of the density, scaled from the peak value within the frame at relative levels 0.5 and 0.1. The contours on the S^{r} plots are of S^{θ} scaled from the peak within the frame at relative levels 0.67 and 0.33. Notice that any contribution from an electrodynamic effect associated with the horizon appears minimal. The white contour is the stationary limit surface. There is no data clipping, so plot values that exceed the limits of the color bar appear white

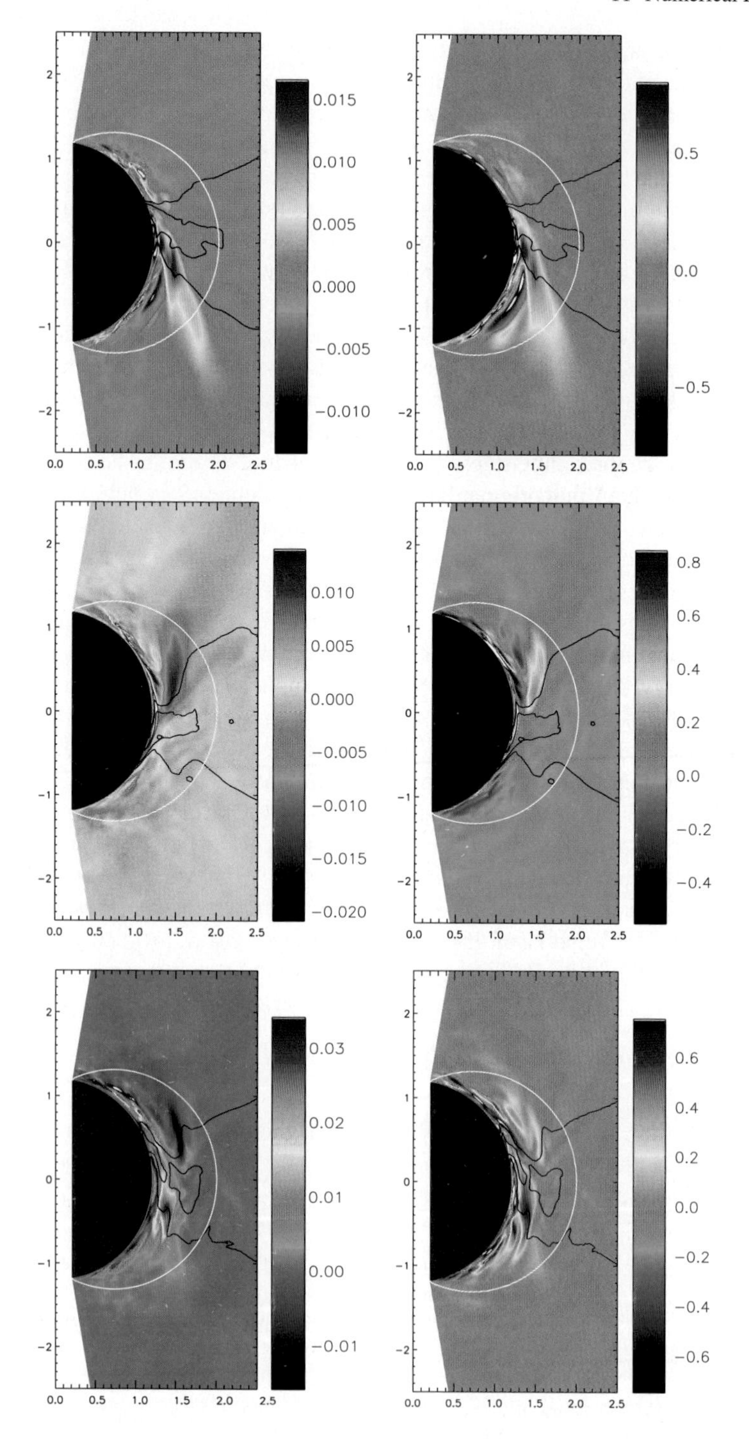
0.015
0.010
0.005
0.000
−0.005
−0.010
0.5
0.0
−0.5
0.010
0.005
0.000
−0.005
−0.010
−0.015
−0.020
0.8
0.6
0.4
0.2
0.0
−0.2
−0.4
0.03
0.02
0.01
0.00
−0.01
0.6
0.4
0.2
0.0
−0.2
−0.4
−0.6

every location in which S^θ is strong in the left frames, there is a pronounced enhancement in B^θ in the right frames in Figs. 11.8. Recall that the sign of S^θ is not determined by the sign of B^θ. These intense flux patches penetrate the inertially dominated equatorial accretion flow in all three frames. The density contours indicate that the regions of enhanced vertical field greatly disrupt the equatorial inflow. As noted in Chap. 7, a GHM interaction is likely to occur when the magnetic field impedes the inflow in the ergosphere. The regions of large B^θ are compact compared to the global field configuration of the jet, only $\sim 1M - 2M$ long. Considering the turbulent, differentially rotating plasma in which they are embedded, these are most likely highly enhanced regions of twisted magnetic loops created by the MRI. The strength of B^θ at the base of the flares is comparable to, or exceeds the radial magnetic field strength. The situation is clearly very unsteady and vertical flux is constantly shifting from hemisphere to hemisphere. The time slice $t = 10{,}000$ M, although primarily a southern hemisphere event, also has a significant contribution in the northern hemisphere (see the blue fan-like plume of vertical Poynting flux in the top frame of Fig. 11.8 right). The GHM interaction is provided by the vertical flux that links the equatorial plasma to the relatively slowly rotating plasma of the magnetosphere within the accretion vortex. The vertical flux transmits huge torsional stresses from the accretion flow to the magnetosphere.

In all three time steps, a strong patch of vertical flux interacts strongly with the equatorial accretion flow. As the accreting plasma is inhibited from flowing inward by the magnetic pressure, torsional stresses move up and down the vertical flux patch into the Poynting jet within the funnel: as evidenced by the strong S^θ that is coincident both spatially and temporally with the strong vertical flux patches. This establishes condition 2 of an ergospheric disk dynamo in the introductory remarks to this section.

In the language of Chap. 7, the Poynting jet plasma and field is rotating slower than the accreting plasma. Thus, this load on the vertical flux tubes provides a torque on the accreting plasma. As a back reaction, the accreting plasma sends angular momentum into the funnel Poynting jet in an attempt to spin it up in this torsional tug of war.

11.3.3 The Field Line Angular Velocity

Further corroboration of this interpretation can be found by looking at the values of Ω_F in the vicinity of the S^r flares. In a non-axisymmetric, non-time stationary flow,

←

Fig. 11.8 The source of Poynting flux. The left frame is S^θ and the right frame is B^θ in KDJ, both averaged over azimuth, at *top*: $t = 9840$ M, *middle*: $t = 9920$ M, *bottom*: $t = 10000$ M. The relative units (based on code variables) are in a color bar to right of each plot. The contours on both the S^θ and B^θ plots are of the density, scaled from the peak value within the frame at relative levels 0.5 and 0.1. The white contour is the stationary limit surface. There is no data clipping, so plot values that exceed the limits of the color bar appear white

there is still a well defined notion of Ω_F: the rate at which a frame of reference at fixed r and θ would have to rotate so that the poloidal component of the electric field, $E^{\perp}$, that is orthogonal to the poloidal magnetic field, B^P, vanishes. This was demonstrated more generally in (9.1). If we expand out (9.1) and assume the frozen-in condition, this relation can be written out in Boyer–Lindquist coordinates in terms of the plasma three-velocity, $\tilde{v}^i$ and the Faraday tensor as

$$\Omega_F = \tilde{v}^\phi - \tilde{F}_{\theta r} \frac{g_{rr}\tilde{v}^r \tilde{F}_{\phi\theta} + g_{\theta\theta}\tilde{v}^\theta \tilde{F}_{r\phi}}{(\tilde{F}_{\phi\theta})^2 g_{rr} + (\tilde{F}_{r\phi})^2 g_{\theta\theta}} \,. \tag{11.6}$$

The right frames of Figs. 11.9 are Ω_F/Ω_H plotted at three different time steps for KDJ. For comparison the left frames of these figures are plots of S^r. The plots of Ω_F are very noisy because equation (11.6) is a complicated function of code variables and numerical noise propagates through the algebraic expressions, especially through the denominator. In spite of this, there are still some clear trends that permeate through the strong numerical noise. Notice that each flare in S^r is enveloped by a region of enhanced Ω_F, typically $0.7\Omega_H < \Omega_F < 1.2\Omega_H$. The regions of the funnel outside the ergosphere that are devoid of large flares in S^r, typically have $0 < \Omega_F < 0.5\Omega_H$. It seems reasonable to associate these large peak values of Ω_F in KDJ with the spatially and temporally coincident flares in S^r that occur in KDJ. Furthermore, this greatly enhanced value of Ω_F indicates a different physical origin for Ω_F in the flares than for the remainder of the funnel. The most straightforward interpretation is that it is a direct consequence of the fact that the flares originate on magnetic flux that is locked into approximate corotation with the dense accreting equatorial plasma (i.e., the inertially dominated equatorial plasma anchors the magnetic flux). In the inner ergosphere, frame dragging enforces $0.7\Omega_H < d\phi/dt < 1.0\Omega_H$ on the accretion flow and therefore the frozen-in magnetic flux. This establishes condition 3 of the introductory remarks.

11.3.4 The Creation of Negative Energy Plasma

Finally we look at the generation of negative energy plasma in the GHM ergospheric disk dynamo. From equation (5.20), the negative redshifted specific mechanical energy condition is

$$\omega = \mu(-\tilde{u}_t) < 0 \,. \tag{11.7}$$

Fig. 11.9 The field line angular velocity, Ω_F. The left frame is S^r and the right frame is Ω_F/Ω_H in KDJ, both averaged over azimuth, at *top*: $t = 9840$ M, *middle*: $t = 9920$ M, *bottom*: $t = 10000$ M. The relative units (based on code variables) are in a color bar to right of each plot. The contours on the S^r plots are of S^θ scaled from the peak within the frame at relative levels 0.67 and 0.33. The white contour is the stationary limit surface. There is no data clipping, so plot values that exceed the limits of the color bar appear white

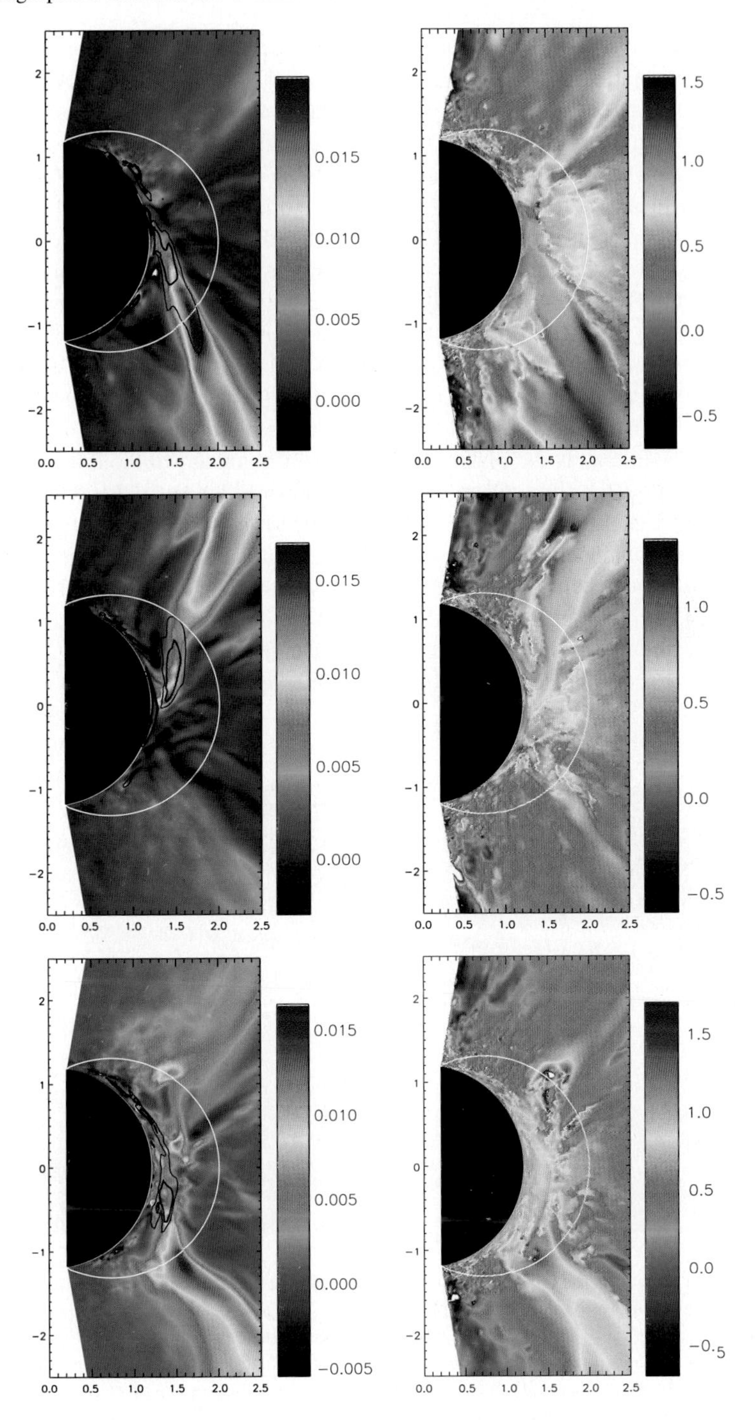
2
1
0
−1
−2
0.0 0.5 1.0 1.5 2.0 2.5
0.015
0.010
0.005
0.000
1.5
1.0
0.5
0.0
−0.5
0.015
0.010
0.005
0.000
1.0
0.5
0.0
−0.5
0.015
0.010
0.005
0.000
−0.005
1.5
1.0
0.5
0.0
−0.5

Recall that when the black hole swallows negative mechanical energy the black hole loses its rotational in energy in the Penrose effect. Note that the definition in equation (11.7) is equivalent to the [16] condition for a Penrose process, i.e., the accretion of particles with $-\tilde{u}_t < 0$. The most dramatic effect of the ideal GHM torsional tug of war that is described in Chap. 7 is that outgoing Poynting flux is made by plasma being forced onto $-\tilde{u}_t < 0$ trajectories. The advantage of using $-\tilde{u}_t$ instead of ω is that its value is not dependent on difficult to interpret code units. Also, the color bar centered on 0 is useful for finding negative energy plasma. The grey color indicates $-\tilde{u}_t = 0$. Slightly negative energy plasma, $-\tilde{u}_t \lesssim 0$, is light blue which is clearly distinct from slightly positive energy plasma, $-\tilde{u}_t \gtrsim 0$, which is yellow. Recall from Chap. 3, that a value of $-\tilde{u}_t = 1$ is equivalent to the energy of a cold particle released from rest at asymptotic infinity. The blue regions of Fig. 11.10 are a very surprising result in that they indicate the incredible power of the GHM interaction. Note that as the plasma accretes radially inward in the dynamo region it crosses the Poynting flux generation region and simultaneously keeps decreasing its mechanical energy per unit enthalpy. This decrease continues as the power is extracted electromagnetically even until the mechanical energy becomes negative. Thus, the extracted energy is more than the energy stored in the plasma. The process creating the Poynting jet is independent of the stored energy within the plasma and is driven by the only other dynamic element available, frame dragging. The blue regions of $-\tilde{u}_t < 0$ indicate that a strong GHM interaction is present. This simulation

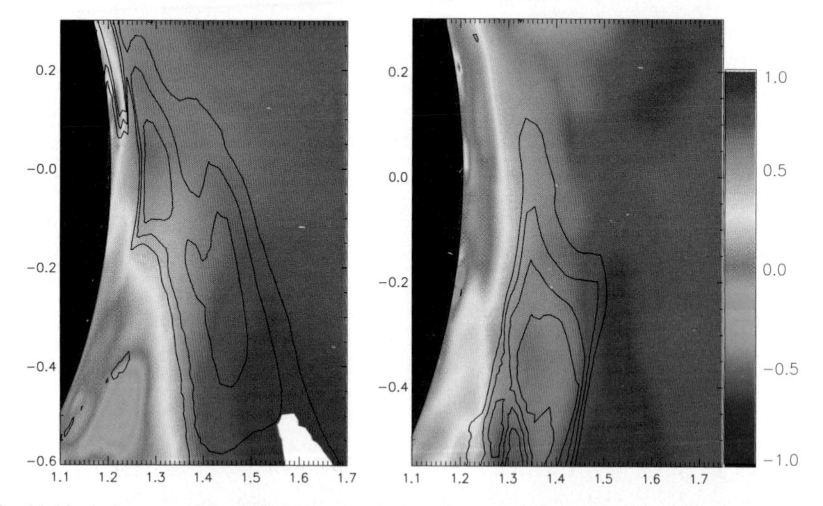

Fig. 11.10 A close-up of the GHM dynamo. A plot of $-\tilde{u}_t$, the mechanical energy per unit enthalpy as viewed from asymptotic infinity of the plasma in KDJ. The data is averaged over azimuth, at *left*: $t = 9840$ M and *right*: $t = 10000$ M. The units are in a color bar to right of the plots. The contours are of S^{θ}. The inside of the inner calculational boundary ($r = 1.203$ M) is black. The calculational boundary near the poles is at 8.1° and 171.9°. There is no data clipping, so plot values that exceed the limits of the color bar appear white

is a high accretion system (like a quasar) with rapid accretion driven by a strong MRI. Thus, there is an enormous flood of positive energy plasma (the large $-\tilde{u}_t \approx 1$ regions) approaching the vertical magnetic flux region of the dynamo. In this high spin simulation, the GHM interaction can actually remove all this energy from the plasma and more! Figures 11.10 clearly demonstrate that KDJ satisfies condition 4 for a GHM ergospheric disk dynamo from the introductory remarks to this section. The corroboration of conditions 1–4 in KDJ is a robust verification of a GHM driven ergospheric disk in operation. Furthermore, the existence of the ergospheric disk within the strongly accreting system justifies the implementation of the ergospheric disk in the AGN model building of Chap. 10.

11.3.5 The Simulation KDE

The simulations of DeVilliers and Hawley seem to indicate a dramatic increase in GHM jet power as the a/M approaches 1, as predicted in Fig. 11.9 left. Consider the very high spin simulation KDE from [181], with $a/M = 0.998$. The source of Poynting flux for this simulation was studied in [187]. The energy source for the powerful Poynting jet was explored in [187] with the aid of Fig. 11.11.

Figure 11.11 is a magnification of the inner region of Fig. 11.8 left of [181]. It is an excision of a region, $0° < \theta < 65°$, $r \gtrsim r_+$ that is a little larger than the ergospheric portion of the magnetically dominated funnel, $0° < \theta < 55°$, $r \gtrsim r_+$. It is a contour plot of S^r. The data is averaged over azimuth and over time from $2{,}000M < t < 8{,}080M$. A data dump occurs every $t = 80$ M. Thus, 76 discrete time slices are averaged in Fig. 11.11 after the large transients have died down. The most striking feature in the figure is that S^r appears to switch on outside the inner calculational boundary at $r = 1.175M$ (and therefore the horizon at $r = 1.1{,}063M$) in a thin layer near $r = 1.3M$-$r = 1.5M$.

The time and azimuthally averaged data in Fig. 11.11 is trivially equivalent to a discrete sum estimate of the total radial Poynting flux emanating from the ergosphere at $t > 2{,}000M$ in the simulation (i.e., multiply the values by $(\pi/2)(80M)(76)$). Thus, it is interesting to investigate the power source with the aid of Poynting's theorem using the Gaussian pillbox drawn in Fig. 11.11. This requires converting equation (11.4) to an integral version of Poynting's theorem by trivial integration. The symmetry of the Gaussian pillbox simplifies the expression. Curves 1 and 3 are semicircular arcs (r = constant). The curves 2 and 4 are radial segments (θ = constant). Curve 4 is chosen to be at the funnel boundary. However, without complete data to analyze, this is a bit uncertain and $\theta = 55°$ is a very conservative lower bound. Looking at the data in [179] of the same simulation, indicates that it could be as large as $\theta = 65°$. Employing the periodic boundary condition on ϕ and integrating over azimuth and time equation (11.4) becomes,

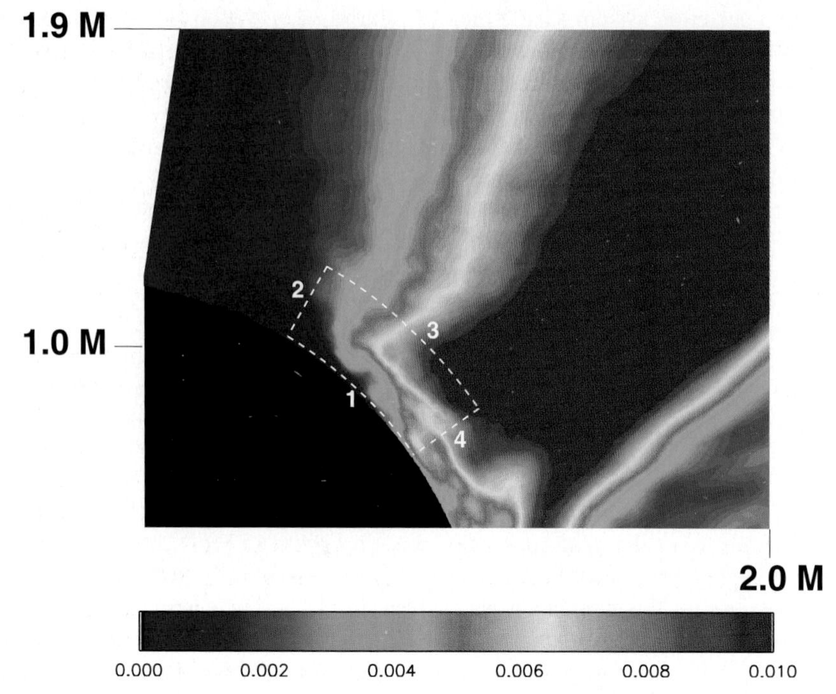

Fig. 11.11 A plot of S^r that is azimuthally averaged and time averaged (over 75% of the simulation that ends at t = 8,080 M) from the simulation KDE. The figure highlights the region, $0° < \theta < 65°, r \gtrsim r_+$ that is a little larger than the ergospheric portion of the magnetically dominated funnel at $r \gtrsim r_+$. The majority of S^r switches-on in a thin layer near $r = 1.3M - r = 1.5M$ (the color bar is in code units). Saturated regions are clipped, so the dark red areas of S^r are stronger than indicated by the color bar. A Gaussian pillbox, $30° < \theta < 55°$, is drawn as a dashed white contour for use in Poynting's Theorem. There are 26 grid zones between the inner boundary, $r = 1.175M$ and $r = 1.5M$. The plot is provided courtesy of John Hawley

$$\begin{aligned}
&\int (-T_t{}^t)_{EM}\, dV + \int (-T_t{}^t)_{fluid}\, dV \\
&= (\pi/2)\left[\int_3 (-\sqrt{-g}T_t{}^r)_{EM}\, d\theta dt - \int_1 (-\sqrt{-g}T_t{}^r)_{EM}\, d\theta dt\right] \\
&+(\pi/2)\left[\int_3 (-\sqrt{-g}T_t{}^r)_{fluid}\, d\theta dt - \int_1 (-\sqrt{-g}T_t{}^r)_{fluid}\, d\theta dt\right] \\
&+(\pi/2)\left[-\int_2 (-\sqrt{-g}T_t{}^\theta)_{EM}\, dr dt + \int_4 (-\sqrt{-g}T_t{}^\theta)_{EM}\, dr dt\right] \\
&+(\pi/2)\left[-\int_2 (-\sqrt{-g}T_t{}^\theta)_{fluid}\, dr dt + \int_4 (-\sqrt{-g}T_t{}^\theta)_{fluid}\, dr dt\right], \qquad (11.8)
\end{aligned}$$

where $dV = \sqrt{-g}\, dr d\theta d\phi$. At the time of [187], it was not clear what the source of S^r was since the raw data was not analyzed. However, with the knowledge gained from studying KDJ, the situation is clear. Strong dissipation and large radiation losses are not allowed by the perfect MHD assumption, thus there is no GHM interaction within the funnel proper. Consequently, in the evacuated funnel with the perfect MHD assumption, the plasma (fluid) stress-energy should be negligible compared to the electromagnetic contribution. Thus, equation (11.8) reduces to

$$\int (-T_t{}^t)_{EM}\, dV \approx (\pi/2)\left[\int_3 (-\sqrt{-g}T_t{}^r)_{EM}\, d\theta dt - \int_1 (-\sqrt{-g}T_t{}^r)_{EM}\, d\theta dt\right]$$
$$+(\pi/2)\left[-\int_2 (-\sqrt{-g}T_t{}^\theta)_{EM}\, drdt + \int_4 (-\sqrt{-g}T_t{}^\theta)_{EM}\, drdt\right]. \qquad (11.9)$$

Furthermore, the amount of energy flux radiated from the ergosphere during the course of the simulation, $(\pi/2)\int_3 (-\sqrt{-g}T_t{}^r)_{EM}\, d\theta dt$ diverges with time while the amount of stored energy in a finite region of spacetime remains bounded, $\int (-T_t{}^t)_{EM}\, dV$. Thus, after a long period of time the source term on the left hand side of equation (11.9) is negligible. Secondly, $(\pi/2)\int_3 (-\sqrt{-g}T_t{}^r)_{EM}\, d\theta dt > 4 \times (\pi/2)\int_1 (-\sqrt{-g}T_t{}^r)_{EM}\, d\theta dt$ based on the contour plot in Fig. 11.11. However, large values of S^r saturate due to the data clipping imposed by [181], so the red color is deceiving and $(\pi/2)\int_3 (-\sqrt{-g}T_t{}^r)_{EM}\, d\theta dt$ is likely significantly larger than this relative estimate. Thus, we can approximate (11.9) as

$$\int_3 S^r\, d\theta dt \approx \int_4 S^\theta\, drdt - \int_2 S^\theta\, drdt\,. \qquad (11.10)$$

The situation is very clear from the analogy to KDJ. Compare the vertical extension in S^r at low latitudes in Fig. 11.11 and the similar extension in Fig. 11.7 bottom showing S^r at $t = 10{,}000$ M in KDJ. The morphology is virtually identical and strongly suggests an ergospheric disk origin for the vast majority of S^r radiated from the ergosphere,

$$\int_3 S^r\, d\theta dt \approx \int_4 S^\theta\, drdt\,. \qquad (11.11)$$

Just as in KDJ virtually all of the Poynting flux in the magnetically dominated funnel is produced by ergospheric disk. In KDE, the preponderance of energy flux from the ergospheric disk is even more dramatic.

11.4 Source of Poynting Flux in Event Horizon Magnetospheres

There are two main concepts that are proposed in this book.

- The primary focus of this effort is the GHM mechanism that can drive powerful jets in AGN and X-ray binaries. In particular, the ergospheric disk is the most energetic manifestation of GHM.
- Secondly, causality arguments have been put forward to try to understand what determines the electromagnetic power on field lines that thread the event horizon.

The book expands on the discussion of [67] that the method of solution in [66, 111] is fundamentally acausal. Therefore, in a realistic astrophysical circumstance, the power extracted from the black hole (even with the perfect MHD assumption) will not be as predicted.

In the previous two sections, the first point was explored with the aid of numerical simulations. In particular, it was shown that the physics conducive to an ergospheric disk existed in the 3-D perfect MHD simulations KDJ and KDE. The ergospheric disk jet is launched from the plasma near the equatorial plane of the ergosphere. The ergospheric disk jet fills the outermost portion of the accretion vortex (see Fig. 11.7). Interior to the ergospheric disk jet is the event horizon magnetosphere (EHM) comprised of poloidal flux that threads the event horizon (see the top frame of Fig. 11.12). The EHM is distinct from the poloidal magnetic flux that threads the equatorial plane of the ergosphere, which forms the ergospheric disk magnetosphere. This section explores the second point above by studying the sources of Poynting flux in the EHM within the context of the 3-D simulations. We continue the analysis of KDJ, a/M = 099, and begin an investigation of a third 3-D simulation, KDH with a/M = 0.95. This analysis parallels and expands the work of [191].

The data presented in this section demonstrates that the boundaries of the EHM should be dynamic and are not likely to be passive boundary surfaces for the magnetic field. It is shown that electrodynamic energy flux can arise in the EHM as a result of sources radiating energy from the lateral boundaries. Poynting flux is injected into the EHM from both the ergospheric disk jet as well as strong flares originating in the accretion disk corona. Even if the EHM can be construed as "force-free," the dynamics of the lateral boundaries are determined by strong inertial forces that should make them strong MHD pistons. This circumstance was not anticipated in theoretical treatments of electrodynamic jets in the EHM [66, 111]. The fact that electromagnetic energy can enter the EHM from the side goes right to the heart of the assumptions in the Blandford–Znajek solution. The Blandford–Znajek solution is the perfect MHD solution in which energy conservation reduces to Poynting flux conservation from the horizon to a relativistic wind at asymptotic infinity [111]. From this condition, the parameters of the field are uniquely determined for a given poloidal field distribution, in particular the field line angular velocity, Ω_F, and the total electromagnetic energy output from the black hole, $\int S^r\, d\theta d\phi$. However, if there are strong sources of Poynting flux along the lateral walls of the EHM, the spacetime near the event horizon can not adjust the system to enforce the Blandford–Znajek field parameters within the EHM. This is a direct consequence of the fact that the plasma near the event horizon in the EHM can not effectively react back on the outgoing wind or jet and modify its electromagnetic properties because of the gravitational redshifting of the MHD characteristics (see Sect. 6.4). The plasma near the horizon in the EHM will passively accept any field parameters imposed by the ergospheric disk and the accretion disk corona (see Sect. 6.3). As such, in a general astrophysical context, the basic parameters such as Ω_F and $\int S^r\, d\theta d\phi$ are indeterminant.

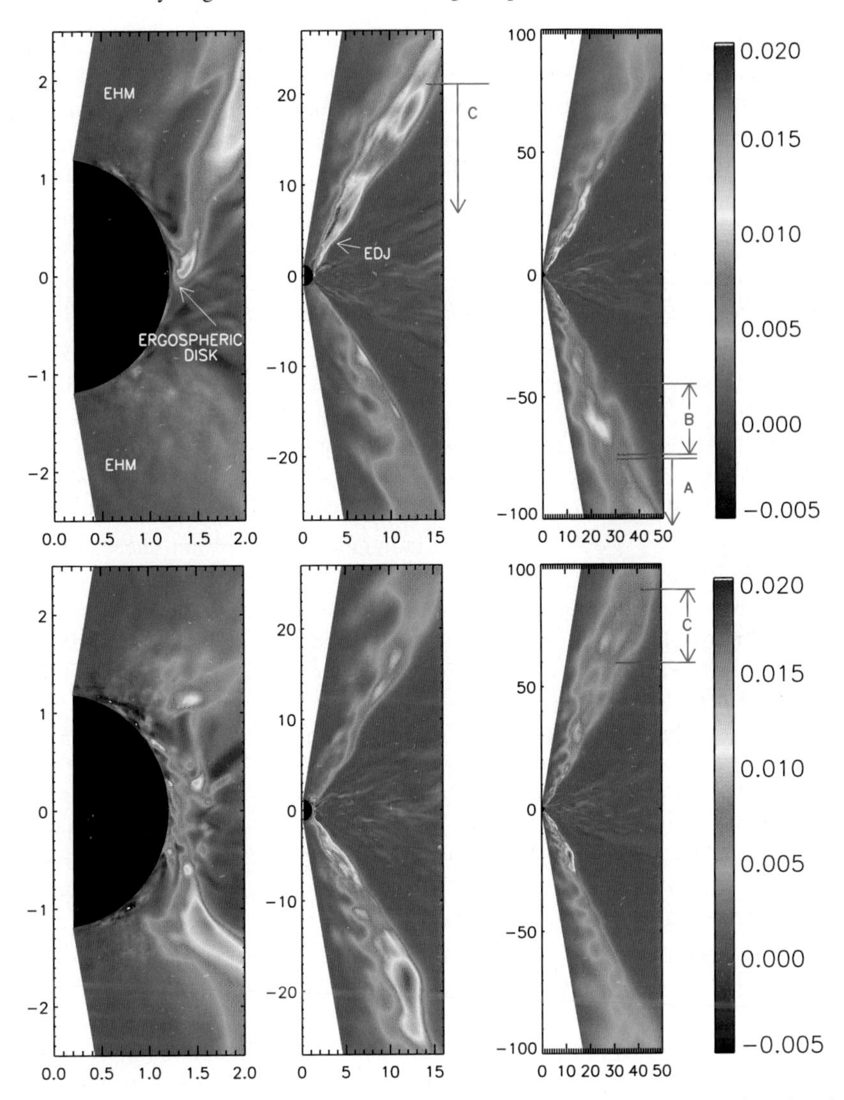

Fig. 11.12 The large scale view of the radial Poynting flux in KDJ. The color bar is in code units. The top row is at $t = 9{,}920$ M and bottom row is $t = 10{,}000$ M. "EDJ" is an abbreviation for the ergospheric disk jet. There is no data clipping, saturated regions are white

In order to demonstrate these points, this section is organized as follows. The first subsection explores the propagation of the ergospheric disk jet in KDJ away from the ergosphere. The ergospheric disk jet is comprised of strong flares of Poynting flux that resistively diffuse to high latitudes within the EHM. By $r \sim 100M$, even the polar regions of the EHM are dominated by the ergospheric disk ejecta. Clearly, the resulting power output is not determined by the Blandford–Znajek solution. In the second subsection, the lower spin simulation KDH is investigated. KDH provides an interesting contrast to KDJ, because the ergospheric disk jet in the available data slices is not nearly as powerful as it is in KDJ. Thus, weaker sources of Poynting flux can be resolved that would otherwise be swamped by a powerful ergospheric disk jet. It is demonstrated that coronal flares act as MHD pistons at the boundary of the EHM that launch substantial Poynting flux. Thus, the total power is significantly different from what one would expect from a Blandford–Znajek solution.

11.4.1 The Propagation of the Ergospheric Disk Jet

Figure 11.12 is a plot of S^r in KDJ viewed at three different levels of magnification, at $t = 9{,}920$ M (top row) and $t = 10{,}000$ M (bottom row). Similarly, Fig. 11.13 is a plot of S^r in KDJ viewed at three different levels of magnification, at $t = 9{,}840$ M (top row) and $t = 9{,}920$ M (bottom row). Each frame is the average over azimuth of each time step. This greatly reduces the fluctuations as the accretion vortex is a cauldron of strong MHD waves. The individual ϕ = constant slices show the same dominant behavior, however it is embedded in large MHD fluctuations. The left hand columns of Figs. 11.12 and 11.13 show strong beams of S^r coming from near the black hole. In this subsection, we turn our attention to the propagation of individual flares from the ergospheric disk out to the outer calculational boundary at $r = 120M$. Even though the time sampling is very coarse in the data dumps ($\Delta t = 80M$), we can understand the propagation of the ergospheric disk jet because of the wide angle views available in the right hand columns of Figs. 11.12 and 11.13. We track the EDJ evolution by identifying the strong knots or flares in Figs. 11.12 and 11.13 in consort with MHD causal constraints. As discussed in Chaps. 5 and 6, the S^r flares in a perfect MHD flares will propagate at the speed of a perfect MHD discontinuity as modified by the plasma bulk flow velocity. The plasma near the edge of the vortex has accelerated to $\tilde{v}^r > 0.9c$ by $r = 30$ M. So the flares of S^r should propagate radially at $V_{flare} \lesssim c$ for $r > 30M$. Without having the benefit of the detailed time evolution, this upper bound is the best estimate that we can make for V_{flare}.

First, consider the strong knot, "C," at $t = 10{,}000$ M in the bottom, right hand frame of Fig. 11.12. Label the outer radial extent of knot C at $t = 10{,}000$ M by $r_{+C}(t = 10{,}000M) = 100.9M$ and inner radial edge by $r_{-C}(t = 10{,}000M) = 65.9M$. Translating this perfect MHD discontinuity back in time to $t = 9{,}920$ M is equivalent to a radial displacement

$$V_{flare}\Delta t = V_{flare}(-80M/c) \gtrsim -80M\,. \tag{11.12}$$

Thus at $t = 9920$ M, knot "C" should extend from the ergospheric disk to $r_{+C}(t = 9{,}920\text{ M}) \gtrsim 20.9M$. The red patches in the middle frame at $t = 9{,}920$ is therefore the early time (past) manifestation of the strong knot "C" in the EHM at $t = 10{,}000$ M.

Next consider the strong knot, "A," at $t = 9{,}840$ M in the middle and right frames in the top row of Fig. 11.13. Label the outer radial extent of knot "A" at $t = 9{,}840$ M

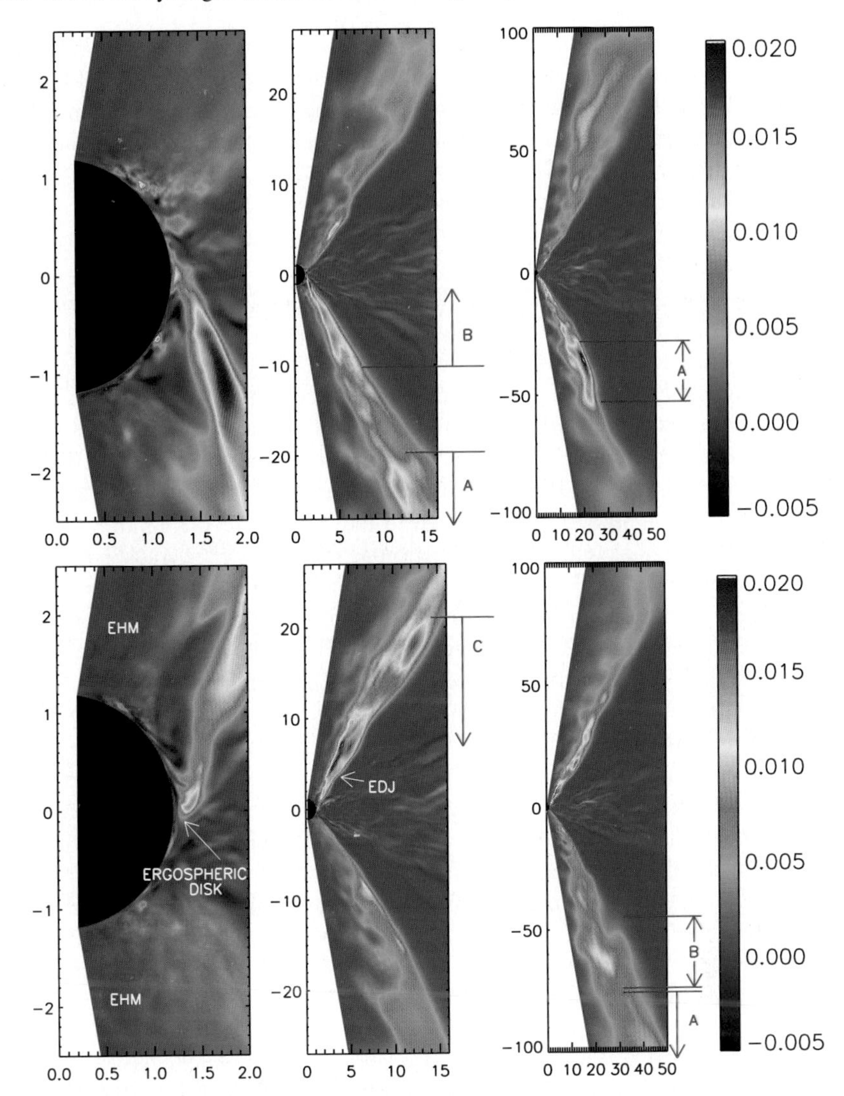

Fig. 11.13 The large scale view of the radial Poynting flux in KDJ. The color bar is in code units. The top row is at $t = 9,840$ M and bottom row is $t = 9,920$ M. There is no data clipping, saturated regions are white. The interior of the inner calculational boundary ($r = 1.203$ M) is black. The calculational boundaries near the poles are at 8.1° and 171.9°

by, $r_{+A}(t = 9,840M) = 58.9M$ and inner radial edge by $r_{-A}(t = 9,840M) = 22.0M$. Time translating this feature to $t = 9,920$ M implies that $r_{-A}(t = 9,920M) \lesssim 102.0M$, so it must be visible near the edge of the right hand frame at $t = 9,920$ M. Furthermore, unless the flare is propagating inordinately slowly, $V_{flare} < 0.75c$, $r_{+A}(t = 9,920M)$ will be beyond the outer boundary of the plot. There is only one plausible feature at $t = 9,920$ M, hence the identification "A" in the bottom, right frame of Fig. 11.13. Secondly, consider the strong flare "B" that is emerging from the ergospheric disk at $t = 9,840$ M in the left and middle frames in the top row. Thus, at $t = 9,920$ M, some portion of the flare must be within 80 M of the black hole, hence the knot labeled "B" in the right hand frame in the bottom row of Fig. 11.13 is uniquely identified as the future time evolution of the ejection from the ergospheric disk at $t \sim 9,840M$.

The dynamics of the ergospheric jet propagation illustrated in Figs. 11.12 and 11.13 can be summarized as follows. At $r \approx 1.5 - -2.0M$, the ergospheric disk jet enters the EHM from the periphery (left hand frames). The ergospheric disk jet gets quickly linked into the EHM because the ergospheric disk magnetosphere in KDJ is comprised of small patches of twisted vertical flux that become intertwined with the large-scale flux in the EHM on scales of ~1–2 M (see Sect. 11.3.2). After this rapid injection, the S^r in the ergospheric disk jet keeps spreading towards the pole as it propagates outward. At the time steps that were made available to this author, the ergospheric disk jet is the predominant source of S^r in the EHM. By $r \approx 100M$, the ergospheric disk jet is flooding the EHM, even close to the polar axis. Furthermore, it should be noted that in the process of S^r migrating towards the pole that prodigious quantities of electromagnetic energy are transferred to the plasma. Therefore the total energy flux is significantly larger than just S^r, $\approx 20\%$ of the total energy flux is in mechanical form in the EHM at $r \approx 100M$. The relevance to this discussion is that the EHM is inundated with S^r (and the bi-product mechanical energy flux) that was not created on field lines that thread the horizon, but on flux entrapped within the equatorial accreting plasma. The slow diffusion of S^r poleward at $r > 30M$ is most likely regulated by numerical diffusion. This might seem like a problem from a numerical point of view, but physically this is not nearly as much of a concern from a qualitative standpoint. Perfect MHD is just a simple tractable method of dealing with the plasma physics. A realistic, high temperature, jet plasma is likely to have anomalous resistivity from a variety of sources, [192, 193], and the diffusion of field energy should naturally occur. The simulation cannot accurately describe the diffusion rate. However, qualitatively speaking, it indicates that if the jet propagates extremely far from the hole ($r \gg 120M$), regardless of the exact details of the diffusion microphysics, the ergospheric disk jet energy flux is likely to get smeared out towards the polar region.

Clearly, KDJ is an example of an EHM magnetosphere in which the Blandford–Znajek wind parameters are irrelevant to the total jet power from the EHM. In this most extreme example (and in KDE), even the notion of electrodynamic energy extraction, that is the fundamental concept behind the Blandford–Znajek mechanism, is merely a weak second order correction to the total output power of the EHM. The plasma in the spacetime near the event horizon is clearly powerless to over-ride the wind parameters imposed on the EHM by the ergospheric disk jet.

11.4.2 The MHD Coronal Piston

J. Krolik and J. Hawley have generously shared the data for the last three time slices of KDH ($a/M = 0.95$), at $t = 9,840$ M, $t = 9,920$ M and $t = 10,000$ M. The strength

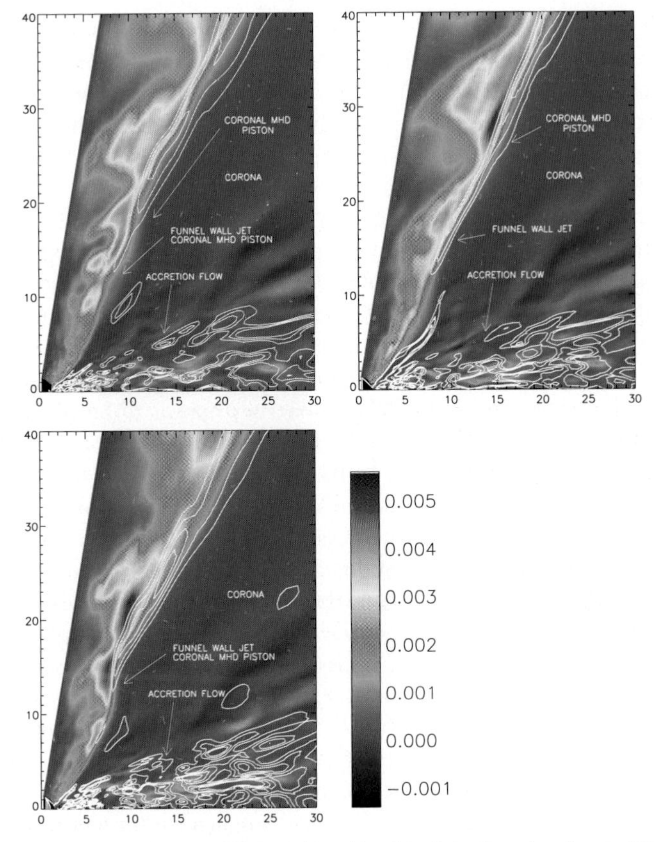

Fig. 11.14 The coronal MHD piston is illustrated by this plot of the Poynting flux in KDH at *top left*: $t = 9840$ M, *top right*: $t = 9920$ M, *bottom*: $t = 10000$ M. The color bar is in code units. There is no data clipping, saturated regions are white. The overlayed white contours represent, P_r, the radial momentum flux described in the text. At each coronal piston location there is a large pressure flare. Inside of $r_{in} = 1.403M$ ($r_+ = 1.312M$) is colored black

of the ergospheric disk jet is variable, it is noticeable at $t = 9,840$ M, but it is negligible otherwise. This circumstance allows for the detection of weaker sources of S^r that would otherwise be swamped by a strong ergospheric disk jet. Figure 11.14 are plots of S^r for these three time slices in chronological order. Each plot is the average over azimuth of the time step. The contours of the radial momentum flux due to mass motion, $P_r \equiv \sqrt{-g}\rho\tilde{u}^r\tilde{u}_r$ (where ρ is the proper mass density), are overlayed in white in order to define the location of the "funnel wall jet," as was done in [178, 183]. The funnel wall jet is a shear layer between the accretion disk corona and the Poynting jet. It is a collimated sub-relativistic flow that transports most of the mass outflow in the jetted system. In [183], it was shown to be driven by the total pressure (gas plus magnetic) gradient in the corona that is oblique to the funnel wall boundary. The gas in this region is constrained from being pushed into the funnel by the centrifugal barrier. The component of pressure gradient that is parallel to the centrifugal barrier forces the flow to be squeezed outward as a shear layer.

In Fig. 11.14, there is an almost one to one correspondence, between locations where P_r of the funnel wall jet increases and sites where S^r increases at the funnel wall boundary in the EHM. In agreement with [183], it was concluded in [191] that it is the coronal pressure and not the Poynting jet that drives the funnel wall jet. In a most general context this must be the case. The results of [183] in Table 4 and Fig. 5 and the analysis in [191] show that the Poynting jet in the funnel is a factor of 3 too weak to drive the funnel wall jet. More graphically Fig. 11.15 shows a strong flare in the total pressure (gas plus magnetic) at $t = 10,000$ M. The pressure gradient seems to provide the accelerating force that drives P_r. The flare appears to be a high pressure loop emerging from the corona, as evidenced by the right hand frame of Fig. 11.15. The loop location and topology are inconsistent with the high pressure feature being injected into the corona from the funnel interior. It was also shown in [191] that the coronal injection sites appear to be required in order to support the large excess (>26%) of $\int S^r\, d\theta d\phi$ that reaches the outer calculational boundary in the EHM compared to the amount of $\int S^r\, d\theta d\phi$ that is created within the ergosphere of the EHM in the simulation KDH. Furthermore, it should be noted that the magnetic pressure in the corona actually exceeds the magnetic pressure in the funnel at these intermediate radii, $10M < r < 30M$! Not coincidentally, this is where the putative MHD coronal pistons are located. There is plenty of magnetic and gas energy in this region to power the injection sites and make up the for the deficit of $\int S^r\, d\theta d\phi$ leaving the ergosphere in the EHM compared to that which reaches the outer calculational boundary.

It is instructive to detail how the total pressure gradient can drive the massive outflow. The total pressure density of a magneto-fluid is defined in terms of the gas pressure, P_g and the Faraday field strength tensor, $F_{\mu\nu}$ as

$$P_{tot} = \sqrt{-g}(P_g + F^{\mu\nu}F_{\mu\nu}/16\pi) \tag{11.13}$$

The radial momentum flux due to mass motion, P_r is linked to the pressure through the stress-energy tensor density of the fluid,

$$\sqrt{-g}T^r_r = hP_r + P_{tot} + \sqrt{-g}[F^{\mu\nu}F_{\mu\nu}/16\pi]u^r u_r - \sqrt{-g}[(*F^{r\nu}u_\nu)(*F_r{}^{\mu}u_\mu)/4\pi]\,. \tag{11.14}$$

The stress energy tensor has been expanded in this unconventional manner as in [178, 183] because it highlights the connection between P_r and P_{tot}. The quantity "h" is the enthalpy per unit mass. In [178,183] and in Figs. 11.14 and 11.15 "h" was left out of the definition of the radial momentum flux, P_r for the following reasons that were discussed in [183]. As pointed out in [183], $h \approx 1$ in the funnel wall jet. However, there are regions of large h in the EHM. The EHM is not force-free even though it is magnetically dominated. There is enormous local heat dissipation as a consequence of the fact that it is an intense cauldron of MHD waves that transfer large amounts of energy to and from the fluid. Not only are there the coronal piston waves discussed here, but there are large fluctuations in general, including intense shocks that heat the plasma to relativistic temperatures [183]. This convention is

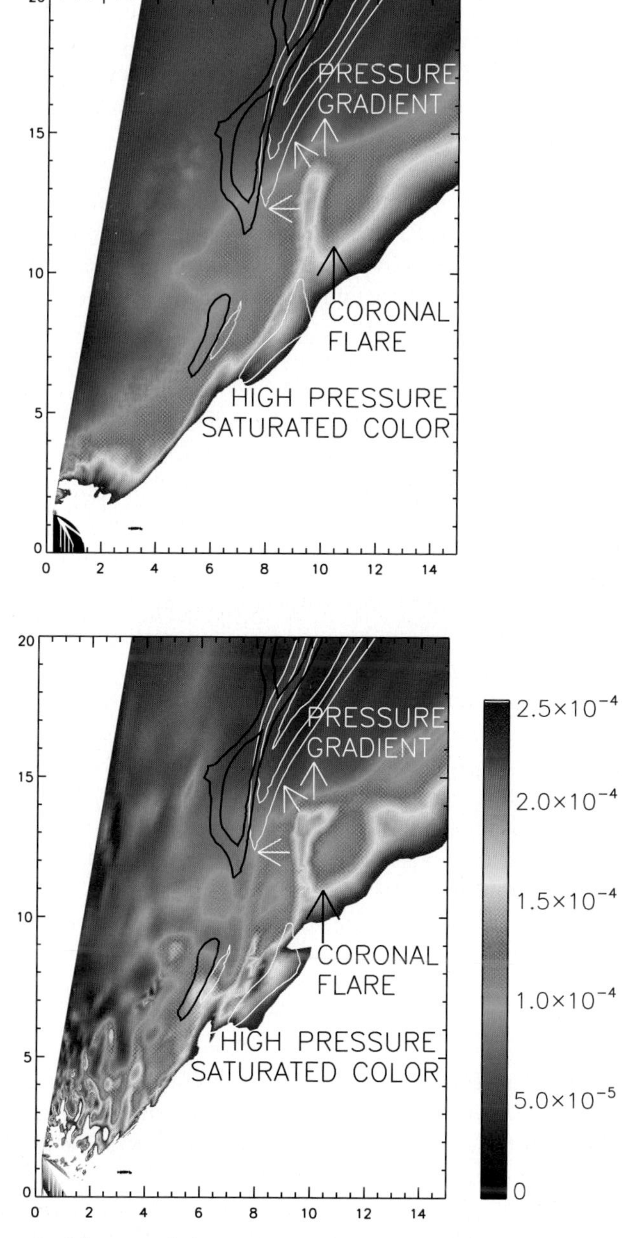

Fig. 11.15 A closeup of the coronal piston at $t = 10,000$ M is depicted in these plots of the total pressure density (gas plus magnetic). The top frame is averaged over ϕ and the bottom frame is at $\phi = 49.2°$. The color bar is the strength of the total pressure in code units. High pressure regions of the corona are saturated and appear white. The force associated with the pressure gradient is indicated by the white arrows. The overlayed white contours represent, P_r, the radial momentum flux as in Fig. 11.14 bottom and the black contours represent S^r

maintained for the sake of clearly distinguishing the location of the larger mass flux of funnel wall jet from hot shock heated gas inside the low density EHM. Applying the relativistic law of the conservation of radial momentum flux, $T^{r\nu}{}_{;\nu} = 0$, to equation (11.14) shows that a decrease in P_{tot} (i.e., the total pressure gradient) can affect an increase in P_r.

Next consider the increase of S^r in locations where P_r increases in Fig. 11.14 and 11.15. The corona does not directly affect an injection of S^r into the funnel as clearly shown in Fig. 11.15. This is because the pressure gradient appears in the radial momentum flux in equation (11.14), but not in the energy flux equation that results from the divergence of equation (11.15) and this is the source equation for S^r.

$$\sqrt{-g}T_t{}^r = \sqrt{-g}(\rho h)\tilde{u}_t\tilde{u}^r - S^r . \tag{11.15}$$

After the plasma is accelerated radially by P_{tot} as a consequence of radial momentum conservation applied to equation (11.14), the funnel wall plasma has attained significant radial momentum. Then by the energy conservation equation, (11.4), and equation (11.15), this radial mechanical momentum can be transferred to field within the EHM and is manifested as S^r. Hence, this intermediate step of plasma acceleration that mediates the flow of coronal internal energy to S^r in the EHM.

We can consider the perfect MHD aspect of the radial momentum being transferred from the funnel wall jet to Poynting flux in the EHM. It has been discussed in Chap. 2 that injecting Poynting flux into a magnetosphere requires an Alfven wave component in addition to the compressive polarizations (slow and fast). In even the most simplified problems, all of these modes are required to meet all the boundary conditions. The Alfven component is particularly important to the Poynting flux because it is the mode that carries an electric charge. However, the coronal piston (via the funnel wall jet) is injecting waves almost perpendicular to the predominantly radial magnetic field in the funnel. Thus, the Alfven speed is slow, it strictly vanishes for perpendicular propagation (see Chap. 2). The plasma in the funnel jet is propagating with a relativistic bulk velocity, so these Alfven waves get swept up in the outgoing wind. Consequently, the coronal piston always appears to be just upstream of the peak of the S^r and the flares in S^r should propagate outward at approximately the bulk velocity of plasma in the EHM. This is depicted in Fig. 11.15.

One might be concerned that there are only $\approx$ 8–10 angular zones between the coronal piston and the EHM and this leads to significant numerical diffusion. From a numerics point of view this is much more of a concern than from a physical point of view. The MHD code is just a simple approximation to any real turbulent plasma state. The turbulent corona is likely to have an anomalous resistivity and diffusion should occur [192, 193]. The rate of diffusion cannot be determined by this simulation. However, the qualitative idea that a strong flare in coronal energy can in principle reach the EHM interior is strongly indicated.

In summary, KDH shows that MHD coronal pistons can inject a modest amount of Poynting flux into the EHM. This provides a dynamic contrast to KDJ and KDE. In KDJ (a/M = 0.99) and KDE (a/M = 0.998), the ergospheric disk jet dominates the energy output from the ergosphere. In the EHM, at large distances from the black hole the jet power is determined primarily by the properties of the ergospheric disk, not any electrodynamic properties associated with the horizon (i.e., any putative Blandford–Znajek effects). In the lower spin simulation, KDH (a/M = 0.95), the ergospheric disk jet is a relatively minor contributor, so the other ergospheric energy sources are highlighted. In this case, the electrodynamic effect associated with the putative Blandford–Znajek mechanism are pronounced with contributions of the same order of magnitude from the coronal pistons within the funnel wall. Clearly, the strength of this coronal piston will depend on the detailed gas dynamics of the magnetized corona.

It should be noted that the very abrupt drop-off of ergospheric disk power with spin, near a/M = 0.95, is a consequence of the nature of the vertical flux generation in the equatorial plane. This result does not contradict the plots in Fig. 10.13. In that figure, it was assumed that large scale vertical flux existed throughout the equatorial plane of the ergosphere. Within the parameters of the simulation, the vertical flux in the ergospheric equatorial plane is comprised of twisted loops generated in the turbulent plasma. These loops seem to form within $r < 1.5M$ in the context of the simulations. The inner calculational boundary in KDH is at $r_{in} = 1.403M$. Thus, there simply is no significant equatorial surface area to support the flares in the vertical magnetic flux. Note that if the inner boundary were truly the event horizon instead of the inner calculational boundary then this argument would indicate that

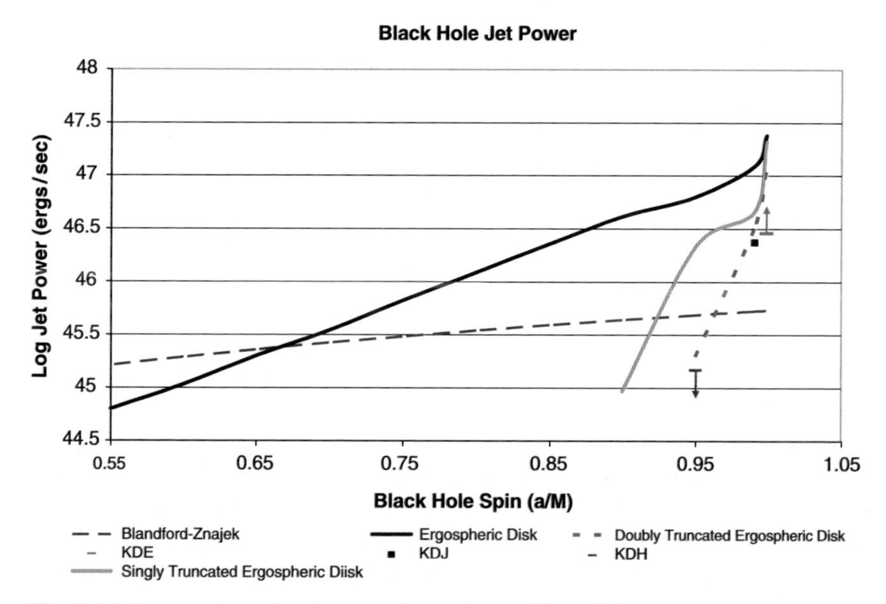

Fig. 11.16 A comparison of the jet power in the theoretical models and the 3-D computer simulations. The figure is in the same relative units as Fig. 10.13, the individual jet powers are expressed in units of $(B_E^P)_4^2$ and M_9^2, where $(B_E^P)_4$ is the strength of the ergospheric magnetic field in units of 10^4 G and M_9 is the black hole mass in units of $10^9\,M_\odot$. The derivations of the individual jet powers from the data are described within the text

the ergospheric disk would likely be powerful even at $a/M = 0.95$ and the switch-on would occur at $a/M \approx 0.9$ (see the Fig. 11.16 and the related discussion in the next subsection).

11.5 Discussion

It is interesting to discuss the relevance of these simulations to the theoretical analysis of this book and astronomical objects. The discrepancy between the theory and the numerics can be used to explore the consequences of the assumptions involved in these disparate treatments and the discrepancies with observations can highlight limitations of the theories and the shortcomings of the numerical methods. In order to explore these issues, it useful to compare the results of the simulations to the quantitative predictions of the theories. Fig 11.16 is an attempt to accomplish this. The figure is patterned after Fig. 10.13. It is assumed that $(B^P_E)_4 = 1$ and $M_9 = 1$, but the scalings with these parameters follows trivially with the vertical axis units in Fig. 10.13. Before analyzing this plot, the generation of each curve and data point in the figure is explained in detail.

11.5.1 The Ergospheric Disk Jet

The power of the ergospheric disk jet from theoretical considerations is plotted as the solid black curve in Fig. 11.16. The plot is slightly different than the corresponding plot in Fig. 10.13. The only change is that B^P is considered to be constant throughout the ergospheric disk. In the model used in Fig. 10.13, there was a build up of B^P at small radii. However, there was no evidence that this happens in the 3-D numerical simulations, so it was dropped for the sake of comparison with the numerical results.

11.5.2 The Truncated Ergospheric Disk Jet

The doubly truncated ergospheric disk jet is shown in the dashed light blue curve. The doubly truncated ergospheric disk represents the region of the ergospheric disk that actually occurs in the 3-D simulations. It is terminated at the inner calculational boundary, r_{in}. Patches of vertical poloidal flux seem to penetrate the ergospheric disk only at $r < 1.5M$, in the turbulent 3-D accretion simulations. Thus, the doubly truncated ergospheric disk is cutoff at $r \lesssim 1.5M$ in the calculations. The dashed light blue curve is the based on the same theoretical calculation as the black curve. However, only the power generated from the full ergospheric disk in the range $r_{in} < r < 1.42M$ is considered in the plot of the doubly truncated disk jet. Furthermore, as in KDJ, the jet was considered to be intermittent, operating in only one hemisphere at a time; so the power was cut in half.

The singly truncated ergospheric disk is the same as the doubly truncated ergospheric disk, except that the numerical artifact of terminating the disk at r_{in} is replaced by extending the inner edge of the ergospheric disk all the way to r_+. The power of the jet that is driven by the singly truncated ergospheric disk is indicated by the solid orange curve. In the singly truncated case, the Poynting jet operates down to $a/M = 0.9$.

11.5.3 The Blandford–Znajek Jet

The Blandford–Znajek jet power is the dark blue dashed curve in Fig. 11.16. The Blandford–Znajek jet power is based on expression (5.50) for B^T in a cylindrical relativistic jet at large distances from the black hole. The total power in the jet is determined by an integration analogous (5.51), over the whole Poynting jet. The parameters that need to be determined are Ω_F and the enclosed poloidal flux, Φ. A split monopole is not a viable choice for these parameters because it forms a spherical relativistic wind and not a jet, so it is not relevant to astrophysics [173]. The only Blandford–Znajek jets found in numerical simulations appear in accretion vortices. Ω_F is determined by averaging the KDH value from [183] with the [184] value, $\Omega_F = 0.4\Omega_H$ and is approximately constant with polar angle. In order to calculate Φ, we consider the distribution of poloidal flux in [183]. In these simulations, the poloidal flux, near the horizon, increases monotonically with polar angle towards the outer boundary of the EHM, more than doubling in magnitude. The opening angle of the Poynting jet at the event horizon is chosen as 70° which is an upper limit based on the analysis of [179, 191]. Thus, the Blandford–Znajek jet power in Fig. 11.16 is an upper limit. The peak value of B^P at this outer boundary at the base of the Poynting jet is the B^P_E value referred to in Fig. 11.16 and this is also the value of B^P that appears in the ergospheric disk calculations. Notice that the total power as a function of spin is relatively flat, not a steep power law in a/M as some authors suggest. The reason is that (5.50) and (5.51) imply that the total jet power is

$$\int S^P dA_\perp \approx \frac{1}{2\pi^2 c}\Omega_F^2\Phi^2 \sim \left(\frac{a(r_+^2+a^2)}{Mr_+}\right)^2 . \tag{11.16}$$

The growth of Ω_H as a/M increases is almost canceled out by the shrinking black hole surface area in the calculation of Φ. Since the result is normalized to $(B^P_E)_4$, the jet power vs. spin must be a relatively flat function at $a/M > 0.5$.

11.5.4 The KDJ Ergospheric Disk Data Point

The KDJ ergospheric disk data point is the black square in Fig. 11.16. The raw simulation data is in code units, so it is not directly interpretable in terms of physical units. Thus, the ergospheric disk jet power must be determined relative to some other parameter in the simulation. The most interesting comparison for our purposes is with the putative power from the Blandford–Znajek mechanism within the simulation. The ergospheric disk jet power is calculated by assuming that the unaccounted for electrodynamic power (non-ergospheric disk power) is a Blandford–Znajek power (no coronal or accretion disk piston power) and that power is represented by the theoretical, dark blue dashed curve. The three available time slices were analyzed and the ratio of the total ergospheric disk jet electromagnetic and mechanical energy flux to the putative Blandford–Znajek power was computed. This resultant ratio was multiplied by the $a/M = 0.99$ value on the dashed blue curve in order to get the black square value. The agreement with the theoretical curve is quite good.

11.5.5 The KDE Ergospheric Disk Data Point

The KDE ergospheric jet power is indicated by the red arrow in Fig. 11.16. The KDE data point is only a lower limit because of the saturated red colors in the false color contour plot in Fig. 11.11, as discussed in Sect. 11.3.5. For consistency with the Blandford–Znajek calculation above, the Poynting jet now is considered to exist all the way to $\theta \approx 65°$. The time average in Fig. 11.11 was used to obtain a lower limit on the ratio of total ergospheric disk jet electromagnetic energy flux to the putative Blandford–Znajek energy flux. As was done for KDJ, this resultant ratio was multiplied by the $a/M = 0.998$ value on the dashed dark blue curve (the theoretical Blandford–Znajek curve) in order to get the red arrow. The agreement with the theoretical curve is likely better than this lower limit.

11.5.6 The KDH Ergospheric Disk Data Point

The KDH ergospheric disk data point is the brick red arrow in Fig. 11.16. It is calculated by assuming that the discrepancy in the electromagnetic and mechanical energy flux at the outer calculational boundary (discussed in Sect. 11.4 and [191]) with the Blandford–Znajek electrodynamic power that is injected into the outgoing Poynting jet forms an upper bound to the energy budget that can be transported by the ergospheric disk jet in KDH. Of course, the coronal piston could be the major contributor to this deficit and this bound could be very loose. The three available time slices were analyzed and an upper limit to the ratio of total ergospheric disk jet electromagnetic and mechanical energy flux to the putative Blandford–Znajek contribution was estimated. This resultant ratio was multiplied by the $a/M = 0.95$ value on the dashed dark blue curve (the theoretical Blandford–Znajek curve) in order to get the brick red upper limit. The KDH estimate falls far below the theoretical prediction even by moving the outer boundary of the vertical flux in from $r = 1.5M$ into $r = 1.42M$ in the calculation of the truncated disk. The reason appears to be the proximity of the inner calculation boundary. Many of the twisted vertical flux loops near the equatorial plane quickly disappear into the adjacent boundary, thus the ergospheric disk power rarely propagates outward, much of it gets terminated by the inner boundary surface.

11.5.7 Constraints Imposed by Observations

There are two striking results that stand out in Fig. 11.16. First of all, at high spin rates the ergospheric disk jet is much more powerful than the electrodynamic power in the EHM, even if the disk is truncated. Secondly, all of the models except an ergospheric disk threaded by large scale flux would have trouble explaining radio sources with a long term time averaged jet power larger than 10^{40} W [189, 190, 194]. This suggests that there might be other modes of accretion that are capable of transporting vertical flux more efficiently to the ergosphere than the MRI driven accretion form thick tori. One possibility that has been suggested is that regions of strong flux should exist as discrete low density islands in the plasma. The plasma is inherently an inhomogeneous two phase accretion flow [195]. In this scenario, the flux tubes are injected into the ergosphere almost vertically in the equatorial plane and already possess there own magneto-centrifugal outflow. They need not be pushed into the horizon by ram pressure in this two phase medium. The plasma can accrete in 3-D by "swimming" around the magnetic islands via interchange instabilities. As such, a magnetosphere of vertical flux could persist in an accreting system. Another possibility is that many modes of thin disk accretion should naturally create enhanced vertical flux distributions in the ergosphere [196].

There seems to be numerical evidence to the effect that significant vertical flux can exist near an accreting compact object. The family of 3-D simulations in the pseudo-Newtonian potential presented in [197] evolve from an endless supply of equatorial gas (from the outer boundary) that is magnetized by weak vertical flux. The flux accretes toward the inner computational boundary at $r = 4M$. As flux accumulates in the inner regions of the accretion flow, large magnetic islands of vertical flux form. These magnetic islands tend to exclude large mass densities and become buoyant. They "swim" outward in the gas (against the strong inward accretion flow) on spiral trajectories (i.e., by interchange instabilities that only exist in 3-D). As the buoyant flux tubes propagate outward, they slowly lose flux by diffusion (numerical) into the low magnetization surrounding gas. The stripped off flux can be dragged inward again until it possibly forms new magnetic islands near the inner boundary, that become buoyant and the process repeats. This seems very reminiscent of the buoyant flux cycle posited for the ergospheric disk in Chaps. 8 and 10, even though the fundamental physics driving the evolution of the buoyant flux might be somewhat different. This family of simulations also included 2-D simulations that highlight the contrast between the 3-D and 2-D assumptions. The 2-D simulations have an entirely different poloidal flux history. The magnetic pressure, in the form of vertical flux, is accreted toward the inner boundary, where it builds-up until there is sufficient pressure to halt the accretion flow. The gas density begins to accumulate, by further accretion, at the interface between the strong field and the accretion flow. This phase continues until the gravitational force of the mass overwhelms the field pressure and pulls the field in radially in close analogy to the field evolution in [174] for the MHD Wald field. However, this phase is short-lived (it ends, once the high density region has accreted into the compact object). The vertical magnetic pressure builds up again and halts the flow. The mass build up is eventually relieved by a brief episode in which the field is pulled in radially, and so on. The contrast between 2-D and 3-D in these simulations strongly indicate the need for 3-D numerical work in order to fully describe black hole magnetospheres.

Alternatively, if the notion that quasar accretion destroys the large scale vertical flux as indicated in the simulations of [178,179,181,183,184] is valid in nature then there are two consequences. First of all, there is no Poynting jet associated with the accretion disk in Fig. 10.13. Secondly, the most powerful FR II radio sources with a long term time averaged jet power $> 10^{40}$ W, could only be explained in terms of a truncated ergospheric disk. The realistic power output is probably somewhere in between the singly and doubly terminated ergospheric jets powers plotted in Fig. 11.16 (i.e., because of the gravitational redshifting of information carried along MHD characteristics, there might not be much power emitted very close to the horizon). However, even in this case, the black hole parameter space is very restricted, most likely to $M > 10^9\,\mathrm{M}_\odot$ and $a/M > 0.99$. Furthermore, the truncated ergospheric disk has a natural switch-on mechanism at $a/M \lesssim 0.95$ for powerful FR II emission. This could explain one of the mysteries of quasars, why are strong FR II radio lobes (which equates to large time averaged powers, [190, 194]) so rare (in about 2% of quasars)?

References

1. K. Kormendy, D. Richstone: Annu. Rev. Astron. Astrophys. **33**, 581 (1995)
2. D. Richstone et al.: Nature **395**, A14 (1998)
3. Y. Tanaka, W.H.G. Lewin: *X-ray Binaries.* ed. by W.H.G. Lewin et al. (Cambridge University Press, Cambridge, 1995)
4. M. Nowak: Publ. Astronom. Soc. Pac. **107**, 1207 (1995)
5. S.L. Shapiro, S.A. Teukolsky: *Black Holes, White Dwarfs and Neutron Stars.* (Wiley, New York, 1983)
6. I.D. Novikov, K.S. Thorne: *Black Holes*, ed. by C. DeWitt and B.S. DeWitt (Gordon and Breach, New York, 1973, p. 344)
7. R. Narayan, M. Garcia, J.E. McClintock: Astrophys. J. **478**, L79 (1997)
8. L. Ferrarese, H.C. Ford: Astrophys. J. **515**, 583 (1999)
9. N. Shakura, R. Sunayev: Astron. Astrophys. **24**, 337 (1973)
10. M. Malkan, W. Sargent: Astrophys. J. **254**, 22 (1982)
11. W-H. Sun, M. Malkan: Astrophys. J. **346**, 68 (1989)
12. R.J. McLure et al.: Month. Notice. R. Astronom. Soc. **301**, 377 (1999)
13. S. Kirhakos et al.: Astrophys. J. **520**, 67 (1999)
14. B. Punsly: Astrophys. J. **527**, 624 (1999)
15. D. Christodoulou, R. Ruffini: Phys. Rev. D. **4**, 3552 (1971)
16. R. Penrose, R. Floyd: Nature **229**, 193 (1971)
17. J. Baldwin, E.J. Wampler, C.M. Gaskell: Astrophys. J. **338**,630 (1982)
18. M. Corbin: Astrophys. J. **391**, 577 (1992)
19. T. Boroson, R. Green: Astrophys. J. Suppl. Ser. **80**, 109 (1992)
20. P.D. Barthel: Astrophys. J. **336**, 606 (1989)
21. R.J. Antonnucci, J.S. Ulvestad: Astrophys. J. **294**, 158 (1985)
22. D. Murphy, I.W.A. Browne, R. Perley: Month. Notice. R. Astronom. Soc. **264**, 298 (1992)
23. R.R.J. Antonnucci: Annu. Rev. Astron. Astrophys. **31**, 473 (1993)
24. P. Hintzen, J. Ulvestad, F. Owen: Astron. J. **88**, 709 (1983)
25. A. Bridle, et al.: Astrophys. J. **108**, 766 (1994)
26. R. Landau, et al.: Astrophys. J. **308**, 78 (1986)
27. W. Gear, et al.: Month. Notice. R. Astronom. Soc. **267**, 167 (1994)
28. C.D. Impey, G. Neugebauer: Astron. J. **95**, 307 (1988)
29. T.J. Pearson, R.A. Perley, A.C.S. Readhead: Astron. J. **90**, 738 (1985)
30. R.I. Kollgaard, et al.: Astron. J. **104**, 1687 (1992)
31. G. Ghisellini, et al.: Astrophys. J. **407**, 65 (1993)
32. P.M. Ogle, et al.: Astrophys. J. **482**, 370 (1997)
33. W. Tucker: *Radiation Processes in Astrophysics.* (MIT Press, Cambridge, 1975)
34. V.L. Ginzburg: *Theoretical Physics and Astrophysics.* Translated by D. Ter Haar (Pergamon, New York, 1979)

35. N.E. Kassim, et al.: In: *Cygnus A - Study of a Radio Galaxy.* ed. by C.I. Carilli, D.E. Harris (Cambridge University Press, Cambridge, 1996, p. 182)
36. A.T. Moffet: In *Stars and Stellar Systems. Volume IX. Galaxies and the Universe.* ed. by A. Sandage, M. Sandage, J.K. Kristian (University of Chicago Press, Chicago, 1975, p. 211)
37. K.I. Kellerman, I.I.K. Pauling-Toth, P.J.S. Williams: Astrophys. J. **167**, 1 (1969)
38. C.L. Carilli, et al.: Astrophys. J. **383**, 554 (1991)
39. K. Arnaud, et al.: Month. Notice. R. Astronom. Soc. **211**, 981 (1984)
40. P.Alexander, M.T. Brown, P.F. Scott: Month. Notice. R. Astronom. Soc. **209**, 851 (1984)
41. D.E. Harris, C.L. Carilli, R.A. Perley: Nature **367**, 713 (1994)
42. S. Rawlings, R. Saunders: Nature **349**, 138 (1991)
43. T.P. Krichbaum, et al.: Astron. Astrophys. **329**, 873 (1998)
44. J-P.Luminet: In: *Black Holes: Theory and Observation.* ed. by F.W. Hehl, C. Kiefer, R.J.K. Metzler (Springer, Heidelberg, 1998, p. 3)
45. A Eckart, R. Genzel: In: *Black Holes: Theory and Observation.* ed. by F.W. Hehl, C. Kiefer, R.J.K. Metzler (Springer, Heidelberg, 1998, p. 60)
46. B. Punsly: AJ. **109**, 1555 (1995)
47. C.L. Carilli, et al.: In: *Cygnus A - Study of a Radio Galaxy.* ed. by C.L. Carilli, D.E. Harris (Cambridge University Press, Cambridge, 1996, p. 76)
48. B. Sorathia, et al.: In: *Cygnus A - Study of a Radio Galaxy.* ed. by C.L. Carilli, D.E. Harris (Cambridge University Press, Cambridge, 1996, p. 86)
49. G. Benford: Month. Notice. R. Astronom. Soc.**183**, 29 (1978)
50. I. Okamoto: Month. Notice. R. Astronom. Soc. **307**, 2530 (1999)
51. T. Sakurai: Astron. Astrophys. **152**, 121 (1985)
52. A. Ferrari: Annu. Rev. Astron. Astrophys. **36**, 539 (1998)
53. P.E. Hardee: In: *Cygnus A - Study of a Radio Galaxy.* ed. by C.L. Carilli, D.E. Harris (Cambridge University Press, Cambridge, 1996, p. 113)
54. K.R. Lind, R.D. Blandford: Astrophys. J. **295**, 358 (1985)
55. P. Padovani, C.M. Urry: Astrophys. J. **387**, 449 (1992)
56. R.C. Vermeulen, M.H. Cohen: Astrophys. J. **430**, 467 (1994)
57. M. Abramowicz, T. Piran: Astrophys. J. **241**, L7 (1980)
58. T. Piran: Astrophys. J. **257**, L23 (1982)
59. F.C. Michel: *Theory of Neutron Star Magnetospheres.* (University of Chicago Press, Chicago, 1991)
60. C. Kennel, F.V Coroniti: Astrophys. J. **283**, 694 (1984)
61. C. Kennel, F.V Coroniti: Astrophys. J. **283**, 710 (1984)
62. E.J. Weber, L. Davis, Jr.: Astrophys. J. **148**, 217 (1967)
63. R. Lovelace: Nature **262**, 649 (1976)
64. R. Blandford: Month. Notice. R. Astronom. Soc. **176**, 465 (1976)
65. R. Ruffini, J. Wilson: Phys. Rev. D. **12**, 2959 (1975)
66. R. Blandford, R. Znajek: Month. Notice. R. Astronom. Soc. **179**, 433 (1977)
67. B. Punsly, F.V. Coroniti: Astrophys. J. **350**, 518 (1990)
68. W. Unruh: Phys. Rev. D. **10**, 3194 (1974)
69. A.M. Anile: *Relativistic Fluids and Magneto-Fluids.* (Cambridge University Press, New York, 1989)
70. A. Lichnerowicz: *Magnetohydrodynamics: Waves and Shock Waves in Curved Space–Time* (Kluwer, Dordrecht, 1994)
71. C. Misner, K. Thorne, J. Wheeler: *Gravitation* (Freeman, San Francisco, 1973)
72. A.P. Lightman, W.H. Press, S.A. Teukolsky: *Problem Book in General Relativity and Gravitation.* (Princeton University Press, Princeton, 1979)
73. T.J.M. Boyd, J.J. Sanderson: *Plasma Dynamics.* (Barnes and Noble, New York, 1969)
74. J. Mathews, R.L. Walker: *Mathematical Methods of Physics* (Benjamin, Menlo Park, 1970)
75. F.V. Coroniti: J. Plasma Phys. **4**, 265 (1970)
76. T.H. Stix: Phys. Rev. **106**, 1146 (1957)
77. T.H. Stix: *Waves in Plasmas.* (American Institute of Physics, New York, 1992)

78. A.R. Kantrowicz, H.E. Petschek: In: *Plasma Physics in Theory and Application.* ed. by W.B. Kunkel (McGraw-Hill, New York, 1966, p. 148)
79. C.C. Johnson: *Field and Wave Electrodynamics.* (McGraw-Hill, New York, 1965)
80. K. Thorne, R. Price, D. Macdonald: *Black Holes: The Membrane Paradigm.* (Yale University Press, New Haven, 1986)
81. P. Goldreich, W.H. Julian: Astrophys. J. **157**, 869 (1969)
82. B. Carter: Phys. Rev. **174**, 1559 (1968)
83. M. Johnston, R. Ruffini: Phys. Rev. D. **10**, 2324 (1974)
84. J. Bardeen, W. Press, S. Teukolsky: Astrophys. J. **178**, 347 (1972)
85. B. Punsly: Phys. Rev. D. **44**, 2970 (1991)
86. A.R. King, J.P. Lasota: Astron. Astrophys. **58**, 175 (1977)
87. J.D. Jackson: *Classical Electrodynamics.* (Wiley, New York, 1975)
88. E. Newman, R. Penrose: J. Math. Phys. **3**, 566 (1962)
89. S. Teukolsky: Astrophys. J. **185**, 635 (1973)
90. J. Bicak, L. Dvorak: Gen. Relativ. Gravit. **7**, 959 (1976)
91. R. Penrose, W. Rindler: *Spinors and Space-time,* Vol. 1. (Cambridge University Press, New York, 1986)
92. S. Chandrasekhar: *The Mathematical Theory of Black Holes.* (Oxford University Press, New York, 1983)
93. J. Petterson: Phys. Rev. D. **12**, 2218 (1975)
94. W. Kinnersley: J. Math. Phys. **10**, 1195 (1969)
95. E. Fackerell, J. Ipser: Phys. Rev. D. **5**, 2455 (1972)
96. A.R. King, J.P. Lasota, W. Kundt: Phys. Rev. D. **12**, 3037 (1975)
97. D.M. Chitre, C.V. Vishveshwara: Phys. Rev. D. **12**, 1538 (1975)
98. H. Bateman, A. Erdélyi: *Higher Transcendental Functions,* Vol. 1 (McGraw-Hill, New York, 1953)
99. T. Regge, J. Wheeler: Phys. Rev. **108**, 1063 (1957)
100. R.H. Price: Phys. Rev. D. **5**, 2419 (1972)
101. R.H. Price: Phys. Rev. D. **5**, 2439 (1972)
102. J. Bicak, V. Janis: Month. Notice. R. Astron. Soc. **198**, 339 (1985)
103. J. Bicak, L. Dvorak: Phys. Rev. D. **22**, 2933 (1980)
104. R. Hanni, R. Ruffini: Phys. Rev. D. **8**, 3259 (1973)
105. R. Ruffini: In: *Physics and Astrophysics of Neutron Stars and Black Holes.* ed. by R. Giacconi and R. Ruffini (North Holland, Amsterdam, 1978)
106. R. Ruffini: *Black Holes, Les Houches 1973* ed. by B. DeWitt and C. DeWitt (Gordon and Breach, New York, 1973, p. 525)
107. R.M. Wald: Phys. Rev. D. **10**, 1680 (1974)
108. R. Ruffini, A. Treves: Astrophys. Lett. **13**, 109 (1973)
109. T. Damour: Phys. Rev. D. **18**, 3598 (1978)
110. R. Znajek: Month. Notice. R. Astron. Soc. **185**, 833 (1978)
111. E.S. Phinney: Ph.D. Dissertation, University of Cambridge, (1983)
112. B. Punsly, F. V. Coroniti: Phys. Rev. D. **40**, 3834 (1989)
113. C. Kennel, F. Fujimura, I. Okamoto: Geophys. Astrophys. Fluid Dyn. **26**, 147 (1983)
114. M. Camenzind: Astron. Astrophys. **156**, 137 (1986)
115. M. Camenzind: In: *Accretion Disks and Magnetic Fields in Astrophysics.* ed. by G. Belvedere (Dordrecht, Kluwer, 1989, p. 129)
116. M. Camenzind: Astron. Astrophys. **162**, 32 (1986)
117. D. Macdonald, K. Thorne: Month. Notice. R. Astron. Soc. **198**, 345 (1982)
118. V.S. Beskin: Physics-Uspekhi. **40**, 659 (1997)
119. M. Takahashi, S. Nitta, Y. Tatematsu, A. Tomimatsu: Astrophys. J. **363**, 206 (1990)
120. F. Michel: Astrophys. J. **157**, 1183 (1969)
121. S. Nitta, M. Takahashi, A. Tomimatsu: Phys. Rev. D. **44**, 2295 (1991)
122. E.T. Scharlemann, R.V. Wagoner: Astrophys. J. **182**, 951 (1973)
123. L.J. Lanzerotti, D.J. Southwood: 'Hydromatic Waves'. In: *Solar System Plasma Physics.* Vol. 3. ed. by E.N. Parker, C.F. Kennel, L.J. Lanzerotti (Amsterdam, North Holland, 1979)

124. V.S. Beskin, Ya.N. Istomin, V.I. Par'ev: Sov. Astron. **36**, 642 (1992)
125. E.S. Phinney: In: *Proc. Torino Workshop on Astrophysical Jets.* ed. by A. Ferrari, A. Pacholczyk (Dordrecht, Reidel, 1982)
126. C.E. Fichtel, J.I. Trombka: *Gamma-Ray Astrophysics.* (Greenbelt, NASA, 1997)
127. S. Chandrasekhar: Proc. R. Soc. London ser. A. **349**, 571 (1976)
128. B. Punsly: Phys. Rev. D. **34**, 1680 (1986)
129. K. Hirotani, A. Tomimatsu, M. Takahashi: Publ. Astron. Soc. Jpn **45**, 431 (1993)
130. B. Punsly: Astrophys. J. **467**, 105 (1996)
131. R. Blandford: In: *Theory of Accretion Disks.* ed. by F. Meyer et al. (Dordrecht, Kluwer, 1989)
132. B. Punsly, F. Coroniti: Astrophys. J. **354**, 583 (1990)
133. V.M. Vasyliunas: Rev. Geophys. Space Phys. **13**, 303 (1975)
134. F. Coroniti: Nucl. Fission **11**, 261 (1971)
135. C. Itzykson, J. Zuber: *Quantum Field Theory* (McGraw-Hill, San Francisco, 1980)
136. B. Punsly: Astrophys. J. **506**, 790 (1998)
137. L. Mestel, P. Phillips, Y-M. Yang: Month. Notice. R. Astron. Soc. **188**, 385 (1979)
138. B. Punsly: Astrophys. J. **372**, 424 (1991)
139. Ya.B. Zel'dovich: Soviet Phys. JETP Lett. **35**, 1095 (1972)
140. C. Kittel: *Elementary Statistical Physics.* (Wiley, New York, 1961)
141. G. Henri, G. Pelletier, J. Roland: Astrophys. J. **404**, L41 (1993)
142. B. Punsly: Astrophys. J. **473**, 152 (1996)
143. S. Koide, K. Shibata, K. Takahiro: Astrophys. J. **522**, 727 (1999)
144. C.D. Dermer, R. Schlickeiser: Astrophys. J. **416**, 458 (1993)
145. G. Ghisellini, L. Maraschi: Astrophys. J. **340**, 181 (1989)
146. L.M.J. Brown, et al.: Astrophys. J. **340**, 129 (1989)
147. S.C. Unwin, M.H. Cohen, M.W. Hodges, J.A. Zensus, J.A. Biretta: Astrophys. J. **340**, 117 (1989)
148. J.A. Biretta, R.L. Moore, M.H. Cohen: Astrophys. J. **308**, 93 (1986)
149. A. Marscher, J. Broderick: Astrophys. J. **290**, 735 (1985)
150. S.C. Unwin, et al.: Astrophys. J. **432**, 103 (1993).
151. E.A. Carrara, Z. Abraham, S.C. Unwin, J.A. Zensus: Astron. Astrophys. **279**, 83 (1993)
152. E.I. Robson, et al.: Month. Notice. R. Astron. Soc. **262**, 249 (1993)
153. B. Punsly: Astrophys. J. **474**, 612 (1997)
154. B. Punsly: Astrophys. J. **473**, 178 (1996)
155. J. Arons, S.M. Lea: Astrophys. J. **235**, 1016 (1980)
156. R. Narayan, I. Yi: Astrophys. J. **428**, L13 (1994)
157. P.S. Smith, T.J. Balonek, R. Elson, P.A. Heckart: Astrophys. J. Suppl. Ser. **64**, 459 (1987)
158. J.R. Webb, et al.: Astrophys. J. **422**, 570 (1994)
159. A.J. Pica, A.G. Smith, J.R. Webb, R.J. Leacock, S. Clements: Astron. J. **96**, 1215 (1988)
160. L. Maraschi, G. Ghisellini, A. Celloti: Astrophys. J. **397**, L5 (1992)
161. H. Spinrad, S. Djorgovski, J. Marr, L. Aguilar: Publ. Astron. Soc. Pac. **97**, 932 (1985)
162. F. Vagnetti, E. Giallongo, A. Cavaliere: Astrophys. J. **368**, 366 (1991)
163. C.M. Urry, P. Padovani, M. Stickel: Astrophys. J **382**, 501 (1991)
164. B.G. Piner, K.A. Kingman: Astrophys. J. **507**, 706 (1998)
165. S.J. Tingay, et al.: Astrophys. J. **497**, 594 (1998)
166. S.J. Tingay, D.W. Murphy, P.G. Edwards: Astrophys. J. **500**, 673 (1998)
167. S. Koide: Phys. Rev. D**67** 104010 (2003)
168. S. Koide, K. Shibata, T. Kudoh, D. Meier: Science **295**, 1688 (2002)
169. M. Camenzind, R. Khanna: Nuovo Cimento **B115**, 815 (2000)
170. S. Komissarov: Month. Notice. R. Astron. Soc. **326** L41 (2001)
171. B. Punsly, D. Bini: Astrophys. J. Lett. **601** 135 (2004)
172. S. Komissarov: Month. Notice. R. Astron. Soc. **350**, 1431 (2004)
173. S. Komissarov: Month. Notice. R. Astron. Soc. **350**, 407 (2004)
174. S. Komissarov: Month. Notice. R. Astron. Soc. **359**, 801 (2005)
175. V. Semenov, et al.: Physica Scripta **65**, 13 (2002)
176. V. Semenov, S. Dyadechkin, B. Punsly: Science **305**, 978 (2004)

177. J.P. De Villiers, J. Hawley: Astrophys. J. **589**, 458 (2003)
178. J-P. De Villiers, J. Hawley, J.H. Krolik: Astrophys. J. **599**, 1238 (2003)
179. S. Hirose, K. Krolik, J. De Villiers, J. Hawley: Astrophys. J. **606**, 1083 (2004)
180. J-P. De Villiers, J. Hawley, J. Krolik, S. Hirose: Astrophys. J. **620**, 878 (2005)
181. K. Krolik, J. Hawley, S. Hirose: Astrophys. J. **622**, 1008 (2005)
182. J-P. De Villiers, J. Staff, R. Ouyed: Astrophys. J. (2006) astro-ph 0502225
183. J. Hawley, K. Krolik: Astrophys. J. **641** 103 (2006)
184. J. McKinney, C. Gammie: Astrophys. J. **611**, 977 (2004)
185. J. McKinney: Astrophys. J. Lett. **630**, 5 (2004)
186. B. Punsly: Astrophys. J. Lett. **661**, 21 (2007)
187. B. Punsly: Month. Notice. R. Astron. Soc. **366**, 29 (2006)
188. C. Fragile: Astrophys. J. **668**, 417 (2007)
189. B. Punsly: Astrophys. J. Lett. **651**, L17 (2006)
190. B. Punsly: Month. Notice. R. Astron. Soc. **374**, 10 (2007)
191. B. Punsly: Month. Notice. R. Astron. Soc. **381**, L79 (2007)
192. B. Somov, A. Oreshina: Astron. Astrophys. **354**, 703 (2000)
193. R. Treumann: Earth Planets Space **53**, 453 (2001)
194. C. Willott, S. Rawlings, K. Blundell, M. Lacy: Month. Notice. R. Astron. Soc. **309**, 1017 (1999)
195. H. Spruitt, D. Uzdensky: Astrophys. J. **629**, 965 (2005)
196. G.S. Bisnovatyi-Kogan, R.V.E. Lovelace: Astrophys. J. Lett. **667**, 167 (2007)
197. Igumenshchev, I. Astrophys. J. Lett. **677** TBD (2008) http://xxx.lanl.gov/abs/0711.4391v1

Index

Printing: Krips bv, Meppel, The Netherlands
Binding: Stürtz, Würzburg, Germany